T0251399

MUNICIPAL SEWAGE SLUDGE MANAGEMENT

WATER QUALITY MANAGEMENT LIBRARY
VOLUME 4
SECOND EDITION

MUNICIPAL SEWAGE SLUDGE MANAGEMENT
A REFERENCE TEXT ON PROCESSING, UTILIZATION AND DISPOSAL

EDITED BY

Cecil Lue-Hing, D.Sc., P.E., D.E.E.
David R. Zenz, Ph.D., P.E., D.E.E.
Prakasam Tata, Ph.D., QEP
Richard Kuchenrither, Ph.D., P.E., D.E.E.
Joseph Malina, Jr., Ph.D., P.E., D.E.E.
Bernard Sawyer, M.S.

LIBRARY EDITORS
W. W. ECKENFELDER, D.Sc., P.E.   J. F. MALINA, JR., Ph.D., P.E., D.E.E.   J. W. PATTERSON, Ph.D.

TECHNOMIC
PUBLISHING CO., INC.
LANCASTER · BASEL

Water Quality Management Library—Volume 4
a TECHNOMIC publication

Published in the Western Hemisphere by
Technomic Publishing Company, Inc.
851 New Holland Avenue, Box 3535
Lancaster, Pennsylvania 17604 U.S.A.

Distributed in the Rest of the World by
Technomic Publishing AG
Missionsstrasse 44
CH-4055 Basel, Switzerland

10 9 8 7 6 5 4 3 2 1

Main entry under title:
   Water Quality Management Library—Volume 4 / Municipal Sewage Sludge Management:
      A Reference Text on Processing, Utilization and Disposal, Second Edition

A Technomic Publishing Company book
Bibliography: p.
Includes index p. 785

Library of Congress Catalog Card No. 98-85169
ISBN No. 1-56676-621-4 (Volume 4)
ISBN No. 1-56676-660-5 (11-Volume Set)

Table of Contents

Foreword

IN 1992 the United States National Committee of IAWQ (International Association on Water Quality) organized eight specialty courses offered in conjunction with the 1992 IAWQ Biennial Conference in Washington, D.C. Designed for the practicing engineer, the specialty courses covered critical topics in environmental quality management water pollution control, wastewater treatment, toxicity reduction, and residuals management. These courses were compiled in an eight-volume series as the Water Quality Management Library. Experts from the United States and many countries contributed their expertise and experience to the preparation of these state-of-the-art texts.

The success of this series prompted the editors to expand the series to include volumes on such timely topics as water reuse, non-point source control and aeration and oxygen transfer. Additional volumes are presently being considered. In addition to the new topics, in order to keep pace with this rapidly developing field, many of the original volumes have been updated to reflect current advances in the field. In addition to providing an up-to-date technical reference for the practicing engineer and scientist as in the first series, these volumes will provide a text for continuing education courses and workshops.

The Water Quality Management Library should provide a unique reference source for professional and education libraries.

W. WESLEY ECKENFELDER
JOSEPH F. MALINA, JR.
JAMES W. PATTERSON

Preface

THROUGH the years, municipal agencies have successfully processed
municipal sewage sludge through various combinations of unit operations. Final disposition of municipal sewage sludge has been a continuing
challenge, but landfilling, incineration, and land application have been and
continue to be successfully implemented by many municipal agencies.
However, the relatively recent public awareness of environmental issues
and the federal, state, and local regulatory framework are forcing a closer
look at the problem of how to effectively process, dispose of, or utilize
the sludge produced by municipal wastewater treatment.

This reference textbook represents an update of the first edition issued
in 1992. This updated edition contains a more comprehensive database
on the metal, nutrient, and pathogen content of municipal sludges throughout the United States. The update reflects the fact that the Part 503 regulations were issued in final form in 1993 by the United States Environmental
Protection Agency (U.S.EPA). New sections have been added that describe
the important sludge processing operations of aerobic and anaerobic digestion, disinfection by ionizing radiation, and a procedure for the selection
and procurement of polymers for centrifugal dewatering. The updated
edition has been significantly revised based upon recent development in
sludge processing technology and changes in various sludge regulations
in the United States and Europe.

This second edition explains and discusses many of the unit operations
used for processing municipal sewage sludge. It also contains valuable
information on the available methods for final disposition of this sludge.
This book can be used for planning, designing, and implementing municipal
sludge management projects.

Because the U.S.EPA issued its final comprehensive regulations (40

CFR Part 503) governing the management of municipal sewage sludge in the United States, some emphasis has been placed on describing and commenting on these regulations as they relate to landfilling, incineration, and land application. This emphasis on these regulations represents a unique aspect of this second edition.

List of Contributors

ROBERT BASTIAN, U.S. Environmental Protection Agency, Washington, District of Columbia

ALLEN BATURAY, P.E., Carlson Associates, Manassas, Virginia

JAMES J. BERTUCCI, PH.D., Metropolitan Water Reclamation District of Greater Chicago, Chicago, Illinois

JAMES W. BRADLEY, Royds Garden Ltd., Dunedin, New Zealand

JEFFREY C. BURNHAM, PH.D., Medical College of Ohio, Toledo, Ohio

JOSEPH CALVANO, Metropolitan Water Reclamation District of Greater Chicago, Chicago, Illinois

ROBERT DOMINAK, Northeast Ohio Regional Sewer District, Cleveland, Ohio

JOHN F. DONOVAN, P.E., D.E.E., Camp Dresser & McKee, Inc., Cambridge, Massachusetts

JOSEPH B. FARRELL, PH.D., U.S. Environmental Protection Agency, Cincinnati, Ohio

JANE FORSTE, Bio Gro Systems, Annapolis, Maryland

THOMAS C. GRANATO, PH.D., Metropolitan Water Reclamation District of Greater Chicago, Chicago, Illinois

ROBERT A. GRIFFIN, PH.D., University of Alabama, Tuscaloosa, Alabama

JOHN GSCHWIND, Metropolitan Water Reclamation District of Greater Chicago, Chicago, Illinois

DONALD W. HARPER, Metropolitan Water Reclamation District of Greater Chicago, Chicago, Illinois

ROGER T. HAUG, PH.D., P.E., Solids Technology Division of the City of Los Angeles, Los Angeles, California

PATRICIA HUNT, HCI, Inc., Philadelphia, Pennsylvania

NABIH P. KELADA, PH.D., Metropolitan Water Reclamation District of Greater Chicago, Chicago, Illinois

RICHARD KUCHENRITHER, PH.D., P.E., Black and Veatch, Kansas City, Missouri

SHUNSOKU KYOSAI, D.E., Ministry of Construction of Japan, Tsukuba-shi, Ibaraki-ken, Japan

TERRY J. LOGAN, PH.D., Ohio State University, Columbus, Ohio

DAVID T. LORDI, PH.D., Metropolitan Water Reclamation District of Greater Chicago, Chicago, Illinois

CECIL LUE-HING, D.SC., P.E., D.E.E., Metropolitan Water Reclamation District of Greater Chicago, Chicago, Illinois

JOSEPH F. MALINA, JR., PH.D., P.E., D.E.E., The University of Texas, Austin, Texas

PETER MATTHEWS, PH.D., FIWEM, Anglian Water Services, Cambridge, England

DAVID OERKE, Black and Veatch, Aurora, Colorado

RICHARD I. PIETZ, PH.D., C.P.S.S., Metropolitan Water Reclamation District of Greater Chicago, Chicago, Illinois

K. C. RAO, PH.D., Metropolitan Water Reclamation District of Greater Chicago, Chicago, Illinois (Deceased)

GEORGE R. RICHARDSON, Metropolitan Water Reclamation District of Greater Chicago, Chicago, Illinois

KAZUAKI SATO, D.E., Ministry of Construction of Japan, Tsukuba-shi, Ibaraki-ken, Japan

BERNARD SAWYER, Metropolitan Water Reclamation District of Greater Chicago, Chicago, Illinois

SALVADOR J. SEDITA, PH.D., SM(AAM), Metropolitan Water Reclamation District of Greater Chicago, Chicago, Illinois

RONALD B. SIEGER, PH.D., CH2M Hill, Dallas, Texas

STANLEY SOSZYNSKI, Metropolitan Water Reclamation District of Greater Chicago, Chicago, Illinois

ROBERT M. SOUTHWORTH, U.S. Environmenal Protection Agency, Washington, District of Columbia

RICHARD C. SUSTICH, Metropolitan Water Reclamation District of Greater Chicago, Chicago, Illinois

PRAKASAM TATA, PH.D., QEP, Metropolitan Water Reclamation District of Greater Chicago, Chicago, Illinois

WARREN R. UHTE, Brown and Caldwell, Seattle, Washington

MEL WEBBER, PH.D., Environment Canada, Burlington, Ontario, Canada

THOMAS E. WILSON, PH.D., P.E., D.E.E., Greeley & Hansen, Chicago, Illinois

DAVID ZENZ, PH.D., P.E., D.E.E., Metropolitan Water Reclamation District of Greater Chicago, Chicago, Illinois

CHAPTER 1

Introduction

MUNICIPAL sewage sludge processing, utilization, and disposal are some of the most difficult and expensive operations conducted by municipalities today. In the United States about 15,300 dry mt (16,900 tons) per day of municipal sewage sludge is produced and this sludge must be managed in an environmentally acceptable way. In addition, the rising costs of energy, labor, and material makes it imperative for municipalities to develop management programs which are reliable and cost effective.

Although reliable technologies are available for processing and ultimate disposition of municipal sewage sludge, proper planning, designing, and implementing solutions for particular site specific situations still remain perplexing problems. This reference textbook contains information compiled by a team of experts who have been working in the field of municipal sewage sludge management for many years. It will be valuable to those who must plan, design, and implement municipal sewage sludge management programs.

Chapter 2 contains a comprehensive discussion of the regulatory framework that the United States Environmental Protection Agency constructed at the national level, within which all existing and future municipal sewage sludge management projects must operate. The regulations themselves are presented along with a working history of their development.

Chapter 3 contains data on the levels of various constituents found in municipal sewage sludges and presents information on how pretreatment regulations, trade waste control, and enforcement can influence these levels. Because of the importance of having accurate and precise data on the

---

David Zenz, Metropolitan Water Reclamation District of Greater Chicago, Chicago, IL.

levels of constituents present in municipal sewage sludge, this chapter includes examples of the methodologies used for municipal sludge analysis.

The microbiological quality of municipal sewage sludge has public health implications and such quality is regulated at the local, state, and federal level. Chapter 4 contains a comprehensive discussion of the microbiological aspects of municipal sludge management. Data is presented on the levels of various microbiological species in municipal sewage sludge and how these levels are affected by sludge processing. In addition, this chapter presents the relevant scientific research work regarding the potential for movement of microbiological species through soils, surface water, groundwater, and the food chain.

As anyone knowledgeable in the field of municipal sewage sludge management can attest, odors from municipal sewage sludge projects can be a vexing problem. Chapter 5 contains valuable information on the sources, measurement, and control of odors from municipal sewage sludge management operations.

Chapter 6 describes municipal sewage sludge processing techniques. This chapter focuses on chemical stabilization techniques, sludge conditioning (for dewatering), composting, lagooning, air-drying, and aerobic and anaerobic digestion. Chapters 7, 8, 9, 10, and 11 deal with the final disposition of municipal sewage sludge; namely, landfilling, incineration, land application, and production and distribution.

Chapter 7 describes the mathematical models that can be used to determine the migration of municipal sewage sludge constituents from a landfill to groundwater. This chapter also contains valuable information on the design and operation of landfills used for the final disposition of municipal sewage sludge. Lastly, Chapter 7 presents a compelling case for using municipal sewage sludge for daily cover, and final cover to support vegetation for municipal solid waste landfills.

Chapter 8 presents two important aspects of municipal sewage sludge incineration. First, there is valuable information about retrofitting existing incinerators for regulatory compliance. Secondly, some of the beneficial uses available for incinerator ash are presented.

Chapter 9 presents a complete description of the planning design and implementation of dedicated sites used for land application of municipal sewage sludge. This is one of the most exhaustive presentations available today.

Land application of municipal sewage sludge to soil reclamation sites is addressed in Chapter 10. This chapter will be invaluable to those who are involved in using municipal sewage sludge for this beneficial purpose.

Chapter 11 contains information on the production and distribution of municipal sewage sludge products.

Public policy represents a very important aspect of municipal sewage sludge management, and Chapter 12 discusses it in depth.

CHAPTER 2

Regulatory Issues

THE intent of this chapter is to review the development of knowledge concerning sludge application to land and the associated development of sludge regulations. This chapter describes the development of knowledge on the utilization of sewage sludge on land during the 1970s and 1980s. The chapter discusses the development of sludge regulations in the 1970s, with the implementation of 40 CFR Part 257 in 1979, and the development of 40 CFR Part 503 in the 1980s and 1990s and its final publication in 1993. Public and peer review comments on the Part 503 regulations are described, and then the final regulations for 40 CRF Part 503 are discussed. Finally, this chapter outlines items of concern in probable future regulations that could also impact the utilization of sewage sludge.

U.S. POLICIES, LAWS AND REGULATIONS IMPACTING THE MANAGEMENT OF MUNICIPAL SEWAGE SLUDGE

Approximately 30 billion gallons of domestic, commercial, and industrial wastewaters are discharged to the collection systems of over 15,500 publicly owned treatment works (POTWs) serving about 75% of the nation's population. These POTWs generally discharge the treated effluent to rivers, lakes, estuaries, groundwater, or coastal ocean waters. They dispose of or recycle their residuals (grit, screenings, scum, and sludges—

Robert Bastian and Robert M. Southworth, U.S.EPA, Washington, D.C.; Joseph B. Farrell, U.S. U.S.EPA, Cincinnati, OH; Thomas C. Granato, Cecil Lue-Hing, Richard I. Pietz, and K.C. Rao, Metropolitan Water Reclamation District of Greater Chicago, Chicago, IL.

an estimated 7.7 million dry metric tons/year) produced while treating the wastewater by a number of practices, including incineration, landfilling, land application (to agricultural and non-agricultural lands), distributing or marketing products (such as heat dried, composted, or alkaline stabilized sludge), and long-term storage. Municipal sewage sludge use and disposal practices are greatly impacted by a number of different federal policies, laws, and regulations.

U.S.EPA POLICY ON BENEFICIAL USES

The official "Policy on Municipal Sludge Management" issued by the U.S. Environmental Protection Agency (U.S.EPA) in 1984 (49 *FR* 24358; June 12, 1984) states that:

> The U.S. Environmental Protection Agency (U.S.EPA) will actively promote those municipal sludge management practices that provide for the beneficial use of sludge while maintaining or improving environmental quality and protecting public health. To implement this policy, U.S.EPA will continue to issue regulations that protect public health and other environmental values. The Agency will require states to establish and maintain programs to ensure that local governments utilize sludge management techniques that are consistent with federal and state regulations and guidelines. Local communities will remain responsible for choosing among alternative programs; for planning, constructing, and operating facilities to meet their needs; and for ensuring the continuing availability of adequate and acceptable disposal or use capacity.

As noted in the policy statement, U.S.EPA continues to prefer wherever possible well-managed beneficial uses of sewage sludge to disposal. Such uses include land application practices using sludge as a soil amendment or fertilizer supplement and various practices that derive energy from sludge or convert it to useful products. Such well-managed beneficial use practices can help reduce the volume of waste to be disposed of in the limited available landfill capacity, and avoid the threat of pollution problems from such disposal practices as landfilling, incineration, ocean dumping, or discharge to coastal waters. Secondary benefits can also be realized from these beneficial use practices of sludge such as a reduction in the use of inorganic fertilizers, improvement in soil properties and plant productivity, and reductions in energy consumption and air emissions.

A number of issues concerning some programmatic and technical considerations needed to implement sewage sludge beneficial use programs on federal lands were the subject of discussions held in 1990 and 1991 by representatives of a number of federal agencies. U.S.EPA worked closely with the Office of Management and Budget, U.S. Department of the Interior, U.S. Department of Agriculture (USDA), Department of Defense, Department of Energy, Food and Drug Administration (FDA), Tennessee Valley Authority (TVA), and other federal agencies that generate policy or have been approached to allow the use of sewage sludge on federal lands to establish a unified federal policy on beneficial use of sludge and to prepare guidelines that federal agency land managers can use in determining the appropriateness of land application of sewage sludge for their facilities. A *Federal Register* notice was issued in July 1991 containing the new "Interagency Sludge Policy on Beneficial Use of Municipal Sewage Sludge on Federal Land." It reaffirms and supplements the existing federal policy (as described in the 1984 U.S.EPA policy promoting beneficial use of sludge and an earlier joint 1981 U.S.EPA/USDA/FDA policy and guidance document addressing the use of sewage sludge in the production of fruits and vegetables) by advocating those municipal sludge management practices that provide for the beneficial use of sludges while maintaining environmental quality and protecting public health.

U.S. LAWS, U.S.EPA REGULATIONS, AND PROGRAMS

For many years, sewage sludge use and disposal practices were regulated under a number of federal statutes, with no comprehensive program at the national level. Federal requirements were authorized under several federal environmental statutes and were developed independently in response to media-specific concerns. The primary role for control of sewage sludge use and disposal practices was left up to the states, and while nearly all states have had some type of sludge management program, these programs have varied widely in the coverage of various practices. A brief description of the historical U.S.EPA regulations, requirements, and federal statutes that have applied to sewage sludge use and disposal follows.

U.S.EPA Regulations Applying to Sewage Sludge Use and Disposal Before 40 CFR Part 503 was Promulgated

Coverage	Reference	Application
Ocean dumping	40 CFR Parts 220–228 (MPRSA[1])	The discharge of sludge from barges or other vessels
Ocean discharge	40 CFR Part 125, Subpart M (CWA[2])	The discharge of sludge from outfalls

[1] MPRSA = Marine Protection, Research and Sanctuaries Act.
[2] CWA = Clean Water Act.

Coverage	Reference	Application
Polychlorinated biphe-nyls (PCBs)	40 CFR Part 761 (TSCA[3])	All sludges containing more than 50 mg/kg
National Ambient Air Quality Standards	40 CFR Part 52 (CAA[4])	Major stationary sources emitting more than 100 tons/yr
New Source Perform-ance Standards for Air Emissions	40 CFR Part 60 (CAA)	Incineration of sludge at rates above 1000 kg/d
Hazardous Air and Emis-sion Standards—Mercury and Beryllium	40 CFR Part 61 (CAA)	Incineration and heat dry-ing of sludge
Cadmium, PCBs, patho-genic organisms	40 CFR Part 257 (RCRA[5] and CWA)	Land application of sludge, landfills and stor-age in lagoons
Toxicity Characteristic	40 CFR Part 261 Appendix II (RCRA)	Test to determine if sludges (or other wastes) are hazardous wastes un-der RCRA
Environmental impacts	40 CFR Part 1500 (NEPA[6])	Environmental impact as-sessments and reviews

Ocean Dumping and Discharge of Sewage Sludge

Ocean dumping is covered under provisions of the Marine Protection, Research, and Sanctuaries Act (MPRSA) of 1972 [PL 92-532] and its amendments (40 CFR Parts 220–228) including the Ocean Dumping Ban Act of 1988, and provisions of the Clean Water Act (CWA), 40 CFR Part 125, Sub-Part M. MPRSA and its amendments focus on regulating the dumping of all types of materials into ocean waters and limiting the ocean dumping of materials that would adversely affect human health and the welfare of the marine environment and its commercial values. The MPRSA prohibits ocean dumping of sludge from barges or other vessels and transpor-tation from the U.S. for purposes of dumping except pursuant to a permit, and permits are not to be issued where dumping would "unreasonably de-grade" the marine environment. Permitting regulations, marine research, and marine sanctuaries establishment provisions are included. The U.S.EPA was required to establish criteria for evaluating ocean dumping permit applications applying specific statutory factors. A 1977 Amendment (PL 95-153) effectively established December 31, 1981 as the deadline for terminating ocean dumping of "sewage sludge" defined as municipal

[3] TSCA = Toxic Substances Control Act.
[4] CAA = Clean Air Act.
[5] RCRA = Resources Conservation and Recovery Act.
[6] NEPA = National Environmental Policy Act.

waste "which may unreasonably degrade or endanger human health, welfare, amenities, or the marine environment, ecological systems, or economic potentialities." A similar amendment was enacted in 1980 for "industrial wastes." In the Ocean Dumping Ban Act of 1988 (Public Law 100-88), Congress amended the MPRSA to prohibit after December 31, 1991, the dumping of sewage sludge or industrial waste in ocean waters. New York City, the last authority to end ocean disposal, closed ocean dumping operations in July 1992 under provision of a court order/enforcement agreement.

Guidelines for issuance of National Pollutant Discharge Elimination System (NPDES) permits for the discharge of pollutants from point sources (outfalls) into the territorial seas, the contiguous zone, and the oceans (i.e., seaward from the baseline) have been issued under Section 403(c) of the CWA (40 CFR Part 125, Subpart M). This provision provides consistency with the MPRSA, which regulates ocean dumping of materials by vessels into the same marine waters, and provides for similar protection from point source discharges as from the dumping of waste materials. However, there is one major difference in these regulatory programs. The holder of an ocean dumping permit must only comply with Section 102(g) of MPRSA [which is similar to the language of Section 403(c) of CWA]. In contrast an NPDES permittee discharging into the territorial seas, the contiguous zone, or the oceans must comply with all applicable CWA provisions, including compliance with effluent limitations, secondary treatment standards and guidelines, and water quality standards, as well as with the Section 403(c) requirements. A controversy that developed over the potential for applying the criteria and standards issued in 40 CFR 125 Subpart G to sludge (which are to be applied by U.S.EPA in acting on Section 301(h) of CWA requests for modifications to the secondary treatment requirements for marine discharges from POTWs) was resolved when Congress amended Section 301(h) in December 1981 to specifically exclude consideration of sludge discharges under this provision.

PCBs in Sewage Sludge

PCBs are addressed under provisions of the Toxic Substances Control Act (TSCA) of 1976 (PL 94-469). TSCA focuses on the need for testing, premanufacture notification, regulating production, and placing necessary use restrictions on certain chemical substances and mixtures that present an unreasonable risk of injury to health or the environment. Along with its many regulatory, testing, and reporting requirements, the U.S.EPA is also required under Section 9 of TSCA to coordinate actions taken under TSCA with actions taken under other federal laws. In addition, TSCA requires that U.S.EPA issue rules restricting the manufacturing, processing, distribution in commerce, and disposal of PCBs.

Sewage sludges contaminated with PCBs at levels greater than 50 ppm must be handled in accordance with 40 CFR Part 761, and disposed of in a hazardous waste incinerator (that complies with 40 CFR Part 761.70), a chemical waste landfill (that complies with 40 CFR Part 761.75), or an alternative method approved by the U.S.EPA. Spills of such PCB-contaminated sludges (including intentional or unintentional spills, leaks, and other uncontrolled discharges) must be cleaned up to stringent levels in accordance with the U.S.EPA's PCBs Spill Cleanup Policy (52 *FR* 10688, Subpart G of 40 CFR Part 761). POTWs that experience PCB spills resulting in the direct contamination of sewage treatment systems must immediately notify U.S.EPA and clean up the spill (40 CFR Part 761.120). POTWs may be excluded from final decontamination standards as determined on a case-by-case basis by the U.S.EPA.

Air Emissions

General air emissions are dealt with under provisions of the Clean Air Act (CAA) Amendments of 1970 (PL 91-604), 1977 (PL 95-95), and 1990 (PL 101-549). The CAA focuses on the protection and enhancement of the quality of the Nation's air resources in order to protect public health and welfare and the productive capacity of the country. A national R&D program, technical and financial assistance, emission standards, and air quality planning assistance program authorities are included. CAA provides for technical and financial assistance to state and local governments in connection with the development and execution of their air pollution and control programs, encourages and assists the development and execution of their air pollution and control programs, and initiates an accelerated national R&D program to achieve the prevention and control of air pollution. It also authorizes the development of state implementation plans (SIPs) for the purpose of meeting minimum federal ambient air quality standards and the issuance of regulations to control hazardous air pollutants and new source performance standards (i.e., emission standards).

POTWs that incinerate sludge or operate sludge dryers are currently regulated by three U.S.EPA programs under the CAA—New Source Review [either Prevention of Significant Deterioration (PSD) or nonattainment are permitted], the New Source Performance Standards (NSPS), and the National Emission Standards for Hazardous Air Pollutants (NESHAPs). POTWs may also be regulated by state standards that are more stringent than the federal CAA requirements. The U.S.EPA's program for PSD is a system whereby new sources of air pollution are subject to special requirements. The requirements are designed to protect and maintain air quality in those areas that meet the National Ambient Air Quality Standards (NAAQS) (40 CFR Part 52). NAAQS have been set for carbon monoxide, particulate matter, lead, nitrogen dioxide, ozone, and sulfur oxides. For areas meeting the NAAQS (referred to as attainment areas), increments in air pollution are specified that would be considered significant in their impact on public health and welfare. These PSD increments serve as limits on allowable additions to existing air emissions. POTWs that plan to construct an incinerator or other air pollution source in the areas meeting the NAAQS must show that they will not create pollution sufficient to violate any of the PSD increments. In addition, these facilities must employ the Best Available Control Technology for controlling the air emissions. The U.S.EPA's PSD program applies to major stationary sources that have the potential to emit or do emit more than 100 tons per year (or 250 tons per year, depending on the source category) of any pollutant covered under the CAA (40 CFR Part 52). For areas not meeting the NAAQS (nonattainment areas), POTWs may be required to achieve even more stringent emissions levels.

U.S.EPA's New Source Performance Standards (NSPS) (40 CFR Part 60) program regulates emissions from sources that threaten the NAAQS. The NSPS program restricts emissions from new industrial facilities or facilities undergoing major modifications. The NSPS standards are uniform national rules for specific industrial categories, such as utility steam boilers. The NSPS standards for sewage treatment plants currently cover only particulate matter and opacity (40 CFR Part 60). The standards specify that no sewage sludge incinerator can discharge particulate matter at a rate in excess of 0.65 g/kg dry sludge input (1.30 lb/ton dry sludge input), or discharge any gases that exhibit 20% opacity or greater. In addition, certain boilers may be subject to NSPS standards for particulate matter, sulfur dioxide, and nitrogen oxides. These standards depend upon the BTU output and type of fuel utilized by the unit. POTWs planning to construct, modify, or expand a sludge incinerator are regulated by the U.S.EPA's NSPS program.

The National Emissions Standards for Hazardous Air Pollutants (NESHAPs) program is applicable to both sludge incinerators and dryers and requires POTWs whose mercury emissions exceed 1600 g per twenty-four-hour period to test their sludge or emissions for mercury at least once every year. Emissions from any combination of sludge incineration and sludge drying cannot exceed 3200 g of mercury per twenty-four-hour period (40 CFR Part 61 Subpart C). Although the U.S.EPA anticipates that few POTWs will exceed NESHAPs standards, all facilities must demonstrate compliance through testing. POTWs that incinerate beryllium-containing waste may be subject to the beryllium NESHAP, which sets daily emission limits from the incinerator. Only POTWs receiving significant volumes of wastewater from a beryllium processing plant could potentially exceed the limit (40 CFR Part 61 Subpart E).

Landfilling and Land Application

Landfilling is covered under provisions of the Resource Conservation and Recovery Act (RCRA) of 1976 (PL 94-580) and its amendments. RCRA emphasizes the regulation of solid waste management practices to protect human health and the environment while promoting the conservation and recovery of resources from solid wastes. Technical and financial assistance, training grants, solid waste planning, resource recovery demonstration assistance, and hazardous waste regulatory program authorities are included. The key focus of RCRA implementation to date has been the comprehensive regulatory system (including a cradle-to-grave permit/manifest system) to ensure the proper management of hazardous waste. RCRA also provides for technical and financial assistance to state, local, and interstate agencies for the development of solid waste management plans and prohibits open dumping of waste. In addition, it promotes a national R&D program for improving solid waste management practices, and calls for a cooperative effort among the federal, state, and local governments and private enterprise to recover valuable materials and energy from solid wastes.

Any sludge produced by a POTW that is a hazardous waste must be stored, shipped, and disposed of in accordance with the RCRA hazardous waste regulations (40 CFR Parts 260–270). Since it is not a listed hazardous waste, sewage sludge would be considered to be a hazardous waste only if it exhibits a hazardous waste characteristic—ignitability, corrosivity, reactivity, or toxicity [most likely if it fails the Toxicity Characteristic Leachate Procedure (TCLP) as described in 40 CFR Part 261, Appendix II RCRA, and it is not further treated to pass TCLP and render it no longer hazardous]—or it is derived from the treatment of U.S.EPA listed hazardous waste received by the POTWs by truck, rail, vessel, or dedicated pipeline. Only RCRA-permitted hazardous waste treatment, storage, or disposal facilities can manage hazardous waste for incineration, land application, or landfilling. Luckily, most POTWs will not generate sludge that exhibits any of the hazardous waste characteristics as currently defined (40 CFR Part 261).

Federal criteria (40 CFR Part 257) identifying ''acceptable solid waste disposal practices''—including landfill and land application—were issued in 1979 under the joint authority of RCRA (and the CWA relative to sewage sludge). These criteria were to be used by state agencies in phasing out ''open dumping'' (i.e., unacceptable solid waste disposal practices), which is specifically prohibited under Section 405 of RCRA. Federal permits for regulating the disposal of nonhazardous waste were authorized under RCRA. Limits for annual and cumulative amounts of cadmium in sludge applied to cropland and the requirement to maintain soil pH at 6.5

or above were specified in the criteria. All sludges containing greater than 10 mg PCB/kg were required to be incorporated into the soil whenever animal feed crops were grown on sludge-amended sites by the criteria. The criteria also imposed pathogen reduction requirements for land application and disposal practices, and defined operating conditions for various sludge treatment processes that "significantly reduced pathogens" (PSRP) and processes that "further reduced pathogens" (PFRP).

In October 1991, new municipal solid waste landfill requirements (40 CFR Part 258; 56 *FR* [196]:50978–51119) were co-promulgated under the CWA and RCRA which apply to sewage sludge and other wastewater solids co-disposed of in municipal solid waste landfills for bulk disposal or used as landfill cover material.

Comprehensive Sewage Sludge Regulations

The passage of new Section 405 CWA provisions in 1977 (PL 95-217) and 1986 (PL 100-4) mandated greater direct Federal involvement in the control of sewage sludge use and disposal practices. With the issuance of new Federal technical regulations under the expanded authority of Section 405 of the CWA—40 CFR Part 258 (new municipal solid waste landfill regulations) in 1991, 40 CFR Part 122–124 and 501 (Federal permitting of sewage sludge use/disposal facilities under expanded National Pollution Discharge Elimination System (NPDES) programs or other approved State programs) in 1989, and 40 CFR Part 503 (technical standards for the use/ disposal of sewage sludge) in 1993—U.S.EPA attempted to address all sewage sludge use/disposal practices comprehensively, including cross-media considerations.

The CWA focuses on the protection, restoration, and maintenance of chemical, physical, and biological integrity of the nation's waters. The goals of the CWA are to eliminate the discharge of pollutants into navigable waters and, in the meantime, to provide for recreation and the protection and propagation of fish, shellfish, and wildlife. These goals are to be achieved through the use of appropriate control technology and management practices, or efficient reuse and reclamation of wastewater effluents. Under the CWA, states designate the uses for water bodies and establish water quality standards and criteria for pollutants to protect those uses, and policies to protect water quality and prevent its degradation. Research, standards and enforcement, water quality planning, and construction grants program authorities are included, which center on the control of both point and nonpoint discharge sources of water pollution. The law authorizes federal funding for the planning, design, and construction of publicly owned treatment works (POTWs), including sludge management facilities. It also authorizes the issuance of comprehensive sewage sludge management guidelines and regulations, the issuance of NPDES permits for point source discharges, and the development of area-wide waste treatment management plans including best management practices (BMPs) for non-point sources of water pollution, and requires the development and implementation of pretreatment standards for industrial discharges into POTWs. The 1977 amendments added several important management provisions including special incentives for greater use of innovative and alternative waste treatment technologies, broader authority to regulate sewage sludge management practices, and pretreatment credits for industrial dischargers to POTWs.

Section 405 of the CWA, entitled "Sewage Sludge," was not used very extensively to regulate the use and disposal of sewage sludge until the amendments in the Water Quality Act of 1987 greatly expanded upon the requirements previously included in this section of the law. Section 405 of the CWA, as amended by the Water Quality Act of 1987, required the U.S.EPA to develop comprehensive technical standards by specific dates which:

- Identify all major sludge use and disposal methods.
- Identify toxic pollutants that may in certain concentrations interfere with each use and disposal method.
- Establish acceptable levels of the identified pollutants for each use and disposal method to protect public health and the environment.
- Establish management practices, where necessary.

In addition, Section 405 of the CWA required that the technical standards be implemented through an NPDES permit (or another permit approved by U.S.EPA). Finally, Section 405 also required that prior to promulgation of the technical standards, the U.S.EPA must incorporate sludge conditions developed on a case-by-case basis into NPDES permits (or take other appropriate measures) to protect public health and the environment.

Development of U.S.EPA's New Part 503 Sewage Sludge Regulations

The initial round of comprehensive sewage sludge technical regulations required by Section 405 of the CWA was published on February 6, 1989, in the *Federal Register* (Vol. 54, No. 23:5746–5902) as "Proposed Standards for the Disposal of Sewage Sludge" for public comment. The technical regulations issued as 40 CFR Part 503 covered the use and disposal of sewage sludge when incinerated, applied to the land, distributed and marketed, placed in sludge-only landfills (monofills) or on surface disposal sites. Co-landfilling of sewage sludge with municipal solid waste was to be covered under the new 40 CFR Part 258 Municipal Solid Waste Landfill regulations. Since Ocean dumping was to be phased out by the end of

1991 under the provisions of the Ocean Dumping Ban Act of 1988 (PL 100-68), it was not covered by the proposed rule.

The proposed Part 503 rule contained standards for each end use and disposal method consisting of limits for twenty-eight pollutants in the form of sludge concentration limits or pollutant loading limits, as well as management practices and other requirements such as treatment works management controls over users and contractors, and monitoring, record keeping, and reporting requirements. As proposed, the requirements would apply to the final use and disposal of sludges produced by both publicly owned treatment works (POTWs), and privately owned treatment works that treat domestic wastewater and septage, but would not apply to sludges produced by privately owned industrial facilities that treat domestic sewage along with industrial waste.

Extensive public comments were received in response to the proposed rule; 650 parties submitted more than 5500 comments identifying some 250 issues. Formal comments on the regulations were received from thirty states and four environmental groups, as well as many POTWs, consultants, equipment vendors, etc. During the 180-day comment period provided on the proposal, experts from both inside and outside the U.S.EPA were involved in thoroughly reviewing the technical basis of the proposal. The review involved experts from the Agency's Science Advisory Board, environmental groups, academia, and various scientific bodies with expertise in areas covered by the proposed rule. The majority of commentators indicated that the proposed rules were overly stringent, used unrealistic conservative assumptions, and at a minimum would discourage beneficial use of sludge. Others raised questions about how to better define the sludge use and disposal categories, terms such as "de minimis" and "clean" sludge, and which models, risk assessment methodologies, and data to use for determining the proposed numeric limitations.

The Agency also conducted a National Sewage Sludge Survey (NSSS) to obtain better information on current sludge quality, use and disposal practices. The survey collected information from 479 POTWs on sludge use and disposal practices and costs, and analyzed sludges from 181 POTWs for 419 analytes—all the metals and inorganics (including pesticides, dibenzofurans, dioxins, and PCBs) for which gas chromatography/mass spectroscopy (GC/MS) standards existed. These data were used in developing regulatory impact analysis and aggregate risk analysis of human health, environmental, and economic impacts and benefits of sludge use and disposal practices to help refine the Part 503 regulations, and to help identify which additional pollutants in sewage sludge should be regulated in the future. As a result of settlement of litigation, the U.S.EPA agreed to identify additional pollutants and a schedule for a second round of sewage sludge rulemaking in June 1992.

The Agency published its analysis of the new NSSS data in the November 9, 1990, *Federal Register* notice (55 *FR* 47210) for public comment. In addition, the notice requested comments on alternative approaches for various sections of the Part 503 regulations that the U.S.EPA was considering based on comments received on the proposed Part 503 regulation and information received since the proposal. These include:

- revised approaches for regulating (1) land application of septage, (2) organic pollutants in emissions from sewage sludge incinerators, (3) the application of sewage sludge to non-agricultural land, and (4) the disposal of sewage sludge on a surface disposal site
- potential changes to the input parameters for the models used to develop pollutant limits for sewage sludge applied to agricultural land or distributed and marketed
- alternative pollutant limits (i.e., "clean sludge concept" for sewage sludge applied to the land or distributed and marketed), and
- the eligibility of a pollutant for a removal credit with respect to the use and disposal of sewage sludge

U.S.EPA utilized the comments received on the November, 1990 notice, the February 6, 1989 proposal, and the recommendations of the peer review panels to craft the final rule. A number of the external scientists involved in the peer review effort assisted the Agency in developing scientifically defendable pollutant limits and in addressing key technical issues raised in public comments. As a result of the comments received, the pollutant limits and management practices included in the proposed regulations, and even some of the basic approaches for regulating sewage sludge, changed significantly.

Recent revisions to U.S.EPA's program for controlling PCBs under TSCA have lead to additional controls being placed on materials with <50ppm PCBs. However, based upon the risk assessment conducted during the development of the Part 503 rule and the NSSS data on current PCB levels in sewage sludge, it was determined that there was no need for further TSCA controls on PCBs to be applied to sewage sludge.

The New Part 503 Regulations and Their Implementation

On February 19, 1993, the final 40 CFR Part 503 rule "Standards for the Use or Disposal of Sewage Sludge" was published in the *Federal Register* (58 *FR* [32]:9248–9415). Relatively minor changes to the Part 503 regulation were published in the *Federal Register* on February 25, 1994 (59 *FR* [38]:9095–9099) and on October 25, 1995 (60 *FR* [206]:54764–54792).

Prior to this, on May 2, 1989, revisions to the 40 CFR Parts 122, 123 and 501 "NPDES Permit Regulations; State Sludge Management Program Requirements" were published in the *Federal Register* (54 *FR* [83]:18716–18796). A "Sewage Sludge Interim Permitting Strategy" was issued in September 1989, describing the Agency's strategy for carrying out the new Section 405 CWA requirements to impose controls on sewage sludge use and disposal practices in NPDES permits issued to POTWs until the new Part 505 technical standards became effective. Pursuant to the "Interim Strategy," U.S.EPA or the states could issue sludge permits as agreed to by the state/U.S.EPA agreements. A "POTW Sludge Sampling and Analysis Guidance Document" (along with a video) was issued in August 1989 to provide technical guidance on the sampling and analysis of municipal sewage sludge. Draft guidance for writing case-by-case permit requirements for municipal sewage sludge was issued in May 1990; it was updated in March 1993, and again in October 1995 as final "Part 503 Implementation Guidance." A "State Sludge Management Program Guidance Manual" was issued in October 1990.

The new technical regulations are risk based in nature, and although Federal permits can now be involved in the control of sludge (biosolids) management practices, they are designed to be self-implementing. Still, the choice of use/disposal practice (with the exception of ocean disposal—ocean dumping was banned Ocean Dumping Ban Act of 1989 and ocean discharge is prohibited by provisions of the Clean Water Act) will continue to be up to the generator. This potentially opens the door to more consistent regulation of biosolids and could well be extended to other non-hazardous residuals in the future as a result of future rulemaking under the RCRA.

The U.S.EPA rule addresses beneficial use practices involving land application as well as surface disposal and incineration of biosolids. They affect generators, processors, users and disposers of biosolids—both public and privately owned treatment works treating domestic sewage (including domestic septage haulers and non-dischargers), facilities processing or disposing of biosolids, and the users of biosolids and products derived from biosolids.

The Part 503 regulation addresses the use and disposal of only biosolids, including domestic septage, derived from the treatment of domestic wastewater. It does not apply to materials such as grease trap residues or other non-domestic wastewater residues pumped from commercial facilities, sludges produced by industrial wastewater treatment facilities, or grit and screenings from POTWs.

Although facilities which dispose of biosolids in municipal solid waste (MSW) landfills or use processed biosolids as a cover material are regulated under the Part 258 rules issued in 1991, all treatment works treating domestic sewage (TWTDS), including non-dischargers and sludge-only facilities, were required to apply for a federal permit from the U.S.EPA (or an approved state program). TWTDS that land apply their biosolids and have existing NPDES permits must apply for their federal "sewage sludge" permit as a part of their next NPDES permit renewal application, while those without existing NPDES permits were required to apply for their permit within one year of publication of the new regulation (i.e., February 19, 1994).

While disposal facilities such as sewage sludge incinerators, monofills and other surface disposal sites are clearly TWTDS and are required to apply for federal permits, the definition does not extend automatically to areas such as farm land where biosolids are beneficially used; only under unusual situations would these areas be considered TWTDS by the permitting authority. The permitting authority has the flexibility to cover both the generator and the treatment, use/disposal facility in one permit or separate permits (including covering one or both under general permits).

The U.S.EPA is working closely with the states to encourage their adoption of biosolids regulatory programs that can be approved to carry out delegated programs and avoid the need for separate U.S.EPA permits, compliance monitoring and enforcement activities. Utah received the first (and to date the only) state program approval in 1996. Until a state applies for and is approved to carry out a delegated program, all TWTDS in the state will be dealing directly with their U.S.EPA Regional Office regarding federal permits, compliance monitoring and enforcement issues associated with the implementation of the Part 503 requirements—*in addition* to dealing with their state regulatory authorities and requirements.

By statute, compliance with the Part 503 standards was required within 12 months of publication of the new regulation (i.e., February 19, 1994). If pollution control facilities needed to be constructed to achieve compliance, then compliance was required within two years (i.e., February 19, 1995) of publication. Under the new regulation, compliance with the monitoring and recordkeeping requirements was required within 150 days (i.e., July 20, 1993) of publication of the rule.

For the most part the Part 503 regulation was written to be "self implementing," which means that citizen suits or U.S.EPA can enforce the regulation even before permits are issued. As a result, treatment works must start monitoring and keeping records of biosolids quality (and in many cases land appliers must start keeping records of loading rates and locations receiving biosolids), and must comply with pollutant limits and other technical standards, even in the absence of a federal permit.

Part 503 is organized into the following subparts: general provisions, land application, surface disposal, pathogens and vector attraction reduction, and incineration. Subparts under each of the use/disposal practices generally address: applicability, general requirements, pollutant limits, management practices, operational standards, frequency of monitoring, recordkeeping, and reporting requirements.

Under Part 503, *Land Application* includes all forms of applying biosolids to the land for beneficial uses at agronomic rates (rates designed to provide the amount of nitrogen needed by the crop or vegetation grown on the land while minimizing the amount that passes below the root zone). These uses include: application to agricultural land such as fields used for the production of food, feed and fiber crops, pasture and range land; non-agricultural land such as forests; disturbed lands such as mine spoils, construction sites and gravel pits; public contact sites such as parks and golf courses; and home lawns and gardens. The distribution and marketing of biosolids derived materials such as composted, chemically stabilized or heat dried products is also addressed under land application, as is land application of domestic septage.

The rule applies to the person who prepares biosolids for land application or applies biosolids to the land. These parties must obtain and provide the necessary information needed to comply with the rule. For example, the person who prepares bulk biosolids that is land applied must provide the person who applies it to land all information necessary to comply with the rule, including the total nitrogen concentration of the biosolids.

The regulation establishes two levels of biosolids quality with respect to ten heavy metal concentrations—pollutant Ceiling Concentrations and Pollutant Concentrations ("high quality" biosolids); two levels of quality with respect to pathogen densities—Class A and Class B; and two types of approaches for meeting vector attraction reduction—biosolids processing or the use of physical barriers. Under the Part 503 regulation, fewer restrictions are imposed on the use of higher quality biosolids. Based upon the results of the NSSS published in November 1991, U.S.EPA anticipates that a large percentage of the biosolids currently being produced will be capable of meeting the "high quality" pollutant concentrations.

To qualify for land application, biosolids or material derived from biosolids must meet at least the pollutant Ceiling Concentration Limits, Class B requirements for pathogens and vector attraction reduction requirements. Cumulative Pollutant Loading Rates are imposed on biosolids that meet the pollutant Ceiling Concentrations but not the Pollutant Concentrations. A number of general requirements and management practices apply to biosolids that are land applied other than an "Exceptional Quality" biosolids or derived material which meets three quality requirements—the Pollutant Concentration limits, Class A pathogen requirements, and vector attraction reduction biosolids processing. However, in all cases the minimum frequency of monitoring, recordkeeping, and reporting requirements must be met.

Under Part 503, *Surface Disposal* addresses biosolids and septage disposal, including sludge-only monofills; dedicated disposal surface application sites (where sludge pollutants are applied at higher than the Cumulative Pollutant Loading Rates allowed under land application); piles or mounds and impoundments or lagoons where sludge is placed for final disposal. It is not intended to include the placement of biosolids in similar locations for storage or treatment; however, the facility operator will need to provide an adequate explanation concerning why it is being stored for longer than two years.

For surface disposal, the regulation establishes requirements for both unlined facilities and those with liners and leachate collection systems. These include concentration limits for three heavy metals (As, Cr & Ni) that apply to biosolids disposed of in lined surface disposal sites, but no specific concentration limits for biosolids going to sites with liners and leachate collection systems. Specific management practice requirements address such areas as the location of surface disposal sites, control of surface runoff, methane gas monitoring, and restrictions on crop production, grazing and public access. A provision allowing for the establishment of site-specific limits and management practices for surface disposal sites is provided.

Part 503 also establishes requirements for sewage sludge *Incineration.* The rule applies to the person who fires sewage sludge in a sewage sludge incinerator, to the sewage sludge fired in the incinerator, and the exit gas from the sewage sludge incinerator stack. It does not apply to incinerators in which hazardous waste (as defined by 40 CFR Part 261) are fired, or to the firing of sewage sludge containing ≥50 ppm concentrations of PCBs. It also does not apply to incinerators that co-fire sewage sludge with other wastes (although up to 30% MSW as auxiliary fuel is not considered "other wastes"). Furthermore, this rule does not apply to the ash produced by a sewage sludge incinerator.

The rule limits concentrations of heavy metals in sewage sludge fed to the incinerator and the concentration of total hydrocarbons (THC) or carbon monoxide (CO) in the exit gas from sewage sludge incinerator stacks, and establishes management practices, as well as frequency of monitoring, recordkeeping and reporting requirements. The rule contains equations to calculate the allowable concentration of metals in the sewage sludge fed to the incinerator. Sewage sludge incinerators are required to conduct performance tests to determine pollutant control efficiencies for heavy metals, and to conduct site-specific air modeling to determine a dispersion factor for their site. Continuous emissions monitoring equipment is also required to be installed.

Pollutant limits for sewage sludge fired in a sewage sludge incinerator are imposed for beryllium, mercury, arsenic, lead, cadmium, chromium, and nickel. The limits for beryllium and mercury are those that already exist under the National Emission Standards for Hazardous Air Pollutants (NESHAPS; 40 CFR Part 261). Pollutant limits for the remaining metals are determined using site-specific performance characteristics and emission dispersion modeling results. Incinerators must also meet a monthly average stack exit gas concentration of 100 ppm for either THC or CO (not both), corrected for moisture level (for zero percent) and oxygen content (to 7%). The allowable THC or DO concentrations are an indicator of the toxic organic compound concentrations in the gas. Parameters, such as percent oxygen, information used to determine moisture content, maximum temperature, and values for the operating parameters for the air pollution control device also must be monitored.

The Part 503 regulation requires minimum frequency of monitoring biosolids pollutant quality (but not septage) based on the annual amount of biosolids used or disposed; maintenance of records (in most cases for a minimum of 5 years) regarding such information as biosolids quality, application sites, application dates, various certification statements and descriptions of management practices, pathogen and vector reduction measures used. Annual reporting is required only for Class I biosolids facilities (the ~1,600 pretreatment POTWs and an estimated 400 other facilities likely to be designated TWTDS), and other POTWs with a design flow of ≥1 MGD or serving a population ≥10,000.

For the most part, Part 503 is a risk-based regulation designed to protect public health and the environment from reasonable worst case situations. Models were established to facilitate the evaluation of fourteen major pathways of exposure (e.g., sludge→soil→plant→human; sludge→soil→ soil biota; etc.) associated with land application practices. Fifty pollutants of potential concern were evaluated in great detail using the available scientific data. Acceptable biosolids pollutant concentration limits and loading rates were calculated based upon conservative assumptions and endpoints previously established as adequate to protect public health and the environment. A summary of the extensive risk assessment effort and documentation was published by U.S.EPA in 1995 [29]. Land application pollutant limits were established for only ten metals (note that Cr was dropped from coverage in October 1995) in the final rule because the Agency determined that it would not establish numerical pollutant limits for any pollutants meeting one of the following criteria:

(1) The pollutant is banned or restricted by the Agency or no longer manufactured or used in manufacturing a product.

(2) The pollutant is not present in biosolids at significant frequencies or detection based on data gathered from the NSSS.

(3) The Agency's risk assessment for the pollutant showed no reasonably anticipated adverse effects on public health or the environment at the 99th-percentile concentration found in biosolids from the NSSS.

Concerns have been raised over some of the scientific data, assumptions and models used in developing the land application numerical limits. As required by the CWA (and several pending law suits), future rounds of rulemaking are expected to address these concerns and additional pollutants that may be present in biosolids. In materials submitted to the court during May 1993 in response to pending law suits on the final Part 503 regulation, U.S.EPA provided a list of 31 additional candidate pollutants that the Agency intended to further evaluate via research and risk assessment, and propose for regulation no later than December 15, 1999. A schedule for developing and issuing a Round II rule has been established. After full evaluation of the Table 14 list, U.S.EPA announced on November 28, 1995 its plans to limit the Round II rulemaking effort to Dioxins/dibenzofurans (all monochloro- to octochloro-congeners), and co-planar PCBs.

Areas such as the long-term fate of some land applied pollutants in biosolids relative to plant uptake rates, surface runoff and groundwater movement, and the potential impacts (both positive and negative) on wildlife and unmanaged ecosystems are ripe for further research due to the limited amount of field data currently available. Future attempts to make the pathogen control portion of the rule more "risk-based" will also require additional research efforts.

Environmental Impact Assessment and Review

Under provisions of the National Environmental Policy Act (NEPA) of 1969 [PL 91-190], if a POTW receives federal funding, the facility is required to plan its policies and actions in light of the environmental consequences under the NEPA. The POTW must prepare an environmental impact statement (EIS) for any major action that will significantly affect the quality of the human environment. Such actions could include expansion of the facility or changing the location of the sludge disposal facility. The EIS must identify and address the environmental impact of the proposed action and identify, analyze, and compare options according to the requirements of 40 CFR Part 1500.

Superfund Cleanup and Right-to-Know

These topics are dealt with under provisions of the Comprehensive Environmental Response, Compensation, Liability Act (CERCLA) of 1980 [PL 96-150] and its amendments under the Superfund Amendments and Reauthorization Act (SARA) of 1986 [PL 99-499]. CERCLA or "Superfund" provides broad federal authority to respond directly to releases or threatened releases of hazardous substances. This law also provides for the cleanup of inactive or abandoned hazardous waste sites. Under CERCLA, U.S.EPA assesses the nature and extent of contamination at a site, determines the public health and environmental threats posed by a site, analyzes the potential cleanup alternatives, and takes action to clean up the site. POTWs that dispose of sludge in impoundments or landfills that become Superfund sites may be required to help pay for cleanup of those sites.

Title III of SARA, which amended CERCLA, created a new program to increase the public's knowledge of and access to information on the presence of hazardous chemicals in their communities and releases of these chemicals into the environment. Title III requires facilities, including POTWs, to notify state and local officials if they have extremely hazardous substances present at their facilities in amounts exceeding certain "threshold planning quantities." If appropriate, the facility must also provide material safety data sheets on hazardous chemicals stored at their facilities, or lists of chemicals for which these data sheets are maintained, and report annually on the inventory of these chemicals used at their facility. The law may also require certain POTWs to submit information each year on the amount of toxic chemicals released by the facilities to all media (air, water, and land).

If the sludge from a POTW contains any CERCLA hazardous substance (there are currently 725 substances on the list of hazardous substances covered under Superfund [40 CFR Part 302.4]) and the sludge is released into the environment in an amount equal to or greater than its reportable quantity within a twenty-four-hour period, the POTW is subject to the reporting requirements under Section 103(a) of CERCLA and Section 304 of SARA [50 *FR* 13462 (April 4, 1985)]. For example, if 1000 pounds of sludge containing 1 pound or more of 1,4 dichloro-2-butene are released to the environment in one day, the release containing this hazardous substance must be reported to the National Response Center and state and local authorities. There is an exception to these reporting requirements for federally permitted releases and the reduced reporting requirements for continuous releases of hazardous substances.

POTENTIAL FOR LAWSUITS AND FINANCIAL LIABILITY UNDER RCRA AND CERCLA ASSOCIATED WITH LAND APPLICATION OF MUNICIPAL SEWAGE SLUDGE

A number of concerns have been raised over the potential for lawsuits and financial liability associated with land application of municipal sewage sludge. The following discusses a number of these questions.

Does the beneficial use of sludge fall within the CERCLA exemption for the "normal application of a fertilizer"? To the extent that the placement of sludge on land is for the purpose of providing fertilizer to the soil, and is applied in the normal manner (i.e., not spilled or dumped on the ground) and in the normal concentrations, that placement would not constitute a "release," and thus should not give rise to CERCLA liability. However, neither the statute nor the legislative history of CERCLA state whether the exclusion for the initial application of the fertilizer extends also to any later contamination of groundwater caused by the chemicals in the fertilizer applied to the ground. The exclusion may only apply to the initial application and not to later contamination from the application. On the other hand, the legislative history indicates a concern that the applicators not be held liable for normal applications of fertilizer, the exclusion may have very little meaning if it covers only the initial application and not any later contamination.

The U.S.EPA has not yet taken a position on this issue. To the extent that the Agency takes the narrow approach to the exclusion, landowners could remain liable for cleanup costs associated with the contamination from chemicals in fertilizer and sludge (e.g., groundwater contamination). If the sludge is used as a fertilizer in accordance with the CWA requirements, the likelihood of a need to respond to residual contamination from the "fertilizer" would at most be quite remote.

Is sewage sludge applied to the land authorized under and in compliance with a permit required by Section 405 of the CWA a "federally permitted release" for the purposes of CERCLA? In certain circumstances, the application of sewage sludge to land in compliance with a permit required under Section 405 of the CWA may be a federally permitted release as defined in CERCLA.

Releases in compliance with sludge permits issued under U.S.EPA (or U.S.EPA-authorized state) NPDES programs *would* qualify as federally permitted releases under CERCLA. In some cases, however, permits may be issued under U.S.EPA-approved state solid waste programs. Such permits would not be NPDES permits and thus releases under such permits would seem not to constitute federally permitted releases for purposes of CERCLA liability based on a strict reading of the limited NPDES exception.

However, standard NPDES permit conditions require compliance with federal sludge use and disposal standards as well as implementation of these standards in the NPDES permit unless included in another permit. Therefore, it could be argued that releases authorized under and in compliance with the non-NPDES permit would effectively constitute a release under the NPDES permit. Such a release consequently could be a federally permitted release even though the specific permitting vehicle was a state solid waste permit so long as the non-NPDES permit was linked to a NPDES permit that required compliance with federal sludge use/disposal standards. The U.S.EPA has not yet taken a position on this issue.

In a substantial part of the universe of sludge land appliers, most of whom are likely to be (or linked to by contract) POTWs with NPDES permits, land application practices *would* be able to qualify as "federally permitted releases" for purposes of CERCLA.

Is sewage sludge applied to the land as a fertilizer a "solid waste" within the meaning of RCRA? Section 1004(27) of RCRA defines "solid waste" as "garbage, refuse sludge from a waste treatment plant, water supply treatment plant, or air pollution control facility and other discarded material. . . ." The use of the term *discarded material* in this definition has led to a long-standing debate about whether secondary materials that are recycled or reused are "solid wastes" and thus within the jurisdictional ambit of RCRA.

While the U.S.EPA has from time to time modified its views as to when recycled secondary materials are "solid wastes" under RCRA, it has been the Agency's consistent position since 1979 that sewage sludge used as a fertilizer *is* a solid waste [40 CFR 257.3–5; 44 *FR* 53438, 53449–53455 (Sept. 13, 1979); 40 CFR 260.10, 261.4—1981; 40 CFR 261.2(c) (1)—1990]. The 1984 amendments to RCRA and their legislative history also indicate that uses of secondary materials constituting disposal involve "solid wastes."

In 1985, the U.S.EPA amended its hazardous waste regulations to better define when secondary materials (such as sludges and by-products) which are recycled are "solid wastes." The U.S.EPA took the position that all sludges used in a manner constituting disposal (i.e., under RCRA this means placed on the land) were "solid waste"; thus, under the amended hazardous waste regulations, sludges used as a fertilizer would continue to be considered a solid waste.

The U.S.EPA's 1985 regulations have been challenged directly or indirectly in a number of cases. There are now three D.C. circuit decisions addressing the issue of when recycled secondary materials are solid wastes although none directly address the issue of whether secondary materials used as fertilizer are "solid wastes." However, one consistent principle that emerges from them is that recycled secondary materials that are "part of the waste disposal problem" are "solid wastes" under RCRA.

It is unlikely that sewage sludge that is allowed to be used as a fertilizer under Section 405 of the CWA would ever be considered a hazardous waste under RCRA. Thus, whether it is a "solid waste" for purposes of the hazardous waste program would appear to be largely irrelevant as a practical matter. As a nonhazardous solid waste, sewage sludge used as a fertilizer may be subject to corrective action (if disposed of at a hazardous waste facility requiring an RCRA Subtitle C permit—which is *not* the case for land application to cropland, rangeland, etc.) or an abatement action under RCRA 7002 and 7003. However, if the application of the sludge is in compliance with the CWA Section 405 guidelines (i.e., Part 503 regulations), the likelihood of its creating a health or environmental problem that would necessitate some type of remedial or abatement action is extremely remote.

Could a land management agency or landowner who owns land upon which sludge has been applied for beneficial purposes be a liable party under CERCLA? Yes. CERCLA Section 107(a) identifies four broad classes of responsible parties that are liable for the response costs incurred at a facility from which there has been a release or threat of release of a hazardous substance, including: (1) the owners or operators of a vessel or facility, (2) persons who owned or operated a facility at the time hazardous substances were disposed of at the facility, (3) persons who arranged for disposal or treatment of the hazardous substances, and (4) persons who transported the hazardous substances and selected the disposal facility [42 U.S.C. Section 9607(a) (1)–(4)].

The fact that the relevant land management agency is a part of the United States government does not absolve it from the potential liability under CERCLA. Nor does the beneficence of the sludge application necessarily release the Federal land management agency (or other landowner) from CERCLA liability. This is because the terms "owner or operator," "disposal," "release," and "hazardous substance" are all very broadly defined under CERCLA. No distinction is made between the beneficial placement of solid wastes (such as treated sewage sludge containing small quantities of hazardous substances) on land and the dumping of leaking barrels of hazardous wastes into an open pit. However, the likelihood is remote that the beneficial use of treated sewage sludge on land in compliance with the Part 503 regulations would result in the incurrence of a response cost for which liability under CERCLA is premised.

Can U.S.EPA exempt sewage sludge containing "de minimis" amounts of hazardous substances or hazardous constituents from the liability provisions of CERCLA or from RCRA requirements? CERCLA does not appear to authorize U.S.EPA to exempt wastes containing "de minimis" amounts of hazardous substances from the definition of "hazardous substance." The United States has repeatedly—and successfully—argued in CERCLA cost-recovery and enforcement cases that the release of even small amounts of hazardous substances triggers CERCLA liability.

While sewage sludges are not a listed hazardous waste under RCRA, they may be regulated as hazardous waste if they exhibit an RCRA characteristic (most likely the toxicity characteristic) or are derived from a listed waste. It would be at least theoretically possible for the U.S.EPA to exempt sewage sludges from RCRA Subtitle C requirements if the Agency could establish that they would not be expected to be identified as a hazardous waste under any RCRA characteristic. Given the variability of sewage sludge and the fact that it generally contains many of the hazardous constituents tested for under the toxicity characteristic, it may be difficult to build the necessary technical record to make this demonstration. Because sewage sludge used as a fertilizer is still a solid waste, it might be subject to an imminent or substantial endangerment action under RCRA when used at a facility that requires a Subtitle C permit. RCRA Sections 7002 and 7003 authorize citizens and the U.S.EPA to seek abatement of imminent and substantial endangerment caused by disposal of solid waste.

If sewage sludge is used as a fertilizer in accordance with CWA Section 405 guidelines (i.e., Part 503 regulations), the likelihood of its creating an environmental or health problem that would trigger a CERCLA response, RCRA corrective action, or an imminent or substantial endangerment is extremely remote.

Does the domestic sewage exemption in Section 1004 of RCRA provide any defense against potential suit for beneficial use of sludge? No. Under RCRA, the term *solid waste* does not include "solid or dissolved materials in domestic sewage" [RCRA Section 1004(27)]. It has been the U.S.EPA's long-standing position that this "domestic sewage exemption" does not extend to residuals from the treatment of domestic sewage, e.g., sewage sludge [44 *FR* 53440 (Sept. 13, 1979) and 53 *FR* 33321–33322 (Aug. 18, 1988)]. Thus, the domestic sewage exemption does not exempt sewage sludge from RCRA solid waste requirements.

RCRA defines "hazardous waste" as a subset of "solid waste" [RCRA Section 1004(5)]. Thus, the domestic sewage exemption does not exempt sewage sludge from RCRA hazardous waste requirements.

CERCLA incorporates the domestic sewage exemption in RCRA Section 1004 through the definition of hazardous substances in CERCLA Section 101(14)(C). This latter provision identifies as one category of hazardous substances "any hazardous waste having the characteristics identified under or listed pursuant to Section 3001 of [RCRA]." Since the domestic sewage exemption does not exclude sewage sludge from the definition of hazardous waste, it also does not exclude it from the list of Section 101(14)(C) hazardous substances.

Even if sewage sludge were excluded from the list of Section 101(14)(C) hazardous substances by virtue of the domestic sewage exemption, sewage sludge would very likely contain other categories of hazardous substances {e.g., CWA Section 311 hazardous substances [CERCLA Section 101(14)(A)], CWA Section 207(a) toxic pollutants [(CERCLA Section 101(14)(D)]}.

In light of the decades of experience in the private sector, is there any record of financial liability actually arising from application of sewage sludge in beneficial use projects? The Agency has not conducted a comprehensive search of reported cases involving the beneficial use of sewage sludge to determine whether any of these cases have authorized the recovery of cleanup costs or damages. However, there are no known lawsuit judgments where plaintiffs have been successful in recovering cleanup costs, or in obtaining compensation for damage to property or persons, resulting from the use of sewage sludge as a fertilizer. There have, however, been a number of cases and actions brought by the U.S.EPA, states, and citizen groups to enforce regulatory requirements for the landspreading of sewage sludge.

What is the potential for liability for the beneficial use of municipal sewage sludge on land? The potential liability to a land management agency or landowner arising from a beneficial use project depends on two factors: the likelihood of harm occurring which would then lead to legal action, and the existence of a legal basis to support such an action. Despite decades of experience with beneficial use in the private sector, and the fact that about half of the nation's sewage sludge is currently being put to beneficial use, there are no known judgments in connection with beneficial use projects.

A hypothetical analysis of the possible legal basis for litigation in connection with a flawed beneficial use project suggests that the incremental liability risk for a land owner, beyond what already exists in connection with practices in more common use on public and privately owned lands, is insignificant, and could generally be managed satisfactorily through conformance with applicable regulations and guidance.

The question of liability need not be an obstacle if the land management agencies and private landowners undertake beneficial use projects in conformance with applicable law and regulations, appropriate permit conditions, and federal policy.

HISTORICAL DEVELOPMENT OF KNOWLEDGE, RESEARCH, AND REGULATIONS FOR SEWAGE SLUDGE UTILIZATION UP TO 40 CFR PART 503

INTRODUCTION

Early Land Application Activities

The disposal of human wastes has probably been a matter of concern since man has congregated in cities, towns, and villages. De Turk [1] reports that the earliest account he found of sewage sludge utilization on land was in Bunzlau, Germany, where in 1559 the sewage was allowed to flow out onto land used for growing crops. In 1869, the practice was implemented in Berlin, Germany, which bought large areas of cropland to be irrigated with raw sludge. The city of Paris, France also purchased farmland to be irrigated with sewage. In 1897, Melbourne, Australia went to land treatment of raw wastewater at Weribee [2]. These land treatment systems have been very successful in treating wastewater with no major environmental impacts, and they were implemented without imposed regulations.

Early workers saw the value of sewage sludge being used as fertilizer on land. Muller [3] obtained satisfactory growth of grasses in sand cultures containing dried sewage sludge. Rudolfs [4] determined the fertilizer value of sludge from different treatment plants throughout the United States. De Turk [1] evaluated the potentials of activated and anaerobically digested (Imhoff) sewage sludges as a fertilizer. American cities such as Milwaukee and Toledo marketed "Milorganite" and "Tol-e-Gro," respectively, as fertilizers with much success. Lunt [5] reported the results of testing to use digested sewage sludge for soil improvement and gave recommendations for its safe, effective use.

Fulton County, Illinois Land Reclamation Site

The Metropolitan Water Reclamation District of Greater Chicago (MWRDGC) by late 1960 was confronted with a serious problem of what to do with solids that were removed from the treatment process. In 1967, the MWRDGC began an extensive program of research and testing [6]. The most significant project initiated was a cooperative effort between the MWRDGC and University of Illinois (Dr. T. D. Hinesly and others) under the sponsorship of a federal research grant from the U.S.EPA. The purpose of this project was to demonstrate the possible agricultural benefits and environmental changes occurring from applying digested sewage sludge to field crops. Dalton and Murphy [7] discuss the history of the MWRDGC's move toward using land reclamation as a means of recycling solids. The concern about solids utilization led to the development of the Fulton County, Illinois site [7]. All of these activities and research projects were developed to demonstrate that sewage sludge could be safely used on agricultural lands and for reclamation of disturbed lands. The MWRDGC conducted these activities and developed the Fulton County site without federal regulations.

DEVELOPMENT OF LAND APPLICATION KNOWLEDGE AND REGULATORY STRATEGIES

Land application of sludges and wastewaters was recognized as an alternative method for effecting stages of wastewater treatment and for ultimate disposal of solid wastes under the Federal Water Pollution Control Amendments of 1972 [8]. For certification and shared-cost funding under this legislation, a waste treatment proposal must include evidence that the plan is based on "the best practicable technology" and "the most cost-effective method(s) over the life of the works." Requirements for compliance were to be phased over all pollutants into navigable surface waters by 1985.

1973 Champaign, Illinois Conference

The recognition that the land application of sludges and wastewaters was a variable alternative for treatment in 1972 led the leaders of U.S.EPA, USDA, and the National Land Grant Universities to create a Coordinating Committee on Environmental Quality. The committee recognized that the utilization of land as a treatment medium required a coordinated effort by multidisciplinary interests. A subcommittee entitled "Recycling Municipal Effluents and Sludges on Land" was created to effectively use the resources available within the U.S.EPA-USDA-University structures for a cooperative and coordinated research, development, and demonstration program [9].

The initial task for this subcommittee was to identify what was known about liquid effluent and sludge application to the land, and what research was needed for successful utilization of land as a soils treatment system from economic, engineering, health, and esthetic points of view [9]. A research needs workshop was held on July 9–13, 1973, in Champaign, Illinois to summarize knowledge currently available and to identify and recommend areas of research.

The impact of the conference was far-reaching in that the most knowledgeable people were brought in to discuss their knowledge about the various aspects of applying effluents and sludges on land. The result of this conference was to focus management in government agencies and university researchers on the need to develop research programs to address the research needs related to effective utilization of sludge on land.

Activities of the MWRDGC in developing the Fulton County, Illinois, site into a resource to recycle sewage sludge on minesoil was already underway with sludge applications at the time the Champaign, Illinois Conference occurred. The MWRDGC sponsored cooperative research using anaerobically digested sewage sludge from its Stickney treatment plant on cropped research plots and lysimeters at the University of Illinois Elwood agricultural experiment station. This research, at the time the conference was held, was into its sixth year.

Metal Loadings

One of the concerns about sludge application to land after the 1973 Champaign, Illinois Conference was the impact of sludge applied metals on crops and the food chain. At this time, no regulations had been developed for land application of sewage sludge. Chaney [10] expressed the opinion that boron, cadmium, cobalt, chromium, copper, mercury, nickel, lead, and zinc were a potential hazard to plants or the food chain and the amounts of these elements applied in sludge needed to be limited.

At about the same time, members of the British National Agricultural Advisory Service (NAAS) investigated sites where poor growth of crops was observed following applications of sewage sludge. Chumbley [11], in preparing the advisory service report, recommended on the basis of advisory experience and pot experiments the "zinc equivalent concept." He proposed a maximum accumulation of 560 kg/ha of zinc equivalent, where the zinc equivalent is 1 zinc, 2 copper, and 8 nickel. The coefficients of the equation reflect the relative phytotoxicities of the three metals. Leeper [12] suggested combining the zinc equivalent approach with the soil cation exchange capacity (CEC). He proposed a maximum zinc equivalent accumulation of 5% of the CEC.

In 1974, U.S.EPA published a draft technical bulletin that proposed to increase the allowable zinc equivalent to 10% of CEC and decrease the relative phytotoxicity of nickel with respect to zinc from eight to four. In addition, the draft bulletin proposed that the cadmium:zinc ratio not exceed 1:100. The cadmium:zinc ratio was proposed by Chaney [10] as a means of preventing excessive cadmium uptake, with the rationale being that zinc would become phytotoxic before cadmium had accumulated to excessive levels.

A Cooperative State Research Technical Committee (NC-118) met in 1974 to review the proposed U.S.EPA guidelines. The committee suggested several items that are described by Logan and Chaney [13]. Among the more significant suggestions were that the annual rate of cadmium application should not exceed 2.24 kg/ha (2 lb/acre) and the cumulative cadmium application should not exceed 11.2 kg/ha (10 lb/acre). This was based on the assumption that after five years at the maximum annual application rate of 2.24 kg/ha, the levels of soil cadmium still be within the range found for natural mineral soils. The committee also indicated that allowable accumulations of lead, nickel, zinc, and copper should reflect the current available knowledge about their relative phytotoxicities. Consequently, maximum accumulations for medium textured soils in the North Central region of the U.S., where soil pH would be ≥ 6.5, were:

Trace Element	lb/acre	kg/ha
Pb	1000	1120
Zn	500	560
Cu	250	280
Ni	100	112
Cd	10	11.2

These loading limits were intended to be lifetime accumulations. The concepts expressed by Chumbley, Leeper, and Chaney concerning metals in land applied sewage sludge resulted in considerable research activity, with a variety of opinions. The MWRDGC sponsored research related to use of sludge for pasture reclamation, with emphasis on parasitology and the occurrence of heavy metals in soil, vegetation, and animal tissue, with Dr. Fitzgerald, College of Veterinary Medicine, University of Illinois at the Fulton County site. In addition, sponsored research at the University of Illinois with Drs. Hinesly, Jones, and Hansen on effects of using sewage sludge on agricultural and disturbed lands was well underway since 1968. Activities included field lysimeter studies, long-term plot studies, and feeding studies with pheasants and swine.

1976 Council of Agricultural Science and Technology (CAST) Report

On June 3, 1976, a proposed U.S.EPA technical bulletin entitled "Municipal Sludge Management: Environmental Factors" was published in the *Federal Register* for public comment. During the development of this document by an interagency workgroup, considerable concern and conflicting opinions were expressed regarding the merits and potential hazards of applying sewage sludge to agricultural lands. The U.S.EPA requested that CAST convene a task force to review the most recent research, especially field research, on the application of sludge to cropland. The CAST was to prepare a consensus statement on the current understanding of the relationships among the metals applied in the sludge, the chemical and physical properties of the soils, the soil and crop management practices, and the plant growth and uptake of these metals by plants. The report was developed by a multidisciplinary task force of scientists who were actively engaged in research on application of sewage sludge to cropland.

The CAST [16] reviewed the currently available suggestions made to limit the application of metals in sewage sludge to land. The suggestions included the "zinc equivalents" equation, a zinc-to-cadmium ratio, and the cumulative amounts of the metals supplied. The report concluded that each method has its limitations and that none of the methods used alone was universally applicable.

The CAST [16] also reviewed all the metals present in sludge and evaluated their potential hazard to plants, animals, or humans. The metals manganese, iron, aluminum, chromium, arsenic, selenium, antimony, lead, and mercury, with correct management practices, pose relatively little hazard to crop production and plant accumulation when sludge was applied to soil. The remaining metals, cadmium, copper, molybdenum, nickel, and zinc, because they can accumulate in plants, may pose a hazard to plants, animals, or humans under certain circumstances.

During this time, the NC-118 Committee was working on the development of suggested maximum rates of metal application to land. They felt that the maximum safe applications of individual metals may differ among soils as a result of differences in CEC of the soils and differences in relative toxicity of the various metals. At the request of the USDA, the NC-118 Committee modified their guidelines to distinguish between fine, medium, and coarse textured soils [13]. The allowable nickel addition was increased to the same level as copper. These guidelines were subsequently adopted by many state regulatory agencies.

The CAST [14] indicated that the concentration of sludge-borne metals in plants was related more closely to the total amount of metals applied than to the concentration of the metals in the sludge. Consequently, a limit on the concentration of metals in sludge would not necessarily protect plants and animals from the problems that might result from applying excessive amounts of metals to land. The CAST [14] also presented several management options that could be used to keep the concentration of cadmium in food and feed crops at a low level on sludge-treated land. One of these options was to maintain the soil pH at or above 6.5.

IMPLEMENTATION OF 40 CFR PART 257 REGULATIONS AND CONTINUING RESEARCH

Research continued on the land application of sewage at many universities and the MWRDGC's Fulton County site. In 1978, the U.S.EPA printed "Solid Waste Disposal Facilities—Proposed Classification Criteria, Part II" in the *Federal Register* on February 6, 1978. The proposed regulations contained minimum criteria for determining which solid waste land disposal facilities should be classified as posing no reasonable probability of adverse effects on health or the environment. Numerous and extensive responses were submitted to the U.S.EPA. The MWRDGC submitted extensive comments based on available research information, sponsored research, and its own activities. In addition, Cooperative State Research Service Technical Committee (W-124) submitted a response to the U.S.EPA.

The U.S.EPA reviewed these comments and published "Criteria for Classification of Solid Waste Disposal Facilities and Practices" in the *Federal Register* [15] under 40 CFR Part 257. Briefly, the regulation required that pH be maintained at 6.5 for as long as food-chain crops were grown. The U.S.EPA established cumulative cadmium application rates on sludge-amended soils based on soil CEC and annual application rates for cadmium which declined with time. Tables 2.1 and 2.2 show these limits, respectively. Limits were also established for PCBs with the requirement that solid waste containing concentrations of PCBs equal to or greater than 10 mg/kg be incorporated into the soil when applied to land used for producing animal feed, including pasture crops for animals raised for milk.

The U.S.EPA provided a second approach that allowed unlimited appli-

TABLE 2.1. Time Period for Specified U.S.EPA Annual Cadmium Application Rates.

Time Period	Annual Cadmium Application Rate (kg/ha)
Present to June 30, 1984	2.0
July 1, 1984 to December 31, 1986	1 25
Beginning January 1, 1987	0.50

cation of cadmium provided that four specific control measures were taken. Briefly, these measures were: (1) crops grown could only be used for animal feed, (2) soil pH must be maintained at 6.5 or above as long as crops were grown, (3) facility operating plan must describe how the animal feed was distributed, and (4) future landowners must be informed that there are high levels of cadmium in the soil and that food chain crops should not be grown. The MWRDGC's Fulton County site was placed in this category.

In 1979, the MWRDGC received a draft copy of sponsored research in cooperation with the U.S.EPA titled "Agricultural Benefits and Environmental Changes Resulting from the Use of Digested Sludge on Field Crops: Including Animal Health Effects" by Hinesly and Hansen, University of Illinois. A notable conclusion in that draft report was that soil CEC does not affect plant metal uptake on sludge-amended soils. Another conclusion was that the plant uptake of sludge-applied metals after the first year or two of annual sludge applications remains fairly constant from year to year for a particular plant species.

1980 CAST Report

In 1979, the U.S.EPA asked the CAST to prepare a report on the effects of sewage sludge on the cadmium and zinc content of plants as a means of compiling the latest published and unpublished information on this

TABLE 2.2. U.S.EPA Maximum Cumulative Application Rates for Cadmium Based on Soil Cation Exchange Capacity and pH.

Soil Cation Exchange Capacity	Maximum Cumulative Applicative[a]	
	Soil pH < 6.5	Soil pH ≥ 6.5
	(kg/ha)	
< 5	5	5
5–15	5	10
> 15	5	20

[a]Rates are based on background soil pH.

subject. The U.S.EPA wanted this information so that it could be used in connection with proposed regulations being developed to control the application of sludge to agricultural soils. A task force of scientists involved in research on sewage sludge was assembled in 1980 to prepare a report.

The CAST [16] concluded that soil CEC does not adequately reflect the properties that control the availability to plants of cadmium and zinc in sludge-treated soils. Other conclusions of note included: (1) the increases in cadmium and zinc concentrations in plants grown on calcareous soils was usually substantially less than those observed under comparable conditions in noncalcareous soils; and (2) at a given pH value, the concentrations of cadmium and zinc in crops after repeated annual sludge additions was essentially the same as, or less than, the same amounts of cadmium and zinc applied at a rate equal to the sum in repeated annual additions. The CAST [16] also indicated that factors such as soil and sludge properties and plant species and cultures influence the concentrations of cadmium and zinc in plants following either a single application or repeated applications of sludge to soils.

1983 Denver, Colorado Workshop

Much progress was made in the land application of sludges during the 1970s. The data base established from research enabled the promulgation of federal and state regulations and guidance concerning municipal sludge use or disposal on land.

In 1979, the USDA-CSRS Technical Committee for ''Optimum Utilization of Sewage Sludge on Agricultural Land'' felt there was a need to assess and evaluate progress made during the last decade. In 1982, an U.S.EPA sludge policy committee was formulated between agency offices, signaling a need to assess information available and to define future research needs for land application of sludge. It was agreed that a second workshop would be held to reassess the research needs enumerated at the 1973 Champaign, Illinois Conference, and to identify existing recommendations for future research. On February 23–25, 1983 in Denver, Colorado, a workshop on ''Utilization of Municipal Wastewater and Sludge on Land'' was convened. The workshop was co-sponsored by the U.S.EPA, USDA-Cooperative State Research Service, the University of California-Kearney Foundation of Soil Science, U.S. Corps of Engineers, and National Science Foundation.

Logan and Chaney [13] presented an extensive discussion of metals in sludge and their impact on soils, crops, and animals. In reviewing the current status of U.S.EPA sewage sludge regulations in the United States, they evaluated the phased reduction of the annual cadmium application rates shown in Table 2.1. Logan and Chaney [13] concluded that, based on the report by CAST [16] and additional research information, there was little justification for restricting annual application rates of cadmium to <2.0 kg/ha as long as the cumulative application limit was followed.

DEVELOPMENT OF 503 SLUDGE REGULATIONS

Sludge Task Force

In 1983, the U.S.EPA Sludge Task Force began to develop a general statement of U.S.EPA's position on the key sludge management issues. The Sludge Task Force also had the responsibility of assisting with the initial phase of regulation development in 1984. In 1984, the Sludge Task Force prepared for review a draft on sludge management guidance for "Municipal Wastewater Sludge Use and Disposal." Briefly, the draft document covered land application, distribution and marketing, landfilling, incineration, and ocean disposal. The signing of the Sludge Task Force policy on sewage sludge management by the administrator in 1984 marked the beginning of the U.S.EPA's commitment to developing a comprehensive sludge management program and the Part 503 regulations.

The U.S.EPA in 1985 presented the agency program for developing sludge regulations at various cities throughout the United States. The agenda covered the overall process for generating technical regulations, development of state management regulations, developing environmental profiles and hazard indices, and development of risk assessment methodologies. The history and recommendations of the Sludge Task Force were also presented.

1985 Las Vegas, Nevada Workshop

In 1985, a workshop was held in Las Vegas, Nevada on November 13–15 entitled "Effects of Sewage Sludge Quality and Soil Properties on Plant Uptake of Sludge-Applied Trace Constituents." The workshop was jointly sponsored by the U.S.EPA, the University of California, Riverside, and Ohio State University, Columbus. The results of the workshop were published by Page et al. [17]. The workshop presented scientific evidence to indicate that certain factors now known to affect metal uptake by plants were not considered in previous guidelines and criteria for land application of sludge. Corey et al. [18] presented a very significant concept that stated that the soil and sludge properties will determine the activity of metals and, consequently, the plant uptake of metals. Trace element concentrations in plants will reach a level that reflects the properties of the soil and applied sludge metal concentrations, and the levels will not rise any higher after continued sludge applications. As a result, metal levels in plants grown on sludge-amended soils will plateau over time. Consequently, the loading limits for trace elements from municipal sewage sludges applied to land should be based on sludge and soil characteristics that affect plant availability of those elements with continuous sludge applications.

Development of Draft Sludge Regulations

Tiered Approach

In December 1986, the First Option Selection meeting for the Part 503 rule was held with the administrator of the U.S.EPA. A decision was made to make the Part 503 rule a risk-based rule. In the 1987 Water Quality Act amendments to the Clean Water Act (P.L. 100-4, February 4, 1987), Congress reaffirmed its mandate for the U.S.EPA to develop comprehensive sewage sludge regulations and set forth stringent deadlines. Dr. Alan B. Rubin, Chief of the Wastewater Solids Criteria Branch, U.S.EPA, in July of 1987, gave to the Sludge Management Committee of the Association of Metropolitan Sewerage Agencies (AMSA) a draft copy of preliminary regulatory limits being developed by the U.S.EPA as a part of the "Comprehensive Municipal Sludge Management Technical Regulations." These draft preliminary limits governed the land application of municipal sludge. The preliminary draft limits furnished by Dr. Rubin were circulated among the AMSA membership for review and comments.

The fifty-nine AMSA member agencies participated in a sludge analysis survey in 1987. The AMSA mean sludge analysis from the survey was compared to the preliminary regulatory limits for each disposal option. There were thirty-seven instances where the AMSA mean sludge analysis for a particular constituent exceeded the draft preliminary regulatory concentration limits. The comparisons indicated that every sludge disposal option would be precluded because at least two constituents in every option exceeded the draft preliminary regulatory limits [19].

The U.S.EPA at this time was considering a three-tiered approach for sludge management technical regulations. The proposed tiers were as follows:

- tier 1—Those sludges having sludge concentrations lower than the numerical limits in this tier would be acceptable for use regardless of operational conditions.
- tier 2—Those sludges that exceeded the sludge concentration limits in tier 1 may be able to be used provided that certain operational conditions (loading rate, depth to groundwater, etc.) were met.
- tier 3—Those sludges that cannot meet either the tier 1 or tier 2 criteria would be acceptable provided that the municipal agency could show through a U.S.EPA Risk Assessment Mathematical Model that the environment would not be adversely affected.

The U.S.EPA proposed draft numerical standards for maximum sludge concentrations of chemicals and maximum application rate for these chemicals in various use categories under the tier 1 and 2 approach. The use categories being considered were agriculture, silviculture, land reclamation, dedicated land, and distribution and marketing under tier 1 and sludges used in growing human food chain crops, and crops not in the human food chain under tier 2.

Risk Assessment and MEI Approach

The U.S.EPA, for a variety of reasons, subsequently dropped the tiered approach to regulating sewage sludge application to land and went to a more concentrated risk assessment approach. On March 10, 1988, the National Resources Defense Council (NRDC) filed suit against the U.S.EPA in the U.S. District Court of Eastern Pennsylvania to force the agency to promptly issue sludge regulations. On May 31, 1988, the U.S.EPA released draft guidance for its interim sludge strategy, as required under the 1987 Water Quality Act, to ensure safe use and disposal of sewage sludge until technical regulations became final. The U.S.EPA decided to conduct a comprehensive national sewage sludge survey to better define the quality of chemical sludge produced by sewage treatment facilities nationwide.

In the summer of 1988, the U.S.EPA had completed the first draft of the proposed sewage sludge regulation (Part 503). At this time, the structure of the proposed regulation consisted of land application (agricultural and nonagricultural), distribution and marketing, monofills, incineration, and ocean disposal. The U.S.EPA made reasonable worst case assumptions, and where there were no numeric limits, risk reference dose for non-carcinogenics, derived from Integrated Risk Information System, and a risk specific dose corresponding to an environmental carcinogenic risk level of 1×10^{-4} were used in the calculations. For incineration, the U.S.EPA used a risk level of 1×10^{-5}. When high levels of pollutant exposure were likely to occur or where there was significant scientific uncertainty, the exposure assessment models (MEI) were used as the basis for numeric limits.

The U.S.EPA continued to make revisions in the draft sludge regulations (Part 503) during 1988 and dropped ocean disposal. Members of AMSA, which had received copies of the August 1988 draft, made numerous comments that were forwarded to the U.S.EPA through AMSA to encourage the agency to modify proposed numerical limits before they were published in the *Federal Register* in 1989. The response from AMSA members indicated that there would be considerable problems and significant impacts for sewage treatment agencies if the draft 503 regulations remained as written by the U.S.EPA.

THE 40 CFR PART 503 RULEMAKING PROCESS: PROPOSAL, PUBLIC COMMENT, PEER REVIEW, REVISION, AND THE QUEST FOR A BALANCED PERSPECTIVE

UNITED STATES ENVIRONMENTAL PROTECTION AGENCY PROPOSES 40 CFR PART 503 STANDARDS FOR THE MANAGEMENT AND DISPOSAL OF SEWAGE SLUDGE

Proposed Rule Governs All Major Sludge Management Practices

The U.S.EPA, the wastewater treatment, and the academic communities expended vast resources in the 1970s and 1980s in efforts to ascertain the impact of sewage sludge management and disposal on public health and the environment. These efforts culminated in the U.S.EPA's proposal, on February 6, 1989, of their 40 CFR Part 503 Standards for the Management and Disposal of Sewage Sludge, which is the most comprehensive, technically based sludge regulation in the world.

The proposed regulation was intended to govern all of the major sludge management and disposal practices utilized by municipalities in the United States. This included application of sludge to agricultural land (land on which human food crops are grown) and to nonagricultural land (land on which vegetation, not directly consumed by humans, is grown), distribution and marketing of sludge (sale or giveaway of sludge to the general public), monofilling of sludge (disposal in sludge-only landfills), sludge incineration and surface disposal of sludge (final disposition in stockpiles or impoundments).

Proposed Rule Dictates Numerical Criteria

The U.S.EPA proposed cumulative soil loading rates (kg/ha) for ten elements (As, Cd, Cr, Cu, Pb, Hg, Mo, Ni, Se, Zn) and annual soil loading rates (kg/ha) for fourteen organic compounds (total aldrin and dieldrin; benzo(a)pyrene; chlordane; total DDT, DDD, DDE; dimethylnitrosamine; heptachlor; hexachlorobenzene; hexachlorobutadiene; lindane; toxaphene; trichloroethylene) and total PCBs, to regulate sludge application to agricultural land and distribution and marketing [20]. The numerical criteria were generated by risk-based mathematical models that calculate the maximum contaminant loading rate that maintains the level of exposure of most exposed individuals (MEI) below predetermined allowable limits for each of fourteen environmental transport pathways (Table 2.3).

The U.S.EPA also utilized risk assessment to establish numerical criteria for sludge incineration and monofilling. For monofilling, the U.S.EPA

TABLE 2.3. Environmental Pathways Through Which Contaminants in Sludge Applied to Agricultural Land or Distributed and Marketed Are Transported to Most Exposed Individuals (MEI).

Pathway	Pathway MEI
1	Sludge-Soil-Plant-Human	Human consuming rural produce
1F	Sludge-Soil-Plant-Human	Human home gardener
2F	Sludge-Soil-Human	Pica child
3	Sludge-Soil-Plant-Animal-Human	Human farmer
4	Sludge-Soil-Animal-Human	Human farmer
5	Sludge-Soil-Plant-Animal	Livestock most sensitive to contaminant
6	Sludge-Soil-Animal	Livestock most sensitive to contaminant
7	Sludge-Soil-Plant	Food crop most sensitive to contaminant
8	Sludge-Soil-Soil Biota	Earthworm
9	Sludge-Soil-Soil Biota-Soil Biota Predator	Birds that consume soil biota
10	Sludge-Soil-Airborne Dust-Human	Tractor operator at land application site
11	Sludge-Soil-Surface Water	Water quality criteria for receiving water
12A	Sludge-Soil-Air-Human	Human living on land application site, breathing fumes
12W	Sludge-Soil-Groundwater-Human	Human drinking groundwater from land application site

proposed sludge-borne contaminant concentration limits for six elements (As, Cd, Cu, Pb, Hg, Ni), eleven organic compounds (benzene; benzo(a)-pyrene; bis(2-ethylhexyl) phthalate; chlordane; total DDT, DDE, and DDD; dimethylnitrosamine; lindane; toxaphene; and trichloroethylene), and total PCBs [20]. The risk assessment consisted of modeling the transport of these regulated contaminants through the groundwater (sludge-soil-groundwater-human) and vaporization (sludge-air-human) pathways to assess exposure to the MEI. Sludges that have contaminant concentrations below the proposed limits can be monofilled and those with one or more contaminants present in concentrations in excess of the limits may not be monofilled.

The U.S.EPA proposed equations that could be used to calculate maximum sludge feed rates for incineration based on the concentrations of Be, Hg, Pb, As, Cd, Cr, and Ni in sludge, the incinerator stack height and related dispersion factors, incinerator control efficiencies (the fraction of total contaminant introduced into the incinerator that is contained in the system and not released to the atmosphere), and risk-specific air concentrations for each contaminant that were determined from mathematical models in the sludge-atmosphere-human pathway. The U.S.EPA also proposed a similar equation to be used to calculate maximum sludge feed rates based on total hydrocarbon emissions [20].

The U.S.EPA proposed sludge contaminant concentration criteria for ten elements (As, Cd, Cr, Cu, Pb, Hg, Mo, Ni, Se, and Zn), fourteen organic compounds (total aldrin and dieldrin; benzo(a)pyrene; chlordane; total DDT, DDE, and DDD; dimethylnitrosamine; heptachlor; hexachloro-benzene; hexachloro butadiene; lindane; toxaphene; trichloroethylene) and PCBs for application of sludge to nonagricultural land. Sludge contaminant criteria were also proposed for six elements (As, Cd, Cu, Pb, Hg, Ni), eleven organic compounds (benzene; benzo(a)pyrene, bis(2-ethylhexyl)phthalate; chlordane; total DDT, DDE, DDD; dimethylnitrosamine; lindane; toxaphene; and trichloroethylene), and PCBs for surface disposal of sludge [20]. Municipalities would be allowed to apply sludge to nonagricultural land or to surface dispose of it if the concentrations of all regulated contaminants in their sludge are below the specified concentration limits. Unlike the regulatory limits derived for the other sludge management and disposal options, which were based on pathway risk assessments, the limits that U.S.EPA derived for sludge application to nonagricultural land and sludge surface disposal were based on 98th percentile national sludge quality data. The U.S.EPA developed this approach because impacts of sludge management by these two options on the human diet, and potential health risks to the MEI were determined to be negligible [20]. The U.S.EPA therefore decided to impose sludge quality criteria to insure that contaminant concentrations in sludges being applied to nonagricultural land or being surface disposed would remain at or below current levels, which they determined to be having negligible impacts on human health (MEI cancer risk was estimated at 2×10^{-8} for sludge application to nonagricultural land [20]).

Proposed Rule Dictates Pathogen and Vector Attraction Reduction Requirements

The U.S.EPA specified three sludge classifications in the proposed rule, based on the level of pathogen reduction achieved [20]. Class A sludges were proposed to have pathogenic bacteria, viruses, protozoa, and helminth ova reduced to below detectable limits. Alternatively, Class A sludges must be heated to prescribed temperatures for stipulated lengths of time and as a result, the density of fecal coliforms and fecal streptococci (enterococci) per gram of volatile suspended solids must be equal to or less than 100 (Table 2.4). There were no use restrictions placed on Class A sludge on the basis of their pathogen reduction level. Class B and C sludges were proposed to have pathogenic bacteria and virus densities that are 100 and 31.6 times lower than the densities in the wastewaters from which the Class B and C sludges were produced, respectively (Table 2.4). Alternatively, fecal coliform and fecal streptococci densities must be less than or equal to 10^6 and $10^{6.7}$ for Class B and Class C sludges, respectively (Table 2.4).

TABLE 2.4. **Pathogen Reduction Criteria for Class A, B and C Sewage Sludges in February 1989—Proposed Part 503 Regulations.**

Class	Regulatory Criteria
A	Pathogenic bacteria, viruses, protozoa and helminth ova below detectable limits. or Sludge heated to 53°C for 5 days or Sludge heated to 70°C for 3 days or Sludge heated to 70°C for 0.5 hours and Density of fecal coliforms and fecal streptococci per gram of volatile suspended solids is less than or equal to 100
B	Density of pathogenic bacteria and viruses per gram of volatile suspended solids is 100 times less in sludge than in influent wastewater or Density of fecal coliforms and fecal streptococci per gram of volatile suspended solids is less than or equal to 10^6.
C	Density of pathogenic bacteria and viruses per gram of volatile suspended solids is 31.6 times less in sludge than in influent wastewater. or Density of fecal coliforms and fecal streptococci per gram of voilatile suspended solids is less than or equal to $10^{6.7}$.

Class B and C sludges had land use and access restrictions associated with their application to land [20]. For Class B sludges, food crops having harvested parts that are above ground but contact the sludge-amended soil could not be grown until eighteen months after sludge application. Food crops with harvested parts below ground could not be harvested until five years after sludge application, and no harvesting of animal feed crops or grazing could occur until thirty days after sludge application. Land receiving Class B sludge application was proposed to have access to the general public restricted for at least twelve months after the applications cease. For Class C sludge,the same use and access restrictions were proposed to apply, except that animal feed crops could not be harvested and animals could not graze on the land until sixty days after sludge application.

The U.S.EPA also proposed several alternative criteria for meeting vector attraction reduction standards [20]. This was intended to diminish the potential for vectors such as rodents, flies, and mosquitos to spread pathogenic diseases. In order to meet the vector attraction reduction standards, sludges would have to have 38% lower volatile solids after processing (anaerobic digestion, etc.) than influent, or they would have to be demonstrated to have an oxygen uptake rate equal to or less than 1 mg per hour per gram of sludge solids. Vector attraction reduction requirements would also be met if the pH of the sludge is raised to 12 or above for two hours and remains at 11.5 or above for an additional twenty-two hours. Alternatively, sludges may be dried to 75% solids or greater, or the sludges may be injected below the soil surface. These measures were thought to reduce the threat of transmission of pathogens by reducing the moisture and putrescible organic compound content of the sludge, which reduces the sludge's tendency to attract vectors.

Proposed Rule Dictates Management Practices

In addition to establishing numerical criteria, the Part 503 regulation, proposed in February 1989, also dictated management practices for each sludge management and disposal option [20].

The management practices proposed for distribution and marketing were required to be presented on a label affixed to the sewage sludge product or on an information sheet that would be required to accompany the product to inform the eventual user of the sewage sludge.

These management practices were thought to complement the numerical regulatory criteria in protecting human health and the environment. These practices were intended as minimum guidelines and it was recognized that individual states may impose a more restrictive set of management practices on municipalities in their jurisdiction.

PUBLIC RESPONSE TO THE PROPOSED U.S.EPA PART 503 SEWAGE SLUDGE REGULATIONS

Following proposal of their 40 CFR Part 503 sewage sludge regulations on February 6, 1989, the U.S.EPA allowed 180 days for submission of written public comments. The Agency received over 5000 pages of written comments from more than 500 municipal sludge generators, local regulatory agencies, environmental consultants, professional environmental lobby organizations, and private citizens. In addition, the U.S.EPA sponsored a peer review of the proposed rule to insure that the complex risk assessments and regulatory approaches would receive critical evaluation by exceptionally well qualified members of the scientific community.

Comments from Municipal Sewage Sludge Generators

The U.S.EPA received comments on their proposed Part 503 regulation from hundreds of municipal sewage sludge generators an from the Association of Metropolitan Sewerage Agencies (AMSA), an organization of approximately 200 of the largest municipal sewage sludge generators in the U.S. The efforts undertaken by municipalities to assess the impacts of, and comment on, the proposed rule were extraordinary. The results of an AMSA survey showed that over $7,000,000 was spent by municipalities to conduct chemical and biological testing of sludges, to evaluate sludge treatment methods and alternative reuse and disposal options, to conduct technical studies, and to conduct detailed evaluations of policy and technical aspects of the proposed rule [21].

Comments from municipalities unanimously expressed shock at the highly conservative approach utilized by the U.S.EPA, which produced extremely restrictive numerical criteria in the proposed rule, for sludge management practices that the municipalities, through years of experience, have come to regard as being quite safe and effective. AMSA conducted a survey of thirty-eight of its member agencies on the impacts of the proposed rule and the results have been analyzed and summarized by Lue-Hing et al. [22].

Lue-Hing et al. [22] reported that two of the thirty-eight agencies surveyed by AMSA were currently monofilling their sludge and that the proposed rule would require both to terminate operations. The regulatory limits for As, Cd, Cu, Pb, Hg, Ni, bis(2-ethylhexyl)phthalate, and toxaphene were found to be responsible for the noncompliance. The proposed regulatory sludge concentrations for Cd (0.04 mg/kg), Cu (8.4 mg/kg), Pb (0.35 mg/kg), and Ni (7.0 mg/kg) were particularly restrictive. Lue-Hing et al. [22] report that three of the thirty-eight agencies surveyed by AMSA are currently practicing surface disposal of sewage sludge, and that one of the agencies would be required to terminate operations by the proposed Part 503 regulation. The regulatory limits for dimethylnitrosamine and toxaphene were found to be responsible for the non-compliance.

Fifteen of the thirty-eight agencies surveyed by AMSA were reported to be applying sludge to agricultural land, nine of which were found to be prohibited from making one-time applications of even 11 Mg/ha (4.91 T/A) [22]. Lue-Hing et al. [22] further concluded that fourteen of the fifteen municipalities would be unable to apply 45 Mg/ha (20.1 T/A) annually for twenty years to the same parcel of land under the proposed criteria for agricultural land (a scenario judged necessary to make agricultural land application programs economically feasible). The cumulative loading limits in the proposed rule that were found to be most restrictive were: 18 kg/ha (16.1 lbs/A) for Cd, 46 kg/ha (41.0 lbs/A) for Cu, 78 kg/ha (69.6 lbs/A) for Ni, and 170 kg/ha (151.6 lbs/A) for Zn, and the most restrictive annual loading limits were found to be: 0.0055 kg/ha (0.0049 lbs/A) for total DDT, DDE, and DDD, and 0.0056 kg/ha (0.0050 lbs/A) for total PCBs.

The proposed Part 503 regulations were found to be severely restrictive to distribution and marketing of sludge as well. Municipalities commented that the proposed labeling requirement would adversely impact long-standing programs by giving the general public the impression that products, which have been proven over the course of time to be relatively innocuous, are hazardous. Lue-Hing et al. [22] report that five of the thirty-eight AMSA members surveyed reported distributing and marketing their sludge and that the proposed regulatory criteria would be so restrictive that all of them would have to abandon the practice. None of the distributed and marketed products produced by the five municipalities could be applied at rates above 10 Mg/ha (4.46 T/A) annually, which is a requirement for all uses of these products. The U.S.EPA proposed a 50 Mg/ha (22.3 T/A) maximum annual application rate, regardless of product composition, for distributed and marketed sewage sludges that municipalities found to be stifling to their programs. In order to be able to apply the maximum 50 Mg/ha (22.3 T/A) of product annually, several extremely restrictive regulatory contaminant concentration limits in the proposed Part 503 regulation would have to be met, including: 1.5 mg/kg benzo(a)pyrene, 18 mg/kg Cd, 530 mg/kg Cr, 46 mg/kg Cu, 130 mg/kg Pb, 76 mg/kg Ni, and 170 mg/kg Zn.

In its review and comments, AMSA was highly critical of the fact that no provision was made for dedicated sites in the proposed Part 503 regulation and that nonagricultural land sites would be restricted from growing animal feed crops, which was inconsistent with the existing 40 CFR Part 257 regulations [21]. This restriction was found to be completely contrary to the U.S.EPA's stated policy of promoting beneficial use of sewage sludge.

By failing to acknowledge dedicated sites and by prohibiting nonagricultural land sites from growing crops for animal feed, the proposed regulation would require municipalities that are currently operating sites dedicated to production of animal feed, a beneficial use of sludge, to convert their sites to nonagricultural land disposal sites where only a vegetative cover would be grown. Lue-Hing et al. [22] report that five of the thirty-eight AMSA members surveyed apply their sludge to nonagricultural land, and that only one of the five agencies would be able to continue its program under the restrictive regulatory sludge concentration criteria for PCBs (0.11 mg/kg) and total DDT, DDE, and DDDs (0.11 mg/kg) in the proposed Part 503 regulation. AMSA was especially critical of these extremely restrictive limits since PCBs and DDT, DDE, and DDD are compounds that have been banned from being manufactured, sold, or distributed in the United States, making it nearly impossible for municipalities to control their inputs into POTWs [21].

The results of the AMSA survey therefore indicated that the proposed Part 503 regulations would have severe impacts on its members and estimated that 100% of the monofilling and distribution and marketing programs would be terminated, 93% of the agricultural land application programs would become impractical or economically unfeasible, 80% of all nonagricultural land application programs would be terminated, and 33% of all surface disposal sites would be forced to close.

Most municipalities that conducted analyses of the impacts of the proposed Part 503 regulations on their existing sludge management programs and on potential alternatives found that they would be forced to abandon tried and true practices, many of which were beneficial uses, for codisposal of sludge with municipal solid waste. This is because codisposal of sludge with other municipal solid wastes was not regulated under Part 503 and all other options would be too severely restricted under the proposed rule. Thus the proposed rule was found to have the effect of restricting municipalities to codisposal of their sludge rather than promoting its beneficial use. The impacts and ramifications of the proposed rule were especially difficult for municipalities to accept in light of the fact that the numerical criteria were not derived from sound scientific information or risk assessment methodologies, which was emphasized in the comments of the Peer Review Committee (PRC) commissioned by the U.S.EPA to review the technical basis for the regulatory criteria in the proposed rule.

Comments and Recommendations of the EPA-Sponsored Peer Review Committee

The U.S.EPA realized that the proposed Part 503 regulations were based on a comprehensive risk assessment, which consisted of complex mathematical models, and its relative accuracy and hence the appropriateness of the numerical limits derived from it were largely dependent on the structure of the models and the quality and suitability of technical data inputs. In order to insure that the technical basis for the proposed regulation was adequately reviewed during the public comment period, the U.S.EPA sponsored a technical review by a Peer Review Committee (PRC).

The nucleus of the PRC was made up of members of the USDA Cooperative States Research Service Regional Research Technical Committee W-170. This committee consisted of the most highly expert and experienced researchers of sewage sludge management in the academic community in the United States. In addition to W-170 committee members, the thirty-five-member PRC was also made up of experts from the regulatory community, municipalities generating sewage sludge, professional environmental lobby organizations, and private consulting firms. The PRC met in Washington, DC from April 15 to April 18, 1989. During this meeting members were assigned to workgroups (monofills, surface disposal, nonagricultural land application, agricultural land application, distribution and marketing, and risk assessment) and each workgroup produced a preliminary draft of comments and recommendations resulting from their review. The draft documents were revised, and the review completed, over the ensuing three months, and the chairmen of the original workgroups met from July 9 through July 12, 1989 to make final revisions to the draft report. The PRC's final comments were submitted to the U.S.EPA prior to the close of the public comment period on August 7, 1989.

The PRC focused its review of the proposed Part 503 regulations on the information contained in the technical support documents (TSDs) that were utilized to derive the regulatory limits. In its final report to the U.S.EPA, the PRC stated that they "found such extensive misinterpretation and errors in the TSDs, that it is imperative that U.S.EPA review and revise them completely" [23]. Like the municipal sewage sludge generators who commented on the proposed rule and its impacts on local sludge management programs, the PRC concluded that "the proposed rule is based on a series of worst case scenarios, which are so stringent and inflexible that local communities are precluded from beneficial use options considered protective of the public health and the environment" and that "in spite of the Agency's [U.S.EPA's] own findings that the aggregate risk for the land-based sludge utilization options are lower than that associated with other disposal options, the proposed regulations encourage non-utilization practices" [23].

These findings indicated that the effect of the proposed rule would be to contradict the U.S.EPA's stated policy of encouraging beneficial use of sludge. The PRC then proceeded to elucidate technical flaws in the proposed rule that were responsible for this apparent lack of consistency between the U.S.EPA's beneficial use policy and the proposed Part 503 regulations.

Risk Assessment Methodology

The PRC was critical of the U.S.EPA's risk assessment methodology. The PRC criticized the U.S.EPA for not keeping policy separate from the scientific conduct of the risk assessment. The PRC found that technical data inputs were often chosen on the basis of bias resulting from the U.S.EPA's conservative policy of protecting human health and the environment rather than being chosen on the basis of being technically sound [23]. Thus, instead of selecting data and models that most accurately reflected the behavior of sludge-borne contaminants and MEIs, and then adjusting the resulting regulatory criteria, to provide a margin of safety consistent with Agency policies, the U.S.EPA attempted to introduce margins of safety throughout the risk assessment by making overly conservative data selections and assumptions about the behavior of the MEI.

This approach was found to result in a proposed rule that was not scientifically sound, and a risk assessment from which it was impossible to ascertain real risk or the magnitude of the overall margin of safety. The PRC recommended that the U.S.EPA replace all of the overly conservative data inputs and models it selected with technically sound models and inputs so as to conduct an accurate risk assessment, which would produce more realistic regulatory criteria [23].

The PRC was most critical of the U.S.EPA's use of the Most Exposed Individual (MEI). The PRC stated that the MEI descriptions utilized by the U.S.EPA "substantially overestimate actual exposure for the sector(s) of the population with the highest potential exposures from sewage sludge" [23].

Data Selection

The PRC specifically criticized the U.S.EPA's selection of data from experiments in which plants were grown in pots rather than in fields, and contaminants were added to soil as metal salts or pure organic compounds rather than as constituents of sewage sludge. The PRC pointed out that plant uptake of metals and organic compounds are greatly exaggerated under these conditions and do not accurately portray the behavior of sludge-borne contaminants at actual field sites [23]. As a result, the U.S.EPA's models grossly overpredicted bioaccumulation of metals and organics by plants. This was recently reiterated by Granato et al. [24] who compared actual metal concentrations in tissues of crops grown at a dedicated site by the MWRDGC with those predicted by models in the Proposed Part 503 regulation. The PRC urged the U.S.EPA to utilize only data from experiments in which plants were field grown in sludge-amended soils, and to modify criteria for generation of regulatory limits so that "no-observed-adverse-effect data from valid field studies with sludge" could be utilized [23].

Selection of Models

The PRC also specifically criticized the models utilized by the U.S.EPA in pathways 7, 11, 12A, and 12W (Table 2.3). The PRC found these models to be inappropriate or technically flawed and observed that they led to generation of regulatory limits that were extremely restrictive and did not coincide with real world observations of safe sludge contaminant concentrations or soil loading rates for the various practices regulated [23]. For instance, the model for pathway 12W (sludge-soil-groundwater-human) produced a regulatory limit of 0.04 mg/kg for cadmium in sludges that are monofilled. This concentration is below that found in many POTW's influents. Another example cited by the PRC is that the model and data inputs for pathway 7 (sludge-soil-plant) "led U.S.EPA to conclude that sludge-borne copper, nickel, and zinc would cause phytotoxicity in crops, a situation never observed in the field" [23]. The conclusion that the Pathway 7 model was flawed was corroborated by Granato et al. [24] who did not observe any phytotoxic effects of sludge on corn and wheat grown at a dedicated site operated for twenty years by the MWRDGC. The PRC recommended that the U.S.EPA utilize more appropriate models in their risk assessment analysis for pathways 7, 11, 12A, and 12W.

Regulatory Approach for Nonagricultural Land

The PRC was highly critical of the use of 98th percentile sludge concentration data to regulate nonagricultural land application practices. The PRC stated that "the proposed 98th percentile-based approach has no technical merit" and concluded that the approach should not be used in regulating nonagricultural land application [23]. The PRC pointed out that the 98th percentile approach will either over- or under-regulate. For the case of organic pollutants, where concentrations in sludge are extremely low, the 98th percentile approach tends to over-regulate because even the highest concentrations may be insignificant with respect to risk of human health effects. In contrast, the approach could allow land application of sludges with high concentrations of pollutants at rates that may pose unacceptable risk of human health effects.

The PRC recognized that a diversity of nonagricultural land application practices exist and recommended that the U.S.EPA account for differences among these practices in the final rule. The PRC recommended that the U.S.EPA categorize nonagricultural land application sites on the basis of their potential for conversion to other land uses (i.e., agricultural land or housing development) and the subsequent human exposure that would result. The PRC recommended that sites having a high potential for conversion to other land uses be restricted by cumulative metal loading limits, while sites having low potential for conversion should be restricted to adhering to management practices to control runoff and public access [23].

The PRC was also critical of the U.S.EPA's proposed approach for regulating nonagricultural land application because they felt it was unclear whether the proposed rule recognized dedicated sites, a management option that was encouraged under 40 CFR Part 257. The PRC stated their conviction that dedicated sites were a preferred mode of operation with many municipalities, and urged the U.S.EPA to offer direction on this issue in the proposed rule so that municipalities could consider dedicated sites in their long-term management plans. The PRC concluded that promoting dedicated site management would increase the opportunity to beneficially use sewage sludge for many municipalities.

Regulatory Approach for Distribution and Marketing

The PRC also reviewed the proposed rule for distribution and marketing of sludge. They were critical of the labeling requirements mandated in the proposed rule and found no technical reason to label distribution and marketing products. The PRC stated that the labeling should be used to provide information to consumers on proper use of the product rather than to provide warnings of the potentially hazardous chemicals it contains. The PRC expressed concern that imposing a requirement that the concentration of all twenty-two regulated contaminants be printed on the label would cause undue confusion and would discourage beneficial use of distributed and marketed sludge products [23].

The PRC concluded that a diversity of distributed and marketed products exist, and that it was not appropriate to set maximum whole sludge application rates that would be applicable to all products. The PRC also criticized the U.S.EPA for attempting to define the applicability of distributed and marketed products based on the percentage of their total mass that consisted of sewage sludge. These criticisms prompted the PRC to recommend that the U.S.EPA propose a set of "clean sludge" standards that would be numerical contaminant concentration criteria [23].

Sludges or sludge products, regardless of the percentage of their mass comprised of sludge, having concentrations for all regulated contaminants below the "clean sludge" standard, could be land-applied as deemed appropriate by the end user without concern about compliance with cumulative loading limits or maximum whole sludge application rates. Only standard management practices designed to protect surface and groundwaters would need to be dictated in the regulation. Sludges and sludge products that could not meet the "clean sludge" criteria would be subjected to maximum whole sludge application rates and cumulative loading rates. The "clean sludge" limits would be derived from the risk assessment analysis such that no adverse effect concentration levels could be determined. These are contaminant concentrations below which the risk of adverse effects on human health or the environment, from exposure through any of the fourteen known pathways, would be negligible. Thus, sewage sludges and sludge products would be treated in a manner analogous to other commercial fertilizers which are distributed and marketed in the United States.

Pathogen Reduction Requirements

The PRC also commented on the pathogen reduction requirements of the proposed Part 503 regulation and indicated that they appeared to be overly restrictive. The PRC stated that they generally favored continued use of the requirements in 40 CFR Part 257. They stated this because there have been no reported incidents of adverse health effects resulting from adherence to the management practice guidelines established in 40 CFR Part 257 [23]. These included a thirty-day waiting period after sludge application before grazing livestock and an eighteen-month waiting period prior to growing crops that are marketed without undergoing processing to reduce pathogens.

The PRC submitted their comments and recommendations to the U.S.EPA on July 24, 1989. The recommended revisions to the proposed Part 503 regulations were very extensive and identified the technical deficiencies and inaccuracies inherent in the proposed rule that were responsible for the extremely restrictive numerical limits that municipalities felt would severely impact their tried and true management and disposal programs. The U.S.EPA carefully reviewed the public comments they received and, having come to the realization that the proposed regulation was indeed seriously flawed, decided to retain consulting experts to assist the Agency in revising its regulatory approach and its selection and use of scientific data and models.

UNITED STATES ENVIRONMENTAL PROTECTION AGENCY PROPOSES REVISED APPROACH FOR REGULATING SEWAGE SLUDGE MANAGEMENT UNDER 40 CFR PART 503

The voluminous public comments and extensive criticism received by the U.S.EPA during the 180-day public comment period prompted the Agency to work with individual consultants, with expertise in various technical fields, to develop a revised regulatory approach. On November 9, 1990, the U.S.EPA published a proposal of their revised approach for regulating sewage sludge management and disposal under 40 CFR Part 503 in the *Federal Register* [25].

Revised Approach for Regulating Nonagricultural Land Application

The U.S.EPA proposed revisions to their approach to regulating nonagricultural land application. The Agency decided to drop the 98th percentile sludge quality approach and to replace it with a pathway risk assessment based approach, similar to that taken for agricultural land application and distribution and marketing.

The Agency proposed dividing nonagricultural land into five categories: forest and range lands, soil reclamation sites, public contact sites (urban parks, golf courses, cemeteries, and highway medians and buffer strips), dedicated beneficial use sites, and dedicated disposal sites. The U.S.EPA proposed pathways, MEIs, and management practices that they felt were appropriate for each nonagricultural land category (Table 2.5). The application of sludge to nonagricultural land would be limited by annual organic contaminant loading limits and cumulative metal loading limits generated by the pathway exposure assessments for each category of land.

The MEIs were similar to those for agricultural land, but they were modified where the Agency felt it was appropriate to reflect conditions more likely to be encountered at nonagricultural land sites. For instance, the exposure of the MEI for pathway 1 to edible plants grown on sludge-amended soil was limited to wild fruits and berries, and mushrooms that could be gathered at sites having no public access restrictions. Pathway exposure assessments were not applied to specific nonagricultural land categories when management practice requirements or general conditions precluded the existence of an MEI.

Revised Approach for Regulating Surface Disposal Sites

The U.S.EPA proposed revisions to its approach for regulating surface disposal of sludge in the November 9, 1990 notice [25]. The Agency changed the definition of surface disposal sites to exclude sites that are part of a POTW's treatment process or that are used for temporary storage prior to final use or disposal.

The U.S.EPA also decided to abandon the 98th percentile sludge quality approach to setting numerical criteria for surface disposal sites as they did for nonagricultural land sites. Instead, the Agency proposed a pathway exposure assessment approach that was based on pathways 11, 12A, and 12W. The Agency also proposed to allow for computation of site specific

TABLE 2.5. Exposure Assessment Pathways and Management Practices Applied to the Five Proposed Nonagricultural Land Categories.

Pathway or Management Practice	Nonagricultural Land Category				
	FRL	PCS	SRS	DDS	DBUS
1	yes	yes	yes	no	no
2	yes	yes	no	no	no
3	yes	no	yes	no	yes
4	yes	no	yes	no	yes
5	yes	yes	yes	no	yes
6	yes	yes	yes	no	yes
7	yes	yes	yes	yes	yes
8	yes	yes	yes	yes	yes
9	yes	yes	yes	yes	yes
10	no	no	no	no	yes
11	yes	no	yes	yes	yes
12A	yes	no	no	yes	yes
12W	yes	no	yes	yes	yes
Buffer strips around wells and surface waters required	yes	yes	yes	yes	yes
Animal grazing allowed	yes	no	yes	no	no
Animal feed crop allowed	yes	no	yes	no	yes
Public access restriction required	no	no	no	yes	yes
Land conversion restriction required	no	no	no	yes	no

FRL = Forest and range land
PCS = Public contact site.
SRS = Soil reclamation site.
DDS = Dedicated disposal site.
DBUS = Dedicated beneficial use site

numerical limits based on soil properties and site design, as an alternative to the national limits that would be calculated using reasonable worst case assumptions. This is very similar to the approach originally proposed for sludge monofills.

Revised Approach for Regulating Distribution and Marketing and Agricultural Land Application

The U.S.EPA stated that, pursuant to their policy of promoting beneficial use of sludge, they gave careful consideration to the public comments on their proposed approach for regulating sludge application to agricultural land and distribution and marketing. As a result, they proposed a revised approach for regulating these practices [25].

The Most Exposed Individual

The U.S.EPA stated that they would reevaluate the MEIs selected for each pathway as well as the assumptions made concerning the behavior of the MEI. The Agency stated that they would consider more realistic human dietary exposure scenarios to compute human food-chain risk. In the rule proposed on February 6, 1989, the U.S.EPA used the consumption data from the age and sex groups having the highest consumption of each food group. As a result, the teenage male diet was used to represent MEI consumption of grain, potatoes, root vegetables, dairy products, and dairy fat, while the diet of the adult female was used to represent MEI consumption of lamb and lamb fat, and so on. The U.S.EPA, responding to public comment, proposed revising the approach so that, for each sex, the consumption rate for each food group was integrated over an individual's life span. The U.S.EPA thus proposed utilizing a time-weighted, lifetime average consumption value for the MEI [25].

The U.S.EPA also stated that they had reevaluated the fraction of animal product food groups in the MEI's diet that were derived from animals raised on sludge-amended soil. The Agency had originally assumed that 44% of meat and 40% of dairy products consumed by the MEI were raised on sludge-amended soil. Public comment led the Agency to reconsider these values. The Agency proposed changing their assumption that for farms receiving sludge applications, 100% of the land receives sludge application each year. Instead, the Agency proposed the assumption that 33% of the land receives sludge in any given year [25].

The U.S.EPA also proposed changing their assumption that 8% of a grazing animal's diet is composed of sludge. The Agency proposed utilizing 1.5% instead. This is to reflect the fact that, at most, only 33% of a farmer's fields will receive sludge in a given year, and livestock in the United States does not graze year-round [25].

Relative Effectiveness of Exposure

The relative effectiveness of exposure is a unitless factor that indicates the relative toxicological effectiveness of a contaminant when ingested in one form relative to another form. This factor is important in extrapolating data obtained from studies in which metals were fed to animals as soluble salts as compared to the actual retention of metals in tissues of animals fed metals endogenous in sludges or crops grown on sludge-amended fields. The Agency had previously assumed that this factor had a value of one for all metals and all pathways, but the Agency proposed on November 9, 1990 that, in fact, the relative effectiveness of exposure was less than one in many cases [25]. That is, less metal is retained in animal tissue when fed as a constituent of sludge or crop tissue then when fed as a soluble salt.

Model Selection

The U.S.EPA acknowledged that the models they had originally chosen for pathways 7, 11, and 12 were either technically flawed or inappropriate for their intended purpose. The Agency thus proposed substituting more appropriate, technically sound models into these pathways. The Agency acknowledged that the limiting cumulative metal loading rates in the originally proposed rule, which were based on potential phytotoxicity (pathway 7) or groundwater pollution (pathway 12W), were too conservative in light of the voluminous scientific data made available during the public comment period, demonstrating that the Agency's models were flawed [25]. Thus, in an effort to promote beneficial use, the U.S.EPA proposed utilizing models that generated exposure assessments that were more in line with scientific observations of actual exposures.

Screening and Selection of Data

The U.S.EPA was severely criticized for utilizing data from experiments where plants were grown in small pots and in soils that were spiked with metals as soluble salts or organic contaminants as pure compounds. The U.S.EPA acknowledged that these data were inappropriate for use in their pathway models for extrapolating to metal and organic compound behavior in sludge-amended field soils. The Agency proposed utilizing only data from experiments in which metals or organics were added to field soils as constituents of sludge [25].

Distribution and Marketing Labeling Requirement

The U.S.EPA carefully considered the comments it received on requiring that the label of distributed and marketed products include a listing of the twenty-two regulated contaminants and their concentrations. The Agency decided that such information was of little value to the general public and would cause undue confusion and apprehension. They therefore decided to drop the requirement and proposed that the label provide information on the nitrogen and phosphorus content of the product, as well as a statement saying that a listing of the trace element concentrations in the product can be obtained from the generator or manufacturer. The name, address, and phone number of the product manufacturer would be required to appear on the label as well [25].

Regulation of Banned Compounds

The U.S.EPA's originally proposed 40 CFR Part 503 sludge regulations contained regulatory limits for many organic contaminants that have been banned, or severely restricted, from use in the U.S. by the Agency. Among these compounds are aldrin, dieldrin, chlordane, DDT, lindane, PCBs, and hexachlorobutadiene. The Agency stated that they were reevaluating the necessity for regulating banned compounds if the compounds were found to be present in sludge at insignificant concentrations, or below detectable limits. The U.S.EPA would utilize the results of their extensive National Sewage Sludge Survey [25] to determine the concentrations of these constituents in municipal sludges generated in the U.S.

This proposal was applauded by municipalities and AMSA members who have very little control over the influx of these banned compounds into their treatment works. Since the compounds are already banned from sale or distribution in the U.S., there is little more that local government agencies could do to lower their concentrations in municipal sewage sludge except to wait for these ubiquitous compounds to "bleed" out of the urban environment. Thus, setting overly restrictive numerical limits for these compounds would not lead to tighter control of industrial discharges, but would discourage beneficial use of sludge. The Agency apparently recognized that this was unwarranted in light of the generally low levels of these compounds in sludge and the minuscule risks associated with their land application.

Preliminary Results of Revised Approach for Regulating Distribution and Marketing and Agricultural Land Application

The U.S.EPA published some preliminary numerical limits for metals that resulted from the revised approaches they proposed on November 9, 1990 [25]. The new numerical limits resulted from use of more technically sound pathway models, data inputs, and MEI assumptions. The new limits were not intended to be a proposal of final criteria, but were published to give some indication of the effects that the revisions were having on the regulatory limits generated by the risk assessment analysis. Final criteria would not be generated until the revision process was complete.

The preliminary results of the revision process on the regulatory limits generated for arsenic, copper, nickel, lead, and zinc are illustrated in Table 2.6. These limits are maximum soil loading rates generated by the most restrictive pathways for each metal. These preliminary results of the revision process indicated that the originally proposed rule was extremely conservative and overly restrictive, as was communicated to the Agency by municipalities, AMSA, and the PRC during the public comment period.

Another major result of the revision process was that the U.S.EPA proposed alternative pollutant limits based on the revised numerical limits for the regulated metals. The alternative pollutant limits were first proposed by the PRC and were intended to set quality standards for no-adverse-

TABLE 2.6 Comparison of Preliminary Revised Cumulative Soil Loading Limits with Originally Proposed Limits for As, Cu, Ni, Pb and Zn.

Pathway		Originally Proposed Soil Loading Limit	Preliminary Revised Soil Loading Limit
		[kg/ha (lbs/A)]	
As			
1	Sludge-Soil-Plant-Human	6960 (6208)	a
1F	Sludge-Soil-Plant-Human	382 (341)	a
2F	Sludge-Soil-Human	14 (12.5)	1628 (1452)
Cu			
5	Sludge-Soil-Plant-Animal	153 (136)	> 2390 (2132)
6	Sludge-Soil-Animal	458 (409)	3930 (3506)
7	Sludge-Soil-Plant	46 (41)	1160 (1035)
9	Sludge-Soil-Soil Biota-Predator	224 (200)	1200 (1070)
Ni			
1F	Sludge-Soil-Plant-Human	206 (184)	≥500 (≥446)
7	Sludge-Soil-Plant	78 (70)	500 (446)
Pb			
1	Sludge-Soil-Plant-Human	1190 (1060)	≥1000 (≥892)
1F	Sludge-Soil-Plant-Human	195 (174)	> 1000 (> 892)
2F	Sludge-Soil-Human (Ag land)	378 (337)	580 (517)
2F	Sludge-Soil-Human (D&M)	378 (337)	300c
7	Sludge-Soil-Plant	b	≥1000 (> 892)
9	Sludge-Soil-Soil Biota-Predator	125 (112)	> 1000 (> 892)
Zn			
7	Sludge-Soil-Plant	172 (153)	2750 (2453)
9	Sludge-Soil-Soil Biota-Predator	452 (403)	2600 (2319)

aRevised soil loading limits not yet determined (as of November 9, 1990)
bNo soil loading limit for these metals and pathways was proposed in the original rule
cRegulatory limit has been changed from soil loading rate to sludge concentration (mg/kg).

effect sludges, which could be managed similar to other commercially available fertilizers. The U.S.EPA proposed deriving the alternative pollutant limits by assuming that sludge application to any given parcel of land, either in one-time applications or cumulatively, would not exceed 1000 Mg/ha (446 T/A). The Agency would then calculate sludge concentration limits for metals such that the total metal loading associated with the 1000 Mg/ha (446 T/A) sludge application does not exceed the regulatory cumulative loading limit derived for each metal from the most restrictive exposure pathway. The Agency would also calculate sludge concentration limits for the regulated organic contaminants such that the organic contaminant loading associated with the 1000 Mg/ha (446 T/A) sludge application does not exceed the regulatory annual loading limit derived for each organic contaminant from the most restrictive exposure pathways [25].

The Agency never presented proposed alternative pollutant limits in their November 9, 1990 *Federal Register* notice. However, the limits for metals could be inferred from the limits in Table 2.6. The alternative pollutant limits for arsenic, cadmium, chromium, copper, mercury, lead, and zinc as calculated from information in the November 9, 1990 notice [25] are presented in Table 2.7. Sludges that have metal concentrations below those in Table 2.7, and organic contaminant concentrations below the then undisclosed organics limits, were proposed to meet the alternative pollutant limits. These sludges would not be subjected to regulation under 40 CFR Part 503, except that pathogen and vector attraction reduction standards must be met, and the N requirement of the crop should not be exceeded. This proposal marked a dramatic step forward in the Agency's endeavor to promote beneficial use of sludge.

Public Response to Revisions to Part 503 Proposed in November 1990

Following the publication of the U.S.EPA's revised approaches for 40 CFR Part 503 on November 9, 1990 [25], there was a sixty-day public comment period. The Agency solicited comments on the many revisions it proposed making to the original regulatory approach, and on the content of the rule that would likely result when the proposed revisions were completed.

AMSA submitted detailed comprehensive comments to the U.S.EPA on the proposed revised approaches and on the likely form of the final rule [26]. The AMSA comments commended the U.S.EPA for undertaking the task of revising their original proposal. AMSA applauded most of the proposed revisions and concluded that the process would result in a vastly improved final rule. However, AMSA's comments remained critical of some aspects of the regulatory approach.

TABLE 2.7. Alternative Pollutant Limits Proposed in November 9, 1990 Notice for No-Adverse-Effect Sludges.

Metal	Alternative Pollutant Limit (mg/kg)
As	382[a]
Cd	18 4[a]
Cr	530[a]
Cu	1160
Hg	14.9[a]
Ni	500
Pb	300
Zn	2600

[a]Alternative pollutant limit is based on unrevised originally proposed pathway limits.

Model Selection and Pathway Assumptions

The comments that the U.S.EPA received from AMSA were very critical of the assumptions utilized in constructing pathway 12A (Sludge-Soil-Vaporization-Human). In the original proposal, the U.S.EPA based its pathway model on the assumption that for volatile contaminants, 100% of the compound would volatilize after land application and that the MEI would reside twenty-four hours per day, every day, for seventy years, downwind of the site where the volatilization was occurring. However, if 100% of the volatile organic contaminant volatilizes, then there is no need to include it in any other exposure pathway, and obviously, wind direction does not remain constant every day for seventy years. AMSA pointed out that these were unreasonable assumptions and urged the U.S.EPA to consider a revised approach [26].

AMSA was also critical of the fact that the U.S.EPA did not specifically propose a revised approach for pathway 12W (Sludge-Soil-Groundwater-Human). In the originally proposed rule, this pathway generated very restrictive limits for both metals and organics. The behavior of the regulated contaminants predicted by this pathway, which utilized overly conservative worst case assumptions and inappropriate models, did not correspond with observed behavior at actual land application sites [27]. AMSA was concerned that the pathway, if left unrevised, would generate unwarranted restrictive limits that would discourage beneficial use. AMSA urged the U.S.EPA to carefully revise this pathway [26].

Revised Approach to Regulating Nonagricultural Land Application

AMSA commended the U.S.EPA for abandoning its originally proposed 98th percentile approach to regulating nonagricultural land application practices by creating five land use categories. However, AMSA was very critical of the U.S.EPA's selection of exposure pathways for the five categories. AMSA found many of the pathways to be inappropriate because the MEIs did not exist, due to management practice requirements precluding their existence or because the MEIs were presumed to exist on the basis of unrealistic, overly conservative, assumptions about their nature and behavior [26]. AMSA recommended that the U.S.EPA drop the inappropriate exposure pathways from the risk assessment for each nonagricultural land category.

AMSA also recommended that the U.S.EPA revise their approach for regulating nonagricultural land to make provision for case-by-case regulatory criteria generation based on site-specific conditions. AMSA pointed out that the risk assessment generated by the pathway exposure models could not be expected to be accurate for all sites across an area as large and diverse as the U.S. Nonagricultural land practices, such as dedicated beneficial use sites, are well suited for generation of site-specific regulatory criteria because they represent parcels of land dedicated to land application of sludge year after year, and site-specific criteria can be generated once, which will last for the lifetime of the site [26]. This would provide incentive for beneficial use and for selecting and managing sites to reduce risk.

The revised approach proposed by the U.S.EPA for dedicated beneficial use sites would dictate cumulative metal loading limits that were the same for all sites regardless of soil type or management practices. The site life would be determined *a priori* by these metal loading limits regardless of impacts on human health or the environment that were actually occurring at the site. These uncertainties in the risk assessment could be circumvented if the site was extensively monitored for environmental impacts through likely exposure pathways. Sites would be allowed to operate until monitoring data indicated that the environment was being adversely impacted and site life would no longer be determined by generic pathway exposure models. This regulatory approach would be more congruent with the nature of dedicated sites, which are parcels of land utilized and managed to safely land-apply sewage sludge for indefinitely long periods of time.

The U.S.EPA carefully considered these and other comments that they received during the sixty-day public comment period and completed the revision process with their consultants.

EFFECTS OF THE PEER REVIEW, PUBLIC COMMENTS AND THE RESULTING U.S.EPA REVISION PROCESS ON 40 CFR PART 503: THE FINAL RULE

Following completion of the revision process by the U.S.EPA's Office of Water, the draft final rule was subjected to months of careful scrutiny throughout the Agency. The regulatory limits were compared for consistency with those of other programs and regulations developed for RCRA, TSCA and CERCLA. The structure of the draft final rule was scrutinized for ease of implementation, permitting and enforcement. The level of protection that the rule actually provided, to the health of the exposed individual, and to the quality of the surrounding ecosystem were vigorously debated.

In the end, most of the U.S.EPA's Office of Water's draft framework and its scientific basis survived this process, although some changes were made on the basis of overriding policy, ease of implementation, or enhancing protection. The final rule was published in the *Federal Register* on February 19, 1993 [30], with requirements for frequency of monitoring, recordkeeping and reporting becoming effective on July 20, 1993 and compliance with the standards becoming effective on February 19, 1994 for situations where compliance can be achieved without construction of new pollution control facilities. In cases where construction of new pollution control facilities were required for compliance, the effective date for compliance with the standards was February 19, 1995.

Major Changes in Part 503 Resulting from the Revision Process

The changes outlined in this section are the result of the evolution of thought and refinement of the risk assessment methodology that occurred over the four years and 13 days that elapsed from the date Part 503 was proposed to the date it was promulgated. The changes were made possible by the cooperative interaction that occurred between the U.S.EPA, municipal wastewater treatment agencies, academic experts, environmental groups and professional organizations, and by a great desire on the part of all parties to produce a protective and effective final rule. The following is a summary of the most important changes that resulted from this process.

Risk Assessment Revisions

As has been discussed, major revisions were made to the risk assessment, including (1) the replacement of models utilized in the groundwater and phytotoxicity pathways; (2) the quality of data input into these models was improved; (3) the descriptions of the exposed organisms and the extent of their exposure to the various pathways were refined; and (4) the regulatory criteria based on 98th percentile sludge quality data were replaced by risk based numbers.

Organic Pollutants Deleted from the Rule

As a result of the revised risk assessment methodology, the U.S.EPA determined that the organic pollutants originally proposed to be regulated in Part 503 did not pose significant risk for the use and disposal options considered. This conclusion was reached by considering the following:

(1) The NSSS indicated that none of the organics proposed for regulation were present in greater than 5 percent of the nation's sludge with the exception of PCB congeners 1248, 1254 and 1260 which were present in 10, 8 and 9 percent of the nation's sludges respectively. Hexachlorobenzene; hexachlorobutadiene; dimethylnitrosamine; PCB congeners 1016, 1221, 1232, 1242; and toxaphene were not detected in any of the sludges sampled (Table 2.8).

(2) Those organic compounds that were detected were present at extremely low concentrations (Table 2.8).

TABLE 2.8.   Data Utilized by U.S.EPA in Decision to Delete Organic
Pollutants from Part 503.

Organic Pollutant	Occurrence in Nation's Sludges %	Mean National Sludge Concentration[a] (mg/kg)	Risk Based Pollutant Concentration Limit[b] (mg/kg)
Aldrin	3	0.0038	2.7
Benzo (a) pyrene	3	0.067	15
Chlordane	< 1	0.0019	86
Dieldrin	4	0.0017	2.7
Heptachlor	< 1	0.0007	7.4
Hexachlorobenzene	0	N.D.	29
Hexachlorobutadiene	0	N D	600
Lindane	< 1	0.0024	84
Dimethylnitrosamine	0	N.D.	2 1
PCB	~ 10	1.177	4.6
Toxaphene	0	N.D.	10
Trichloroethene	1	0 0177	10,000
DDT/DDE/DDD	3	0.0268	150

[a]From Reference [26] as calculated from NSSS data
[b]From the Technical Support Document for Land Application [29].
N D —Not detected in any NSSS samples.

(3) The risk assessment produced annual pollutant loading limits that were extremely high relative to the concentrations of the pollutant found in the NSSS sludges (Table 2.8).

(4) Many of the compounds such as DDT/DDD/DDE, PCB, toxaphene and chlordane are banned from manufacture, sale and use in the United States and so their concentrations in sludge should approach zero with time.

Revised Regulatory Framework

In the interest of simplifying the implementation of Part 503, the U.S.EPA streamlined the regulatory framework. One set of criteria were established for land application which, in the final rule, covers application to agricultural land, distribution and marketing, and application to all of the non-agricultural land categories except dedicated disposal sites and dedicated beneficial use sites. The U.S.EPA had conducted a separate risk assessment for sludge application to each category of non-agricultural land and for application to agricultural land, and had found that non-agricultural land risk assessments generally produced numerical limits that were only 25 to 50 percent higher than those derived from assessment of risk at agricultural land application sites. This difference was not deemed large enough to justify complicating the rule with separate criteria for non-agricultural and agricultural land [31].

Since Part 503's promulgation in 1993, the Land Application Subpart of the rule and the various practices it governs have become synonymous with beneficial use. Unfortunately the U.S.EPA excluded dedicated beneficial use sites from this subpart. In the final Part 503 framework no distinction is made between dedicated disposal sites and dedicated beneficial use sites (see Chapter 9 Introduction for a discussion of these sites) and both are referred to collectively as dedicated sites and have been placed in the subpart intended by the U.S.EPA to govern all land based disposal options.

In the final Part 503 framework the practices of monofilling, surface disposal and land application at dedicated sites are governed in the Surface Disposal Subpart by a single set of regulatory criteria which includes a site specific option. This subpart and all of the practices it governs, including sludge application at dedicated beneficial use sites, have come to be synonymous with disposal. AMSA, as well as the United States Department of Agriculture and the Water Environment Federation, has urged the U.S.EPA to move regulatory control of dedicated beneficial use sites to the Land Application Subpart so that it will receive the recognition it deserves as a beneficial use practice.

No Adverse Effect Sludge Concept Adopted

In the final Part 503, the U.S.EPA adopted the concept of the No-Adverse-Effect Sludge and produced a set of sludge pollutant concentration limits and pathogen and vector attraction reduction criteria to define it. Sludges that meet the No-Adverse-Effect standards are currently referred to as exceptional quality (EQ) sludges and are treated as equivalent to a commercial fertilizer product allowing nearly unrestricted use.

Deletion of Chromium from Criteria for Land Application

The originally promulgated Part 503 rule contained regulatory criteria for Cr in the Land Application Subpart. However the criteria was based on 99th percentile NSSS sludge quality data which the U.S.EPA apparently utilized despite the fact that the risk assessment indicated that Cr concentrations far in excess of the 99th percentile would be protective.

The Cr criteria for land application was challenged in the United States Court of Appeals by the Leather Industries of America and AMSA in March 1993; and on November 15, 1994 the Cr limits were remanded to U.S.EPA for modification or additional justification by the Washington D.C. Circuit Court [32]. The Court concluded that section 405 of the Clean Water Act mandates a risk-based regulation and that U.S.EPA did not have the statutory authority to adopt pollutant concentration limits based on the 99th percentile because they are not risk-based [33]. Following this ruling, the U.S.EPA reexamined its risk assessment for Cr and called upon AMSA and the public for additional data. The U.S.EPA concluded that there is an insufficient basis at this time for the regulation of Cr in land applied sludge. This is because all of the existing literature and unpublished data on behavior of Cr in land applied sludge indicate no adverse effects are produced even at Cr loading rates as high as 3,000 kg/ha [33]. The U.S.EPA therefore decided to delete Cr from the Land Application Subpart of the rule [33].

Regulatory Framework of Final 40 CFR Part 503

The final rule is organized into five major sections referred to as subparts with each subpart further divided into sections. While it is beyond the scope of this textbook to provide a detailed description of each section, highlights of each subpart will be outlined below.

Subpart A—General Provisions

This subpart is organized into nine sections including:
- 503.1 Purpose and applicability
- 503.2 Compliance period
- 503.3 Permits and direct enforceability
- 503.4 Relationship to other regulations
- 503.5 Additional or more stringent requirements
- 503.6 Exclusions
- 503.7 Requirements for a person who prepares sewage sludge
- 503.8 Sampling and analysis
- 503.9 General definitions

The first section, 503.1 states that the purpose of the rule is to establish standards, including general requirements, pollutant limits, management practices and operational standards for the final use or disposal of sewage sludge generated during treatment of domestic sewage in a treatment works. Reporting requirements must be met by all Class I sludge management facilities (as defined in 40 CFR 501.2, 40 CFR 403.8, 40 CFR 122.2) and all POTWs with either design flow greater than 1 MGD or servicing 10,000 people or more.

While the U.S.EPA intended to enforce the rule for sludge use and disposal through issuance of permits, a unique feature of Part 503 is established in section 503.3b where it is stated that "No person shall use or dispose of sewage sludge through any practice for which requirements are established in this part except in accordance with such requirements.'' This statement makes Part 503 self implementing.

Section 503.6 defines sludges and sludge management practices that are excluded from Part 503. Sewage sludge that is co-fired in an incinerator with other wastes, and sewage sludge that is landfilled with other wastes is excluded from Part 503. Sludges excluded from Part 503 include: sludge generated at an industrial facility during treatment of industrial wastewater, hazardous sewage sludges as defined in 40 CFR Part 261, sewage sludges with greater than 50 mg/kg of PCB, incinerator ash from sewage sludge incinerators, grit and screenings generated during preliminary treatment of domestic sewage at POTWs, drinking water treatment sludge and commercial and industrial septage.

Section 503.8 establishes sampling and analysis requirements for sludge monitoring under Part 503. This includes methods for analysis of enteric viruses, fecal coliform, Helminth ova, inorganic pollutants, *Salmonella* bacteria, specific oxygen uptake rate and total, fixed, and volatile solids.

Subpart B—Land Application

This subpart is organized into nine sections including:

- 503.10 Applicability
- 503.11 Special definitions
- 503.12 General requirements
- 503.13 Pollutant limits
- 503.14 Management practices
- 503.15 Operational standards—pathogens and vector attraction reduction
- 503.16 Frequency of monitoring
- 503.17 Recordkeeping
- 503.18 Reporting

In the final Part 503, all of the practices described in previous drafts of the Part 503 regulations as application of sludge to forest land, use of sludge at soil reclamation sites, application of sludge to agricultural land, application of sludge to public contact sites, and distribution and marketing were grouped together and are governed by Subpart B. The U.S.EPA actually conducted a separate risk assessment for each practice and, during the Part 503 revision process considered regulating application to agricultural land (including distribution and marketing) separately from application to non-agricultural land (forest land, soil reclamation sites, public contact sites). However in their technical support document for land application, the Agency states that a policy decision was made to produce only one set of criteria for all land application to simplify compliance for preparers and appliers of sewage sludge [31]. As a consequence, Subpart B governs land application of sludge at agricultural land, soil reclamation, forest, and public contact sites, as well as distribution and marketing with a single set of criteria.

Section 503.13 contains numerical pollutant concentration limits that define three categories of sludge quality for all land applied sludge. The first category of sludge quality defines sludges of such poor quality that they cannot be land applied under Part 503. In order for a sludge to be prohibited from land application it must exceed the pollutant concentration ceiling limits listed in Table 1 of Section 503.13 (Table 2.9). The second category of sludge quality includes sludges that meet the pollutant concentration ceiling limits in Table 1 of Section 503.13, and are thus deemed suitable for land application, but do not meet the pollutant concentration limits of Table 3 of Section 503.13 (Table 2.9). The sludges in this category are therefore subject to the cumulative pollutant loading limits of Table 2 of Section 503.13 (Table 2.9) which limit potential risk to human health and the environment from land application of these sludges to acceptable levels. The third category is synonymous with high quality sludge as defined by the pollutant concentration limits in Table 3 of Section 503.13. All sludges meeting these limits are exempt from cumulative loading limits.

This third category is the final form of the APL or No-Adverse-Effect-Sludge limits discussed earlier in this chapter. The U.S.EPA developed this final regulatory structure to encourage municipalities to produce "cleaner" sludges with lower metal concentrations. The incentive offered is that regulatory control is progressively relaxed as sludge quality is increased through the three categories described above. The U.S.EPA encouraged this improvement in sludge quality hoping that it would lead to increases in beneficial use through land application.

Section 503.13 also contains a set of annual pollutant loading rates which are applicable to sludges that are "sold or given away in a bag or other container for application to the land". The intent of the U.S.EPA in issuing these annual pollutant loading rates (Table 2.10) was to insure that the home gardeners, the most sensitive distribution and marketing customer, would not apply excessive amounts of material to their land. The annual loading limits ensure that even high quality sludge, meeting the pollutant concentration limits in Table 3 of Section 503.13, would not be applied in excess of the cumulative loading limits of Table 2 of section 503.13 for twenty years if sludge was applied every year.

Since promulgation in 1993, the numerical limits in the four tables of Section 503.13 have been revised. As was discussed earlier, Cr was deleted altogether due to the risk assessment failing to show that it posed any significant risk in land applied sludge [33]. The limit for Se in Table 2 of Section 503.13 was increased from 36 kg/ha to 100

TABLE 2.9 Pollutant Concentration Ceiling Limits, Cumulative Loading Limits and Pollutant Concentration Limits for Land Application from Section 503.13 of Part 503.

Regulated Pollutant	Pollutant Concentration Ceiling Limit[a] (mg/kg)	Cumulative Pollutant Loading Limit[b] (kg/ha)	Pollutant Concentration Limit[c] (mg/kg)
As	75	41	41
Cd	85	39	39
Cu	4,300	1,500	1,500
Hg	57	17	17
Mo	75	—	—
Ni	420	420	420
Pb	840	300	300
Se	100	100	100
Zn	7,500	2,800	2,800

[a]From Table 1 of Section 503 13 Cr was originally regulated with a limit of 3,000 mg/kg but was deleted as per *Federal Register* 60 54764–54770, October 25, 1995
[b]From Table 2 of Section 503 13 Cr was originally regulated with a limit of 1,200 kg/ha but was deleted as per *Federal Register* 60· October 25, 1995 Se was originally regulated with a limit of 36 kg/ha but was increased to 100 kg/ha as per *Federal Register* 60 54764–54770, October 25, 1995 Mo was originally regulated with a limit of 18 kg/ha but has been temporarily deleted until new criteria are developed as per *Federal Register* 59·9095–9099, February 25, 1994
[c]From Table 3 of Section 503 13 Cr was originally regulated with a limit of 1,200 mg/kg but was deleted, and Se was originally regulated with a limit of 36 mg/kg which was increased to 100 mg/kg as per *Federal Register* 60 54764–54770, October 25, 1995 Mo was originally regulated with a limit of 18 mg/kg but was temporarily deleted until new criteria are developed as per *Federal Register* 59· 9095–9099, February 25, 1994

TABLE 2.10. Annual Pollutant Loading Limits Applicable to Distribution and Marketing of Sludge from Section 503.13 of Part 503.

Regulated Pollutant	Annual Pollutant Loading Limit[a] (kg/ha per 365 day period)
As	2 0
Cd	1.9
Cu	75
Hg	0.85
Ni	21
Pb	15
Se	5.0
Zn	140

[a]From Table 4 of Section 503 13 Cr and Mo were originally regulated with limits of 150 and 0.90 kg/ha per 365 day period, respectively. Cr has been deleted as per *Federal Register* 60·54764–54770, October 25, 1995, Mo has been temporarily deleted until new criteria are developed as per *Federal Register* 59.9095–9099, February 25, 1994

kg/ha and in Table 3 of the same section from 36 mg/kg to 100 mg/kg. This was the result of another remand by the U.S. Court of Appeals due to U.S.EPA basing the original limits on the 99th percentile sludge concentrations from the NSSS. The 100 mg/kg and 100 kg/ha limits for Se are risk-based [33]. The limits for Mo in Tables 2, 3, and 4 of Section 503.13 have also been deleted due to U.S.EPA deciding that the risk assessment for Mo was flawed by lack of sufficient data. The U.S.EPA is trying to acquire additional data on transport of Mo through terrestrial pathways to properly reassess the risk for Mo in land applied sludge. New limits for Mo will eventually be inserted in these tables [34]. It is thought that the new risk assessment will produce new Mo limits of approximately 55 kg/ha, 55 mg/kg, and 2.75 kg/ha per 365 day period for Tables 2, 3, and 4 of Section 503.13, respectively.

Section 503.14 outlines management practices that must be met at land application sites. These include:

(1) Bulk sewage sludge cannot be applied to land where it is likely to adversely affect a threatened or endangered species or its designated critical habitat.

(2) Bulk sewage sludge shall not be land applied to sites that are frozen, flooded, or snow covered if this will cause sludge to enter a wetland or other waters of the United States.

(3) Bulk sewage sludge shall not be applied to land that is 10 meters or less from waters of the United States.

(4) Bulk sewage sludge shall not be applied to land at rates in excess of the agronomic rate unless, in the case of a reclamation site, otherwise specified by the permitting authority.

(5) Sewage sludge that is sold or given away in a bag or other container shall have a label attached to the bag or an information sheet shall be provided to the person who receives the sludge. The label or information sheet shall contain the name and address of the sludge preparer, a statement that land application is prohibited except in accordance with the instructions, and the annual whole sludge application rate that does not cause any of the annual pollutant loading rates in Table 4 of 503.13 to be exceeded.

Section 503.15 contains operational standards with respect to pathogens and vector attraction reduction. It states that all land applied sludge must meet either Class B or Class A pathogen standards and that sludges that are distributed and marketed, either in bulk for use on lawns or home gardens or in bags or other containers, must meet Class A pathogen standards. It also states that all sludges that are land applied must meet at least one vector attraction reduction standard.

Section 503.16 outlines requirements for frequency of monitoring. This section states that the frequency of monitoring of all pollutants in Tables 1 through 4 of Section 503.13, for pathogen density as required in Section 503.32 and for vector attraction reduction requirements as required in 503.33 shall be once per year for POTWs producing less than 290 metric tons of sludge per 365 days, once per quarter (4 times per year) for POTWs producing 290 to 1499 metric tons of sludge per 365 days, once per 60 days (six times per year) for POTWs producing 1,500 to 14,999 metric tons of sludge per 365 day period, and once per month for POTWs that produce greater than 15,000 metric tons of sludge per 365 day period. The permitting authority is granted the right to reduce the frequency of monitoring after two years but it may not be reduced to less than once per year for land applied sludge.

The U.S.EPA further states in Subpart B that if a sludge meets the no-adverse-effect limits in Table 3 of 503.13, the Class A pathogen requirements in Section 503.32 and the vector attraction reduction requirements of Section 503.33 the sludge is exempt from the general requirements of Section 503.12 and the management practices of Section 503.14. These sludges have come to be referred to as Exceptional Quality or EQ sludges. The U.S.EPA has deemed them to be equivalent to commercial fertilizer products and may thus be used virtually without restriction. Again, this was done to encourage beneficial utilization of sludge by encouraging production of exceptional quality material with the incentive that such sludge is virtually exempt from regulation.

Subpart C—Surface Disposal

This subpart is organized into nine sections including:

- 503.20 Applicability
- 503.21 Special definitions
- 503.22 General requirements
- 503.23 Pollutant limits
- 503.24 Management practices
- 503.25 Operational standards—pathogens and vector attraction reduction
- 503.26 Frequency of monitoring
- 503.27 Recordkeeping
- 503.28 Reporting

This subpart represents a merging of the regulatory requirements for the practices of surface disposal, monofilling, and land application to dedicated sites, including both disposal sites and beneficial use sites. These practices were all regulated by separate criteria in all of the prior proposed drafts of the Part 503 regulation. However, to simplify the regulatory framework and implementation of the rule, the U.S.EPA decided to regulate all of these practices with a single set of criteria and they are all classified as surface disposal sites in the final Part 503.

Section 503.23 summarizes pollutant limits. This section was written to apply to monofills that don't have a liner and leachate collection system. It would also apply to any surface disposal site where sludge is placed for final disposition and to dedicated land application sites since these sites are usually not built with liners and leachate collection systems.

Section 503.23 sets numerical limits on pollutants in sludge and also has provision for site specific limits. Sludge concentration limits for As, Cr, and Ni are established in Tables 1 and 2 of Section 503.23. The allowable concentration of these three metals is dependent on the distance from the edge of the "sewage sludge unit" (meaning the monofill cell, surface disposal impoundment, or dedicated site field) to the surface disposal site property line (Table 2.11). Maximum allowable limits are achieved when the sewage sludge unit is 150 meters or greater from the surface disposal site property boundary.

The U.S.EPA had utilized a groundwater and a vapor inhalation pathway to derive the criteria for surface disposal sites. The vapor inhalation pathway was utilized for organic pollutants and produced minimum pollutant concentration limits of thousands of mg/kg [29]. The groundwater pathway was utilized for both organics and metals and assumed a distance of 150m from the edge of the monofill unit cell, or surface impoundment to the property boundary. It produced limits of 73 mg/kg for As, 600 mg/kg for Cr, 46,000 mg/kg for Cu, and 690 mg/kg for Ni [33]. The U.S.EPA then chose to delete Cu from this subpart since the regulatory limit derived

TABLE 2.11. Pollutant Concentration Limits for Sludge Management at Surface Disposal Sites Including Monofills, Dedicated Sites and Surface Disposal Sites From Section 503.23 of Part 503.

Distance to Property Line[a] (m)	Sludge Concentration Limit (mg/kg)		
	As	Cr	Ni
> 150	73	600	420
125–150	62	450	420
100–125	53	360	390
75–100	46	300	320
50–75	39	260	270
25–50	34	220	240
0–25	30	200	210

[a]Distances are from edge of monofill active cell, surface disposal impoundment, or dedicated site application field, to the surface disposal site property boundary

was far in excess of even the highest reported Cu concentrations in the NSSS. Organics were deleted for the same reason.

The U.S.EPA made a policy decision and decided to "cap" the Ni limit at 420 mg/kg which is the 99th percentile sludge concentration from the NSSS. Coincidentally this corresponds to the risk-based pollutant concentration limit for Ni in Table 3 of Section 503.13 for land application. The concentration limits for these metals decrease as the monofill, impoundment or dedicated application field becomes closer than 150m to the property boundary to compensate for decreased dilution of pollutant in waters leaving the site.

Section 503.23 also makes provision for site specific limits to be developed. These may be utilized if a site has a monofill, impoundment or dedicated application field that is less than 150m from the property boundary so that actual distances may be used rather than the distance ranges of Table 2 of Section 503.23 (Table 2.11). It may also be used if the concentration of one or more of the pollutants in the sludge exceeds the regulatory limits. Then the owner or operator of the site can submit documentation showing that site-specific data should be used in recalculating the pollutant concentrations. All surface disposal sites with liners and leachate collection systems are exempt from the pollutant concentration limits in Section 503.23.

Section 503.24 outlines management practices for Surface Disposal Sites. These management practices include:

(1) Sludge cannot be placed in a surface disposal site if it is likely to adversely affect a threatened or endangered species or its critical habitat.

(2) Sewage sludge units cannot block the flow of a base flood.

(3) Sewage sludge units must be built to withstand maximum recorded horizontal ground level acceleration if they are located in a seismic impact (earthquake) zone.

(4) Sewage sludge units must be at least 60m from a geological fault that has displacement in Holocene time, unless specified by the permitting authority.

(5) Sewage sludge units cannot be located in an unstable area.

(6) Sewage sludge units cannot be located in a wetlands unless specially permitted.

(7) Runoff from an active sewage sludge unit shall be collected and disposed of in accordance with National Pollutant Discharge Elimination System permits and the system must have capacity to capture runoff from a 24 hour, 25 year storm event.

(8) Leachate collection system for active sewage sludge units having

such systems must be operated and maintained while the sewage sludge unit is active and for 3 years after it closes.

(9) Leachate from active sewage sludge units that have liners and leachate collection systems must be collected and disposed in accordance with applicable requirements while the sewage sludge unit is active and for 3 years after the unit closes.

(10) The concentration of methane gas must not exceed 25 percent of the lower explosive limit in air in any structure on the surface disposal site and shall not exceed the lower explosive limit in air at the property line at any time while the sewage sludge unit is active or for 3 years after it closes.

(11) No human food, animal feed or fiber crop can be grown on an active sewage sludge unit unless the owner/operator of the surface disposal site can demonstrate to the permitting authority that management practices can protect public health and the environment from reasonably anticipated adverse effects (this is a provision for site specific permitting to allow crops to be grown at dedicated beneficial use sites).

(12) Animals cannot be grazed on an active sewage sludge unit unless the owner/operator of the surface disposal site can demonstrate that management practices can protect public health and the environment from reasonably anticipated adverse effects. (This is another provision for site specific permitting to allow grazing at dedicated beneficial use sites.)

(13) Public access shall be restricted at surface disposal sites for the entire time it contains active sewage sludge units and for 3 years after final closure.

(14) Sewage sludge placed on an active sewage sludge unit shall not contaminate the aquifer, this shall be verified by a ground water monitoring program or site assessment conducted by a qualified professional.

Section 503.25 summarizes operational standards for pathogens and vector attraction reduction which require that the sludge managed at a surface disposal site meet either the Class A or B pathogen requirements of Section 503.32 and the vector attraction reduction requirements of Section 503.33.

Section 503.26 contains a summary of frequency of monitoring requirements. They are essentially the same as those of Section 503.16 described above except that air in structures at surface disposal sites, and at the property boundary, is required to be monitored continuously for methane gas for the entire period that the site contains an active sewage sludge unit and for 3 years following final closure.

Subpart D—Pathogens and Vector Attraction Reduction

This subpart contains all of the requirements for classifying a sludge as class A or B with respect to pathogen content, as well as vector attraction reduction requirements. Class A is the no-adverse-effect category for pathogen content and when the class A pathogen standards are met in conjunction with the no-adverse-effect metal limits (from Table 3 of Section 503.13), the sludge is considered to be exceptional quality and is exempt from nearly all regulatory requirements except pollutant and pathogen content monitoring. The subpart is organized into subsections as follows:

- 503.30 Scope
- 503.31 Special definitions
- 503.32 Pathogens
- 503.33 Vector attractions reduction

Section 503.32 lists the criteria for Class A and Class B pathogen content. Since the original proposal of Part 503, the pathogen criteria were dramatically revised. The original proposal criteria contained three quality classes as presented in Table 2.4. The final Part 503 contains only two quality classes and a variety of criteria for meeting each.

The final Part 503 provides six alternatives for meeting the class A pathogen standards; they are discussed in detail in another section of this chapter but may be briefly summarized as follows:

(1) The sludge must be heat treated for a prescribed time and temperature, and must meet class A fecal coliform limits (<1,000 most probable number/g total dry solids) or class A *Salmonella* limits (<3 most probable number/4g total dry solids) at the time it is used or disposed or prepared for use or disposal.

(2) The sludge must be lime treated and the pH and temperature must be elevated to prescribed levels for prescribed time periods and the sludge must meet class A fecal coliform or *Salmonella* limits at the time of use or disposal.

(3) The sludge must meet the class A fecal coliform or *Salmonella* limits, at the time of use or disposal and the class A enteric virus limit (<1 plague forming unit/4g total dry solids) and the class A viable helminth ova limit (<1 viable ova/4g total dry solids) prior to pathogen treatment.

(4) The sludge must meet the class A fecal coliform or *Salmonella* limits and the class A enteric virus limit and viable helminth ova limit at the time of use or disposal.

(5) The sludge must meet the class A fecal coliform or *Salmonella* limits at the time of sludge use or disposal and the sludge must have been

processed through a "process to further reduce pathogens" which are described in appendix B of the rule. Processes to further reduce pathogens (PFRP) include composting, heat drying, heat treatment, thermophilic aerobic digestion, beta ray irradiation, gamma ray irradiation and pasteurization.

(6) The sludge must meet the class A fecal coliform or *Salmonella* limits at the time of sludge use or disposal and the sludge must be treated in a process that is equivalent to a PFRP. In this option the equivalent process must be documented and critical operating parameters for the process must be monitored and maintained above threshold limits. A detailed discussion of this alternative is presented in Chapter 4 of this book. The alternative was included to allow for utilization of innovative and unique pathogen treatment processes that achieve adequate levels of pathogen destruction.

Sludges that meet the class A pathogen standards have no management practice requirements with respect to pathogens, however, they are still subject to other management practices prescribed for the specific use or disposal option in Subpart B or C. As discussed previously, if the sludge meets the class A pathogen requirements and the Table 3 limits in Section 503.13, then no management practices are required whatsoever.

Section 503.32 also outlines three alternatives for meeting the class B pathogen requirements. These alternatives are discussed in more detail in a later section of this chapter but are summarized as follows:

(1) The geometric mean density of fecal coliform in seven samples of sludge collected at the time of sludge use or disposal must be <2,000,000 most probable number/g total dry solids or <2,000,000 colony forming units/g total dry solids.

(2) The sludge must be treated by a "process to significantly reduce pathogens" as described in appendix B of the final rule, including: aerobic digestion, air drying, anaerobic digestion, composting, and lime stabilization.

(3) The sludge must be treated by a process equivalent to the processes that significantly reduce pathogens.

For sludges meeting only class B pathogen requirements, the final Part 503 dictates a series of management practices that must be met. These management practices include:

(1) Waiting periods to harvest food crops grown on sludge amended soil. Above ground crops that touch the sludge soil mixture cannot be harvested until 14 months after sludge application; corps with harvested parts below ground cannot be harvested for 38 or 20 months after sludge application if the land applied sludge remains on the

soil surface for less than 4 months or greater than 4 months prior to incorporation, respectively.

(2) Food crops with harvested parts that never contact the sludge amended soil, feed crops and fiber crops cannot be harvested until 30 days after sludge has been applied to the land.

(3) Animals cannot graze on sludge amended soil until 30 days after the sludge was applied.

(4) Turf grown on sludge amended soil cannot be harvested until one year after the sludge was applied if the turf is to be placed on a public contact site or home lawn.

(5) Public access to land with sludge amended soil will be restricted for 1 year for public contact sites and for 30 days for land with a low potential for public exposure.

While these pathogen standards are not risk based, the management practices prescribed for class B sludges were included to reduce risk of exposure to sludge-borne pathogens from sludges that had not received the rigorous and harsh pathogen treatments of class A sludges and are thereby assumed to harbor higher pathogen populations.

Section 503.33 outlines requirements for vector attraction reduction which are meant to insure that a stabilized product will be land applied or surface disposed to minimize odor potential and reduce proliferation of pathogens via vectors. Sludges must meet at least one of the outlined requirements in this section. The vector attraction reduction requirements are summarized as follows:

(1) The mass of volatile solids in the sludge must be reduced by a minimum of 38 percent during the stabilization process.

(2) In anaerobically digested sludge that does not undergo 38 percent reduction of volatile solids, if a portion of the digested sludge is anaerobically digested for an additional 40 days at bench-scale at between 30 and 37°C, less than 17 percent additional volatile solids destruction must occur for the sludge to be considered in compliance with the vector attraction reduction standards.

(3) In aerobically digested sludge that does not undergo 38 percent reduction of volatile solids, if a portion of the digested sludge, with <2 percent solids, is digested aerobically at bench scale for an additional 30 days at 20°C, less than 15 percent additional volatile solids destruction must occur for the sludge to be considered in compliance with the vector attraction reduction standards.

(4) It must be demonstrated that the specific oxygen uptake rate for sludge treated in an aerobic process is ≤ 1.5 mg O_2/hr per gram of total dry solids at 20°C.

(5) Sludge must be treated in an aerobic process for at least 14 days

with mean temperature >45°C and minimum temperature >40°C for the entire process.

(6) Sludge pH must be raised to 12 or higher where it must remain for 2 hrs and then must not drop below 11.5 for an additional 22 hrs.

(7) Sludge that does not contain unstabilized primary wastewater treatment solids must be dried to ≥75 percent solids.

(8) Sludge that contains unstabilized primary wastewater treatment solids must be dried to ≥90 percent solids.

(9) Sludge must be injected below the surface of the land.

(10) Sludge that is surface applied to the land must be incorporated into the soil within six hours after application.

(11) Sludge applied to an active sewage sludge unit must be covered with soil or other material at the end of each operating day.

The last option was written specifically for monofills. Sludge meeting at least one of the summarized vector attraction reduction requirements listed above are deemed suitable for land application or surface disposal.

Subpart E—Incineration

This subpart regulates all sludges that are incinerated except those that are co-fired in multiple waste incinerators. The subpart contains the following sections:

- 503.40 Applicability
- 503.41 Special definitions
- 503.42 General requirements
- 503.43 Pollutants limits
- 503.44 Operational standards—total hydrocarbons
- 503.45 Management practices
- 503.46 Frequency of monitoring
- 503.47 Recordkeeping
- 503.48 Reporting

Section 503.43 specifies pollutant limits for seven inorganic pollutants including As, Be, Cd, Cr, Hg, Ni, and Pb. All the pollutant limits are site specific. For Be and Hg, Section 503.43 states that "firing of sewage sludge in a sewage sludge incinerator shall not violate the requirements in the National Emissions Standards of subpart C of 40 CFR part 61". Hence Hg and Be may be present in incinerated sludge to the extent that their concentration does not cause the stack gas to violate the National Emission Standard.

Section 503.43 sets a pollutant limit for Pb in the incinerator feed sludge. The daily concentration limit of Pb in sewage sludge (mg Pb/kg total dry solids) C, can be computed as follows:

$$C = [0.1 \times NAAQS \times 86,400]/[DF \times (1 - CE) \times SF]$$

where NAAQS is the National Ambient Air Quality Standard for Pb (1.5 μg Pb/m^3), DF is the dispersion factor (μg Pb/m^3 per g per s), CE is the incinerator control efficiency for lead (expressed as a decimal where 1.00 represents 100 percent retention of Pb in the incinerator), and SF is the sludge feed rate (Mg dry weight/day). The constant, 86,400, is the number of seconds in one day. The dispersion factor is dependent on the incinerator's stack height. The actual stack height is used in an air dispersion model specified by the permitting authority to determine the dispersion factor for stacks not greater than 65m in height. For stacks greater than 65m in height a creditable stack height is determined and utilized in the air dispersion model to determine the dispersion factor. The 0.1 factor in the equation allocates 10 percent of the NAAQS for lead to the firing of sewage sludge in a sewage sludge incinerator, that is, sludge incineration may not contribute more than 10 percent of the NAAQS for Pb.

Pollutant limits were set for As, Cd, Cr and Ni using risk assessment based on the inhalation pathway. The daily concentration limit for these pollutants, C, may be calculated from the following equation:

$$C = [RSC \times 86,400]/[DF \times (1 - CE) \times SF]$$

Here, SF is the sludge feed rate (dry Mg/day) and RSC is the risk specific concentration of pollutant in air to which the MEI is exposed, as determined from the risk assessment inhalation pathway. The RSC determined for As, Cd, Cr and Ni is presented in Table 2.12 (Table 1 and 2 of Section 503.43, [30]). These risk specific concentrations were derived assuming that 10 percent of the total Ni emitted from the sludge incinerator is in the form of Ni subsulfide and that the RSC for Cr is dependent on the fraction, r, of total Cr in the emissions that is in the hexavalent form, where

$$RSC = 0.0085/r$$

The RSC values for Cr listed in Table 2 of Section 503.43 for various incinerators and scrubbers (Table 2.12) are calculated using r values that were observed to be typical for that technology in field tests conducted by the U.S.EPA. Operators are given the option of measuring the actual fraction of hexavalent Cr in their emissions and using the site specific equation above instead of the values in Table 2 of Section 503.43.

Section 503.44 specifies operational standards for sludge incinerators

TABLE 2.12. Risk Specific Concentrations of As, Cd, Cr, and Ni to Be Utilized in Determining Daily Concentration Limits for Incinerated Sludge.

Pollutant	Risk Specific Concentration[a] $(\mu g/m^3)$
As	0 023
Cd	0 057
Ni	2.0
Cr (FBWS)[b]	0.65
Cr (FBEP)[b]	0.23
Cr (OIWS)[b]	0.064
Cr (OIEP)[b]	0.016

[a]Risk specific concentrations for As, Cd, and Ni are from Table 1 of Section 503 43, and from Table 2 of Section 503 43 for Cr
[b]FBWS is fluidized bed incinerator with wet scrubber, FBEP is fluidized bed incinerator with wet scrubber and wet electrostatic precipitator, OIWS is other incinerator types with wet scrubber, and OIEP is other incinerator types with wet scrubber and wet electrostatic precipitator.

which consists of the requirement that the monthly average concentration of total hydrocarbons in the exit gas of the incinerator stack be equal to or less than 100 ppm on a volumetric basis after correction to zero percent moisture and 7 percent oxygen. Subsequent to the promulgation of the final Part 503 the U.S.EPA added an alternative operational standard in Section 503.44 as a result of a petition filed with the Washington D.C. Circuit Court seeking a review of the Part 503 requirement for total hydrocarbon emissions monitoring. The U.S.EPA decided to allow an operational standard of 100 ppm CO on volumetric basis as an alternative to the 100 ppm total hydrocarbon limit [34]. These operational standards were set to insure adequate combustion to destroy harmful organic emissions that would otherwise result from sewage sludge incineration.

Section 503.45 specifies management practices that must be met by sludge incineration operators to insure that the operational standard is complied with and thereby that the incinerator is operated properly and effectively. The incinerator management practices may be summarized as follows:

(1) An instrument must be installed, calibrated, operated and maintained to continuously monitor either total hydrocarbons or CO in the stack exit gas of each incinerator.

(2) An instrument that measures and records O_2 concentration shall be installed, calibrated, operated and maintained to continuously monitor the stack exit gas of each incinerator.

(3) An instrument that measures and records moisture content shall be installed, calibrated, operated and maintained to continuously monitor the stack exit gas of each incinerator.

(4) An instrument that measures and records combustion temperature shall be installed, calibrated, operated and maintained for continuous monitoring of each incinerator.

(5) The maximum combustion temperature for an incinerator will be specified by the permitting authority based on information obtained during the performance test to determine pollutant control efficiencies.

(6) Operating parameters for incinerator air pollution control devices will be specified by the permitting authority, based on information obtained during the performance test to determine pollutant control efficiencies.

(7) Sludge shall not be incinerated if it is likely to adversely effect a threatened or endangered species or its designated critical habitat.

The final Part 503 rule is the product of the arduous three-year revision process that involved the successful cooperation of the U.S.EPA, the academic community, municipalities, and environmental lobbyists through public comment, peer review, and consultation. The metamorphosis that has occurred as a result of these cooperative efforts, and of litigation of unresolved issues in the Washington, D.C. Circuit Court, has resulted in a rule that is amply protective of human health and the environment and that promotes beneficial use of sludge. The major accomplishments of the revision process were:

(1) Faulty pathway models and overly conservative data inputs were replaced with more technically sound alternatives.

(2) The 98th percentile approach was abandoned and replaced by risk based numerical limits.

(3) Unnecessary requirements on labeling sludge products for distribution and marketing were dropped.

(4) Pollutant limits were derived that define no-adverse-effect sludges that can be managed like commercial fertilizers.

All of these accomplishments confirm the U.S.EPA's commitment to promoting beneficial use of sludge.

THE FUTURE OF SEWAGE SLUDGE REGULATIONS IN THE UNITED STATES: ROUND II AND BEYOND

The U.S.EPA is currently in the final phase of revising Round I rulemaking. The Agency published a notice in the Federal Register [33] outlining issues to be "cleaned up" in the final Round I rule. For land application, the revisions proposed are minor except that a final pollutant concentration limit must still be developed for Mo.

For incineration the proposed changes mainly concern removing the burden on the permitting authority to dictate air dispersion models or methods of installing, calibrating operating and maintaining monitoring equipment. It is expected that a follow up notice will be published in early 1997 clarifying these issues, which will end the Round I rulemaking process.

At the time of this writing the U.S.EPA has set a schedule to propose Round II rules by December 15, 1999 and to promulgate them by December 15, 2001. The U.S.EPA appears to be primarily concerned with assessing risk due to the tetrachlorodibenzodioxins (TCDD), tetrachlorodibenzofurans (TCDF) and co-planar PCBs in Round II.

Ideally, the Round II rule-making process should entail identifying contaminants that are ubiquitous in sludge as determined by an updated NSSS, and running a preliminary risk screening to determine if there is adequate information to justify conducting a full-scale risk assessment. If the best available environmental toxicological data and environmental fate data on the contaminants indicate a potential cause for concern, then the contaminant should be subjected to the same risk assessment as was conducted in Round I. If scientific information is lacking on some contaminants that are found to be ubiquitous in sludge, the Agency should sponsor research targeted at producing data that can be utilized in their pathway risk assessments.

After it has been established that there is just cause for concern regarding a particular contaminant, and if adequate scientific data exists, the U.S.EPA should conduct a detailed risk assessment analysis to generate numerical limits to be included in 40 CFR Part 503. The Agency should then propose the addition of this contaminant to the rule and should subject its reasoning and risk assessment methodology to public review and comment. The Agency should also solicit information during the Round II public comment period on the environmental toxicology and fate of contaminates from the Round I rule that has been generated subsequent to its promulgation. Any significant new scientific information that becomes available should be carefully considered and where appropriate, should be used to adjust the Round I risk assessment. The Agency should consider deregulating Round I contaminants if their adjusted risks become insignificant as was the case for Cr in land applied sludge. The Agency should repeat this process periodically as the generation of additional scientific and sludge composition information dictates.

Future rounds of rule making may also include developing risk assessments for pathogens and gross radioactivity and for sludge management practices that may be developed or innovated after promulgation of the Round II rule.

Regulation of sewage sludge management and disposal is a journey and not a destination. It is an ongoing process that must be dynamic, to account for new scientific information as it is continuously generated, and to adapt to changing practices that are being constantly innovated. Above all, the regulatory process must maintain a balanced perspective that not only considers the potential threat of sludge-borne contaminants to public health and the environment based on the best scientific data available, but that also considers practical constraints, which may be fiscal, institutional, sociological or technological, faced by municipalities that generate the sludge. Overly conservative, or unduly restrictive regulations will not stop the flow of sewage sludge from POTWs and the sludge produced will not disappear. Regulatory agencies should identify those sludge management practices that pose the lowest risk of adversely impacting public health and the environment and should promulgate regulations that promote such practices that provide an incentive for municipalities to further reduce risk through site management and production of cleaner sludges.

PATHOGENS AND VECTOR ATTRACTION REDUCTION—PART 503 REGULATION

INTRODUCTION

On February 19, 1993, the U.S. Environmental Protection Agency (U.S.EPA) published the Standards for the Use or Disposal of Sewage (40 CFR Part 503) [30]. Part 503 contains the requirements that have to be met when sewage sludge is applied to the land, placed on a surface disposal site, placed in a municipal solid waste landfill, or fired in a sewage sludge incinerator. A Part 503 standard contains the following seven elements: general requirements, pollutant limits, management practices, operational standards, and frequency of monitoring, recordkeeping, and reporting requirements. The general requirements in a Part 503 standard contain what often are called "administrative requirements" (e.g., notification for interstate transfer of land-applied sewage sludge), while the pollutant limits and management practices contain the requirements that protect public health and the environment from the reasonably anticipated adverse effects of pollutants in sewage sludge. An operational standard is a technology-based requirement that in the judgement of the U.S.EPA Administrator protects public health and the environment. For land application, the pathogen and vector attraction reduction requirements are operational standards and for incineration, the allowable concentration of total hydrocarbons in the emissions from the incinerator stack is an operational standard.

The frequency of monitoring, recordkeeping, and reporting requirements in a Part 503 standard help make Part 503 self-implementing (i.e., the requirements have to be met even if a federal permit has not been issued). These requirements indicate how often a representative sample of sewage sludge has to be collected and analyzed; who has to keep what records and for how long; and who has to report information to the permitting authority annually.

As mentioned above, the operational standards in Part 503 are technology-based. This means they are based on the performance of a technology, not on the results of a risk assessment. U.S.EPA's conclusion that the operational standards protect public health and the environment is based on the judgement of the U.S.EPA Administrator. This section of Chapter 2 discusses the pathogen and sector attraction operational standards for land application of sewage sludge.

OVERVIEW OF THE LAND APPLICATION PATHOGEN AND VECTOR ATTRACTION REDUCTION REQUIREMENTS

General Approach

Two approaches were taken in the Part 503 land application operational standards for pathogens and vector attraction reduction. In the first approach, sewage sludge can be treated to reduce pathogens and to reduce the characteristics of sewage sludge that attract vectors. If specified treatment-related requirements are met, nothing has to be done at the application site with respect to pathogens and vector attraction reduction.

The second approach in the Part 503 rules requires a combination of sewage sludge treatment and management practices that must be met at the application site. For pathogens, some reduction must be achieved through treatment of the sewage sludge, and in addition management practices have to be met at the application site. The management practices prevent exposure to the sewage sludge for a period long enough to allow the environment to further reduce the pathogens to below detectable levels, which is the goal in both approaches.

The vector attraction reduction requirements that are met at the application site (i.e., injection below the land surface and incorporation after being surface-applied) place a barrier of soil between the sewage sludge and the vectors. This prevents contact between the sewage sludge and the vectors.

Part 503 contains several treatment-related alternatives for pathogen reduction and several alternatives for the combined treatment and site-related pathogen reduction. This provides the person who prepares the sewage sludge (i.e., the generator or a person who derives a material from sewage sludge) flexibility to choose the alternative that best fits a particular situation.

Part 503 also contains several treatment-related options for vector attraction reduction and two barrier options for vector attraction reduction. The person who prepares the sewage sludge can choose any of the treatment options or the applier of the sewage sludge can choose any of the barrier options.

EPA concluded that in the judgement of the Administrator, all of the pathogen destruction alternatives and all of the vector attraction reduction options protect public health and the environment. Note that the pathogen requirements always are called "alternatives" and the vector attraction reduction requirements are called "options" in the Part 503 Standards.

What Is a Pathogen?

Part 503 defines a pathogen as a disease causing organism. This includes, but is not limited to, certain bacteria, protozoa, viruses, and viable helminth ova.

The Part 503 regulation contains requirements for the reduction of three pathogens (i.e., *Salmonella* sp. bacteria, enteric viruses, and viable helminth ova). In some cases, fecal coliform is used as an indicator of those organisms. U.S.EPA concluded that, in the judgement of the Administrator, when these three pathogens are reduced to below detectable levels, public health and the environment are protected from pathogens in sewage sludge. More details on each organism (e.g., densities and survival) are presented in the technical support document for the Part 503 pathogen and vector attraction reduction requirement [28].

Part 503 contains requirements for two classes of pathogen reduction (i.e., Class A and Class B). The Class A pathogen requirements have to be met if the first approach mentioned above is used (i.e., treat the sewage sludge). The Class B pathogen requirements plus restrictions at the application site have to be met under the second approach. Public health and the environment are protected for both classes of pathogen reduction.

Which Pathogen Requirements Should Be Met?

The decision about whether to meet the Class A pathogen requirements or the Class B requirements with site restrictions is based on the type of land onto which the sewage sludge is applied. It also depends on whether the sewage sludge is applied to the land in bulk, or sold or given-away in a bag or other container for application to the land (hereafter called bagged sewage sludge).

Either the Class A pathogen requirements or the Class B pathogen requirements can be met when bulk sewage sludge (i.e., any sewage sludge that is not bagged sewage sludge) is applied to agricultural land, forest, a reclamation site, or a public contact site with applicable site restrictions.

The Class A pathogen requirements must be met, however, when bulk sewage sludge is applied to a lawn or home garden or when sewage sludge is bagged. The reason for this is there is no way to implement the site restrictions (i.e., there is no control of the application site) for a Class B sewage sludge when bulk sewage sludge is applied to a lawn or home garden, or when sewage sludge is bagged. There are no site restrictions when the Class A requirements are met.

The Class A Pathogen Alternatives

The Part 503 regulation contains six alternatives that can be met for a sewage sludge to be considered Class A with respect to pathogens. There are two requirements that apply to all of the Class A alternatives.

The first requirement is that the Class A pathogen requirements be met either prior to or at the same time as the vector attraction reduction requirements. The purpose of this requirement is to ensure that enough competitive organisms remain in the sewage sludge. This reduces the potential for *Salmonella* sp. bacteria to regrow under certain conditions (i.e., available nutrients and removal of a stress such as elevated temperature). If vector attraction reduction occurs before pathogen reduction, not enough competitor organisms remain in the sewage sludge and regrowth of *Salmonella* sp. bacteria can occur.

The second requirement is what is called the "regrowth" requirement. This requirement addresses the density of either fecal coliform or *Salmonella* sp. bacteria in the sewage sludge at the time the sewage sludge is used or disposed, or at the time the sewage sludge is prepared for sale or give-away in bags. The purpose of this requirement is to ensure that *Salmonella* sp. bacteria have not regrown between the time the sewage sludge is treated and the time it is used or disposed (i.e., during storage).

To meet the "regrowth" requirement, either the density of fecal coliform in the sewage sludge has to be less than 1000 Most Probable Number (MPN) per gram of total solids or the density of *Salmonella* sp. bacteria in the sewage sludge has to be less than three MPN per four grams of total solids at the time the sewage sludge is used or disposed, or at the time the sewage sludge is prepared for sale or give-away in bags. This means that the sewage sludge has to be tested for either fecal coliform or *Salmonella* sp. bacteria (not both) after storage, but prior to when it is used or disposed.

In some cases, the density of fecal coliform in a sewage sludge sample may exceed 1000 MPN per gram of total solids even though the *Salmonella* sp. bacteria density is less than three MPN per four grams of total solids. In this case, the regrowth requirement is met because the purpose of that requirement is to insure *Salmonella* sp. bacteria have not regrown. Thus, if the density of fecal coliform in the sewage sludge sample exceeds 1000 MPN per gram of total solids, that sample or another sample can be analyzed for *Salmonella* sp. bacteria to determine whether the regrowth requirement is met.

The first Class A pathogen alternative addresses percent solids, time, and temperature. If the temperature of the sewage sludge with a certain percent solids is raised to the value determined using equations in Part 503 for the time indicated by the equations, the sewage sludge is Class A. This alternative applies to sewage sludge where an external source is used to raise the temperature. It does not apply to sewage sludge that is composted.

The second alternative addresses pH adjustment, time/temperature, and percent solids after the pH and time/temperature requirements are met. Any material can be used to adjust the pH. The percent solids requirement has to be met by air drying the sewage sludge.

The third alternative is a demonstration alternative. If the sewage sludge contains enteric viruses and viable helminth ova prior to treatment and if the treatment process reduces the density of those organisms to below detectable levels, the demonstration requirements in this alternative are met. After the demonstration is made, the process has to be operated like it was during the demonstration. No further analysis of the sewage sludge for enteric viruses and viable helminth ova is required after the demonstration. The regrowth requirement (i.e., measure either fecal coliform or *Salmonella* sp. bacteria density) also has to be met.

The fourth alternative is called the "storage pile" alternative. Under this alternative, a representative sample of the sewage sludge has to be analyzed for fecal coliform or *Salmonella* sp. bacteria, enteric viruses, and viable helminth ova. If the density requirements for those organisms are met, the sewage sludge is Class A with respect to pathogens.

The last two alternatives address treatment of sewage sludge in a Process to Further Reduce Pathogens (PFRP). In Alternative 5, if the sewage sludge is treated in one of the PFRPs described in Appendix B to Part 503, the sewage sludge is Class A with respect to pathogens. In Alternative 6, if the sewage sludge is treated in a process that is equivalent to a PFRP, as determined by the permitting authority, the sewage sludge is Class A. The regrowth requirement also has to be met for both of these alternatives.

The Class B Pathogen Alternatives

Part 503 contains three alternatives that can be met for a sewage sludge to be Class B with respect to pathogens. For a Class B sewage sludge, there is no order in which pathogen reduction and vector attraction reduction have to be achieved and there is no "regrowth" requirement.

The first Class B alternative is that seven representative samples of the sewage sludge have to be analyzed for fecal coliform during each monitoring episode. The geometric mean of the fecal coliform densities in those samples has to be less than two million MPN per gram of total solids or less than two million Colony Forming Units (CFU) per gram of total solids. The samples have to be collected at the time the sewage sludge is used or disposed (e.g., long enough to analyze samples and receive results prior to use or disposal). The two million MPN or CFU value is based on the performance of a well-operated anaerobic digester.

The second and third Class B alternatives address treatment of the sewage sludge. If the sewage sludge is treated in one of the Processes to Significantly Reduce Pathogens (PSRP) described in Appendix B to Part 503 or if the sewage sludge is treated in a process that is equivalent to a PSRP, as determined by the permitting authority, the sewage sludge is Class B with respect to pathogens.

Site Restrictions for Class B Sewage Sludge

When a Class B sewage sludge is applied to the land, restrictions have to met at the application site. These site restrictions provide the environment time to reduce remaining pathogens in the sewage sludge to below detectable levels.

The first three site restrictions address harvesting of food crops. These restrictions prevent exposure to viable helminth ova, which can survive on the land surface for up to a year and beneath the soil surface for up to three years after sewage sludge application. The restriction vary depending on whether the harvested parts touch the sewage sludge/soil mixture or are below the surface of the land. The restriction when harvested parts are below the land surface also varies depending on how long the sewage sludge remains on the land surface before it is incorporated into the soil. The time periods for the restrictions include time for viable helminth ova to be reduced to below detectable levels (i.e., the density level in Class A, Alternative 4) plus an assumed two month growing season. Note that the restrictions address harvesting of food crops, not growing of food crops.

The fourth and fifth restrictions prohibit the harvesting of food, feed, and fiber crops for 30 days after sewage sludge application and the grazing of animals for 30 days after sewage sludge application, respectively. These restrictions provide the time for the environment to reduce *Salmonella* sp. bacteria and enteric viruses in the sewage sludge to below detectable levels.

The sixth restriction addresses harvesting of turf. It ensures that turf grown on land where Class B sewage sludge is applied is not harvested for placement on land with a high potential for exposure (e.g., a ball field) or a lawn for one year after sewage sludge application.

The last two site restrictions require that public access to land on which Class B sewage sludge is applied be restricted for certain periods. If sewage sludge is applied to land with a low potential for exposure (e.g., land in a remote area), the public access restriction is 30 days. If the land has a high potential for exposure (e.g., a ball field or a construction site in a city), the public access restriction is one year. Part 503 does not specify how to restrict public access. These restrictions prevent exposure to viable helminth that may remain on the land surface.

What Is Vector Attraction Reduction?

Vector attraction reduction is the reduction in the characteristics of sewage sludge that attract organisms capable of transporting infectious agents. This concept was first introduced in the publication of 40 CFR Part 257 in 1979 [15]. Prior to Part 257, it was common to "stabilize" sewage sludge before it was used or disposed. The primary purpose of stabilization was to make sewage sludge acceptable aesthetically (e.g., reduce odors). Secondary purposes of stabilization included reduction in pathogens and in the characteristics of sewage sludge that attract vectors.

The Part 257 regulation reversed this approach. The primary purposes of the Part 257 disease requirements are to reduce the pathogen densities in sewage sludge and to reduce the characteristics of the sewage sludge that attract vectors. While the reduction of pathogens and vector attraction is necessary to protect public health and the environment, the reduction of odor is not. Odor is a nuisance, but not a public health and environment concern.

The Part 503 Vector Attraction Reduction Options

Part 503 contains 10 vector attraction reduction options for sewage sludge that is land-applied. The first eight options are treatment-related, while the last two are barrier options.

The first three options address percent volatile solids reduction in the sewage sludge. The first option requires that the mass of volatile solids be reduced by a minimum of 38 percent. This percent reduction is based on the reduction achieved by an anaerobic digester with a temperature maintained at 35 degrees Celsius for 15 days. When this reduction is achieved, most of the biodegradable material in the sewage sludge is degraded to lower activity forms. The remaining biodegradable material degrades so slowly that vectors are not attracted to the sewage sludge [28].

In some cases, the volatile solids content of the sewage sludge entering the treatment process is so low that a 38 percent reduction can not be achieved. In those cases, either Option 2 or Option 3 can be met. Both options require further digestion of a portion of the digested sewage sludge in bench-scale units for certain periods. Option 2 is for an anaerobically digested sewage sludge and Option 3 is for an aerobically digested sewage sludge. If the percent volatile solids reduction during the extra digestion period is less than the specified value, vector attraction reduction is achieved.

The fourth and fifth options address aerobic digestion processes. In Option 4, if the Specific Oxygen Uptake Rate (SOUR) for a liquid sewage sludge is equal to or less than 1.5 milligrams of oxygen per hour per gram of total solids at a temperature of 20 degrees Celsius, vector attraction reduction is achieved.

Option 5 addresses composting of sewage sludge and other aerobic processes. If the specific conditions described in this option are met, vector attraction reduction is achieved.

Option 6 requires the pH of the sewage sludge to be raised to 12 or higher and remain at 12 or higher for two hours. After two hours, the pH must remain at 11.5 or higher for an additional 22 hours. Adjusting the pH of the sewage sludge does not change the nature of the substances in the sewage sludge. Instead, it causes stasis in biological activity. If the pH should drop, the surviving bacterial spores could become active biologically, and the sewage sludge could putrefy and attract vectors [28]. This is an important consideration when this option is chosen for sewage sludge that is stored before it is land-applied.

Options 7 and 8 address the percent solids of the sewage sludge. The difference between the two options is that in Option 7 the sewage sludge does not contain unstabilized solids generated in a primary wastewater treatment process, and in Option 8, the sewage sludge does contain those types of solids. When unstabilized solids are present, the percent solids of the sewage sludge has to be higher (i.e., sewage sludge has to be drier) to ensure that vectors are not attracted to the sewage sludge. These options are valid as long as the sewage sludge does not become wet. If the percent solids drops between the time the percent solids is achieved and the time the sewage sludge is used or disposed, vector attraction reduction is not achieved.

Options 9 and 10 are ''barrier'' options. In both of these options, vectors are not attracted to the sewage sludge because of the layer of soil between the sewage sludge and the vectors. Option 9 is injection of the sewage sludge below the land surface and Option 10 is incorporation of the sewage sludge into the soil after it is surface-applied.

CONCLUSIONS

U.S.EPA's Standards for the Use or Disposal of Sewage Sludge (40 CFR Part 503) contain several alternatives for pathogen reduction and several options for vector attraction reduction for land-applied sewage sludge. Neither the pathogen requirements nor the vector attraction reduction requirements are risk-based. Instead, they are technology-based, which is why they are called operational standards.

Two approaches were taken in developing the Part 503 pathogen and vector attraction reduction requirements. One approach is to treat the sewage sludge to reduce pathogens and the attractiveness of the sewage sludge to vectors. The second approach for pathogen reduction is a combination of treatment and restrictions at the application site. The site restrictions allow the environment time to further reduce remaining pathogens in the sewage sludge to below detectable levels. The second approach for vector attraction reduction is to place a barrier of soil between the sewage sludge and the vectors at the application site. This prevents vectors from contacting the sewage sludge. In the judgement of the U.S.EPA Administrator, the pathogen alternatives and the vector attraction reduction options under both approaches protect public health and the environment.

REFERENCES

1 De Turk, E. E. 1935. "Adaptability of Sewage Sludge as a Fertilizer," *Sewage Works J.* 7:597–610.

2 Seabrook, B. L. 1975. *Land Application of Wastewater in Australia, the Werribee Farm System, Melbourne, Victoria,* U.S.EPA Rep. EPA-430/9-75-017, Washington, DC: U.S. Environmental Protection Agency.

3 Muller, J. F. 1929. "The Value of Raw Sewage Sludge as Fertilizer," *Soil Sci.* 28:423–432.

4 Rudolfs, W. 1928. "Sewage Sludge as Fertilizer," *Soil Sci.* 26:455–458.

5 Lunt, H. A. 1959. *Digested Sewage Sludge for Soil Improvement,* Connecticut Experiment Station Bulletin 622.

6 Lynam, B. T., B. Sosewitz and T. D. Hinesly. 1972. "Liquid Fertilizer to Reclaim Land and Produce Crops," *Water Research* 6:545–549.

7 Dalton, F. E. and R. E. Murphy. 1973. "Land Disposal IV: Reclamation and Recycle," *J. Water Pollut. Control Fed.* 45:1489–1507.

8 Knezek, B. and R. H. Miller, eds. 1976. *Application of Sludges and Wastewaters on Agricultural Land: A Planning and Educational Guide,* Research Bulletin 1090, Ohio Agricultural Research and Development Center, Wooster, OH., pp. 1.1–1.2.

9 NASULGC. 1974. *Proceedings of the Joint Conference on Recycling Municipal Sludges and Effluents on Land,* July 9–13, 1973, Champaign, Illinois. Washington, DC: National Association of State Universities and Land Grant College.

10 Chaney, R. L. 1974. "Crop and Food Chain Effects of Toxic Elements in Sludges and Effluents," in *Proceedings of the Joint Conference on Recycling Municipal Sludges and Effluents on Land,* July 9–13, 1973, Champaign, Illinois. Washington, DC: National Association of State Universities and Land-Grant Colleges, pp. 129–141.

11 Chumbley, C. G. 1971. *Permissible Levels of Toxic Metals in Sewage Used on Agricultural Land,* Agric. Dev. Advis. Serv. Paper No. 10. 12 pp.

12 Leeper, G. W. 1972. *Reactions of Heavy Metals with Soils with Special Regard to*

Their Application in Sewage Wastes, Dept. of the Army, Corps of Engineers. Contract No. DACW 73-73-C-0026. 70 pp.

13 Logan, T. J. and R. L. Chaney. 1983. "Utilization of Municipal Wastewater and Sludge on Land—Metals," in *Proceedings of the 1983 Workshop on Utilization of Municipal Wastewater and Sludge on Land*, A. L. Page, T. L. Gleason, III, J. E. Smith, Jr., I. K. Iskandar and L. E. Sommers, eds., University of California, Riverside, CA., pp. 235–326.

14 CAST. 1976. *Application of Sewage Sludge to Cropland: Appraisal of Potential Hazards of the Heavy Metals to Plants and Animals*. Report No. 64, Council for Agricultural Science and Technology, Ames, IA., 63 pp.

15 U.S.EPA. 1979. "Environmental Protection Agency, 40 CFR Part 257, Criteria for Classification of Solid Waste Disposal Facilities and Practices," *Federal Register* 44(179):53438–53464.

16 CAST. 1980. *Effects of Sewage Sludge on the Cadmium and Zinc Content of Crops*, Report No. 83, Council for Agricultural Science and Technology, Ames, IA. 77 pp.

17 Page, A. L., T. G. Logan and J. A. Ryan, eds. 1987. *Land Application of Sludges; Food Chain Implications*. Chelsea, MI: Lewis Publishers, Inc. 168 pp.

18 Corey, R. B., L. D. King, C. Lue-Hing, D. S. Fanning, J. J. Street and J. M. Walker. 1987. "Effects of Sludge Properties on Accumulation of Trace Elements by Crops," in *Land Application of Sludge, Food Chain Implications*, A. L. Page, T. G. Logan, J. A. Ryan, eds., Chelsea, MI: Lewis Publishers, Inc., pp. 25–51.

19 AMSA. 1987. *Comments of the Association of Metropolitan Sewerage Agencies on the Preliminary Draft Regulatory Limits Disseminated by Dr. Alan B. Rubin in July of 1987*, Association of Metropolitan Sewage Agencies, Washington, DC.

20 U.S.EPA. 1989. "40 CFR Parts 257 and 503 Standards for the Disposal of Sludge; Proposed Rule. Monday, February 6, 1989," *Federal Register* 54(23):5746–5902.

21 AMSA. 1989. *Comments of the Association of Metropolitan Sewerage Agencies on the Proposed Standards for the Disposal of Sewage Sludge. (Federal Register, February 6, 1989, pages 5746–5902)*, Association of Metropolitan Sewerage Agencies, Washington, DC.

22 Lue-Hing, C., T. C. Granato, D. R. Zenz, J. Gschwind and G. R. Richardson. 1991. *Impact of the Proposed U.S.EPA Part 503 Sludge Management Technical Regulations on POTWs*, Metropolitan Water Reclamation District of Greater Chicago, Department of Research and Development.

23 Cooperative State Research Service Technical Committee W-170. 1989. *Peer Review. Standards for the Disposal of Sewage Sludge U.S.EPA Proposed Rule 40 CFR Parts 257 and 503 (Federal Register February 6, 1989, pp. 5746–5902)*, United States Department of Agriculture.

24 Granato, T. C., G. R. Richardson, R. I. Pietz and C. Lue-Hing. 1991."Prediction of Phytotoxicity and Uptake of Metals by Models in Proposed U.S.EPA 40 CFR Part 503 Sludge Regulations. Comparison with Field Data for Corn and Wheat," *Water, Air, and Soil Pollution*, 57–58:891–902.

25 U.S.EPA. 1990. "National Sewage Sludge Survey: Availability of Information and Data, and Anticipated Impacts on Proposed Regulations," *Federal Register* 55(218):47210–47283.

26 AMSA. 1991. *Comments of the Association of Metropolitan Sewerage Agencies on the Notice of Availability of Information and Data from the National Sewage Sludge Survey and Request for Comments" (Federal Register, November 9, 1990, pp. 47210–47283)*, Association of Metropolitan Sewerage Agencies, Washington, DC.

27 Metropolitan Water Reclamation District of Greater Chicago. 1989. *Comments of the Metropolitan Water Reclamation District of Greater Chicago on the Proposed Standards for the Disposal of Sewage Sludge (Federal Register, February 6, 1989, pp. 5746–5902)*, Department of Research and Development.

28 U.S.EPA. 1989. *Technical support document for pathogen reduction in sewage sludge*, NTIS No. PB 89-13600.

29 U.S.EPA, September 1995. *A Guide to the Biosolids Risk Assessments for the U.S.EPA Part 503 Rule*, U.S.EPA Rep. EPA 823-B-93-005, Office of Wastewater Management, Washington, D.C. 144 pp.

30 U.S.EPA. 1993. "40 CFR Part 503—Standards for the Use or Disposal of Sewage Sludge. Friday, February 19, 1993," *Federal Register* 58 (32):9387–9413.

31 U.S.EPA. 1992. *Technical Support Document for Land Application of Sewage Sludge*, Office of Water, November, 1992.

32 U.S. Court of Appeals. 1994. *Leather Industries of America, Inc., Petitioner v. Environmental Protection Agency; Carol M. Browner, Administrator, United States Environmental Protection Agency, Respondents*. District of Columbia Circuit Docket No. 93-1187, November 15, 1994.

33 U.S.EPA. 1995. "40 CFR Parts 403 and 503 Standards for the Use or Disposal of Sewage Sludge," *Federal Register* 60 (206):54764–54792.

34 U.S.EPA. 1994. "40 CFR Part 503 Standards for the Use or Disposal of Sewage Sludge," *Federal Register* 59 (38):9095–9099.

CHAPTER 3

Chemical Constituents Present in Municipal Sewage Sludge

IN this chapter, a summary of information concerning the occurrence and distribution of pollutants in raw municipal sewages and municipal sewage sludge is presented. The distribution of heavy metals, cyanide, and various organic priority pollutants through activated sludge treatment processes is reviewed. Concentration factors for various constituents in sludges are estimated. Regulatory requirements for the pretreatment of industrial discharges to Publicly Owned Treatment Works (POTWs) are reviewed. A discussion of analytical techniques for specific pollutants in municipal sludge is presented.

OCCURRENCE AND DISTRIBUTION OF POLLUTANTS IN SLUDGE AT POTWs

CATEGORIES OF PRIORITY POLLUTANTS

The term priority pollutants results from the 1976 consent decree between the U.S.EPA and the Natural Resources Defense Council, which established a list of sixty-five toxic compounds or classes of compounds considered as having the greatest potential to harm human health or to be detrimental to the environment and for which controls were to be established [1]. The original list was subsequently divided, modified, and listed under Section 307 of the Clean Water Act and incorporated into the U.S.EPA's pretreatment regulations. The list of priority toxic pollutants

John Gschwind, Donald W. Harper, Nabih P. Kelada, David T. Lordi, George R. Richardson, Stanley Soszynski, and Richard C. Sustich, Metropolitan Water Reclamation District of Greater Chicago, Chicago, IL.

includes 127 compounds, inorganic and organic, and is presented in Table 3.1 [2].

The major categories and the number of compounds in each category are as follows:

- metals and cyanide—14
- semivolatile organics—58
- volatile organics—28
- pesticides and PCBs—25
- others—2
- total—127

The semivolatile organics are further subdivided into acid extractable and base/neutral extractable compounds. These categories for the organic compounds were established according to the analytical procedures used for identification and quantification.

OCCURRENCE OF PRIORITY POLLUTANTS IN RAW SEWAGE

The occurrence of priority pollutants in raw sewage is influenced by the type of area served by the POTW. A highly industrialized area would contribute greater quantities of metals, cyanide, and organic pollutants than would be found in sewage from residential and/or commercial areas.

Metals

The occurrence of heavy metals in raw sewage results from a variety of sources, which can include industrial operations such as metal finishing and printed circuit boards, corrosion of water pipes, and urban storm water in combined sewered areas [3–6]. The metals of major concern because of their widespread occurrence in POTW influents and effluents and their possible adverse effects on POTWs are cadmium, chromium, copper, lead, nickel, and zinc. Other metals to be considered are arsenic, silver, and mercury.

The influence of industrial contributions on raw sewage metals concentrations is shown in Table 3.2, which summarizes data collected from an U.S.EPA survey of 239 POTWs and from treatment plants operated by the MWRDGC. The data from the 239-POTW survey include the mean concentrations grouped according to the percentage of the industrial flow [7]. The POTWs having less than 4% industrial flow had lower concentrations than the POTWs with greater than 4% industrial flow. The two MWRDCG plants with large industrial inputs (Calumet WRP and North Side WRP) had generally higher raw sewage metals concentrations than the smaller plants with lower industrial inputs (Lemont WRP and Hanover

TABLE 3.1. Categories of Priority Pollutants.

Metals and Cyanide	
Antimony	Arsenic
Beryllium	Cadmium
Chromium	Copper
Lead	Mercury
Nickel	Selenium
Silver	Thallium
Zinc	Cyanide (total)

Volatile Organic Compounds	
Acrolein	Acrylonitrile
Benzene	Bromoform (tribromomethane)
Carbon tetrachloride (tetrachloromethane)	Chlorobenzene
Chlorodibromomethane	Chloroethane (ethyl chloride)
2-Chloroethyl vinyl ether (mixed)	Chloroform (trichloromethane)
1,2-Dichloroethane	Dichlorobromomethane
1,2-Dichloropropane	1,1-Dichloroethane
Ethylbenzene	1,1-Dichloroethylene
Methyl chloride (chloromethane)	1,3-Dichloropropylene
1,1,2,2-Tetrachloroethane	Methyl bromide (bromomethane)
Toluene (methyl benzene)	Methylene chloride (dichloromethane)
1,1,1-Trichloroethane	Tetrachloroethylene
Trichloroethylene	1,1,2-Trichloroethane
Vinyl chloride (chloroethylene)	1,2-Trans-dichloroethylene

Semivolatile Organic Compounds	
Acid extractable	
2-Chlorophenol	2,4-Dichlorophenol
2,4-Dimethylphenol	4,6-Dinitro-o-cresol
2,4-Dinitrophenol	4-Nitrophenol
2-Nitrophenol	p-Chloro-m-cresol
Pentachlorophenol	Phenol (4-APP method)
2,4,6-Trichlorophenol	
Base/neutral extractable	
Acenaphthene	Acenaphthylene
Anthracene	Benzidine
Benzo(a)anthracene	Benzo (a) pyrene
Benzo(b)fluoranthene	Benzo (ghi) perylene
Bis(2-chloroethyl) ether	4-Bromophenyl ether
Bis(2-ethylhexyl) phthalate	2-Chloronaphthalene
Butyl benzyl phthalate	Chrysene
4-Chlorophenyl phenyl ether	1,2-Dichlorobenzene
Dibenzo(a,h)anthracene	1,4-Dichlorobenzene
1,3-Dichlorobenzene	Diethyl phthalate
3,3-Dichlorobenzidine	Di-n-butyl phthalate
Dimethyl phthalate	2,4-Dinitrotoluene
2,6-Dinitrotoluene	1,2-Diphenylhydrazine

(continued)

TABLE 3.1. (continued).

Di-n-octyl phthalate	Hexachlorobenzene
Fluoranthene	Hexachlorocyclopentadiene
Fluorene	Indeno(1,2,3-cd)pyrene
Hexachlorobutadiene	Naphthalene
Hexachloroethane	N-nitrosodimethylamine
Isophorone	N-nitrosodiphenylamine
Nitrobenzene	Pyrene
N-nitrosodi-n-propylamine	1,2,4-Trichlorobenzene
Phenanthrene	
Pesticides and PCBs	
Aldrin	Alpha-BHC
Beta-BHC	Delta-BHC
Gamma-BHC (lindane)	Chlordane (mixed)
4,4'-DDT	4,4'-DDE
4,4'-DDD	Dieldrin
Alpha-endosulfan	Beta-endosulfan
Endosulfan sulfate	Endrin
Endrin aldehyde	Heptachlor
Heptachlor epoxide	PCB-1242
PCB-1254	PCB-1221
PCB-1232	PCB-1248
PCB-1260	PCB-1016
	Toxaphene
Miscellaneous	
2,3,7,8-Tetrachlorodibenzo-p-dioxin (TCCD)	Asbestos (fibrous)

Park WRP). The Hanover Park service area, which has mainly copper water piping, showed a higher influent copper concentration than the Calumet WRP and North Side WRP, which include areas of the city of Chicago where copper water pipe is not permitted.

The greatest mean concentrations from the 239-POTW survey were zinc and copper [0.354 and 0.151 mg/L, respectively] followed by chromium, nickel, and lead (0.145, 0.140, 0.130 mg/L, respectively). Although the mean concentrations for the MWRDGC plants are lower than for the 239-POTW survey a similar pattern in relative metal concentrations is observed.

The implementation of an industrial waste pretreatment program as well as changes in industrial operations influences the resulting metals concentrations in raw sewage. Historical influent metals data from the North Side WRP of the MWRDGC has been summarized in terms of annual averages in Table 3.3. The North Side WRP serves an area with a large number of metal finishing and electroplating operations. The area

TABLE 3.2. Occurrence of Heavy Metals (mg/L) in the Raw Sewage of POTWs.

| Metal | 239-Plant Survey[a] | | MWRDGC[b] | | | |
	< 4% Industrial Flow	> 4% Industrial Flow	Lemont	Hanover Park	Calumet	North Side
Cadmium	0 0095	0.0166	0.00	0.001	0.00	0.001
Chromium	0.025	0 145	<0.03	0.008	<0.03	0.015
Copper	0 095	0.151	0.05	0.073	0.05	0 050
Lead	0.042	0.103	0.00	0.00	0.00	0.00
Mercury	0 0006	0.0012	0.0001	0.0002	0.0001	0.0001
Nickel	0.037	0.140	<0.02	<0.03	<0.02	<0.03
Zinc	0.238	0 354	0.096	0.080	0.243	0 101

[a]Minear, R A et al 1981. "Data Base for Influent Heavy Metals in Publicly Owned Treatment Works" EPA-600/2-81-220, Municipal Environmental Research Laboratory Cincinnati, OH
[b]Average of influent daily values collected in 1995.

is also served by a combined sewer system. The MWRDGC has had an ongoing industrial waste control program since the late 1960s. Generally the metals levels in the influent have declined since 1971. From 1971 to 1995, cadmium declined from a range of 0.03–0.04 mg/L to 0.001 mg/L. Copper levels have declined from 0.22 mg/L to 0.050 mg/L from 1971 to 1995. Chromium decreased from 0.37 mg/L in 1971 to 0.015 mg/L in 1995. Zinc levels declined from 0.40–0.60 mg/L to a level of 0.101 mg/L over the period of 1971 to 1995. Lead, which ranged between 0.12 and

TABLE 3.3. Yearly Average Raw Sewage Metals Concentrations at North Side WRP.

| Year | Average Concentrations (mg/L)[a] | | | | | |
	Cadmium	Chromium	Copper	Lead	Nickel	Zinc
1971	0.04	0.37	0.22	0.20	0.2	0.6
1973	0.02	0 22	0.20	0.12	0.1	0 4
1975	0.01	0.13	0.13	0 07	0 0	0.2
1977	0.02	0.16	0.12	0.06	0.1	0.2
1983	0.01	0 09	0 11	0 04	0 0	0 2
1985	0 01	0.07	0.10	0.03	0.0	0.3
1987	0.01	0 08	0 13	0 04	0.0	0.2
1989	0 00	0.06	0.07	0.01	0.0	0.1
1991	0.00	0.02	0.04	0 00	0 0	0.1
1993	0 003	0.017	0.047	0.00	0.00	0.133
1995	0.001	0.015	0 050	0.00	0.00	0.101

[a]Average calculated upon values less than analytical detection limit taken as zero.

0.20 mg/L in the early 1970s, decreased to <0.05 mg/L in the 1990s. Part of the decrease in lead concentrations could be attributed to the discontinuance of gasoline lead additives.

Cyanide

Yearly average cyanide levels observed in the influent of four MWRDGC plants over the period of 1983 through 1995 are presented in Table 3.4. The highest influent levels were observed at the Calumet WRP, which receives steel manufacturing and coking wastes. All of the plants have exhibited decreasing influent cyanide levels. North Side WRP influent cyanide decreased steadily by approximately 79% from 0.090 mg/L in 1983 to 0.019 mg/L in 1995. Calumet WRP had an average cyanide of 0.581 mg/L in 1985 and then decreased to 0.211 mg/L in 1995 a 64% decrease. Both John Egan and Hanover Park WRPs had decreasing cyanide influent concentrations over the 1983–1995 period with 1995 average concentrations of 0.023 mg/L and 0.021 mg/L, respectively.

The cyanide content of wastewaters may be divided into two broad species, namely simple and complex cyanide. Simple cyanides are the free cyanide plus the medium strength metal cyano-complexes such as nickel. The complex cyanides represent the difference between total and simple cyanides and include the iron and cobalt complexes. The cyanide in raw wastewaters has been found to be predominantly in the complex form, with the complex cyanide ranging between 64 and 83% of the total cyanide [9,10].

Organics

The occurrence of the organic priority pollutants in raw sewage has not been as completely characterized as the heavy metals. Table 3.5 pre-

TABLE 3.4. Annual Average Raw Sewage Cyanide Concentrations at MWRDGC Plants.

| Year | Concentration in mg/L | | | |
	North Side	Calumet	Hanover Park	John Egan
1983	0.090	0.278	0.037	0.086
1985	0 078	0.581	0.024	0.040
1987	0.071	0.427	0.022	0.056
1989	0 047	0.334	0.019	0.025
1991	0.019	0.306	0.012	0.020
1993	0.019	0.264	0.015	0.023
1995	0.019	0.211	0.021	0.023

TABLE 3.5. Occurrence of Priority Organic Compounds in Influent Wastewaters to POTWs.

	Forty-POTW Study[a]		MWRDGC Plants[b]	
Compound	Times Detected (%)	Maximum Conc. (μg/L)	Times Detected (%)	Maximum Conc. (μg/L)
Acenaphthene	3	21	0	—
Acrylonitrile	<1	82	0	—
Benzene	61	1,560	19	15.5
Carbon tetrachloride (tetrachloromethane)	9	1,900	0	—
Chlorobenzene	13	1,500	0	—
1,2,4-Trichlorobenzene	10	4,300	0	—
Hexachlorobenzene	1	20	0	—
1,2-Dichloroethane	15	76,000	0	—
1,1,1-Trichloroethane	85	30,000	31	1.9
Hexachloroethane	<1	12	0	—
1,1-Dichloroethane	31	24	0	—
1,1,2-Trichloroethane	7	135	0	—
1,1,2,2-Tetrachloroethane	7	52	0	—
Chloroethane	1	38	0	—
2-Chloroethyl vinyl ether	<1	10	0	—
2-Chloronaphthalene	<1	7	0	—
2,4,6-Trichlorophenol	5	11	0	—
Parachlorometacresol	3	41	0	—
Chloroform (trichloromethane)	91	430	88	5.5
2-Chlorophenol	3	5	0	—
1,2-Dichlorobenzene	23	440	0	—
1,3-Dichlorobenzene	7	270	0	—
1,4-Dichlorobenzene	17	200	0	—
1,1-Dichloroethylene	26	243	0	—

(continued)

TABLE 3.5. (continued).

Compound	Forty-POTW Study[a]		MWRDGC Plants[b]	
	Times Detected (%)	Maximum Conc. (µg/L)	Times Detected (%)	Maximum Conc. (µg/L)
1,2-Trans-dichloroethylene	62	200	69	2.6
2,4-Dichlorophenol	7	25	0	—
1,2-Dichloropropane	7	2,600	0	—
1,3-Dichloropropylene	2	100	0	—
2,4-Dimethylphenol	10	55	6	70.7
Ethylbenzene	80	730	31	4.6
2,4-Dinitrotoluene	1	8	0	—
2,6-Dinitrotoluene	<1	5	0	—
1,2-Diphenylhydrazine	1	5	0	—
Fluoranthene	7	5	31	4.3
4-Bromophenyl phenyl ether	<1	5	0	—
Bis(2-chloroethoxy)methane	1	5	0	—
Methylene chloride (dichloromethane)	92	49,000	25	11.7
Methyl chloride (chloromethane)	11	1,900	0	—
Methyl bromide	3	164	0	—
Bromoform (tribromomethane)	2	81	0	—
Dichlorobromomethane	8	22	6	1.6
Trichlorofluoromethane	9	190	0	—
Dichlorodifluoromethane	2	1,000	0	—
Chlorodibromomethane	8	3	0	—
Hexachlorobutadiene	<1	5	0	—
Isophorone	2	23	0	—
Naphthalene	49	150	12	11.0
2-Nitrophenol	<1	64	0	—

TABLE 3.5. (continued.)

	Forty-POTW Study[a]		MWRDGC Plants[b]	
Compound	Times Detected (%)	Maximum Conc. (µg/L)	Times Detected (%)	Maximum Conc. (µg/L)
2,4-Dinitrophenol	<1	7	0	—
N-nitrosodimethylamine	2	14	0	—
Pentachlorophenol	29	640	0	—
Phenol	79	1,400	69	302.2
Bis(2-ethylhexyl)phthalate	92	670	38	44.0
Butylbenzylphthalate	57	560	50	15.1
Di-n-butylphthalate	64	140	44	9.2
Di-n-octylphthalate	7	210	6	3.9
Diethylphthalate	53	42	75	8.9
Dimethylphthalate	11	110	0	—
Benzo(a)anthracene (1,2-benzanthracene)	3	15	19	2.9
Benzo(a)pyrene	1	10	0	—
3,4-Benzofluoranthene	<1	5	0	—
Benzo(k)fluoranthene	<1	5	0	—
Benzo(ghi)perylene	1	35	0	—
Chrysene	3	5	19	2.8
Acenaphthylene	<1	5	0	—
Anthracene	18	93	0	—
Dibenzo(a,h)anthracene	1	5	0	—
Indeno(1,2,3-cd)pyrene	1	5	0	—
Fluorene	4	5	0	—
Phenanthrene	20	93	31	2.9
Pyrene	7	84	6	6.5

(continued)

99

TABLE 3.5. (continued.)

Compound	Forty-POTW Study[a]		MWRDGC Plants[b]	
	Times Detected (%)	Maximum Conc. (µg/L)	Times Detected (%)	Maximum Conc. (µg/L)
Tetrachloroethylene	95	5,700	81	19.9
Toluene	94	13,000	100	35.4
Trichloroethylene	90	18,000	69	37.7
Vinyl chloride (chloroethylene)	6	3,900	0	—
Aldrin	1	5	0	—
Dieldrin	<1	0.04	0	—
4,4'-DDE	<1	1.20	0	—
4,4'-DDD	<1	2.70	0	—
Alpha-endosulfan	1	2.70	0	—
Heptachlor epoxide	<1	0.5	0	—
Heptachlor	5	0.5	0	—
Alpha-BHC	8	4.4	0	—
Beta-BHC	<1	1,000	0	—
Delta-BHC	3	1,400	0	—
Gamma-BHC (Lindane)	26	3.9	6	0.21
PCB-1242 (Aroclor 1242)	5	46.6	0	—
PCB-1254 (Aroclor 1254)	1	5.5	0	—
PCB-1248 (Aroclor 1248)	0	—	0	—
PCB-1260	0	—	19	0.45

[a] 1982. "Fate of Priority Toxic Pollutants in Publicly Owned Treatment Works," final report Volume 1, EPA 440/1-82/303, Effluent Guidelines Division, Washington, DC.
[b] Results from seven MWRDGC plants monitored in 1995.

sents data taken from a survey of forty POTWs nationwide [8] and data from seven treatment plants operated by the MWRDGC. The data in Table 3.5 is summarized as to the percent of the samples in which a specific organic pollutant was detected along with the maximum sample concentration found. The organic pollutants with the most frequent occurrence (80 to 100% of the time) included the volatile organic compounds—1,1,1-trichloroethane, chloroform, ethyl benzene, methylene chloride, tetrachloroethylene, toluene, and trichloroethylene—and the semivolatile organic compounds—phenol and bis(2-ethylhexyl)phthalate. Only fourteen organic pollutants were measured in 50% or more of the samples. Out of the 111 priority organic pollutants analyzed, 22 were not detected in any sample.

The concentrations of the organic pollutants vary greatly. The majority of the organics in most samples was found to be less than 10 μg/L. The highest concentration observed in a single sample from the forty-POTW survey was 1,2-dichloroethane at 76,000 μg/L. This compound was found in only 15% of the samples [8]. The highest single sample concentration observed at the seven MWRDGC facilities was phenol at 302 μg/L. Other compounds detected at relatively high concentrations (greater than 3,000 μg/L maximum observed concentration) included methylene chloride, tetrachloroethylene, toluene, 1,1,1-trichloroethane, 1,2,4-trichlorobenzene, and vinyl chloride from the forty-POTW survey and for the MWRDGC treatment plants (30–70 μg/L maximum observed concentration) were toluene, trichloroethylene, bis(2-ethylhexyl)phthalate, and 2,4-dimethylphenol.

DISTRIBUTION OF PRIORITY POLLUTANTS THROUGH THE WASTEWATER TREATMENT PROCESSES

Pollutants entering a wastewater treatment facility are either removed in the physical and biological treatment processes or pass through the treatment facility into the effluent. The distribution of priority pollutants in the wastewater treatment facility is the result of physical, chemical, and biological phenomena. The characteristics of the pollutant, the concentrations in the wastewater, and the type of treatment process all influence the resulting distribution.

Metals

A partial removal of the metals occurs in the primary settling tank but the principal removal mechanisms of the metals from wastewater is by adsorption on biological flocs and subsequent settling. The removal of different metals varies widely. Data reported in the literature on metals removal in waste treatment processes indicate that about 5 to 50% removal occurs in primary sedimentation [8,11–14]. Overall removals of 15 to 80%

may occur in secondary treatment, with removals of 100% also being reported [7,8,11–17].

Data as to the percent removals obtained for six selected metals from three pilot plant scale studies of the fate and distribution of metals in secondary treatment systems is summarized in Table 3.6 [18,19,20]. The removal by primary sedimentation ranged from zero to 46%. The activated sludge system alone gave removals between 15 and 81%. The overall removals for the total secondary treatment process varied from 24 to 88%. Although there is a substantial variation between investigators, certain generalizations can be made. The degree of removal of metals for the secondary treatment process varied according to the specific metal with the following general order:

$$(Cu,Pb) > (Cr,Cd) > (Zn) > (Ni)$$

According to Neufeld and Herman, the affinity for metals by activated sludge decreases as Hg > Cd > Zn [21]. Cheng et al. found the preferred order of uptake to be Pb > Cu > Cd > Ni [22].

Table 3.7 shows the annual average metals removals and influent and effluent concentrations observed at five MWRDGC treatment facilities for 1995. Cadmium, chromium, lead, and nickel were below detection in both the influent and effluent of the Lemont and Calumet Water Reclamation Plants. Removals were expressed as 0 percent in these cases. In those cases where there was detectable influent concentration but effluent was below detection, removal was calculated as 100 percent. Cadmium, lead, and nickel were not detected in effluents from any of the five water reclamation plants. Copper removals ranged from 60 to 95 percent while zinc removal ranged from 56 to 84 percent, excluding the Lemont Water Reclamation Plant in which similar influent and effluent zinc concentrations resulted in a calculated negative removal.

Metals removed from the wastewater are conserved in the resulting process sludges. Concentration factors for sludge metals taken from the three pilot plant studies previously cited are summarized in Table 3.8 [18–20]. The sludge concentration factors were calculated as the metal concentration in the sludge (on a wet basis, mg/L) divided by the metal concentration (mg/L) in the raw wastewater. The concentration factors for the primary sludge are generally higher than those for the secondary sludges. This probably is a result of higher solids concentration in the primary sludges and lower influent concentrations to the aeration tanks. Although there is great variation between the studies, the concentration factors vary with the type of metal. Cadmium and nickel showed the lowest concentration factors and chromium and copper generally had the highest concentration factors.

TABLE 3.6. Removal of Heavy Metals in Wastewater Treatment Pilot Plant Scale Studies.

Metal	Primary Sedimentation			Activated Sludge Only			Total Secondary Treatment		
	Ref. [18]	Ref. [19]	Ref. [20]	Ref. [18]	Ref. [19]	Ref. [20]	Ref. [18]	Ref. [19]	Ref. [20]
Cadmium	19	20	12	54	35	14	62	48	24
Chromium	19	29	7	33	5	80	46	33	82
Copper	29	20	19	72	37	78	80	49	82
Lead	34	42	30	81	36	49	88	55	65
Nickel	40	46	4	33	−18	40	60	36	43
Zinc	0	18	—	64	15	—	63	31	—

TABLE 3.7. Annual Average Metals Removals at MWRDGC Plants.[a]

Parameter	North Side	Calumet	Lemont	Hanover Park	John Egan
Cadmium					
Influent mg/L	0.001	0	0	0 001	0.003
Effluent mg/L	0.000	0	0	0.000	0.000
Removal %	100%	—	—	100%	100%
Chromium					
Influent mg/L	0.015	0.00	0	0.008	0.037
Effluent mg/L	0.003	0	0	0.002	0.002
Removal %	80%	—	—	75%	95%
Copper					
Influent mg/L	0.050	0.05	0 05	0.073	0.181
Effluent mg/L	0.008	0.02	0.02	0 013	0.010
Removal %	84%	60%	60%	82%	95%
Lead					
Influent mg/L	0.00	0.00	0.00	0 00	0.01
Effluent mg/L	0.00	0.00	0.00	0.00	0.00
Removal %	—	—	—	—	100%
Nickel					
Influent mg/L	0.00	0	0	0 00	0 01
Effluent mg/L	0.00	0	0	0.00	0.00
Removal %	—	—	—	—	100%
Zinc					
Influent mg/L	0.101	0.243	0.096	0.080	0.230
Effluent mg/L	0.041	0 096	0.116	0 035	0.038
Removal %	59%	60%	−21%	56%	84%

[a]Annual average of 24-hour composite samples collected in 1995.

Table 3.8 also includes concentration factors for combined primary-secondary sludge from the MWRDGC's North Side WRP. The factors ranged from 9.0 (copper) to 16.2 (lead). Two metals (cadmium and nickel) were not detected in the influent and the concentration factors would theoretically be infinity.

On a mass-based concentration (mg of metal per kg of dry sludge solids) the secondary sludges generally have higher concentrations than the primary sludges. Further information with regard to the metals concentrations occurring in various types of sludges prior to disposal is presented in a later section of this chapter.

Cyanides

Cyanide influent and effluent concentrations observed at five MWRDGC treatment plants in 1995 and the minimum and maximum

TABLE 3.8. Heavy Metal Concentration Factors in Primary and Secondary Sludges.[a]

Metal	Primary Sludge			Activated Sludge			Primary Plus Activated Sludge
	Ref. [18]	Ref. [19]	Ref. [20]	Ref. [18]	Ref. [19]	Ref. [20]	Ref.[c]
Cadmium	6.7	7.4	24.0	4.2	7.6	16.0	b
Chromium	32.4	4.2	70.6	25.1	7.2	57.9	9.2
Copper	38.8	14.5	68.1	22.5	14.9	43.2	9.0
Lead	36.6	24.4	61.2	17.2	23.8	32.1	16.2
Nickel	32.4	10.8	36.8	10.0	7.9	24.8	b
Zinc	37.4	42.4	—	15.3	28.3	—	10.7

[a]Concentration factor taken as the metal concentration in the sludge (on a wet basis) divided by the metal concentration in the raw wastewater.
[b]The metal concentration in the raw wastewater was below detection or zero and thus a calculated concentration factor would be infinity.
[c]Source: 1988. Data from MWRDGC North Side WRP

average values from the forty-POTW study are presented in Table 3.9 along with the corresponding removal. The percent removals generally increase with increasing influent concentrations. The Calumet WRP, which had the highest average influent concentration (0.211 mg/L), also showed the highest percent removal (90.0). The average cyanide removals observed in the forty-POTW study ranged between 7 and 98%.

Organics

The removal mechanisms of the priority organic pollutants passing through a secondary wastewater treatment plant include:

- volatilization
- sedimentation
- adsorption onto the biological floc
- biodegradation

Volatile organic compounds will be stripped from the wastewater during the aeration process [20,23,24]. Sedimentation and adsorption are the primary removal mechanisms for the higher molecular weight organic compounds. Many of the organics such as acrylonitrile, benzene, dichloro-benzene, 2,4-dichlorophenol, ethyl benzene, naphthalene, phenol, and toluene are generally biodegradable to varying degrees. Organics such as the PCBs and many of the pesticides (aldrin, chlordane, DDT, endrin, and heptachlor) show very little or no degradation [24–26].

Several studies of the fate and distribution of organic compounds in wastewater treatment have been reported in the literature [8,20,24,25].

TABLE 3.9. Average Cyanide Removals through POTWs.

Treatment Facility	Influent Concentration (mg/L)	Effluent Concentration (mg/L)	Removal (%)
MWRDGC[a]			
Calumet	0 211	0.021	90.0
North Side	0.019	0.008	57 9
John E. Egan	0.023	0.011	63.2
Lemont	0.012	0.006	50.0
Hanover Park	0 021	0.013	39.1
40-POTW Study[b]			
Minimum	0.003	0.002	7
Maximum	7.58	2.14	98

[a]Annual averages of 24-hour composite samples collected in 1995
[b]Range of average values from 40-POTW study [8] Excludes data of four POTWs which showed negative removals.

Table 3.10 presents occurrence data and maximum observed concentrations of the detected pollutants for secondary effluents from the forty-POTW survey [8] and from seven treatment plants operated by the MWRDGC. The data in Table 3.10 indicate that seventy-four priority organic pollutants were detected in at least one sample of secondary effluent. Only seven priority organic pollutants were detected in more than 50% of the effluent samples from the forty-POTW study and the most frequently detected were methylene chloride (86%), bis(2-ethylhexyl) phthalate (84%), and chloroform (82%). In the case of the MWRDGC plants, there were eight organic pollutants detected in at least one of the plant effluents with only one organic pollutant being detected in more than 50% of the effluents. Dichloromethane, bis(2-ethylhexyl) phthalate, and di-*n*-butyl phthalate were detected in all of the samples analyzed.

Data on removals by primary sedimentation are limited. Based upon pilot plant studies by Hannah et al. [20], twenty-one listed priority pollutant organics showed an average removal for volatiles of 7% and for semivolatiles ranging from −4% (lindane) to 45% (2,4-dichlorophenol). Petrasek et al. [24], in a pilot plant study of twenty-two semivolatile organics, calculated an average removal of 40%. The median percent removals of selected organic pollutants in the forty-POTW study range from zero to 62%.

Removal data collected from pilot studies of activated sludge systems operated on sewage spiked with priority organic pollutants conducted by Hannah et al. [20] and by Petrasek et al. [24] are summarized in Table 3.11. The removals of priority organics in the pilot plant studies were between 18 and 99% with the majority of the compounds having removal greater than 90%. The lowest removals were for lindane (18% reported by Petrasek), and pentachlorophenol (19% reported by Hannah). There is a wide variation in the results for pentachlorophenol with other investigators having reported removals of zero to 96% of pentachlorophenol [28–30].

Table 3.11 also presents the median percent removals of selected pollutants for those POTWs having secondary activated sludge treatment from the forty-POTW study. The removal calculations only included those plants with average influent concentrations greater than three times the detection limit of each pollutant. Secondary treatment removed from 62% [chloroform and bis(2-ethylhexyl) phthalate] to 92 and 93% (naphthalene and toluene) of the priority organic pollutants.

Sludge concentration factors (concentration found in the sludge on a wet basis divided by the concentration in the influent for a specific compound) calculated from the pilot plant data presented in the studies of Petrasek [24] and Hannah [20] are given in Table 3.12 for primary and secondary sludges. Typically a 30- to 200-fold increase occurred for the semivolatile organics in the primary sludge. Concentration

TABLE 3.10. Occurrence of Priority Organic Pollutants in Effluents from POTWs.

Compound	Forty-POTW Study[a]		MWRDGC Plants[b]	
	Times Detected (%)	Maximum Value (μg/L)	Times Detected (%)	Maximum Value (μg/L)
Acenaphthylene	1	5	0	—
Benzene	23	72	0	—
Carbon tetrachloride (tetrachloromethane)	6	67	0	—
Chlorobenzene	3	9	0	—
1,2,4-Trichlorobenzene	4	310	0	—
Hexachlorobenzene	1	10	0	—
1,2-Dichloroethane	8	13,000	0	—
1,1,1-Trichloroethane	52	3,500	7	1.3
1,1-Dichloroethane	8	6	0	—
1,1,2-Trichloroethane	3	6	0	—
1,1,2,2-Tetrachloroethane	0	—	0	—
Chloroethane	<1	5	0	—
2,4,6-Trichlorophenol	3	3	0	—
Parachlorometacresol	<1	4	0	—
Chloroform (trichloromethane)	82	87	64	20.1
2-Chlorophenol	1	5	0	—
1,2-Dichlorobenzene	8	27	0	—
1,3-Dichlorobenzene	2	5	0	—
1,4-Dichlorobenzene	3	9	0	—
3,3'-Dichlorobenzidine	<1	5	0	—
1,1-Dichloroethylene	10	11	0	—
1,2-Trans-dichloroethylene	13	17	0	—
2,4-Dichlorophenol	4	3	0	—
1,2-Dichloropropane	4	8	0	—

TABLE 3.10. (continued.)

Compound	Forty-POTW Study[a]		MWRDGC Plants[b]	
	Times Detected (%)	Maximum Value (µg/L)	Times Detected (%)	Maximum Value (µg/L)
2,4-Dimethylphenol	4	10	0	—
2,4-Dinitrotoluene	<1	2	0	—
1,2-Diphenylhydrazine	<1	2	0	—
Ethylbenzene	24	49	0	—
Fluoranthene	1	5	0	—
Methylene chloride (dichloromethane)	86	62,000	14	1.1
Methyl chloride (chloromethane)	7	540	0	—
Methyl bromide	1	220	0	—
Bromoform (tribromomethane)	3	5	0	—
Dichlorobromomethane	16	6	14	4.6
Trichlorofluoromethane	4	14	0	—
Dichlorodifluoromethane	<1	58	0	—
Chlorodibromomethane	8	5	0	—
Isophorone	1	12	0	—
Naphthalene	6	24	0	—
2-Nitrophenol	3	14	0	—
4-Nitrophenol	2	220	0	—
4,6-Dinitro-o-cresol	<1	2	0	—
Pentachlorophenol	27	440	0	—
Bis(2-ethylhexyl)phthalate	84	370	0	—
Butyl benzyl phthalate	11	34	0	—
Di-n-butyl phthalate	52	97	0	—
Di-n-octyl phthalate	4	13	7	42.6
Diethyl phthalate	13	7	0	—

(continued)

TABLE 3.10. (continued).

Compound	Forty-POTW Study[a]		MWRDGC Plants[b]	
	Times Detected (%)	Maximum Value (µg/L)	Times Detected (%)	Maximum Value (µg/L)
Dimethyl phthalate	2	5	0	—
Benzo(a)anthracene (1,2-benzanthracene)	2	11	0	—
Chrysene	2	11	0	—
Acenaphthene	2	7	0	—
Anthracene	3	32	0	—
Benzo(ghi)perylene (1,12-benzoperylene)	<1	4	0	—
Fluorene	<1	5	0	—
Phenanthrene	3	32	0	—
Dibenzo(a,h)anthracene	<1	5	0	—
Indeno(1,2,3-cd)pyrene	1	5	0	—
Pyrene	<1	5	0	—
Tetrachloroethylene	79	1,200	21	47.1
Toluene	53	110	7	2.1
Trichloroethylene	45	230	14	3.1
Vinyl chloride (chloroethylene)	2	200	0	—
Aldrin	3	6.0	0	—
Dieldrin	<1	0.04	0	—
Chlordane (tech. mixture and metabolites)	<1	0.2	0	—
4,4'-DDD(p,p'-TDE)	1	0.3	0	—
Heptachlor	2	1.5	0	—
Heptachlor epoxide	2	0.5	0	—
Alpha-BHC	8	0.74	0	—
Beta-BHC	40	1.7	0	—
Gamma-BHC (lindane)	33	1.40	0	—

TABLE 3.10. (continued).

Compound	Forty-POTW Study[a]		MWRDGC Plants[b]	
	Times Detected (%)	Maximum Value (µg/L)	Times Detected (%)	Maximum Value (µg/L)
Delta-BHC	3	1.3	0	—
PCB-1242 (Aroclor 1242)	1	2.6	0	—
PCB-1254 (Aroclor 1254)	<1	0.5	0	—
PCB-1260 (Aroclor 1260)	0	—	0	—

[a]Reference [8], 1982 "Fate of Priority Pollutants in Publicly Owned Treatment Works," Final Report, Volume 1, EPA 440/1-82/303, Effluent Guidelines Division.
[b]Results from seven MWRDGC plants monitored in 1995

increases of 5- to 50-fold occurred in the secondary sludges depending upon the specific semivolatile compound. The volatile organics showed no concentration increases for most compounds in both the primary and secondary sludges.

Table 3.12 also includes concentration factors calculated for the secondary activated sludge from data presented in a study that measured the priority pollutants in a POTW (Moccasin Bend Wastewater Treatment Plant in Chattanooga, Tennessee) for thirty consecutive days [27]. The sludge concentration factors ranged from zero to 17 (fluoranthene).

In many cases the influent concentration may be too low to be measured, but the pollutant may be concentrated sufficiently in the sludge to be quantified. Data from the forty-POTW study showed eighteen priority organic pollutants occurring in sludges when not detected in the corresponding influents.

TABLE 3.11. Organic Priority Pollutant Removals by Secondary Treatment Systems.

Compound	Pilot Plant Activated Sludge % Removal		Median of Forty-POTW Study % Removal
	Ref. [24]	Ref. [20]	Ref. [8]
Acenaphthene	97	—	—
Benzene	—	—	77
Carbon tetrachloride (tetrachloromethane)	—	74	—
1,2-Dichloroethane	—	84	—
1,1,1-Trichloroethane	—	—	88
1,1-Dichloroethane	—	94	—
Bis(2-chloroethyl) ether	—	80	—
Chloroform (trichloromethane)	—	86	62
1,2-Dichlorobenzene	—	94	—
1,1-Dichloroethylene	—	92	—
1,2-Trans-dichloroethylene	—	—	80
2,4-Dichlorophenol	—	99	—
2,4-Dimethylphenol	99	—	—
Ethylbenzene	—	93	90
Fluoranthene	94	95	—
Methylene chloride (dichloromethane)	—	—	48
Bromoform (tribromomethane)	—	65	—
Isophorone	—	98	—
Naphthalene	99	97	92
Pentachlorophenol	19	96	—

TABLE 3.11. (continued).

Compound	Pilot Plant Activated Sludge % Removal		Median of Forty-POTW Study % Removal
	Ref. [24]	Ref. [20]	Ref. [8]
Phenol (4-APP method)	95	86	—
Bis(2-ethylhexyl) phthalate	79	87	62
Butyl benzyl phthalate	96	—	94
Di-*n*-butyl phthalate	94	88	68
Di-*n*-octyl phthalate	83	—	—
Diethyl phthalate	97	—	91
Dimethyl phthalate	98	—	—
Benzo(a)anthracene (1,2-benzanthracene)	98	—	—
Chrysene	97	—	—
Acenaphthylene	97	—	—
Fluorene	98	—	—
Phenanthrene	97	95	—
Pyrene	94	95	—
Tetrachloroethylene	—	—	82
Toluene	—	—	93
Trichloroethylene	—	—	90
Heptachlor	93	65	—
Gamma-BHC (lindane)	45	18	—
PCB-1254 (Aroclor 1254)	98	—	—

OCCURRENCE OF PRIORITY POLLUTANTS IN SLUDGE AT POTWs

Ultimately, most of the heavy metal priority pollutants entering a municipal wastewater treatment plant will accumulate in the sludge produced. A very small percentage will be found in the plant effluent. Most of the organic priority pollutants will either be biodegraded or volatilized. Adsorption of organics on to sludge solids is not considered a major mechanism, and hence, only a small amount of the organic priority pollutants accumulate in sludge.

There is no national reporting system in the United States (U.S.) for sludge quality that could be used to characterize sludge on a national basis. There were, however, several national surveys during the past decade that attempted to determine the national quality of sludge.

In 1979–1980, the U.S.EPA conducted a study of forty POTWs [8], Mumma et al. in 1984 reported on sludge quality from twenty-three U.S. cities [31], and in 1987 the Association of Municipal Sewerage Agencies (AMSA) conducted a survey of fifty-nine member sewerage agencies [32].

TABLE 3.12. Organic Priority Pollutants Sludge Concentration Factors.*

Compound	Concentration Factors				
	Primary		Secondary		
	Ref. [24]	Ref. [20]	Ref. [24]	Ref. [20]	Ref. [26]
Acenaphthene	84	—	1.0	—	—
Benzene	—	0.3	—	0.2	0.11
Carbon tetrachloride (tetrachloromethane)	—	—	—	—	6.8
1,2,4-Trichlorobenzene	—	0.8	—	0.18	—
1,2-Dichloroethane	—	—	—	—	0.20
1,1,1-Trichloroethane	—	0.9	—	0.06	—
1,1-Dichloroethane	—	0.9	—	0.9	—
Bis(2-chloroethyl) ether	—	0.9	—	0.08	—
Chloroform (trichloromethane)	—	6.2	0	0.5	0.25
1,3-Dichlorobenzene	—	0.7	—	0.07	0.0
1,1-Dichloroethylene	—	0.8	0	0.0	0.0
2,4-Dichlorophenol	—	—	—	0.0	0.0
2,4-Dimethylphenol	0.3	—	0.3	—	—
Ethylbenzene	173	2.8	—	0.03	0.39
Fluoranthene	—	38.5	5.0	4.2	17
Methylene chloride (dichloromethane)	—	0.3	—	0.2	0.37
Methyl chloride (chloromethane)	—	0.8	0	0.2	0.0
Bromoform (tribromomethane)	45	5.7	0.1	0.3	—
Isophorone	—	—	—	—	—
Naphthalene	54	1.4	2.0	1.4	3.8
4-Nitrophenol	—	—	—	—	13
Pentachlorophenol	—	—	—	—	2.5

TABLE 3.12. (continued).

| Compound | Concentration Factors | | | | | |
| | Primary | | Secondary | | | Secondary |
	Ref. [24]	Ref. [20]	Ref. [24]	Ref. [20]	Ref. [26]
Phenol (4-AAP method)	9.0	3.7	3.0	5.0	0
Bis(2-ethylhexyl) phthalate	130	33.0	19.0	48.3	15
Butyl benzyl phthalate	244	15.8	3.0	4.2	—
Di-n-butyl phthalate	79	—	4.0	—	1.8
Di-n-octyl phthalate	187	—	17.0	—	—
Diethyl phthalate	15	—	3.0	—	0
Dimethyl phthalate	1 0	—	1.0	—	—
Benzo(a)anthracene (1,2-benzanthracene)	136	—	8.0	—	2
Chrysene	154	—	7.0	—	—
Anthracene	138	—	3.0	—	—
Fluorene	103	—	1.0	—	—
Phenanthrene	122	30	1.0	2.4	—
Pyrene	218	38.5	3.0	4.3	6
Tetrachloroethylene	—	—	—	—	0.19
Heptachlor	68	30.2	19	25.3	—
Gamma-BHC (lindane)	25	25.9	4.0	5.1	—
Toxaphene	173	—	19	—	—

* Concentration factor taken as the concentration in the sludge (on a wet basis) divided by the concentration in the raw wastewater.

The U.S.EPA in 1988 conducted a National Sewage Sludge Survey [62] to determine the quality of sludge in the U.S. prior to the development of the Part 503 sludge regulations. In 1996, the AMSA membership were contacted to participate in a survey to determine the quality of sludge in the U.S. after implementation of the Part 503 sludge regulations [63].

Metals

Table 3.13 shows a comparison of selected priority pollutant metals based on the three surveys conducted in the 1980s. As a result of the treatment processes, influent metals become part of the sludge mass by precipitation or coprecipitation in the form of sulfides, oxides, bicarbonates, by adsorption onto hydrous oxides, by chelation, by solid phase organic compounds, or by partitioning in soluble form between effluent and sludge during solids separation processes.

Comparison of median metal values in Table 3.13 shows similar orders of magnitude, despite the varying purposes of the three surveys. The forty-POTW survey was actually designed to study the fate of priority pollutants in POTWs using secondary or tertiary treatment. It concentrated on pollutant concentrations in influent and effluent. Most sludges sampled were raw (unprocessed) primary and secondary sludges in order to measure organic constituents so as to produce a mass balance. It was thought that sludge processing (such as anaerobic digestion) would reduce organic constituents by microbial degradation and thus distort the mass balance the U.S.EPA was attempting. Consequently, median priority pollutant levels derived from the forty-POTW survey are primary and secondary raw sludge concentrations.

The Mumma survey was based on analysis of thirty sludge samples

TABLE 3.13. Selected Priority Pollutant (Metals) Concentrations of U.S. Sludges.

Analyte	Forty-POTW Survey Median	Mumma Survey Mean	Mumma Survey Median	AMSA Survey Mean	AMSA Survey Median
	mg/dry kg				
Cd	11.2	41	20	26	11
Cr	248	2132	1275	432	222
Cu	411	1546	991	712	526
Pb	266	327	305	303	220
Ni	70	259	195	167	72
Zn	980	2181	1813	1526	1200
Hg	1.7	7 0	4.8	3.3	2.5

from twenty-three cities out of sixty-five solicited to participate. The sludges were processed sludge products, and the purpose of the study was to determine national sludge quality in order to formulate a means of sludge utilization that is environmentally sound.

The 1987 AMSA survey was based on analysis of sludge provided by sixty-three AMSA member agencies. All three surveys showed that metal concentrations follow the order Zn > Pb > Ni > Cd > Hg. Relative levels of Cu and Cr vary among the three surveys.

In 1996, 152 AMSA member agencies were contacted to provide information on their final sludge product after implementation of the Part 503 sludge regulations. Data on sludge quality was provided by 120 AMSA member agencies. Table 3.14 shows the mean and median metal concentrations in the final sludge products from this survey. Concentrations of most metals were the lowest in alkaline stabilized sludge, which includes brand name products such as Bio Gro and N-Viro sludges, and the highest in incinerator ash. Metal concentrations in all the final sludge products in Table 3.14 are considerably lower than the concentrations from the previous surveys (Table 3.13), and they are well below the limits required for exceptional quality sludge established by the U.S.EPA in the Part 503 sludge regulations.

Organics

Table 3.15 shows a comparison of selected priority pollutant organic compound concentrations in sludges based on the USEPA's NSSS conducted in 1988–89 and data from the MWRDGC's seven WRPs collected in 1995. The data indicates that significant reductions in many organic priority pollutants of environmental concern have occurred over the past seven years, but some of these differences may be the result of improvements in analytical detection limits as opposed to actual differences. The values for total dioxin are of particular interest, as dioxin is currently under study by the USEPA for possible regulatory action relative to acceptable sludge concentrations.

MWRDGC/U.S.EPA Cyanide Study

In August 1987, the MWRDGC began a U.S.EPA-funded study to provide a complete profile of the cyanide content of all its wastewaters and sludges. Samples were collected of raw sewage, final effluent, and various sludges from each WRP once per week for eight weeks. Table 3.16 indicates the types of samples collected at each WRP. The samples

TABLE 3.14. Metal Concentrations of U.S. Sludges from a 1996 Survey of AMSA Member Agencies.

| | Final Sludge Product (mg/dry kg) | | | | | | | |
| Metal | Air Dried | | Alkaline Stabilized | | Incinerator Ash | | Cake[a] | |
	Mean	Median	Mean	Median	Mean	Median	Mean	Median
As	12	10	5.8	5.2	19	13	14	5.1
Cd	8.0	5.0	3.1	2.0	15	11	7.0	4.7
Cr	125	105	50	38	266	51	117	67
Cu	447	308	176	122	554	365	559	459
Pb	126	77	62	70	210	196	129	85
Hg	1.7	1.0	0.8	0.5	0.5	0.2	2.0	1.7
Mo	23	15	8.2	7.5	29	32	23	13
Ni	60	45	39	23	129	42	69	36
Se	5.3	4.0	2.4	1.7	3.8	3.0	6.1	4.5
Zn	825	700	306	250	1540	1256	866	764

TABLE 3.14. (continued).

	Final Sludge Product (mg/dry kg)					
	Compost		Heat Dried		Liquid[b]	
Metal	Mean	Median	Mean	Median	Mean	Median
As	5.4	5.0	5.6	4 5	8.0	5.0
Cd	4.6	4.5	8.9	9.0	4 8	4.0
Cr	73	53	153	95	54	35
Cu	317	380	472	267	486	421
Pb	80	73	93	92	72	55
Hg	1.9	2.0	1.6	1.0	2 7	2.3
Mo	14	13	18	19	12	11
Ni	26	28	35	33	33	30
Se	3.7	3 9	8.3	5.8	6.0	3.1
Zn	878	820	906	817	788	752

[a]Processed sludge with a total solids ≥13 percent.
[b]Processed sludge with a total solids <13 percent.

119

were analyzed for total solids, volatile solids, total cyanide, acid dissociable cyanide, and thiocyanate. The cyanide analyses were performed using an automated method developed by the MWRDGC for differentiating cyanide species [33,34].

The cyanide concentrations found during the eight-week sampling period from September 15 to November 3, 1987 are presented in Tables 3.17 and 3.18. Table 3.17 summarizes the concentrations on a wet basis (mg/L), of total cyanide, acid dissociable cyanide, and thiocyanate by the automated method. The number of observations, means, ranges, and standard deviations are given for all sludge samples at MWRDGC facilities. Table 3.18 presents the same data on a dry weight basis (mg CN/kg dry solids).

As can be seen in Table 3.17, mean total cyanide concentrations on a wet basis, ranged from a low of 0.461 mg/L in the waste-activated sludge to a high of 14.112 mg/L in the centrifuge cake. Mean thiocyanate concentrations ranged from 0.216 mg/L as CN in the waste-activated sludge to 1.622 mg/L as CN in the centrifuge cake. As Table 3.18 indicates, mean total cyanide concentrations on a dry weight basis ranged from a low of 66.1 mg CN/kg dry solids in the waste-activated sludge to a high of 92.9 mg CN/kg dry solids in the centrifuge cake. Mean thiocyanate concentrations ranged from 10.4 mg/kg in the centrifuge cake to 42.0 mg/kg in the primary sludge.

TABLE 3.15. Selected Priority Pollutant (Organics) Concentrations of U.S. Sludges.

Analyte	National Sewage Sludge Survey 1990 Mean Values (μg/kg dry wt)	MWRDGC Treatment Plants 1995 Mean Values (μg/kg dry wt)
Benzidine	< 769,231	< 5,000
Benzo(a)pyrene	12,393	41 9
Bis(2-ethylhexyl)phthalate	104,101	648
Dimethylnitrosamine	< 769,231	< 140
Hexachlorobenzene	< 153,846	< 240
Benzene	122	< 200
Carbon tetrachloride	NA	< 180
Chloroform	3,702	0 22
Tetrachloroethylene	1,358	0.90
Trichloroethylene	848	4.65
Vinyl chloride	< 31,250	< 140
Total PCB	966	12.3
Total Dioxin	924	27.0

NA = No analysis

TABLE 3.16. Wastewater and Sludge Samples Collected from Seven Facilities for Cyanide Analysis for the United States Environmental Protection Agency, 1987.

Sample Stream	Calumet	Hanover Park	John E. Egan	Lemont	North Side	O'Hare	Stickney
Raw sewage	X	X	X	X	X	X	X*
Final effluent	X	X	X	X	X	X	X
Primary sludge	X	X	X		X		X
Imhoff sludge							X
Waste-activated sludge	X	X	X	X	X	X	X
Anaerobic digester feed							X
Anaerobic digester draw-off	X	X	X				X
Centrifuge cake	X		X				X

*Two raw sewage streams were sampled and analyzed from the West Side and the Southwest plants at the Stickney WR

TABLE 3.17. Summary of Cyanide/Thiocyanate Concentrations on a Wet Basis in Wastewaters and Sludges of Seven MWRDGC Facilities, 1987.

| | MWRDGC Automated Method (mg/L) | | | |
| | Total | Acid Dissociable | Thiocyanate As | |
Sample Stream	CN	CN	CN	SCN
Imhoff Sludge				
N	8	8	8	8
Mean	7.583	0 811	1.267	2 826
Minimum	6.170	0 655	0 000	0.000
Maximum	8 074	1.420	2.519	5.617
Std. Dev.	0.596	0 258	0 760	1 695
Primary Sludge				
N	40	40	40	40
Mean	0.776	0.112	0 432	0 964
Minimum	0 029	0 012	0.000	0.000
Maximum	1.880	0 482	1 130	2 520
Std. Dev	0 541	0.084	0.323	0.719
Waste-Activated Sludge				
N	56	56	56	56
Mean	0 461	0.055	0.216	0.482
Minimum	0.042	0.004	0 008	0 018
Maximum	1 246	0.264	0 497	1.108
Std Dev.	0.294	0.042	0 131	0 291
Anaerobic Digester Feed				
N	8	8	8	8
Mean	3.789	0.468	1 221	2 723
Minimum	2 283	0.341	0.1061	0.236
Maximum	5.174	0.793	3 224	7 190
Std Dev.	0 885	1.155	0.993	2.215
Anaerobic Digester Drawoff				
N	32	32	32	32
Mean	2.079	0 224	0 506	1.128
Minimum	0 042	0.018	0 000	0 000
Maximum	4.268	1 010	1 690	3.771
Std Dev.	1.539	0.197	0 416	0.927
Centrifuge Cake				
N	22	22	22	22
Mean	14 112	1.137	1.622	3 617
Minimum	4.542	0 587	0 000	0.000
Maximum	22 288	1.869	5.927	13.217
Std. Dev.	5.633	0 327	1 757	3 917

N = Number of samples
Std Dev = Standard deviation

TABLE 3.18. Summary of Cyanide/Thiocyanate Concentrations on a Dry Basis in Wastewaters and Sludges of Seven MWRDGC Facilities, 1987.

| | MWRDGC Automated Method (mg/L) | | | |
| | Total | Acid Dissociable | Thiocyanate As | |
Sample Stream	CN	CN	CN	SCN
Imhoff Sludge				
N	8	8	8	8
Mean	120.0	12 8	20.2	45.1
Minimum	97.9	10.0	0 0	0.0
Maximum	129 3	21 8	40.0	89 2
Std Dev	9.9	3.8	12 2	27.2
Primary Sludge				
N	40	40	40	40
Mean	91.0	47.7	42 0	93.6
Minimum	4.3	1.1	0.0	0.0
Maximum	1014.1	973 7	192 8	429 1
Std. Dev.	159.3	155.6	38.1	84.9
Waste-Activated Sludge				
N	56	56	56	56
Mean	66.1	7.9	31.9	71.1
Minimum	20.0	1.3	0 9	2.1
Maximum	296.7	32.2	67.2	149.7
Std Dev.	45.1	5 8	16 7	37 2
Anaerobic Digester Feed				
N	8	8	8	8
Mean	78.7	9.8	24.9	55 5
Minimum	43 9	6.7	2.3	5.0
Maximum	103.5	16 9	63 2	141 0
Std. Dev.	17 8	3.4	19.3	43 1
Anaerobic Digester Drawoff				
N	32	32	32	32
Mean	81.7	8.5	21.5	47 9
Minimum	3.6	1.4	0 0	0.0
Maximum	342.8	26.6	62.6	139 7
Std. Dev.	67.6	6.0	15 9	35 4
Centrifuge Cake				
N	22	22	22	22
Mean	92.9	7.4	10.4	23 1
Minimum	24 8	3 5	0.0	0.0
Maximum	169.9	14.6	33.1	73.8
Std Dev	42.0	2 6	11 1	24 9

N = Number of samples
Std. Dev = Standard deviation

NUTRIENTS IN FINAL SLUDGE PRODUCTS AT POTWs

The final sludge products from POTWs contain beneficial nutrients which can be utilized by agricultural, horticultural, and forest crops and vegetation. The major nutrients in sludge are nitrogen, phosphorus, and potassium. The final nutrient content of sludge varies with the processing procedures used to produce the final sludge product. During sludge processing operations water-soluble forms of nitrogen, phosphorus, and potassium are lost, resulting in lower concentrations of these elements in the final product. In addition, nitrogen as ammonia is lost through volatilization during sludge processing. As a result, the nutrient content in the final sludge product is considerably lower than that observed in primary sludges.

The mean nutrient content in final sludge products from the 1996 AMSA survey is shown in Table 3.19. Alkaline stabilized sludge, which includes brand name products such as Bio Gro and N-Viro sludges, had the lowest nitrogen and phosphorus concentrations, and the highest amount of potassium. Heat dried sludges had the highest levels of nitrogen and phosphorus.

EFFECTS OF PRETREATMENT ON CONSTITUENTS PRESENT IN SLUDGE

SCOPE OF PRETREATMENT REGULATIONS

Recognizing the impact of industrial discharges on treatment plant operations and on the quality of the resultant sludge, many publicly owned treatment works (POTW) have implemented industrial pretreatment programs.

TABLE 3.19   **Mean Nutrient Concentrations in Final Sludge Products from a 1996 Survey of AMSA Member Agencies.**

Final Sludge Product	Nutrient (mg/dry kg)			
	Total Kjeldahl-N	NH_3-N	Total P	K
Air dried	24,780	2,780	21,050	2,260
Alkaline stabilized	17,010	2,270	5,220	7,080
Incinerator ash	NA	NA	NA	NA
Cake[a]	39,480	6,160	19,500	2,940
Compost	31,240	11,780	14,640	4,470
Heat dried	58,500	28,700	32,990	2,330
Liquid[b]	49,200	17,050	20,640	3,360

[a]Processed sludge with a total solids ≥13 percent.
[b]Processed sludge with a total solids < 13 percent.
NA = Not available

Pretreatment programs prevent pass-through of pollutants into receiving streams and interference with physical and biological treatment processes, including sludge management, by imposing specific limits on the concentration of pollutants in industrial discharges to public sewer systems. From the mid 1960s through the 1980s, many POTWs established general discharge limits applicable to the dischargers from all industrial users tributary to their treatment facilities.

Parameters Limited

In establishing general industrial discharge limitations, POTWs regulated those parameters that caused physical interference with the operation of the treatment works, such as pH or fats, oils, and greases, those pollutants that passed untreated through the treatment works and for which the discharge from the treatment works was itself regulated, and those pollutants, such as heavy metals and cyanide, which accumulated in the sludge from the treatment works and which adversely impacted sludge disposal options. The general industrial discharge limits established by several POTWs during the period from 1969 to 1990 are indicated in Table 3.20.

Categories—Subcategories Covered

In 1972, the United States Congress enacted the Clean Water Act, which directed the United States Environmental Protection Agency (U.S.EPA) to conduct comprehensive studies of the impact of industrial pollution on the nation's waterways, and to implement nationwide regulations limiting the discharge of pollutants from those industries found to be significant sources of pollution.

The U.S.EPA identified thirty-four industrial categories as significant sources of water pollution, and has established (categorical) pretreatment regulations governing twenty-nine of these categories. The categories for which regulations have been established are indicated in Table 3.21.

In establishing these categorical regulations, the U.S.EPA studied both the industrial processes and the pretreatment options available to the industrial dischargers. The discharge limits established by the U.S.EPA were based upon the best available technology for the pretreatment of pollutants from each industrial category. In this manner, industrial categories were regulated for the pollutants of concern from particular processes and to the highest degree of reduction possible. The regulated pollutants and the discharge limits for the Metal Finishing Category and the Organic Chemicals, Plastic and Synthetic Fibers Category (Phase 1) are indicated in Tables 3.22 and 3.23, respectively.

TABLE 3.20. General Industrial Discharge Limitations at Several POTWs.[a]

Constituent	Metropolitan Water Reclamation District of Greater Chicago [35]	Hampton Roads Sanitation District [36]	Metropolitan Waste Control Commission Twin Cities Area [37]	Municipality of Anchorage [38]	Dallas Water Utilities [39]	East Bay Municipal Utility District [40]
pH (units)	5.0 to 10.0	>5.0	5.0 to 10.0	5.0 to 10.0	5.5 to 10.5	>5.5
Fats, oils and greases	250	100	100	100	100	100
Arsenic	NL	0.1	NL	10	0.5	2.0
Cadmium	2.0	0.1	2.0	1.0	1.0	1.0
Copper	3.0	5.0	6.0	0.3	4.0	5.0
Cyanide (total)	5.0	1.0	4.0	0.3	NL	5.0
Cyanide (free)	NL	NL	NL	NL	1.6	NL
Sulfide	NL	NL	NL	NL	10.0	NL
Lead	0.5	2.0	1.0	5.0	1.6	2.0
Nickel	10.0	2.0	6.0	1.5	9.0	5.0
Silver	NL	0.5	NL	0.02	4.0	1.0
Chromium (total)	25.0	5.0	8.0	10.0	5.0	2.0
Chromium (hexavalent)	10.0	NL	NL	4.0	NL	NL
Zinc	15.0	5.0	8.0	9.0	5.0	5.0
Barium	NL	NL	NL	NL	NL	NL
Boron	NL	NL	NL	NL	NL	NL
Manganese	NL	NL	NL	NL	1.0	NL
Mercury	0.0005	0.02	0.1	0.002	0.01	0.05
Selenium	NL	NL	NL	NL	0.2	NL
Iron	50.0	NL	NL	NL	NL	NL
Phenol	NL	2.0	NL	NL	149.0	100.0
Benzene	NL	1.0	NL	NL	1.0	NL

TABLE 3.20. (continued).

Constituent	Metropolitan Water Reclamation District of Greater Chicago [35]	Hampton Roads Sanitation District [36]	Metropolitan Waste Control Commission Twin Cities Area [37]	Municipality of Anchorage [38]	Dallas Water Utilities [39]	East Bay Municipal Utility District [40]
Toluene	NL	1.0	NL	NL	3.0	NL
Isopropyl alcohol	NL	NL	NL	NL	26,250 0	NL
Acetone	NL	NL	NL	NL	21,000.0	NL
Methylene chloride	NL	1.0	NL	NL	21.0	NL
Ethyl benzene	NL	1.0	NL	NL	1.6	NL
Methyl alcohol	NL	NL	NL	NL	20,000.0	NL
Methyl ethyl ketone	NL	NL	NL	NL	249.0	NL
Xylene	NL	1.0	NL	NL	2.0	NL
Benzene, toluene, ethyl benzene, xylene	NL	1.0	NL	NL	NL	NL
Total toxic organics	NL	2.13	NL	NL	NL	NL

aAll limitations are in mg/L except where noted
NL = Denotes no limitation

Future Categories to Be Regulated

In addition to the twenty-nine industrial categories for which the U.S.EPA has established pretreatment regulations, additional categories are currently under consideration for regulation. These categories are indicated in Table 3.24.

Surveillance Program

Essential to the effectiveness of any industrial pretreatment program are a surveillance program to detect instances of noncompliance by industrial users and the ability of the local public agency to react promptly and effectively to remedy such instances of noncompliance.

Federal regulations require that all significant industrial users (industrial categories listed in Table 3.21 and other industrial users discharging greater than 25,000 gallons per day of process wastewater) conduct self-monitoring of their discharges at least once every six months to verify continued compli-

TABLE 3.21. Industrial Categories with Established Regulations [41].

Industrial Category	40 CFR Part	Proposed Rule Date	Final Rule Date	Existing Sources Compliance Date
Aluminum forming	467	11/22/82	10/24/83	10/24/86
Battery manufacturing	461	11/02/82	03/09/84	03/09/87
Builders' paper and board mills	431	01/06/81	11/18/82	07/01/84
Coil coating I	465	01/12/81	12/01/82	12/01/85
Coil coating II (canmaking)	465	02/10/83	11/17/83	11/17/86
Copper forming	468	11/12/82	08/15/83	08/15/86
Electrical and electronic components I	469	08/24/82	04/08/83	07/14/86
for total toxic organics				07/01/84
for arsenic				11/08/85
Electrical and electronic components II	469	03/09/83	12/14/83	07/14/86
for total toxic organics				07/01/84
for arsenic				11/08/85
Electroplating	413	02/14/78	01/28/81	
for total toxic organics only				07/15/86
for nonintegrated facilities				04/27/84
for integrated facilities				06/30/84
Inorganic chemicals I	415	07/24/80	06/29/82	08/12/85
Inorganic chemicals II	415	10/25/83	08/22/84	08/22/87
Iron and steel	420	01/07/81	05/27/82	07/10/85
Leather tanning and finishing	425	07/02/79	11/23/82	11/25/85

TABLE 3.21. (continued).

Industrial Category	40 CFR Part	Proposed Rule Date	Final Rule Date	Existing Sources Compliance Date
Metal finishing	433 413	08/31/82	07/15/83	02/15/86
for interim total toxic organics only				06/30/84
Metal molding and casting (foundries)	464	11/15/82	10/30/85	10/31/88
Nonferrous metal forming	471	03/05/84	08/23/85	08/23/88
Nonferrous metal manufacturing I	421	02/17/83	03/08/84	03/08/87
Nonferrous metal manufacturing II	421	06/27/84	09/20/85	09/20/88
Organic chemicals, plastic and synthetic fibers	414 416	03/21/83	11/05/87	11/05/90
Pesticides	455	11/30/82	10/04/85	10/04/88
Petroleum refining	419	12/21/79	10/18/82	12/01/85
Pharmaceutical manufacturing	439	11/26/82	10/27/83	10/27/86
Plastics molding and forming	463	02/15/84	12/17/84	01/30/88
Porcelain enameling	466	01/27/81	11/24/82	11/25/85
Pulp, paper and pasteboard	430	01/06/81	11/18/82	07/01/84
Rubber processing	428	12/18/79	no date	no date
Steam electric power generating	125 423	10/14/80	11/19/82	07/01/84
Timber products processing	429	10/31/79	01/26/81	01/26/84

ance, and that the industrial users report the results of their self-monitoring to the POTW. In addition, the POTW is required to inspect and sample each significant industrial user at least once annually to verify continued compliance, independent of information supplied by the industrial user.

While the inspection and monitoring requirements indicated above represent the statutory minimum, many POTWs have found that more frequent surveillance is necessary to ensure continued compliance. For these industrial users, monthly self-monitoring and quarterly monitoring by the POTW are not uncommon. Finally, for industrial users who are known to cause interference or pass-through or who are particularly recalcitrant in resolving a condition of noncompliance, continuous self-monitoring or continuous monitoring by the POTW may be warranted.

To ensure the validity of surveillance samples, and thus their usefulness in enforcement proceedings, it is essential that all sampling and analyses, both by the industrial user and the POTW, be performed in accordance with U.S.EPA-approved methodologies.

TABLE 3.22. Categorical Pretreatment Standards for Metal Finishing Point Source Category [42] (40 CFR 433.15–433.16).

Pollutant	PSES[a]		PSNS[b]	
	1-Day Max.	Monthly Avg.	1-Day Max.	Monthly Avg.
Cyanide (total)	1.20	0 65	1.20	0.65
Copper	3.38	2.07	3 38	2.07
Nickel	3.98	2.38	3.98	2.38
Cadmium	0 69	0 26	0 11	0.07
Chromium	2.77	1.71	2.77	1.71
Zinc	2 61	1 48	2 61	1.48
Lead	0.69	0.43	0 69	0.43
Silver	0 43	0 24	0.43	0.24
Total toxic organics	2 13	—	2.13	—

[a]Pretreatment standards for existing sources
[b]Pretreatment standards for new sources
All limitations are in mg/L

TABLE 3.23. Categorical Pretreatment Standards for Organic Chemicals and Plastics and Synthetic Fibers Category (414.85) [43].

Pollutant	Maximum for Any One Day	Maximum for Monthly Average
Acenaphthene	47	19
Benzene	134	57
Carbon tetrachloride	380	142
Chlorobenzene	380	142
1,2,4-Trichlorobenzene	794	196
Hexachlorobenzene	794	196
1,2-Dichloroethane	574	180
1,1,1-Trichloroethane	59	22
Hexachloroethane	794	196
1,1-Dichloroethane	59	22
1,1,2-Trichloroethane	127	32
Chloroethane	295	110
Chloroform	325	111
1,2-Dichlorobenzene	794	196
1,3-Dichlorobenzene	380	143
1,4-Dichlorobenzene	380	142
1,1-Dichloroethylene	60	22
1,2-Trans-dichloroethylene	66	25
1,2-Dichloropropane	794	196
1,3-Dichloropropylene	794	196
2,4-Dimethylphenol	47	19
Ethylbenzene	380	142
Fluoranthene	54	22

TABLE 3.23. (continued).

Pollutant	Maximum for Any One Day	Maximum for Monthly Average
Methylene chloride	170	36
Methyl chloride	295	110
Hexachlorobutadiene	380	142
Naphthalene	47	19
Nitrobenzene	6402	2237
2-Nitrophenol	231	65
4-Nitrophenol	576	162
2,6-Dinitro-o-cresol	277	78
Phenol	47	19
Bis(2-ethylhexyl) phthalate	258	95
Di-n-butyl phthalate	43	20
Diethyl phthalate	113	46
Dimethyl phthalate	47	19
Anthracene	47	19
Fluorene	47	19
Phenanthrene	47	19
Pyrene	48	20
Tetrachloroethylene	164	52
Toluene	74	28
Trichloroethylene	69	26
Vinyl chloride	172	97
Total cyanide	1200	420
Total lead	690	320
Total zinc	2610	1050

All limitations are in μg/L

Enforcement Strategies

When industrial users are found to be in noncompliance, it is essential that the POTW act promptly and forcefully to remedy the condition of noncompliance. Generally, enforcement action against a noncomplying industrial user consists of two components: remediation and deterrence. For remediations, the industrial user is required to conduct an immediate investigation into the cause of the condition of noncompliance and to develop a formal plan and schedule, with specific milestone events, to attain compliance. The industrial user's progress toward attaining compliance is monitored through the submittal of compliance progress reports to the POTW.

Deterrence against repeated or prolonged instances of noncompliance is achieved through a system of escalated enforcement actions and increasing monetary penalties. In the event of severe instances of noncompliance that constitute an immediate threat to life or the environment or to the operation of the treatment works, the POTW must have authority to immediately halt the noncomplying discharge or other activity. Finally, for industrial users who willfully violate applicable regulations, the POTW should have authority, either directly or through appropriate state or federal agencies, to seek criminal penalties, including fines and imprisonment.

An example of the types of enforcement actions available to a POTW is detailed in Table 3.25.

MWRDGC SLUDGE MONITORING PROGRAM

The sludge monitoring program at the MWRDGC includes the process streams of raw sewage and final effluent in addition to sludge. This monitoring program has been developed for specific MWRDGC needs and can vary for other agencies depending on their needs. The MWRDGC's program has been successful in monitoring MWRDGC operations for over twenty years. Currently raw sewage and final effluent are sampled daily with metal analyses performed daily except for the John E. Egan Water Reclamation Plant (WRP). This plant's raw sewage is analyzed every eight days and the final effluent is analyzed daily for metals. The raw sewage and final effluent of all MWRDGC plants are taken by automatic samplers and composited on a twenty-four-hour basis. Information about the MWRDGC's sludge monitoring program is displayed in Tables 3.26 and 3.27.

Digested sludge samples are taken every eight hours and composited on a twenty-four-hour basis. Prior to 1975, metal analyses were performed at a frequency dependent on the ultimate disposition of the digested sludge. From 1975 through 1991, metals were analyzed for all sludges every sixteen days. Beginning in 1992, sludges were analyzed weekly for metal content.

The raw sewage, final effluent, and sludge of MWRDGC plants are analyzed for a large variety of constituents on a daily basis depending

TABLE 3.24. **Industrial Categories for Which Categorical Pretreatment Regulations Are under Consideration [44].**

Industrial Category	Promulgation Date
Drum reconditioning	NE
Hazardous waste treatment	1996
Industrial laundries	NE
Metal products and machinery	1997
Solvent recycling	NE
Transportation equipment cleaning	NE
Used oil reclamation and refining	NE

NE = Not established.

TABLE 3.25. Enforcement Actions Available to a Typical POTW [45].

Action	Application	Description	Industrial User Response	Penalty
1. Notice of noncompliance	Nonsignificant instance of noncompliance	Letter advising industrial user of instance of noncompliance	Investigation, report and statement of corrective action	None
2. Notice of violation	Significant instance of noncompliance	Cease and desist order requiring compliance within 90 days	Formal compliance plan and schedule, interim and final compliance progress reports	$100.00 to $10,000.00 per day, to be assessed if industrial user fails to comply with subsequent Board Order
3. Show cause action	Failure to comply with cease and desist order	Formal hearing before designee of Board of Commissioners, Board Order requiring compliance by date certain	Formal compliance plan and schedule interim and final compliance monitoring and progress reports	$100.00 to $10,000.00 per day stipulated penalties through Board Order compliance date
4. Civil court action	Failure to comply with Board Order or imminent threat to life, the environment and the POTW	Civil petition for injunctive relief and imposition for fines in Circuit Court	Compliance with Court Order	$1000.00 to $10,000.00 per day assessed by court
5. Criminal court action	Willful violation of an applicable regulation	Criminal proceedings seeking fines and imprisonment brought in state or federal court states' attorney or U.S.EPA	Compliance with Court Order	$2500.00 to $25,000.00 and up to 3 years imprisonment per count

TABLE 3.26. Current Method of Sampling and Frequency of Metal Analysis in Raw Sewage and Final Effluent at MWRDGC Plants.

Plant	Raw Sewage	Final Effluent	Raw Sewage	Final Effluent
Calumet WRP	24-hour composite automatic sampler	24-hour composite automatic sampler	Daily	Daily
North Side WRP	24-hour composite automatic sampler	24-hour composite automatic sampler	Daily	Daily
Stickney WRP	24-hour composite automatic sampler	24-hour composite automatic sampler	Daily	Daily
Hanover Park WRP	24-hour composite automatic sampler	24-hour composite automatic sampler	Daily	Daily
John E. Egan WRP	24-hour composite automatic sampler	24-hour composite automatic sampler	Daily	Daily

TABLE 3.27. Method of Sampling and Frequency of Metal Analysis for Digested Sludge at MWRDGC Plants.

Plant	Method of Sampling	Frequency of Metal Analysis	
		1975–1991	1992–1995
Calumet WRP	24-hour composite grab sample every 8 hours	Once/16 days	Weekly
North Side WRP[a]	24-hour composite grab sample every 8 hours	Once/16 days	Weekly
Stickney WRP	24-hour composite grab sample every 8 hours	Once/16 days	Weekly
Hanover Park WRP	24-hour composite grab sample every 8 hours	Once/16 days	Weekly
John E. Egan WRP	24-hour composite grab sample every 8 hours	Once/16 days since 1976	Weekly

[a]Waste-activated sludge.

135

on National Pollutant Discharge Elimination System (NPDES) permits, process control requirements, and industrial waste pretreatment enforcement needs. These sixty constituents include metals, minerals, nutrients, oxygen demand parameters, and organics.

Those parameters monitored in the treatment plant process streams in relation to the pretreatment program include organics, total cyanide, oil and grease, and metals. Emphasis was placed on cadmium, chromium, copper, lead, nickel, and zinc in the monitoring of the raw sewage, final effluent, and sludges. Flow is continuously monitored at the MWRDGC water reclamation plants. This provides a firm basis for accurately computing the various metal loadings to the plants. Operational records contain totalized daily flows at each of the monitored plants. This data allows the study of the distribution of the metals in the various unit treatment processes and sludges. The solids and moisture components of the sludges are determined to provide metals data expressed on a dry weight basis.

The chronology of this monitoring program goes back to 1971 for raw sewage and final effluent except for the John E. Egan WRP which began operations in 1976. Monitoring of sludge also was initiated in 1971 for the Stickney and Calumet WRPs. The Hanover Park WRP sludge was initially monitored in 1972. The North Side WRP, which transfers its waste activated sludge to the Stickney WRP, began sludge monitoring in 1975. The John E. Egan WRP began monitoring sludge in 1976 with the beginning of operations. This provides twenty to twenty-five years of raw sewage, final effluent, and sludge data to assess the effects of the industrial waste pretreatment program.

Information on the quality of sludges is valuable in developing strategies for proper sludge management practices. Also, this information provides a firm basis for selecting cost-effective sludge management practices.

METHODOLOGIES FOR ANALYSIS OF SLUDGE CONSTITUENTS

PHYSICAL MEASUREMENTS

Total and Volatile Solids

Total solids is the sum of the dissolved and suspended solids in the sludge and is determined by drying a sample at 105°C and measuring the residue. The determination of the total solids of a sludge sample is required for determining the moisture content of the sludge, usually expressed as percentage of wet weight, and for expressing other constituents on a dry weight basis.

The volatile solids content of sludge, generally expressed as a percentage, is used as a measure of the organic content of sludge. Volatile solids are used in the determination of loading rate to a digester. Also, volatile solids reduction is used as a measure of the performance of the digester. The volatile solids are determined by the loss in weight when the sludge is combusted by heating a sample to 500°C.

The analytical methods for total and volatile solids are described in more detail in Section 2540 of the 18th Edition of *Standard Methods for the Examination of Water and Wastewater.*

Caloric Content

The caloric content of sludge is a function of moisture content and elemental composition. Primary combustible elements in sludges are carbon, hydrogen, and sulfur. The contribution of sulfur to the caloric content in sludges can often be neglected in calculations (as may the oxidation of metals). Sludges that contain large fractions of combustibles, such as grease and scum, have high caloric content, while sludges that contain large fractions of inert material, such as grit or chemical precipitates, have low caloric content.

The caloric content of unprocessed (raw) sludges (per unit of dry volatile solids) ranges from 11.16 MJ/kg (4800 BTU/lb) to 23.24 MJ/kg (10,000 BTU/lb) depending on the sludge matrix and volatile content, while a digested sludge has a caloric content ranging from 5.81 MJ/kg (2500 BTU/lb) to 12.78 MJ/kg (5500 BTU/lb) [46]. A typical value of caloric content for grease and scum from a POTW is 388 MJ/kg (167,000 BTU/lb) and for grit material is 9.30 MJ/kg (4000 BTU/lb) [47]. Often, as a rule of thumb, the caloric content for a generic wastewater sludge is assumed to be 23.24 MJ/kg (10,000 BTU/lb) and for a thermally conditioned sludge is assumed to be 27.89 MJ/kg (12,500 BTU/lb) [47–50]. For comparison, the fuel value of high grade coal is approximately 32.54 MJ/kg (14,000 BTU/lb) and of a petroleum product such as oil is approximately 46.48 MJ/kg (20,000 BTU/lb) [48,49].

It is important to realize that these typical caloric content values of sludges are on a dry weight basis. The caloric content for wet sludge (as is) is considerably lower and may be calculated as follows [51]:

$$[BTU/lb \ (Wet \ Basis)] = [BTU/lb \ (Dry \ Basis)]$$
$$\times [1 - (\% \ Moisture/100)] \qquad (3.1)$$

Various methods for obtaining caloric content of sludges have been proposed. In general, they can be categorized as calculation methods and calorimetric methods.

Caloric Content by Calculation

One of the earliest ways proposed to calculate caloric content was by the Dulong formula [46,48,49]:

$$BTU/lb = 145.4[C] + 620 \left([H] - \frac{[O]}{8}\right) + 41[S] \qquad (3.2)$$

where
[C] = Carbon as percent by mass
[H] = Hydrogen as percent by mass
[O] = Oxygen as percent by mass
[S] = Sulfur as percent by mass

It has been reported, however, that the Dulong formula can lead to erroneous results when applied to sludges [31–33]. A well-known alternative to the Dulong formula based on a statistical study of caloric content of vacuum filtered sludges of various types, is as follows [46,48,49]:

$$BTU/lb = a\left(\frac{100C}{100 - D}\right) - b\left(\frac{100 - D}{100}\right) \qquad (3.3)$$

where
C = Volatile solids as percent by mass
D = Dosage of inorganic conditioning agent as percent by mass ($D = 0$ for organic conditioning agents)
a = Empirical constant (107 for activated sludge and 131 for raw domestic or primary sludge)
b = Empirical constant (5 for activated sludge and 10 for raw domestic or primary sludge)

An empirical equation has been suggested for caloric content based upon ignition of sludge at 500°C [48]:

$$BTU/lb = 1.8(83.3P - 1089) \qquad (3.4)$$

where
P = Loss at ignition as percent by mass of total solids

An empirical equation has also been obtained for caloric content of sludges as a function of volatile solids content [48]:

$$BTU/lb = 122(VS) - 660 \qquad (3.5)$$

where
(VS) = Percent volatile solids

This last equation was determined by correlating caloric content from waste activated sludges, digested sludges, primary sludges with lime, and digested sludges with both alum and ferric chloride.

Caloric Content by Calorimetry

This is a direct experimental method to determine caloric content by burning a weighed sample of sludge in an adiabatic oxygen bomb calorimeter under controlled conditions. The caloric content is computed from temperature observations made before and after combustion, making allowances for thermometer and thermochemical corrections. A bomb calorimeter is a massive gun metal cylinder fitted with a screwed cover, capable of withstanding very high pressures. Approximately one gram of material is burned in this sealed cylinder, in an atmosphere of pure oxygen (to assure complete combustion), while the cylinder is immersed in an insulated water bath. The heat of combustion is determined from the temperature rise of the water bath. Some bomb calorimeters can use samples of up to 25 grams, which lower sampling errors. The procedure is available in detail, as an ASTM standard method designated: D 2015-66 [52].

CHEMICAL CONSTITUENTS

Cyanide

Cyanides are known to be toxic to man, but moreso to fish and other aquatic life. The complexity of the chemistry of cyanides has led to the coexistence of several cyanide species in the environment. The great toxicity of cyanide is due to molecular HCN and, to a lesser extent, cyanide ion CN⁻. Analytical distinction between HCN and other cyanide species in solutions of complex cyanides is very important and possible. The degree of dissociation of various metallocyanide complexes increases with decreased concentration and decreased pH, and is inversely proportional to their stability constants. Zinc and cadmium, weak cyano-complexes, dissociate almost totally in dilute solutions, which can result in acute toxicity. On the other hand, strong complexes such as those of cobalt and iron (ferro and ferri-cyanides) are stable and not toxic as such, unless in solutions that are not very dilute and have been aged for a long time. Exposure to ultraviolet radiation of sunlight cause photolysis and yield CN^- and HCN. Losses of HCN to the atmosphere and its destruction by bacterial and chemical interactions, concurrent with its production, tend to decrease the possibilities of reaching harmful levels. Therefore, regulatory distinction between free CN^-, HCN, weak complexes, and strong complexes, can be justified. Several regulations and standards require continuous monitoring of cyanides in water, wastewater, and sludge.

Two cyanide parameters are regulated by the U.S.EPA, namely cyanide amenable to chlorination and total cyanide. The cyanide amenable to chlorination measures CN^-, HCN, and weak cyanide complexes. Part of the sample is chlorinated to decompose the oxidizable cyanides. Total cyanide determinations are required for both the chlorinated portion and the original sample. The difference represents the cyanides amenable to chlorination. Frequently this parameter is subjected to some not understood interferences that result in negative cyanide values.

There are other manual methods that measure the same cyanide species as those amenable to chlorination, and which are not subjected to similar interferences. In addition, only one cyanide determination is needed (thus no subtraction). The nomenclature of these methods is somewhat confusing, i.e., releasable cyanide, weak acid dissociable, or weak and dissociable cyanide. It should be realized that they all practically measure the same components, and a simpler name applicable to all of them is dissociable cyanide.

Manual total cyanide methods are subjected to several types of positive and negative interferences. Among the most difficult to overcome are interferences of oxidants, sulfide, and aldehydes [53]. All of those have more complications in the presence of nitrate and even moreso in the presence of nitrite. Also, thiocyanate interferes with the total cyanide and is variably included in the measurement (thus cannot be determined and subtracted).

The MWRDGC R&D Methodology Laboratory developed a multichannel automated system that overcomes most of the above-mentioned interferences, and does not include any thiocyanate in the measurement [34]. Each of the total cyanide and dissociable cyanide is determined by one direct measurement (no subtraction). In addition, the automated method has a lower limit of detection (0.5 $\mu g/L$), and the rate of analysis is twenty samples per hour, compared to about 1.5 hours or more per sample for manual methods. The automated system utilizes some segmented flow automated modules (Bran Lube/Technicon Autoanalyzer II modules) in addition to an online, thin film distillation and ultraviolet (UV) irradiation. Three factors control separation of cyanides from samples, namely UV irradiation, pH, and thin film distillation ratio. For total cyanide, the breakdown of the strong cyanide complexes (Fe and Co), is achieved by controlled UV irradiation using Pyrex filter, which prevents any breakdown of thiocyanate. The gentle thin film distillation, alone without UV, determines only dissociable cyanide. In all cases, adsorption of the liberated HCN is carried out using sodium hydroxide solution and a glass coil. Colorimetric determination of the recovered cyanide is made by pyridine-barbituric acid reagent. The developed color is measured at 578 nm.

Phenols

Phenols in wastewater and sludge are analyzed either by colorimetric or gas chromatographic techniques. The colorimetric method is much older and determines only an estimate of the total phenol content. The gas chromatographic technique determines values of individual phenols rather than an estimate of the total content (see "Non-Mass Spectrometric Methods" under "Semivolatiles").

Colorimetric Methods

The U.S.EPA-approved method for phenols is given in *Standard Methods for the Examination of Water and Wastewater,* 18th Edition. In this method, the phenols are first separated from most interferences by a manual acidified distillation. Because of the difficulty of bumping from high solids contents, sludges are diluted ten- to fifty-fold before distillation. The phenols in the distillate are coupled with 4-aminoantipyrine (4-AAP) and oxidized in alkaline buffer (pH 10) with potassium ferricyanide. The resultant colored dye is then measured spectrophotometrically. If high levels are expected, an aqueous measurement at 500 nm is made. For better detection (down to 2 µg/L), the dye in 500 mL is extracted in 25 mL of chloroform and measured at 460 nm.

The 4-AAP reagent does not react with all phenols, especially those para-substituted phenols where the substitution is an alkyl, aryl, nitro, benzoyl, nitroso, or aldehyde group. Simple phenol is used as the calibration standard, and because it distills and reacts so well, values found in samples are regarded as the minimum possible value.

A second colorimetric method involves the use of 3-methyl-2-benzothiazolinone hydrazone (MBTH). Although this reagent is not as well known or used as 4-AAP, MBTH reacts with many more substituted phenols (including many parasubstituted). An example is *p*-cresol, which yields 2% of the response of simple phenol using 4-AAP, but 70% using MBTH. The method is listed under Method 420.3 of *U.S.EPA Methods for Chemical Analysis of Water and Wastes.* In this method, the same manual distillation procedure is used as for 4-AAP. The phenols in the distillate are coupled with MBTH and oxidized under acidic conditions with ceric ammonium sulfate. The method has about the same limit of detection as 4-AAP, both with and without solvent extraction. The MBTH method has been automated using the on-line, thin film distillation unit that was developed by MWRDGC for automated cyanide and modified for phenols [54].

Metals

The number of metals contained in the sample, the solids content, and the complex matrix of sludges make the determinations of trace metals by wet chemical analytical techniques very difficult and time-consuming. Proper sampling, preservation, and digestion procedures must be observed even before considering analytical techniques.

Samples of sludge taken for trace metal analyses should be placed in either a plastic bottle or a borosilicate glass bottle, preserved with nitric acid to a pH less than 2, and then refrigerated at 4°C until the time of analysis, which should be as soon as possible, but may be delayed up to six months. Analyses for metals are usually for total concentrations; however, if the dissolved metals are to be determined, the sample is filtered before preservation with nitric acid.

Analyses of sludge may be determined on the sample as received or on a dry weight basis. The latter means is preferred, since comparisons of other samples are more readily demonstrated when the values are all expressed as a dried weight.

To determine the metal concentrations, sludges must be digested with nitric acid to place them in a soluble form. A measurable amount of sludge is weighed in a pre-weighed beaker and calculated by subtracting the beaker weight, the difference is the weight of the sludge. Distilled water is added to the beaker to a volume of 50 to 100 milliliters and 5 milliliters of nitric acid. The beaker is placed on a hot plate and allowed to boil slowly until the sample becomes a light-colored, clear solution. It is then filtered, the beaker carefully washed with distilled water, and transferred to another glass or plastic container. The sample is now ready for analysis. The preferred analytical technique for metals currently is atomic spectroscopy: atomic absorption and atomic emission.

This technique relies on the physical properties of the atoms of each element. These atoms can absorb or emit electromagnetic energy (visible and ultraviolet light) at certain specific wavelengths. Each element's atoms absorb or emit at a different wavelength. The amount of light emitted or absorbed is proportional to the concentration. This property is utilized by two types of instruments (atomic absorption spectrometer and inductively coupled plasma spectrophotometer) to determine the concentration of various elements.

Atomic Absorption Spectrophotometer

The atomic absorption spectrophotometer (AAS) measures the amount of light absorbed by the atoms of the element of interest to determine the concentration. Three techniques are available for the AAS to accomplish this task.

Flame AAS

The instrument must produce a sufficient quantity of ground-state atoms (not ions). This is usually done by aspirating a liquid sample into a burner chamber, mixing the sample with fuel and oxidant, then burning it in a flame. The flame is hot enough to break apart molecules in the sample into individual atoms. It is these ground-state atoms that are quickly and continuously passing through the flame that are capable of absorbing light energy.

The instrument must produce a beam of light energy that is passed through the flame. The light beam must contain the proper wavelength and must be of sufficient intensity. This is produced by means of a hollow cathode lamp (HCL). A HCL is a lamp having a cylindrical shaped cathode made of the metal being analyzed. This lamp produces light rich in the wavelengths that the element of interest absorbs.

The instrument must be able to measure the amount of light absorbed. This is done by using a monochromator to isolate the specific wavelength being absorbed. This specific wavelength is focused onto a photodetector, which measures the decrease in light intensity caused by atoms that are absorbing at that wavelength.

Graphite Furnace AAS

The use of a burner chamber and a flame is the most common and most widely used method for producing the ground-state atoms. But there are other ways. The graphite furnace AA uses a hollow graphite tube that is heated to incandescence by an electric current. Prior to heating, a liquid or solid sample is placed inside the tube. The temperature of the tube is raised in several discrete steps. First, the temperature is raised to about 130°C and held there for about 30 seconds to dry the sample if it is a liquid. Next, the temperature is raised to about 1000°C to char the sample to drive off interferences such as smoke and volatile organics. Finally, the temperature is raised to about 2000°C (charring and atomizing temperatures vary depending on the element) to atomize the sample. At this temperature, the graphite tube is glowing brightly. The sample, which was placed inside the tube, is quickly vaporized into atoms, which fill the tube and remain there for several seconds. The position of the tube is such that it is centered in a beam of light from a hollow cathode lamp. The atoms filling the tube absorb some of the light passing through the tube. This temporary entrapment of atoms inside the tube gives the graphite furnace methods better sensitivity and detection limits than flame AA—approximately a 10× improvement. The graphite furnace is useful in determining concentrations in the parts per billion range.

Cold Vapor AAS

The cold vapor method for determining the concentration of mercury in the sub-parts per billion range is somewhat similar to the graphite tube furnace technique. Instead of the graphite tube, a glass tube with windows is centered in the beam of light from a HCL. Instead of using thermal energy to break down the molecules of the sample to produce ground-state atoms, a chemical reaction is used to reduce the mercury in the sample to elemental mercury. These mercury atoms can then be bubbled out of the sample and pumped into the glass flow cell. The mercury atoms in the cell then absorb a part of the light passing through the cell proportional to the concentration of mercury.

Inductively Coupled Plasma

A second class of instrumentation used in trace metal determinations is the atomic emission spectrophotometer. The most commonly used emission source is an inductively coupled argon plasma (ICP or ICAP). The ICP source is powered by a radio frequency (RF) generator. The output from the RF generator is coupled to a water-cooled copper induction coil that is wrapped around the outside of a quartz torch. Argon flow through the torch assembly provides for plasma discharge. During plasma ignition, the gas stream is seeded with electrons from an external source, such as a spark. These electrons are accelerated by the RF electromagnetic field and they collide with argon atoms to form more electrons and argon ions. These in turn are also accelerated. This process continues until the gas becomes highly ionized (a plasma), at which point the discharge is stable and self-sustaining as long as RF power is applied. The plasma temperature is as high as 10,000 Kelvin.

Liquid samples are introduced into the plasma discharge as an aerosol suspended in argon gas. The resultant elements of the sample are excited. After excitation, the atoms that comprise the sample emit light at their characteristic wavelengths. Light from the plasma emission source is focused onto an entrance slit of the optical system. The light is then dispersed by a grating and passes through an exit slit. The amount of light emitted is then measured by a photomultiplier tube (PMT). By measuring the amount of light emitted, a quantitative determination of the element concentration can be made.

An ICP instrument can also measure the concentration of more than one element at a time. This is accomplished by taking a measurement at one wavelength and moving the grating to another element-specific wavelength. The light detector is now looking at the emission of light from another element. By moving the grating from one wavelength to another, many determinations can be done sequentially. In addition to sequential instruments, there are those that have many PMT detectors in an array. This provides a light path for each element determination so all the element concentrations can be determined simultaneously.

ORGANIC COMPOUNDS SPECIFIC METHODS

General Theory of Gas Chromatography

Gas chromatography is the method of separating mixtures by means of relative volatility, by passing them through a tube (the column) filled or coated with a suitable absorber (the stationary phase) by some gas (carrier, usually helium or nitrogen). The compounds separate because of their differing ability to be coated in the stationary phase. If the column has a relatively wide bore (2 mm to 4 mm i.d.), the stationary phase is normally coated on an inert packing (packed column). If the column is smaller, ≤0.75 mm i.d., the column is considered a capillary column, and the stationary phase is bonded directly to the capillary wall. Capillary columns give greater separation between compounds (higher resolution), while packed columns can handle larger samples.

Detectors

GC detectors may be grouped into two types—universal and specific. Universal detectors respond to all or almost all compounds, and include flame ionization, thermal conductivity, etc. Because universal detectors respond to everything, they are not as useful for detecting trace components in complex mixtures such as sludge. The detectors involved with trace pollutant analysis are described as follows:

(1) The Flame Ionization Detector (FID) utilizes a flame produced by the combustion of hydrogen air to ionize compounds in the GC effluent. It responds to all compounds that burn or ionize in the hydrogen/air flame, which is essentially all pollutants except the heavily halogenated ones.

(2) The Electron Capture Detector (ECD) is highly sensitive to compounds that have electron capturing groups attached, such as halogens, nitro, and sulfur. It is several orders of magnitude less sensitive to hydrocarbons. The ECD is probably the most sensitive detector available for halogenated compounds (such as PCBs and pesticides). It is nondestructive.

(3) The Hall Electrolytic Conductivity Detector (HECD) can be set up to respond only to halogens, nitrogen, or sulfur. It converts the compounds to HX, NH_3, etc. It is less sensitive than the ECD, but also less subject to interferences.

(4) The Photoionization Detector (PID) uses a small UV lamp to ionize compounds. It is most useful for aromatics, olefins, etc. Aliphatic hydrocarbons are not detected. Note: the PID and HECD are normally used in series for monitoring volatile organic compounds (VOCs).

(5) The Nitrogen Phosphorus Detector (NPD) uses a flame as a FID, but with less air flow and an alkali salt present in the flame. The cooler flame minimizes hydrocarbon ionization while the alkali salt enhances N and P compound ionization by an unknown mechanism. The NPD is probably the most sensitive detector for both N and P.

Mass Spectrometer (MS)

Mass spectrometry is both a universal and selective GC detector. By focusing on a particular atomic mass characteristic of a compound, this detector can be quite specific. However, by scanning the full mass spectral range and summing the responses at each mass unit, a total ion chromatogram can be generated, in which any compound eluting from the GC will be detected. Due to the much higher information content of mass spectra, identifications made by GC/MS generally have much greater certainty than those made by other detectors.

Mass spectrometers contain three major components: a region where ions are generated (source), a mass analyzer (magnet or quadrupole), and an ion detector. Molecules introduced to a mass spectrometer via a GC are generally ionized by a beam of high-energy electrons ("electron impact"). The ionized molecule and/or its fragments are then directed to the analyzer section. In a magnetic sector instrument, the strength of the applied magnetic field controls the mass (actually mass-to-charge ratio, m/z) of the ions that can pass through the curved flight path. A quadrupole analyzer has a flight path down the center of four rods. Oscillating radio frequency (RF) and direct current (DC) fields of opposing polarities are applied to the rods. Only those ions of the desired mass pass through the quadrupole; the others oscillate out of the path and strike the rods. Ions passing through either type of analyzer are focused on an electron multiplier for detection. The polarity of the applied voltages determines whether positive ions (usually) or negative ions are transmitted and detected. Associated electronics control all of the components and rapidly change the mass analyzer conditions to scan a mass range (in a second or less when using capillary columns). If all masses within a specified range are measured, a mass spectrum can then be generated for each scan. There are several scans for each GC peak eluted. Organic compounds analyzed by GC are categorized as volatile or semivolatile according to the procedure of separation/concentration from the water/solids matrix.

Volatile Organic Compounds (VOCs)

Sampling and Preservation

Samples must be collected in 40 mL screw cap vials with zero headspace and sealed with Teflon-lined septa. If aqueous samples contain residual chlorine, sodium thiosulfate (10 mg/40 mL) is added to empty sample vials prior to shipment to the sample site. Samples should be refrigerated at 4°C and analyzed within seven days. If aqueous samples have to be held more than seven days, a separate sample should be collected, acidified to pH 2 and analyzed within fourteen days. This acidification is to prevent biological degradation of some aromatic compounds under certain environmental conditions.

Analysis

Volatile compounds are analyzed by U.S.EPA Method 624 or 1624 (isotope dilution), purge and trap packed column GC/MS using 1–5 grams sludge, based on percent solids [55,56]. Method 624 uses internal standard compounds for quantitation. These compounds are similar in analytical behavior to target compounds. Method 1624 employs stable isotopically labeled analogs of the compounds of interest as internal standards in the analysis. An inert gas (helium) is bubbled through the sample (after dilution) contained in a specially designed purging chamber at ambient temperature. The volatiles are transferred from the aqueous phase to the vapor phase and trapped on a sorbent column, which is then heated and backflushed with the GC carrier gas to desorb the volatile compounds onto a gas chromatographic column. The GC is temperature-programmed to separate the volatiles, which are then detected and quantitated with MS.

Identification and Quantitation

Qualitative identification of the compounds of interest is based on manual comparison of the resultant GC retention times and mass spectra. Quantitation of the identified purgeable compounds is based on the mass spectral response (area) of a selected characteristic mass. The internal standard method of quantitation is used to calculate the concentration of each purgeable compound found in the sample according to the following formula:

$$C_s = \frac{(A_s)(C_{is})}{(A_{is})(RF)} \qquad (3.6)$$

where
C_s = Concentration of the compound of interest
A_s = Area of the characteristic mass of the compound of interest
A_{is} = Area of the characteristic mass of the internal standard
C_{is} = Concentration of the internal standard
RF = Response factor calculated from a separate standards run, where C_s is known

Using the same equation algebraically rearranged:

$$RF = \frac{(A_s)(C_{is})}{(A_{is})(C_s)} \qquad (3.7)$$

A computer file and a quantitation program are made to obtain peak areas and calculate concentrations using the selected characteristic masses, retention time windows, and RFs for each compound from updated reference calibration curves.

Quality Assurance

A quality assurance/quality control program has to be performed following U.S.EPA protocol [55]. The method detection limit (MDL) according to the U.S.EPA is based on analyzing seven replicates of 5 ml reagent water spiked with the analytes under investigation. The spiked level is to be roughly in the range of 2.5 to 5 times the signal-to-noise level of the instrument or at a level expected to be near but above the limit of detection.

The MDL for each analyte is calculated based on measurement of precision in terms of standard deviation as follows:

$$MDL = S \times t \qquad (3.8)$$

where
S = the standard deviation of the replicate analyses
t = is the student's value appropriate for a 99% confidence level for n − 1 degrees of freedom

However, MDLs for real samples are usually dependent on sample size and the level of interferences.

Non-Mass Spectrometric Methods

Volatile compounds can also be analyzed with GC/HECD (U.S.EPA Method 601 for halocarbons) and with GC/PID (U.S.EPA Method 602 for aromatics) [55]. When these methods are used, compound identification should be supported by a second GC column of different polarity.

Main Problems Encountered

Samples are contaminated by diffusion of volatile organic compounds (particularly methylene chloride) through the vial seal. This could be checked by a field blank prepared from reagent water and carried through the sampling and handling protocol. Carryover can occur when high level and low level samples are analyzed sequentially; this can be overcome by analyzing a reagent water blank after an unusually concentrated sample is encountered.

Semivolatile Organic Compounds

All the semivolatile organic compounds are listed in U.S.EPA Method 625 [55]. They are classified as either base/neutrals or acids according to the pH of solvent extraction. Although the pesticides and PCBs are included in the base/neutral fraction of the MS Methods, they are analyzed separately according to U.S.EPA Method 608 or 1618 [55,56], because several of the pesticides are unstable at extreme pH conditions. Furthermore, Methods 608 and 1618 provide lower detection limits needed for pesticide and PCB levels typically found in sludge.

Sampling and Preservation

Sludge samples are collected in wide mouth jars with Teflon-lined caps and refrigerated at 4°C. If residual chlorine is present in aqueous samples, add 80 mg sodium thiosulfate per liter. Samples should be extracted within seven days of collection and extracts should be analyzed within forty days of extraction.

Extraction

In U.S.EPA Method 625, base/neutrals are extracted at pH ≥12. The same aqueous phase is pH adjusted to ≤2 and re-extracted to give the acid extractables. A separate sample is needed for extraction of pesticides and PCBs at pH 5–9, according to Method 608. Methylene chloride is used as solvent in all cases.

Separatory Funnel Extraction

Sludge may be diluted with reagent water, and extracted in a separatory funnel, as described in EPA Methods 608, 1618, and 625.

Direct Agitation with Solvent

The sludge sample may be taken as is in a Teflon or glass bottle, mixed with solvent, and shaken by hand or a shaker, or agitated by an ultrasonic mixer.

Continuous Liquid/Liquid Extraction

The sludge samples may be mixed with reagent water and extracted in a continuous liquid/liquid extractor (usually overnight).

Soxhlet

The sample is mixed with sodium sulfate or magnesium sulfate monohydrate, ground into fine particles, and Soxhlet extracted as required. This method is not used for base/neutral and acid extractables (BNAs) as there is no good way to control or change the pH during the extraction.

Steam Distillation

The sample is mixed with reagent water and extracted using a Nielsen-Kryger continuous steam distillation apparatus. This method is only used for PCBs, which are stable to these conditions [57].

Problems in Extractions

Emulsion formation is a significant problem in liquid/liquid extractions, especially with the extremes of pH required for acid and base-neutral extractables. Centrifugation can help break these emulsions. Some heat-sensitive analytes may be lost during Soxhlet extraction or steam distillation, due to the high refluxing temperatures required. Steam distillation does have the advantage of discriminating against the oil and grease materials which have high molecular weight.

Cleanup

Cleanup of extracts is usually required to remove interferences, which can range from oil and grease materials having high molecular weight, nonvolatile compounds, particulates, and substances that will coelute from the GC columns with desired analytes, such as elemental sulfur. Some cleanup methods include gel permeation chromatography, polar-nonpolar separations, and elemental sulfur. Gel permeation chromatography (GPC) separates compounds by molecular size, larger molecules eluting before smaller [58] . It is normally done on columns packed with styrene-divinyl-benzene resins, with pore sizes (for GPC cleanup) of 50 μm or 100 μm. The mobile phase is methylene chloride [59]. The elution order is triglycerides—high MW oils—phthalate esters—PAHs—pesticides and PCBs—small molecules. Elemental sulfur interacts with the resin and is retained past the small molecules, giving an effective cleanup of this interference.

Polar–nonpolar separations may be used, as it is extremely difficult to get complete separation of organochlorine pesticides from each other, and from PCB mixtures. It is useful to divide the PCB/pesticide extract into nonpolar (PCB, aldrin, DDE, DDT, heptachlor) and polar (other organochlorine pesticide) fractions. This can be accomplished using disposable Florosil or silica gel columns, or HPLC silica gel columns in a conventional HPLC instrument [55]. Elution of the desired analyte fractions is done by increasing the polarity of eluting solvent.

Elemental sulfur is naturally found in domestic sewage sludge. It is extremely electron-capture sensitive, and will obscure pesticide and PCB peaks unless it is removed. If GPC on styrene-divinylbenzene columns is not used, sulfur may be removed with metallic mercury or activated copper powder, or by reacting with tetrabutylammonium sulfite.

Problems in cleanup include particulates that will clog cleanup columns unless they are removed by filtration or centrifugation. Each batch of Florosil must be calibrated before use, and the material must be protected from ambient moisture.

Identification and Quantitation, Base/Neutral and Acid Extractables (BNAs), GC/MS Methods

BNAs are analyzed using U.S.EPA Method 625 or 1625 isotope dilution, capillary GC/MS. Base/neutral and acid extracts are separately dried over sodium sulfate, concentrated, and cleaned up using gel permeation chromatography. The acid and base/neutral extracts are combined and injected into the GC after the addition of internal standard. The compounds are separated by GC, detected, and quantitated by the MS as mentioned under volatile compounds.

Occasionally, the acid extracts and the base/neutral extracts have to be injected separately into the GC/MS. This is because injecting combined acid and base/neutral fractions gives a very complicated chromatogram due to heavy matrix interferences. Consequently, each extract is separately quantitated for all the compounds. Final quantitation is based on the sum of values found in both fractions. This is because some base/neutrals are found in the acid fraction and some acids in the base/neutral extract.

Non-Mass Spectrometric Methods

Acid extractables can also be analyzed by GC/FID (U.S.EPA Method 604). The eleven phenols in this method are individually separated and quantitated, unlike colorimetric methods where a total value is estimated. Also, no correlation exists between the colorimetric methods and the sum of the GC obtained phenol values.

Most of the base/neutrals can be analyzed as subgroups in specific GC methods (if not, then by liquid chromatography). Certain phthalate esters can be analyzed by GC/ECD (U.S.EPA Method 606); some nitroaromatics by GC/ECD and GC/FID (U.S.EPA Method 609); polynuclear aromatic hydrocarbons by GC/FID (U.S.EPA Method 610); halo ethers by GC/HECD (U.S.EPA Method 611); and chlorinated hydrocarbons by GC/ECD (U.S.EPA Method 612). When these GC methods are used, compound identification is confirmed by retention time in a second GC column having a different stationary phase.

Organochlorine Pesticides and PCBs

Analysis may be done using either packed (2 M × 2 mm i.d.) or capillary (15 to 30 M) columns. If the MS detector is not used, identification of the single component pesticides is done by elution time, and confirmed by rerunning on a column of different polarity. Multi-component pesticides (chlordane, toxaphene) and PCBs may be identified by pattern matching with standards [55,56]. The ECD has the highest sensitivity, but is subject to interferences from nonhalogenated, electron-capture, sensitive species. Required limits of detection usually mandate this detector.

The HECD will respond only to halogens, but is at least an order of magnitude less sensitive than the ECD. Optimum conditions for converting PCBs are not the same as for other halogenated species. The MS detector has the advantage that identification and quantitation are done simultaneously, but sensitivity is poorer than the ECD.

Quantitation of single-component pesticides is normally done by either internal or external standardization methods. Choice of an internal standard is dependent on the sample matrix and must be left to the analyst.

PCBs may be quantitated either as total PCB or individual congeners. Identification of a particular PCB pattern can be problematic if the sample has been subject to environmental degradation. Summing individual peaks really requires use of a high-resolution capillary column, and a knowledge of the response factor for each congener. Summing the total area under the "Aroclor area" for a particular Aroclor type or the Webb-McCall method may be used effectively for PCB type [60].

Quality Assurance

The quality assurance/quality control program has to be performed following U.S.EPA protocol as detailed in the various methods. MDL is based on analyzing seven replicates of 1 L reagent water, spiked with the analytes under investigation (calculated as mentioned under "Volatiles"). However, MDL for real samples are usually dependent on sample size and the level of interferences.

TCLP TESTING

In 1976, the U.S. Congress enacted the Resource Conservation and Recovery Act (RCRA) to provide a federal jurisdiction over solid waste and resource management and recovery. Its primary goals were to control hazardous waste to protect the environment and human health and to protect and preserve the nation's natural resources through conservation and recovery. In order to control hazardous waste, RCRA mandated that systems be established to identify, trace, and control hazardous waste movement from initial generation to transportation, treatment, storage, and final disposal. These systems were mandated under various parts of RCRA. Part 261, for instance, directs the U.S.EPA to establish ways of determining what waste materials are considered hazardous for regulation. Part 262 then requires solid waste generators to determine if their wastes are indeed hazardous. Solid waste is defined as any material that is abandoned or disposed of, burned, incinerated, or simply stored. It includes all forms of waste: solids such as wastewater treatment sludges, liquids, semi-solids, and contained gaseous materials. It does not, however, include domestic sewage or any mixture of domestic sewage and other wastes.

Every POTW producing a sludge final product must test that sludge to determine if it is a hazardous waste. There are four characteristics that define a hazardous waste, any one of which can cause a sludge to be classified as a hazardous waste.

(1) Ignitability, defined as (for sludges) capable of causing fire through friction, adsorption of moisture, or spontaneous chemical changes under standard temperature and pressure conditions

(2) Corrosivity, meaning an aqueous waste with pH less than or equal to 2 or greater than or equal to 12.5, or a liquid that corrodes steel at a specified rate

(3) Reactivity, the generation of toxic gases, vapors, or fumes when mixed with water; this included cyanide and sulfide, which are present in some sludges

(4) Extraction procedure toxicity (EP), a test used to determine if leachate from disposing of a waste will pollute groundwaters

Most municipal wastewater treatment sludges will have no difficulty passing the first three characteristics. Experience has shown that the EP test generally has caused no problems, either. It is based on analysis of a sludge extraction for eight materials and six pesticides, none of which can exceed specified concentrations.

In 1984, the U.S. Congress enacted the Hazardous and Solid Waste Amendments to RCRA, which required U.S.EPA to examine and revise the EP test. U.S.EPA proposed a revised leachate procedure test known as the Toxicity Characteristic Leachate Procedure (TCLP) which added thirty-eight chemical compounds to the EP list. The final rule was enacted on March 29, 1990 with twenty-five of the thirty-eight proposed additional compounds included. The remaining thirteen constituents will likely be added at a later date.

The regulatory levels are based on human health concentration thresholds and a dilution/attenuation factor specific for each constituent. The concentration threshold indicates adverse affects on human health, while the dilution/attenuation factor indicates the leachate potential into groundwater used for drinking water supplies.

The regulatory levels were determined by multiplying the health based number by a generic dilution/attenuation factor of 100. The TCLP procedure is shown schematically in Figure 3.1 [61].

Approximately 20% of the sludge distributed by the MWRDGC is sent to landfills. Generally this sludge originates from lagoons and is air-dried to remove free water before transportation to a landfill. Table 3.28 shows the regulatory limits for TCLP constituents and the results of the TCLP test on typical sludges from the MWRDGC in 1994. The sludges originated from treatment plants with moderate to heavy industrial input. As may be noted, the constituents were all well within the TCLP regulatory limits. The MWRDGC has never had a sludge fail either the older EP test or the recent TCLP test.

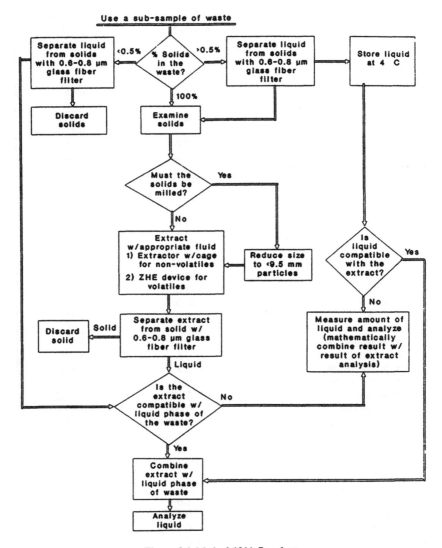

Figure 3.1 Method 1311 flowchart.

TABLE 3.28. TCLP Regulatory Limits and TCLP Analysis of MWRDGC Sludges Sampled in 1994.

Constituent	TCLP Regulatory Limit (mg/L)	MWRDGC Sludge Analysis TCLP Extract (mg/L)
As	5.0	<0.1
Ba	100.0	<0.3
Cd	1.0	0.14
Cr	5.0	0.05
Pb	5.0	<0.08
Hg	0.2	<0.0003
Se	1.0⁻	<0.1
Ag	5.0	<0.01
Endrin	0.02	<0.001
Methoxychlor	10.0	<0.005
2,4-D	10.0	<0.005
Lindane	0.4	<0.0004
Toxaphene	0.5	<0.005
2,4,5-TP	1.0	<0.005
Benzene	0.5	<0.001
Carbon tetrachloride	0.5	<0.001
Chlordane	0.03	<0.003
Chlorobenzene	100.0	<0 001
Chloroform	6.0	<0.001
o-Cresol	200 0	<0.05
m-Cresol	200.0	<0.05
p-Cresol	200.0	<0.05
1,4-Dichlorobenzene	7.5	<0.02
1,2-Dichloroethane	0.5	<0.001
1,1-Dichloroethylene	0 7	<0.001
2,4-Dinitrotoluene	0.13	<0.01
Heptachlor	0 008	<0.0002
Heptachlor epoxide	0.008	<0.0002
Hexachlorobenzene	0.13	<0.02
Hexachloro-1,3-butadiene	0.5	<0.01
Hexachloroethane	3.0	<0 02
Methyl ethyl ketone	200.0	0.006
Nitrobenzene	2.0	<0.04
Pentachlorophenol	100.0	<0.14
Pyridine	5.0	<0.05
Tetrachloroethylene	0.7	<0.002
Trichloroethylene	0.5	<0.001
2,4,5-Trichlorophenol	400.0	<0.04
2,4,6-Trichlorophenol	2.0	<0.01
Vinyl chloride	0.2	<0.001

REFERENCES

1 Natural Resources Defense Council, Inc. v. Train, 8 ERC 2120, 2122–29, 1976.

2 U.S. Code of Federal Regulations. Title 40, 40 CFR, Chapter 2, Part 123.21, App. D, July 1986.

3 Davis III, J. A. and J. Jacknow. 1975. "Heavy Metals in Wastewater in Three Urban Areas," *Jour. WPCF*, 47:2293.

4 Gurnham, C. F. et al. 1979. *Control of Heavy Metal Content of Municipal Wastewater Sludge*. Chicago, Illinois: Gurham and Associates.

5 Klein, L. A. et al. 1974. "Sources of Metals in New York City Wastewaters," *Jour. WPCF*, 46:2653.

6 Bryan, E. H. 1974. "Concentrations of Lead in Urban Stormwaters," *Jour. WPCF*, 46:2419.

7 Minear, R. A., R. O. Ball and R. L. Church. 1981. *Data Base for Influent Heavy Metals in Publicly Owned Treatment Works*. U.S.EPA, Municipal Environmental Research Laboratory, Cincinnati, Ohio, EPA-600/2-81-220.

8 U.S.EPA. 1982. *Fate of Priority Pollutants in Publicly Owned Treatment Works—Final Report, Vol. 1*. Effluent Guidelines Division, Washington, DC, EPA 440/1-82/303.

9 Lordi, D. T. et al. 1980. "Cyanide Problems in Municipal Wastewater Treatment Plants," *Jour. WPCF*, 3:597.

10 Feeney, S. et al. 1988. *Cyanide in Wastewater and Sludges of the Metropolitan Sanitary District of Greater Chicago*, Research and Development Department, Chicago, Illinois.

11 U.S.EPA, *Federal Guidelines—State and Local Pretreatment Programs*. Municipal Construction Division, Washington, DC, EPA 430/9-76-017a.

12 Lester, J. N., R. M. Harrison and R. Perry. 1979. "The Balance of Heavy Metals through a Sewage Treatment Works—I Lead, Cadmium and Copper," *Science of Total Environment*, 12:13.

13 Mytelka, A. I. et al. 1973. "Heavy Metals in Wastewater and Treatment Plant Effluents," *Jour. WPCF*, 45:1859–1864.

14 Barth, E. F. et al. 1965. "Summary Report on the Effects of Heavy Metals on Biological Treatment Processes," *Jour. WPCF*, 37:86.

15 Oliver, B. G. and E. G. Cosgrove. 1974. "The Efficiency of Heavy Metal Removals by a Conventional Activated Sludge Treatment Plant," *Water Research*, 8:869.

16 Kurz, G. E., D. A. Summers and E. G. Wright. 1981. *Proceedings of the 35th Industrial Waste Conference*, May 13–15, 1980, Purdue University, Lafayette, Indiana.

17 Barth, E. F. et al. 1964. "Effects of a Mixture of Heavy Metals on Sewage Treatment Processes," *Proceedings of the 18th Industrial Waste Conference*, April 30–May 2, 1963, Purdue University, Lafayette, Indiana, p. 48.

18 Petrasek, A. C. and I. J. Kugelman. 1983. "Metals Removals and Partitioning in Conventional Wastewater Treatment Plants," *Jour. WPCF*, 55:1183.

19 Patterson, J. W. and P. S. Kodulkula. 1984. "Metals Distribution in Activated Sludge Systems," *Jour. WPCF*, 56:433.

20 Hannah, S. A. et al. 1986. "Comparative Removal of Toxic Pollutants by Six Wastewater Treatment Processes," *Jour. WPCF*, 58:27.

21 Neufeld, R. D. and E. R. Herman. 1975. "Heavy Metal Removal by Acclimated Activated Sludge," *Jour. WPCF*, 47:310.

22 Cheng, M. H., J. W. Patterson and R. A. Minear. 1975. "Heavy Metals Uptake by Activated Sludge," *Jour. WPCF, 47*:363.

23 Anthony, R. M. and L. H. Breimhurst. 1981. "Determining Maximum Influent Concentrations of Priority Pollutants for Treatment Plants," *Jour. WPCF, 53*:1457.

24 Petrasek, A. C. et al. 1983. "Fate of Toxic Organic Compounds in Wastewater Treatment Plants," *Jour. WPCF, 55*:1286.

25 U.S.EPA. 1983. *Treatability Manual.* EPA 600/2-82-00/a.

26 Tabak, H. H. et al. 1981. "Biodegradability Studies with Organic Priority Pollutant Compounds," *Jour. WPCF, 53*:1503.

27 U.S.EPA. 1982. *Fate of Priority Toxic Pollutants in Publicly Owned Treatment Works 30-Day Study.* EPA 440/1-82/302, Effluent Guidelines Division, Washington, DC.

28 Kusch, E. J. and J. E. Etzel. 1974. "Microbial Decomposition of Pentachlorophenol," *Jour. WPCF, 45*:359.

29 Fank, B. E. and P. J. A. Fowlie. 1980. "Treatment of a Wood Preserving Effluent Containing Pentachlorophenol by Activated Sludge and Carbon Adsorption," *Proceedings of the 34th Industrial Waste Conference,* May 1979, Purdue University, Lafayette, Indiana, p. 63.

30 Stover, E. L. and D. F. Kincannon. 1983. "Biological Treatability of Specific Organic Compounds in Chemical Industry Wastewater," *Jour. WPCF, 55*:97.

31 Mumma, R. O., et al. 1984. "National Survey of Elements and other Constituents in Municipal Sewage Sludges," *Archives Environmental Contamination and Toxicology, 13*:75.

32 AMSA. 1987 (unpublished report). "Sludge Analysis Survey of 59 Member Agencies," Association of Municipal Sewerage Agencies, Washington, DC.

33 ASTM. 1990. *Annual Book of Standards, Vol. 11.01,* D-4374 Standard Test Method, "Cyanide in Water—Automated Methods for Total Cyanide and Dissociable Cyanide," Philadelphia, PA: American Society for Testing and Materials.

34 Kelada, N. P. 1989. "Automated Direct Measurement of Total Cyanide Species and Thiocyanate and Their Distribution in Wastewater and Sludge," *Jour. WPCF, 61*:350.

35 Metropolitan Water Reclamation District of Greater Chicago. 1989. Sewage and Waste Control Ordinance, Chicago, IL, pp. 8–9.

36 Hampton Roads Sanitation District. 1990. *Industrial Wastewater Discharge Regulations,* Virginia Beach, VA. pp. 6–8, Appendix D.

37 Metropolitan Waste Control Commission. 1981. *Waste Discharge Rules for the Metropolitan Disposal System.* St. Paul, MN, pp. 7–10.

38 Municipality of Anchorage. 1989. Anchorage Municipal Code, Anchorage, Alaska, 26.50, 022–023.

39 Dallas Water Utilities. 1993. Dallas City Code Chapter 49 "Water and Wastewater," Dallas, TX, pp. 15–18.

40 East Bay Municipal Utility District. 1990. An Ordinance Establishing Regulation for the Interception, Treatment, and Disposal of Wastewater and Industrial Wastes and the Control of Wastewater, Requiring Charges to Be Made Therefore, and Fixing Penalties for the Violation of Said Regulations, pp. 6–7.

41 Metropolitan Water Reclamation District of Greater Chicago. 1989. Sewage and Waste Control Ordinance, Chicago, IL, p. 8.

42 United States Environmental Protection Agency. 1991. 40 Code of Federal Regulations, 433.15–433.16, Washington, DC.

43 United States Environmental Protection Agency. 1991. 40 Code of Federal Regulations, 414.85, Washington, DC.

44 United States Environmental Protection Agency. 1990. "Effluent Guidelines Plan," *Federal Register,* 55(i):30–103.

45 Metropolitan Water Reclamation District of Greater Chicago. 1989. *Enforcement Response Procedure,* Chicago, IL, pp. 2–6.

46 Metcalf & Eddy, Inc. 1972. *Wastewater Engineering: Treatment, Disposal, Reuse,* Revised Edition. New York: McGraw-Hill Publisher.

47 U.S.EPA. 1979. *Process Design Manual for Sludge Treatment and Disposal.* Center for Environmental Research Information, Cincinnati, OH, EPA 625/1-79/011.

48 Vesilind, P. 1979. *Treatment and Disposal of Wastewater Sludges,* Revised Edition. Ann Arbor, MI.

49 Weber, W. 1972. *Physiochemical Processes.* New York: John Wiley & Sons, Inc.

50 WPCF/MOP OM-8. 1987. *Operation and Maintenance of Sludge Dewatering Systems Manual of Practice.* Water Pollution Control Federation, Alexandria, VA.

51 Tchobanoglous, G., H. Theisen, R. Eliassen. 1977. *Solid Wastes.* New York: McGraw-Hill Publisher.

52 ASTM. 1971. *Annual Book of ASTM Standards, Part 19.* Philadelphia, PA: American Society for Testing and Materials.

53 ASTM. *Annual Book of Standards, Vol 11.01,* D-2036 Standard Test Method, "Cyanide in Water," Philadelphia, PA: American Society for Testing and Materials, Philadelphia, PA.

54 Clayton, F. J., N. P. Kelada and C. Lue-Hing. 1991. "Methodology for Automated Phenol Analysis," R&D Report No. 91-26, Metropolitan Water Reclamation District of Greater Chicago.

55 U.S.EPA. 1984. *Federal Register* 49, No. 209, U.S.EPA 40 CFR Part 136, "Guidelines Establishing Test Procedures for the Analysis of Pollutants under the Clean Water Act: Appendix A to Part 136—Methods for Organic Chemical Analysis of Municipal and Industrial Wastewater (Methods 601–625, 1624A and 1625A); Appendix B to Part 136—Definition and Procedure for the Determination of the Method Detection Limit—Revision 1.11."

56 U.S.EPA. 1991. "Analytical Methods for the National Sewage Sludge Survey," Contract No. 68-C9-0019, Methods 1618, 1624C and 1625C.

57 Veith, G. D. and L. M. Kieus. 1977. "An Exhaustive Steam-Distillation and Solvent-Extraction Unit for Pesticides and Industrial Chemicals," *Bull. Environmental Contamination and Toxicology,* 17:631.

58 Stalling, D. L., R. C. Tindle and J. L. Johnson. 1972. "Cleanup of Pesticide and Polychlorinated Biphenyl Residues in Fish Extracts by Gel Permeation Chromatography," *J. Assoc. Offic. Anal. Chem.,* 55:32.

59 Kuehl, D. W. and E. N. Leonard. 1978. "Isolation of Xenobiotic Chemicals from Tissue Samples by Gel Permeation Chromatography," *J. Anal. Chem.,* 50:182.

60 Webb, R. G. and A. C. McCall. 1973. "Quantitative PCB Standards for Electron Capture Gas Chromatography," *J. Chromatogr. Sci.,* 11:366.

61 Environmental Protection Agency, 40 CFR Part 261, et al., Hazardous Waste Management System; Identification and Listing of Hazardous Waste; Toxicity Characteristic Revisions; Final Rule. *Federal Register,* March 29, 1990: 11864.

62 U.S.EPA. 1990. "National Sewage Sludge Survey: Availability of Information and

Data, and Anticipated Impacts on Proposed Regulations,'' *Federal Register* 55(48): 47210–47283.

63 AMSA. 1997. A Sludge Analysis Survey of Member Agencies in 1996. Association of Metropolitan Sewerage Agencies, Washington, DC.

CHAPTER 4

Microbiology of Sludge

THIS chapter will describe the microbial content of sewage. The levels of microorganisms will be followed through the various treatment processes resulting in the production of sludge. The sludge treatment processes will then be examined with respect to their effect on the levels of pathogenic microorganisms prior to the ultimate disposal or utilization of the sludge. Finally, human risk will be assessed based upon a modeling of the environmental movement of the pathogens, data on the doses required for human infection and disease, and exposure to sludge-affected environments by humans during or after the ultimate disposition of municipal sludge. Human risk will also be examined in terms of the available epidemiological data.

INTRODUCTION

Wastewater or sewage is produced by everyday human activities, and by a variety of industrial and agricultural practices. This wastewater is essentially the used water of a community, plus infiltration water and storm water. It may contain organic compounds, inorganic nutrients such as carbon, nitrogen, and phosphorus, trace elements, heavy metals, toxic organic and inorganic substances, disease-producing (pathogenic) microorganisms and other substances that without appropriate treatment, can impair the quality of receiving streams and lakes, making them toxic to aquatic flora and fauna and vehicles for the transmission of disease.

James J. Bertucci and Salvador J. Sedita, Metropolitan Water Reclamation District of Greater Chicago, Chicago, IL.

The objectives of wastewater or sewage treatment are: (1) the immediate and nuisance-free removal of the sewage from its sources of generation, (2) appropriate treatment to remove undesirable materials, and (3) disposal. Ultimately, the goal is the protection of the environment, including the public health, in a manner commensurate with economic, social, and political concerns.

Treatment and disposal of domestic sewage involve a variety of physical, chemical, and biological unit processes. Among the most important of the biological processes are: (1) the activated sludge process, (2) nitrification denitrification processes, and (3) anaerobic digestion. The main driving force in all of these processes is the stabilization (oxidation) of organic substances by microorganisms.

Microorganisms such as bacteria, fungi, algae, helminths, protozoa, and viruses can be present in wastewaters. Many types are capable of proliferating in wastewater during some stage(s) of the treatment train, or in suitably polluted receiving waters.

MICROORGANISMS FOUND IN WASTEWATER

The general types of microorganisms that can be found in most wastewaters reflect three general categories of microorganisms, those which: (1) generally participate in the treatment process or "biofloc" formation, (2) may cause problems in the treatment process, and (3) do not participate in the treatment process and are just present or "along for the ride." The latter category includes most pathogenic bacteria, viruses, and higher parasites.

Microorganisms found in wastewater usually can be characterized by their hardiness since they must survive a variety of hostile environments including the digestive tracts of man or animals, and the fluctuations of temperature, pH, ionic strength, sunlight, nutrient limitations, etc., in the soils or waters from which they originate.

FUNCTIONS OF MICROORGANISMS IN MAJOR BIOLOGICAL TREATMENT PROCESSES

Table 4.1 lists the principal biological treatment processes and their major uses or functions in the treatment train. Table 4.2 lists the major function or use and the types of microorganisms most often associated with these functions.

Most biological treatment processes remove the biomass or microorganisms present, not so much by destruction, although some lysis of biomass does occur, but by concentration into the sludge process streams. The concentrations of all types of microorganisms in sludge will thus be much

TABLE 4.1. Major Biological Treatment Processes Used for Wastewater Treatment.

Type	Common Name	Use[a]
Aerobic processes:		
Suspended growth	Activated sludge processes	
	Conventional (plug flow)	
	Continuous-flow stirred tank	
	Step aeration	
	Pure oxygen	
	Modified aeration	Carbonaceous BOD removal (nitrification)
	Contact stabilization	
	Extended aeration	
	Oxidation ditch	
	Suspended growth nitrification	Nitrification
	Aerated lagoons	Carbonaceous BOD removal (nitrification)
	Aerobic digestion	
	Conventional air	
	Pure oxygen	Stabilization, carbonaceous BOD removal
	High-rate aerobic algal ponds	Carbonaceous BOD removal
Attached growth	Trickling filters	
	Low-rate	Carbonaceous BOD removal (nitrification)
	High-rate	Carbonaceous BOD removal
	Roughing filters	Carbonaceous BOD removal (nitrification)
	Rotating biological contactors	Nitrification
	Packed bed reactors	Carbonaceous BOD removal (nitrification)
Combined processes	Trickling filter, activated sludge, trickling filter	
Anoxic processes.		
Suspended growth	Suspended growth denitrification	Denitrification
Attached growth	Fixed film denitrification	

(continued)

TABLE 4.1. (continued).

Type	Common Name	Use[a]
Anaerobic processes:		
Suspended growth	Anaerobic digestion	
	Standard-rate, single-stage	
	High-rate, single-stage	Stabilization, carbonaceous BOD removal
	Two-stage	
Attached growth	Anaerobic contact process	Carbonaceous BOD removal
	Anaerobic filter	Carbonaceous BOD removal, stabilization (denitrification)
	Anaerobic lagoons (ponds)	Carbonaceous BOD removal (stabilization)
Aerobic anoxic or anaerobic processes:		
Suspended growth	Single-stage nitrification-denitrification	Carbonaceous BOD removal, nitrification, denitrification
Attached growth	Nitrification-denitrification	Nitrification, denitrification
Combined process	Facultative lagoons (ponds)	Carbonaceous BOD removal
	Maturation or tertiary (ponds)	Carbonaceous BOD removal (nitrification)
	Anaerobic-facultative lagoons	
	Anaerobic-facultative aerobic lagoons	Carbonaceous BOD removal

[a]Major use is presented first; other uses are identified in parentheses (Metcalf & Eddy, 1979 [1]).

164

TABLE 4.2. Microorganisms Associated with Function(s) of Major Biological Treatment Processes.

Type of Process and Function	Types of Microorganisms Most Often Associated with Function or Use
Aerobic: Carbonaceous BOD removal	*Zooglea ramigera, Pseudomonas species, Alcaligenes faecalis, Achromobacter species, Microbacterium, Brevibacterium, Flavobacterium, Nocardia, Bdellovibrio, Acinetobacter, Sphaerotilis, Beggiatoa, Thiothrx, Leucothrix,* protozoans, algae, yeast, fungi
Nitrification	Nitrifiers (*Nitrisomonas, Nitrobacter, Nitrosolobus, Nitrosococcus*)
Anoxic processes: Denitrification	*Paracoccus denitrificans, Denitrobacillus, Pseudomonas species, Aquaspirillum iterosonii, Bacillus licheniformis, Alcaligenes species, Thiobacillus denitrificans, Thiobacillus thioparus,* Enterobacteriaceae, *Azospirillum brasilene, Agrobacterium species, Bacillus species, Chromobacterium species, Corynebacterium species*
Anaerobic processes: Stabilization, carbonaceous BOD removal	Heterotrophic facultative bacteria, methane bacteria (*Archaebacteria*)
Denitrification	(See denitrification under anoxic processes, above)
Aerobic/anoxic or anaerobic processes: Carbonaceous BOD removal	Heterotrophic facultative bacteria, methane bacteria (*Archaebacteria*) protozoans, algae, yeast, fungi
Nitrification-Denitrification	*Nitrosomonas, Nitrosolobus, Nitrosococcus, Nitrobacter*—variety of dentrifiers (see above)

Source: Metcalf & Eddy, 1979; Jones et al., 1987, Doelle, 1975, Moat, 1979, Curds, 1982; Taber, 1976 [1–6]

greater than in the original wastewater because the sludge volume is so much smaller than the volume of the wastewater from which it is derived. Raw sludge will therefore contain the same infectious agents found in sewage, but in significantly higher concentrations. Table 4.3 is a list of some of the pathogenic organisms that may be found in sewage and sludge and the symptoms or disease entities with which they are associated. It is the presence of these types of organisms that elicits such concern for sludge utilization or disposal practices among prudent public health practitioners.

To facilitate further handling, primary and secondary sludge is usually treated by one or more water removal processes to concentrate the solids, thus also further concentrating any microorganisms present. These processes also stabilize the sludge to inhibit, reduce, or eliminate the potential for putrefaction and offensive odors, and to reduce pathogens. In fact, the commonly used processes such as anaerobic digestion, aerobic digestion, composting, heat treatment, and lagooning may dramatically reduce the pathogen content of sludge. Certain sludge treatment or management practices, however, may encourage the colonization of the sludge by pathogenic fungi such as *Aspergillus fumigatus.*

Several methods of ultimate municipal sludge utilization or disposal are currently practiced. These include: (1) land application, (2) distribution and marketing, (3) landfilling, (4) incineration, and (5) ocean disposal. In all of these except incineration, there is some slight risk to humans from exposure to the pathogenic microorganisms that these sludges may contain.

Several modes of pathogen transmission are possible when considering the landfilling of sludge or its application to land. Agents such as the hepatitis A virus, released by ocean disposal, may be concentrated from seawater by shellfish, such as oysters, which may be consumed by humans. Aerosols, which may contain pathogens, could be generated if sludge is applied to land by spraying. Runoff and percolation from landfilled sludge or sludge-treated fields may convey pathogens to surface water or groundwater. People may come into direct contact with sludge in the process of landfilling or field application. Finally, pathogens could become associated with edible crops grown on sludge-amended fields. Mechanisms of pathogen transmission from sludges to man will be considered in more detail in the following sections of this chapter.

MICROBIAL CONTENT OF RAW SEWAGE AND FINAL EFFLUENTS

Sewage by its very nature and origin acts as a repository for microorganisms of different types. Feces brings with it the intestinal flora; the soil, its autochthonous and zymogenous flora; the air, its noncharacteristic flora; and water, its own typical flora. Not all of the microorganisms that enter sewage survive, however, and sewage soon acquires a characteristic flora

TABLE 4.3. Principal Pathogens of Concern in Municipal Wastewater and Sludge.

Organism	Disease/Symptoms
Bacteria	
Salmonella spp	Salmonellosis (food poisoning), typhoid fever
Shigella spp.	Bacillary dysentery
Yersinia spp	Acute gastroenteritis (including diarrhea, abdominal pain)
Vibrio cholerae	Cholera
Campylobacter jejuni	Gastroenteritis
Escherichia coli (pathogenic strains)	Gastroenteritis
Viruses	
Poliovirus	Poliomyelitis
Coxsackievirus	Meningitis, pneumonia, hepatitis, fever, common colds, etc
Echovirus	Meningitis, pneumonia, encephalitis, fever, common colds, diarrhea, etc.
Hepatitis A virus	Infectious hepatitis
Rotavirus	Acute gastroenteritis with severe diarrhea
Norwalk agents	Epidemic gastroenteritis with severe diarrhea
Reovirus	Respiratory infections, gastroenteritis
Protozoa	
Cryptosporidium	Gastroenteritis
Entamoeba histolytica	Acute enteritis
Giardia lamblia	Giardiasis (including diarrhea, abdominal cramps, weight loss)
Balantidium coli	Diarrhea and dysentery
Toxoplasma gondii	Toxoplasmosis
Helminth worms	
Ascaris lumbricoides	Digestive and nutritional disturbances, abdominal pain, vomiting, restlessness
Ascaris suum	May produce symptoms such as coughing, chest pain, and fever
Tichuris trichiura	Abdominal pain, diarrhea, anemia, weight loss
Toxocara canis	Fever, abdominal discomfort, muscle aches, neurological symptoms
Taenia saginata	Nervousness, insomnia, anorexia, abdominal pain, digestive disturbances
Taenia solium	Nervousness, insomnia, anorexia, abdominal pain, digestive disturbances
Necator americanus	Hookworm disease
Hymenolepis nana	Taeniasis

Source U S EPA/625/10-89/006 [7]

TABLE 4.4. Typical Microbial Composition of Domestic Wastewater.

Water Quality	Bacteria (#/mL)	Viruses (PFU or MPN/L)	Other Agents
Raw wastewater	Total $(6.8 \times 10^8)^a$ Viable $(1.4 \times 10^7)^a$ TC $(10^5–10^6)^b$ FC $(10^4–10^5)^b$	$35,000^c$ 5000^d $6000–60,000^e$ $25–2600^f$	
Primary effluent	TC $(10^5–10^6)^b$ FC $(10^4–10^5)^b$	$52,000^c$	
Activated sludge tank (MLSS)	Tank $(6.6 \times 10^9)^a$ Viable $(5.6 \times 10^7)^a$ 2° Total $(5.2 \times 10^7)^a$ Viable $(5.7 \times 10^5)^a$		(Protozoans cells/mL) $50,000^g$
Final effluent	2° Filtered TC $(10^2–10^3)^b$ FC $(10^1–10^3)^b$ 2° Nitrified TC $(10^2–10^3)^b$ FC $(10^1–10^3)^b$ 3° Total $(3.4 \times 10^7)^a$ Viable $(4.1 \times 10^4)^a$ Filtered/Nitrified TC $(10^2–10^3)^b$ FC $(10^1–10^5)^b$	2000^c $0.5–10,000^b$ $0.026–0.50^f$	

[a] Pike and Curds, 1971 [8]
[b] U.S.EPA/625/1-86/021 [9]
[c] Berg, 1966 [10]
[d] Clarke et al 1964 [11]
[e] Buras, 1974 [12]
[f] Riggs and Spath, 1983 [13]
[g] Curds, 1973 [14].
2° = secondary
3° = tertiary

168

of its own. The nature of this flora will be conditioned by the character of the sewage, with industrial wastes having a flora different from purely domestic sewage. This selection process goes a step further within the confines of the various biological methods of sewage processing (purification), resulting in a highly specialized flora, able to function properly only under the conditions that favored its selection. We have, thus, a striking example of natural selection at work.

It is difficult to identify, accurately, the general characteristics of municipal wastewater due to regional differences, water uses, seasonal and diurnal variations, etc. Typical microbial compositions of wastewater through various levels of treatment are summarized in Table 4.4. As can be seen, the concentrations of microorganisms decrease by approximately three orders of magnitude through a typical process train.

The total bacterial population of human feces has been estimated at 10^{12} organisms per gram [10]. Within this enormous number of organisms, the density range of fecal coliforms was estimated at 10^6 to 10^9 per gram, and total coliforms at 10^7 to 10^9 per gram. The concentration ranges of certain organisms in domestic wastewater, and their reductions through primary and secondary treatment, and their estimated concentrations in treated secondary effluents are summarized in Tables 4.5, 4.6, and 4.7.

Pike and Curds [8] (Table 4.4), estimated the total and viable bacterial counts in raw sewage, activated sludge mixed-liquor, and secondary and tertiary treated nonchlorinated effluents. The total counts were 6.8×10^8, 6.6×10^9, 5.2×10^7, and 3.4×10^7 organisms per milliliter, respectively. The viable counts were 1.4×10^7, 5.6×10^7, 5.7×10^5, and 4.1×10^4 organisms/ milliliter, respectively, for the same set of samples. These results indicate that viable counts of the bacterial populations in wastewater may underestimate the actual numbers present by two orders of magnitude or more. This, coupled with the fact that these estimates only account for aerobic heterotrophic organisms means that the anaerobic population has been totally ignored.

Table 4.5 shows typical influent or raw sewage concentration ranges

TABLE 4.5. Typical Concentration Ranges for Viruses and Indicator Organisms in Raw Sewage.

Organism	Number/100 mL	
	Minimum	Maximum
Total coliforms	1,000,000	—
Fecal coliforms	340,000	49,000,000
Fecal streptococci	64,000	4,500,000
Viruses	0.5	10,000

Source: From U S EPA/625/1-86/021 [9]

TABLE 4.6. Microorganism Reductions by Conventional Treatment Processes.

Microorganisms	Primary Treatment Removal (%)	Secondary Treatment Removal (%)
Total coliforms	< 10	90–99
Fecal coliforms	35	90–99
Shigella sp.	15	91–99
Salmonella sp.	15	96–99
Escherichia coli	15	90–99
Viruses	< 10	76–99
Entamoeba histolytica	10–50	10

Source From U S EPA/625/1-86/021 [9]

for some indicator bacteria and viruses and Table 4.6 shows the removal efficiencies for similar organisms by primary and secondary treatment processes. It must be remembered that one of the major mechanisms for microorganism removal is adsorption and settling; so that while many of these organisms do die, many are also still viable and become part of the end use or disposal problem that is the primary focus of this book, namely sludge.

Table 4.7 shows the range of concentrations of some pathogenic and indicator organisms in secondary effluents prior to disinfection. Disinfection will reduce these concentrations to values acceptable to the local or state permitting authority.

MICROBIAL CONTENT OF RAW SLUDGES

The sludge resulting from wastewater treatment operations and processes is usually in the form of a liquid or semisolid, containing from 0.25 to 12% solids, depending on the operations and processes used. Sludge is

TABLE 4.7. Secondary Effluent Ranges for Pathogenic and Indicator Organisms Prior to Disinfection.

Organism	Number/100 mL	
	Minimum	Maximum
Total coliforms	45,000	2,020,000
Fecal coliforms	11,000	1,590,000
Fecal streptococci[a]	2,000	146,000
Viruses	0 05	1000
Salmonella sp	12	570

[a]Assuming removal efficiencies for fecal streptococci similar to the fecal coliform removal efficiencies
Source. From U S EPA/625/1-86/021 [9]

by far the greatest in volume of all the constituents removed by treatment; processing and disposing of it are among the most complex problems that engineers in the field of wastewater treatment must resolve [1].

Wastewater sludge may contain beneficial plant nutrients, and have desirable soil conditioning properties. It may also contain bacteria, viruses, protozoa, parasites, and other microorganisms, some of which can cause disease in humans. Land application of sludges thus creates the potential for human exposure to these organisms. The actual species and density of microorganisms present in the sludge produced from a particular municipality, especially pathogens, depends to a large extent on the health status of the local community and may vary substantially with time.

There is not much reliable data available on the density levels of indicator organisms, pathogenic bacteria, viruses, protozoa, or parasites in primary or secondary sludges. Pedersen, in 1981 [15], prepared an extensive literature review on the density of pathogenic organisms in municipal wastewater sludges. This review covered the pertinent domestic and foreign literature from 1940 to 1980. We have drawn liberally from this review to summarize what is known about the levels of microorganisms in primary, secondary, and mixed municipal wastewater sludges.

The density levels of indicators, pathogens, bacteriophages, viruses, protozoans, and parasites in primary, secondary, and mixed sludges are presented in Tables 4.8 and 4.9.

PRIMARY SLUDGE

The total coliform densities in primary sludges range from 1×10^6 to 1.2×10^8 organisms per gram dry weight (GDW). Bacteriophages occur at a density of 1.3×10^5 plaque forming units (PFU) per GDW, with a range of 10^3–10^6 PFU/GDW. The use of bacteriophage has been suggested as a surrogate for the survival of enteric viruses in water and wastewater systems, primarily due to the ease of quantitation compared to eucaryotic viruses. North and Johnson (1979; cited in Pedersen, 1981 [15]) computed the ratio of coliphage to enteroviruses in primary sludge to determine if it was relatively constant. They found that the ratio varied from 28,000:1 to 13:1, indicating no relationship between the numbers of the two types of viruses. This is not too surprising since the host cells for bacteriophage are bacteria that can grow in this milieu and support the replication of the phage particles, whereas enteric viruses cannot increase in numbers since their host cells (mammalian cells) cannot reproduce in the sludge or wastewater environment. Also, there is no relationship between numbers

TABLE 4.8. Levels of Indicator and Pathogenic Organisms in Primary, Secondary and Mixed Sludges: Bacteria and Viruses. [a]

Sludge Type	Total Coliforms	Fecal Coliforms	Fecal Streptococci	Bateriophage	Salmonella Species	Pseudomonas aeruginosa	Enteric Virus	Reference
Primary	$1\ 2 \times 10^8$	2.0×10^7	8.9×10^5	1.3×10^5 PFU	$4\ 1 \times 10^2$	$2\ 8 \times 10^3$	3.9×10^2 PFU	[15]
Primary							$1{-}10^3$ TCID$_{50}$/mL	[22]
Primary							$1\ 2{-}576$ PFU	[16]
Primary							$0.002{-}0.004$ MPN	[21]
Primary							$2{-}1660$ PFU/mL	[22]
Primary							5.7 IU	[17]
Primary	$10^6{-}10^7$	$10^6{-}10^7$					$6.9{-}1400$ PFU	[22]
Secondary	$8 \times 10^6{-}7 \times 10^8$	$8 \times 10^6{-}7 \times 10^8$						[18]
Secondary	$7\ 0 \times 10^8$	8.3×10^6	1.7×10^6		8.8×10^2	1.1×10^4	3.2×10^2 PFU	[15]
Secondary							$3.4{-}49$ PFU	[19]
Secondary							$0.015{-}0\ 026$ MPN	[20]
Mixed	1.1×10^9	1.1×10^5	3.7×10^6		2.9×10^2	$3\ 3 \times 10^3$	3.6×10^2 TCID$_{50}$	[15]
Mixed	$3\ 8 \times 10^7$	1.9×10^6	1.6×10^6		7.0	4.4×10^5		[19]

[a] Organisms, plaque forming unit (PFU), most probable number (MPN), infectious unit (IU), or 50% tissue culture infective dose (TCID$_{50}$) per gram dry weight (GDW) or milliliter (mL).

TABLE 4.9. Levels of Indicator and Pathogenic Organisms in Primary, Secondary and Mixed Sludges: Parasites.[a]

Sludge Type	Parasites Ova/Cysts	Ascaris Species Ova	Trichuris trichiura	Trichuris vulpis	Hymenolepis diminuta	Toxocara Species	Reference
Primary	2.1×10^2 ova	0.72	1×10^{-2}	1.1×10^{-1}	6×10^{-3}	2.4×10^{-1}	[15]
Primary							
Primary	2–100	0.1–2	1			0.2–0.5	[23]
Primary							
Primary	10–10^3						[18]
Secondary	10–10^3						[18]
Secondary		1.36	$<1 \times 10^{-2}$	$<1 \times 10^{-2}$	2×10^{-2}	2.8×10^{-1}	[15]
Secondary							
Mixed	0–50 ova	2.9×10^{-1}	0	1.4×10^{-1}	0	1.3	[15]
Mixed							

[a]Organisms, ova or cysts per gram dry weight (GDW).

of bacteriophage and human viruses in feces which originally contributes the agents to the sewage and sludge.

Two pathogenic bacteria, *Salmonella* sp. and the opportunist pathogen *Pseudomonas aeruginosa* occur in primary sludges at lower densities than the indicator bacteria. The average *Salmonella* and *Pseudomonas aeruginosa* densities were 4.1×10^2 and 2.8×10^3 per GDW, respectively (Table 4.8).

Enteric viruses in samples from five wastewater treatment plants averaged 3.9×10^2 PFU per GDW (Table 4.8).

The density levels of parasite ova in raw primary sludge are shown in Table 4.9. Most of these data are from a study of twenty-seven municipal wastewater treatment plants by Reimers et al. [24]. The ova of *Ascaris* sp. were found in the greatest numbers (720 ova per kilogram dry weight (DW) or 0.72 per GDW). Other studies have found a range of 2 to 40. *A. lumbricoides* per GDW (Theis et al., 1978; and Bond, 1958; cited in Pedersen [15]) in U.S. treatment plant sludges, and *Ascaris* ova in primary sludge from Johannesburg, South Africa approached 7800 per GDW. In the study by Reimers et al. [24], *Trichuris trichiura* (human whipworm) occurred at an average density of 10 per kg DW; *T. vulpis* (dog whipworm) occurred at a mean density of 110 per kg DW; and *Toxocara* sp. (mostly *T. canis* or dog roundworm) at a mean density of 240 per kg DW. The frequency of occurrence of *Hymenolepis diminuta* (rat tapeworm) in primary sludge probably reflects the occurrence of rats in and around treatment plants. *Entamoeba coli*-like cysts were commonly found in all sludges by Reimers et al. [15]. Viable cysts, however, were found in only one sample. Cysts of *Giardia* sp. were found in some samples, but none appeared viable.

SECONDARY SLUDGE

Levels of indicator and pathogenic bacteria, enteroviruses, and parasites are presented in Tables 4.8 and 4.9. Despite the smaller data base, it appears that the concentrations are similar to those reported for primary sludges [15]. Total coliforms and fecal coliforms in secondary sludges both occur in the density range of from 8×10^6 to 7×10^8 per GDW, and fecal streptococci occur in the range of 10^6 to 10^7 per GDW. *Salmonella* sp. averaged 8.8×10^2 and *Ps. aeruginosa*, 1.1×10^4 per GDW. Enterovirus densities from several studies were as follows: 3.2×10^2 PFU per GDW [15], 3.35 to 48.9 PFU per GDW [19], 0.015 to 0.026 MPN per GDW [20].

Data on density levels of parasite ova in secondary sludge were mostly from Reimers et al. [24]. *Ascaris* averaged 1360 per kg DW; *T. trichiura* averaged <10 per kg DW; *T. vulpis*, <10 per kg DW; *Toxocara* sp., 280 kg DW; and *H. diminuta*, 20 per kg DW.

MIXED SLUDGE

The levels of total coliform, fecal coliform, and fecal streptococci in mixed sludges were in the ranges: 3.8×10^7 to 1.1×10^9 per GDW, 1.9×10^5 to 1.9×10^6 per GDW, and 1.6×10^6 to 3.7×10^6 per GDW, respectively. Care should be used in evaluating some of these data since some of the sample preparation methods have a bias to low numbers because organisms associated with solids were discarded [15]. Levels of *Salmonella* sp. ranged from 7 to 290 per GDW. *Pseudomonas aeruginosa* averaged 3.3×10^3 per GDW, and enteric virus averaged 3.6×10^2 TCID$_{50}$.

Total parasite ova in mixed sludge ranged from undetectable to 50 per GDW. *Ascaris* sp. occurred at a density of 290 per g DW; *Toxocara* species at a density of 1300 per kg DW, and *Trichuris vulpis* at 140 per kg DW. By comparison neither *T. trichiura* nor *H. diminuta* were isolated from eight samples examined [24].

REDUCTION IN MICROBIAL CONTENT OF RAW SLUDGE DUE TO VARIOUS TREATMENTS OR PROCESSES

Primary and secondary sludges are important by-products of wastewater treatment processes. The volume of sludge generated by wastewater treatment processes, and hence the volume of pathogenic agents, is less by a factor of one hundred twenty-five or more than the original volume of sewage. While the primary objectives of further treatment of the sludge are to reduce the water and organic content, significant reduction of pathogenic agents also occurs during various sludge treatment processes.

The major processes employed for the stabilization and dewatering of sludge, in sludge management schemes, include anaerobic digestion, aerobic digestion, and composting. Anaerobic digestion involves the biological decomposition of organic and inorganic matter in the absence of molecular oxygen. Aerobic digestion involves the aeration of wastewater sludge and provides a biochemical and oxidative means of sludge stabilization. Composting is an aerobic, microbiological process by which organic matter may be decomposed into a well-stabilized humus-like product.

Each of these processes also produces significant reductions in the levels of various indicator and pathogenic agents. Table 4.10, which lists the levels of indicator bacteria and pathogenic agents in raw primary, secondary, and mixed sludges, was compiled from data collected, for the most part, in the United States. The table was derived using data from a number of separate studies and reports, and the values represent averages from the various studies [15]. In a separate study, Reimers et. al. [24], surveying several southern United States wastewater treatment plants,

TABLE 4.10. **Levels of Indicator Bacteria and Pathogens in Raw Primary, Secondary, and Mixed Sludge.**

Agent	Range of Levels Reported (number/g dry wt)	Average Level Reported (number/g dry wt)
Total coliforms	$1\ 1 \times 10^1\text{--}3\ 4 \times 10^9$	6.4×10^8
Fecal coliforms	$\text{ND--}6\ 8 \times 10^8$	$9\ 5 \times 10^6$
Fecal streptococci	$1\ 4 \times 10^4\text{--}4\ 8 \times 10^8$	2.1×10^6
Salmonella sp	$\text{ND--}1\ 7 \times 10^7$	7.9×10^2
Shigella sp.	ND	ND
Pseudomonas aeruginosa	$1.5 \times 10^1\text{--}9.4 \times 10^4$	5.7×10^3
Enteric virus	$5\ 9\text{--}9.0 \times 10^3$	$3\ 6 \times 10^2$
Parasite ova/cysts	$\text{ND--}1.4 \times 10^3$	$1\ 3 \times 10^2$

ND = none detected
Table adapted from Pedersen (1981) [15]

found the levels of *Ascaris* sp., *Trichuris trichiura, Trichuris vulpis, Toxocara* sp., and *Hymenolepis diminuta* to range from 0 to 32,000, 16,000, 3900, 4700, and 5500 respectively.

Pedersen [15], in an exhaustive literature review of the levels and fate of indicator and pathogenic agents in sludge and during sludge treatment processes, found the following:

For anaerobic digestion:

- Anaerobic digestion, regardless of the specific process type, consistently effected a minimum of a tenfold to more than a hundredfold reduction in total coliforms, fecal coliforms, and fecal streptococci.
- Levels of *Salmonella* sp., and *Pseudomonas aeruginosa* were most often reduced more than tenfold by the various types of anaerobic digestion processes.
- Virus reduction in continuously mixed mesophilic anaerobic digestion processes ranges from approximately 0.25 log to greater than 1.0 log per day.
- Most ova or cysts of parasitic tapeworms, flatworms and roundworms survive anaerobic digestion to a greater extent than indicator and pathogenic bacteria and viruses.

For aerobic digestion:

- Some bacteria including *E. coli* and *Salmonella* sp. may be reduced by tenfold or more after several days of detention.
- Tenfold reduction of enterovirus density is possible after ninety days of detention.
- Parasites seem to be more resistant than other agents.

For composting:

- Indicator bacteria are reduced by mesophilic composting by three to four orders of magnitude with up to a twenty-eight-day period of composting.
- Pathogenic enteric bacteria are reduced during composting to an extent greater than for indicator bacteria.
- Most viruses of concern in sludge appear to be quite vulnerable to the temperature conditions of composting. A tenfold reduction in virus levels per day seems the minimum to be expected during composting.
- *Ascaris lumbricoides* ova may be reduced three orders of magnitude within one hour of mesophilic composting at 51°C.

More recently, Martin et al. [25] mathematically described the relation between temperature and rate of reduction of enteric bacteria and viruses during aerobic digestion. They predicted at least a two $\log_{10}$ reduction in indicator organisms and viruses in sixty days at 15°C, forty days at 20°C, twenty-nine days at 30°C, or twenty-five days at 40°C.

1996 AMSA SEWAGE SLUDGE SURVEY—PATHOGEN CHARACTERISTICS OF VARIOUSLY PROCESSED SLUDGES

The Association of Metropolitan Sewage Agencies (AMSA) conducted a survey of its members in 1996 concerning the nutrient, metal and pathogen characteristics of the members sludge products [74].

The sludge pathogen characteristics related to the pathogen reduction requirements detailed in ''The Standards for the Use or Disposal of Sewage Sludge'' which was published on February 19, 1993 by the U.S.EPA (Part 503 Rule).

As described by the Part 503 Rule, a ''Class A'' sludge is one which has been microbiologically tested or treated by an U.S.EPA approved process to ensure the following numerical limits:

(1) Less than 1000 fecal coliform bacteria per gram or less than three *Salmonella* sp. per four grams dry weight

(2) Less than one viable helminth ovum per four grams dry weight

(3) Less than one virus per four grams dry weight

A ''Class B'' sludge, as described by the Part 503 Rule, has been microbiologically tested or treated by an U.S.EPA approved process to ensure that it contains less than two million fecal coliform bacteria per gram dry weight.

Table 4.11 lists the various types of microbiologically tested sludges that were produced at the facilities reporting in the AMSA survey. One hundred fifty of the 202 survey respondents provided microbiological data on their sludges. The number who reported microbiological data and the number meeting "Class A" or "Class B" criteria are listed in Table 4.11. The reported concentration ranges for fecal coliforms, *Salmonella* sp., viable helminth ova and viruses for each type of sludge product are listed in Table 4.12.

REGROWTH OF MICROORGANISMS IN DISINFECTED AND PASTEURIZED SLUDGES

Despite the considerable reduction in the levels of various indicator and pathogenic agents that occurs during the various sludge treatment processes, certain of these agents may regrow in the sludges after the treatment is complete. Also, potentially pathogenic agents, not originally in the pretreated sludge, may colonize the treated product and thus constitute a potential health hazard. Fortunately, there are only a few microbial agents that are both of potential health concern and capable of regrowth or colonization in treated sludge. They include certain *Salmonella* species, enteropathogenic and drug-resistant coliforms, fungi such as *Aspergillus fumigatus*, and other molds. Human pathogenic viruses and eucaryotic parasites will not regrow or colonize treated sludge for lack of an appropriate host. Other bacterial and mycotic pathogens found in pretreated sludge do not readily grow in treated sludge because of nutritional requirements that are not supplied by the stabilized sludge [26]. The *Salmonella* and enteropathogenic coliforms may cause gastroenteritis if they are ingested in high enough numbers. Drug-resistant coliforms may transfer their drug resistance factor to other bacteria, including pathogens, and render them resistant to the antibiotics normally used for clinical treatment.

Aspergillus fumigatus is an ubiquitous mold that may infect and cause pathology in traumatized tissues, and in immunodeficient or immunosuppressed persons. Pulmonary aspergillosis resulting from either an infective or allergic mechanism may also be caused by this organism. Although ubiquitous in the environment, high levels of *Aspergillus fumigatus* resulting from its colonization and growth in treated sludge obviously will increase its infectiousness [27]. Overall, however, actual human risk from the regrowth of these organisms in sludge is relatively low.

Regrowth of organisms in sludge may follow inoculation of the sludge with the agent. Inoculation of treated, and relatively pathogen-free, sludges with pathogenic organisms may occur by airborne particulates, contaminated equipment, bird droppings, etc. Treated sludge may also be seeded

TABLE 4.11. 1996 AMSA Sewage Sludge Survey: Respondents Producing Class A or Class B Sludge Products.

Final Treatment Process or Product	Number Responding with Microbiological Data	Number Testing for and Meeting "Class A" Requirements	Number Testing for and Meeting "Class B" Requirements
Liquid sludge	37	1	36
Sludge cake	89	5	84
Composted sludge	7	1	6
Alkaline stabilized	5	1	4
Air-dried	7	3	4
Heat dried	5	0	5

Source AMSA, 1997 [74]

TABLE 4.12. 1996 AMSA Sewage Sludge Survey: Concentration Ranges of Fecal Coliforms and Pathogens Reported.

Final Treatment Process or Product	Range of Fecal Coliform Values Reported (#/G)	Range of Salmonella Concentrations Reported (#/4G)	Range of Helminth Ova Concentrations Reported (#/4G)	Range of Virus Concentrations Reported (#/4G)
Liquid sludge	<4 to 6,981,500	none detected	none detected	none detected
Sludge cake	<1 to 776,000	none detected	none detected	none detected
Composted sludge	<3 to 97,500	none detected	none detected	none detected
Alkaline stabilized	<1 to 1,000	<1 to 3	<1 to 1	none detected
Air-dried	<2 to 27,000	<2 to 0.29	<1 to 0.05	<1 to 0.09
Heat dried	<1 to 8	none detected	no analyses	no analyses

Source: AMSA, 1997 [74].

with pathogenic organisms from untreated sludge. Once inoculated, several factors influence the colonization of the sludge by these organisms. Growth in sludge of salmonellae and other possibly pathogenic bacteria may occur when the moisture level is greater than 20%, pH is from 5.5 to 9.0, and the temperature is between 10 and 45°C; while fungi such as *A. fumigatus* may grow at moisture levels as low as 2% and remain active at a pH of 1.5 [26,28]. *A. fumigatus* would grow at the surface of treated sludge because of its oxygen requirement, and because the normally antagonistic bacteria would be inhibited by lack of moisture. Sunlight would tend to be inhibitory to both bacteria and fungi at the surface of treated sludge.

Nutrients in the treated sludge are not likely to be limiting to the regrowth of either enterobacteriaceae, such as the salmonellae, or to fungi [26,29]. Antagonism to potential pathogens by the nonpathogenic microflora of sludges treated by processes that do not kill the nonpathogenic organisms has been demonstrated [30]. On the other hand, several studies have shown significant repopulation by pathogenic bacteria in sludges that were treated in a manner that completely sterilizes or disinfects the sludge [28,31]. The possible mechanisms of pathogen suppression in sludges that are not completely sterilized, and soils treated with sludges, include predation by protozoa and *Bdellovibrio*-like organisms, competition for nutrients, and inhibition by antibiotics produced by the nonpathogenic bacteria.

Conventional sludge treatments reduce levels of pathogenic agents to relatively safe levels. Regrowth or colonization of the sludge by the originally reduced organisms or newly introduced organisms presents a potential public health threat. The magnitude of the problem, however, is much less than that associated with the untreated sludge. Although other agents might also be involved, regrowth of *Salmonella* and colonization with *Aspergillus* likely would present the major problems.

The presence and levels of *Salmonella* and *Aspergillus* in sludge depend upon many factors, some of which were discussed previously. In sludges that have not been treated to eliminate nonpathogenic organisms antagonistic to pathogens, regrowth or colonization by enteric pathogens is not likely to be a significant problem. Considering the fact that the infective dose of various *Salmonella* sp. for man ranges from 10^5 to 10^{10} organisms [32], the possibility of a person ingesting an infective dose is extremely low. Sludges treated by irradiation or high heat are likely to contain low populations of antagonistic organisms. Regrowth of enteric *Salmonella* might therefore occur to a greater extent, but the public health threat would be minimal. Although *A. fumigatus* may colonize the surface of treated sludge to a level of 10^3 organisms per gram, the health threat is not likely to be any higher than that associated with the use of potting soils and similar commercial products [26].

MICROBIAL CONTENT OF SOILS

GENERAL

Few environments on earth provide so great a variety of microorganisms as fertile soil that has never had sludge applied. Bacteria, fungi, algae, protozoa, and viruses make up this microscopic menagerie, which may reach a total of billions of organisms per gram (Table 4.13). The diversity of the microbial flora and fauna makes it difficult to enumerate the total population of a soil sample. Most cultural methods will reveal only those physiological and nutritional types that are compatible with the environment used to recover them [33–35]. This section is being devoted to the microbial content of soils not receiving sludge so that the data on sludge-amended soils can be placed in proper perspective.

Soil is distinguished from subsoil chiefly by the presence of organic matter in the soil. The organic matter is composed of dead plant and animal tissues, and the products of their decomposition by microorganisms.

Microorganisms in the soil include protozoa, fungi, slime molds (*Myxomycetes*), green algae, diatoms, blue-green algae, bacteria, and viruses. The bacteria are a heterogeneous group that includes the procaryotic mycelial forms known as actinomycetes, and the simpler unicellular forms known as eubacteria (true bacteria).

The bacteria, actinomycetes and fungi are the three groups most active in decomposing organic residues, and in rendering inorganic nutrients such as carbon, nitrogen, phosphorus, potassium, and sulfur soluble and available to plants.

BACTERIA

Eubacteria or true bacteria exceed all other soil organisms in numbers and in the variety of their activities (Tables 4.13 and 4.14). In numbers,

TABLE 4.13. Microbial Populations in Fertile Agricultural Soils.

Type	Number per Gram
Bacteria	
Direct count	2,500,000,000
Dilution plate	15,000,000
Actinomycetes	700,000
Fungi	400,000
Algae	50,000
Protozoa	30,000

After Pelczar and Reid, 1972 [33]

TABLE 4.14. Physiological Groups of Bacteria in Various Types of Soil (numbers of bacteria per gram of soil).

Soil Type	Garden	Field	Meadow	Marshland
Moisture content (in percent) of moist soil	17.9	18.1	17.0	37.2
Percent calcium carbonate	4.7	5.0	11.4	7.6
Bacteria developing on nutrient-gelatin plates	8,400,000	8,100,000		1,500,000
Bacteria developing on nutrient-agar plates	2,800,000	3,500,000	3,000,000	1,700,000
Bacteria growing in deep cultures of glucose agar (anaerobes)	280,000	137,000	620,000	2,180,000
Urea-decomposing bacteria	37,000	8500	5200	2500
Denitrifying bacteria	830	400	850	370
Pectin-decomposing bacteria	535,000	70,000	235,000	3700
Anaerobic butyric acid bacteria	368,000	50,300	83,500	235,000
Anaerobic protein-decomposing bacteria	35,000	22,000	36,800	2000
Anaerobic cellulose-decomposing bacteria	367	350	367	1.1
Aerobic nitrogen-fixing bacteria	2350	1885	18	17
Anaerobic nitrogen-fixing bacteria	5500	700	370,000	67
Nitrifying bacteria	880	1701	37	34

After Waksman, 1927 [39].

183

the bacteria may exceed 1.5×10^7 per gram of soil by the plate count, or 2.5×10^9 per gram by microscopic count (Table 4.13) [33,36]. The bacteria vary in size, shape, growth requirements, energy utilization, and function. The members of most taxonomic groups of bacteria, excepting certain animal and human parasites, occur in soil, and some groups are characteristic of soil alone [36].

Nutritionally, soil bacteria can be divided into two groups: autotrophs and heterotrophs. The autotrophs are able to use carbon dioxide as their sole carbon source; heterotrophs must obtain their carbon from organic sources.

The autotrophs can be further subdivided into photosynthetic and chemosynthetic groups, according to their source of energy. Purple and green sulfur bacteria are photosynthetic due to the presence of bacteriochlorophyll or chlorobium chlorophyll pigments. The photosynthetic bacteria, like the chlorophyll-containing algae and higher plants, obtain energy from sunlight.

Other autotrophs are chemosynthetic, deriving their energy from various oxidation reactions. All of their requirements for food and energy are met by inorganic sources. As shown in Table 4.15, the members of this group are able to derive their energy and reducing power from the oxidation of a reduced inorganic compound, and use carbon dioxide as the principal or sole source of carbon. The range of inorganic energy sources is limited for any given organism. Taxonomic subdivisions are based upon strict autotrophy versus ability to utilize organic compounds both as energy and carbon source (facultative). Heterotrophic bacteria, which constitute the

TABLE 4.15. Principal Groups of Chemoautotrophs.

Inorganic Energy Source	Physiological Group	Facultative	Chemoautotrophy	
			Obligate	Genera[a]
H_2	Hydrogen bacteria	+	−	Pseudomonas Algaligenes
NH_3	Nitrifying bacteria	−	+	Nitrosolobus Nitrosomonas Nitrocystis
NO_2	Nitrifying bacteria	+	+	Nitrobacter Nitrospina Nitrosococcus
H_2S, S, $S_2O_3^-$	Sulfur oxidizing bacteria	+	+	Thiobacillus
Reduced iron salts	Iron bacteria	(?)	+	Gallionella, Sphaerotilus

[a]The genera designated are those which have most frequently been investigated. There may be a number of other genera recognized
Source· From Moat, 1979 [37]

vast majority of soil bacteria, derive both food and energy from the decomposition of organic substrates. The majority of the heterotrophs require combined nitrogen sources to build cell substance.

Nitrogen-fixing bacteria utilize elemental nitrogen from the air, and include symbiotic organisms such as *Rhizobium* sp. which live symbiotically with leguminous plants, and nonsymbiotic organisms, such as species of the aerobic genera *Azotobacter, Beijerinckia,* and *Azotomonas,* and the anaerobic genus *Clostridium* (Table 4.16). Indigenous heterotrophic soil bacteria also may be classified according to their nutritional requirements. All require a source of energy, such as a simple sugar, and while the additional needs of some can be satisfied by inorganic salts, others require amino acids or more complex substrates; some may even require growth factors present in soil extracts. The proportion of each nutritional group is fairly constant in soil of a given type, but enrichment, e.g., addition of fertilizer, is reflected in an increase in the proportion of bacteria with complex requirements over those with simple nutritional needs.

ACTINOMYCETES

The actinomycetes are next to the bacteria in numbers in soil, ranging from hundreds of thousands to several million per gram of soil. More actinomycetes are present in dry, warm soils than in wet, cold soils. Their numbers are reduced proportionately less with depth than are the bacteria. Actinomycetes are abundant in grassland soil. The three genera occurring most commonly in soil are *Nocardia, Streptomyces,* and *Micromonospora. Thermoactinomyces* are active in rotting manure, and may be present but inactive in normal soils.

Streptomyces species are dominant in the soil, and though largely saprophytic, a few species are parasitic for plants (e.g., causative agent of potato scab, *Streptomyces scabies*).

Not as much is known of the function of the actinomycetes as of the bacteria. They are heterotrophic organisms, and are nutritionally a very adaptable group. They participate in the decomposition of a wide range of carbon and nitrogen compounds, including the more recalcitrant celluloses and lignins, and play an important role in the formation of humus. The actinomycetes are responsible for the earthy or musty odor characteristic of humus rich soil [36].

Many actinomycetes (~60%) may produce substances that are antagonistic to bacteria or fungi in artificial culture (antibiotics). Such substances can only rarely be demonstrated in soil, but they may play an important role in microenvironments in which intense microbial activity is occurring [36].

FUNGI

Fungi are present in numbers ranging from several thousand to several

TABLE 4.16. **Examples of Nitrogen-Fixing Genera.**

Root-nodulated legumes (associated with various species *Rhizobium*)
  Trifolium, Crotolaria (clovers), *R. trifolii*
  Glycine max (soy bean), *R. japonicum*
  Medicago sativa (alfalfa), *R. meliloti*
  Desmodium (cowpea), *R. leguminosarum*
  Various lupines, vetches, and lotus, *R. lupini*
Root-nodulated angiosperms
  Alnus, alder tree (associated with an actinomycete)
  Myrica, bog myrtle or sweet gale (associated with an actinomycete)
  Shepherdia, Elaeagnus, Hippophae (oleasters)
  Caenothus, New Jersey tea
Root-nodulated gymnosperms
  Cerotozamia
  Macrozamia, Encephalartos, Cycads or seed ferns
  Polocarpus (associated with a phycomycete)
Leaf-nodulated plants
  Rubiaceae (*Psychotria* is associated with *Klebsiella*)
Pavetta
  Myrsinaceae (*Ardisis*)
Free-living bacteria and cyanobacteria
  Aerobic, heterotrophic
    Azotobacteriaceae (Azotobacter, Azomonas, Beijerinckia, Derxis),
    Nocardia, Pullalaria, Pseudomonas
  Aerobic phototrophic
    Nostoc, Calothrix, Anabaena
  Facultative, heterotrophic
    Klebsiella pneumoniae, Bacillus polymyxa, Desulfovibrio
    desulfuricans, Achromobacter
  Anaerobic, heterotrophic
    Clostridium pasteurianum
  Anaerobic, photoautotrophic
    Chromatium, Rhodospirillum rubrum, Rhodomicrobium
    vannielli, Chlorobium limocola
  Nonphotosynthetic, facultatively autotrophic
    Methanobacterium sp.

Source: From Moat, 1979 [37]

hundred thousand per gram of soil. They occur extensively in the mycelial and spore form, and growth may be initiated from both. They most commonly occur near the soil surface, and are more abundant in lighter well-aerated soils than in the heavier soils. Due to their optimal pH range of 4.5 to 5.5 they are more prevalent in acid soils than are the bacteria and the actinomycetes [36].

Ecologically, two broad groups of fungi may be recognized, the soil-inhabiting and the root-inhabiting type. The former are able to survive indefinitely as saprophytes and are widely distributed in the soil, while the latter types are specialized parasites that invade living root tissues. The distribution of the root-inhabiting types is localized and depends on the presence of the host plant. Upon the death of the host plant their activity diminishes, and they persist in the soil only as resting spores or sclerotia [36].

All known fungi are chemoheterotrophs, and nutrients enter the fungal cell or hypha in solution. Some fungi (e.g. the common molds *Aspergillus* and *Penicillium*) can utilize any of a wide range of compounds as sources of carbon and energy; others may have highly specific nutritional requirements (e.g., certain obligate pathogens). A number of fungi require specific growth factors such as the vitamins thiamine and biotin. Carbon sources utilized by fungi include aliphatic hydrocarbons, cellulose, lignins, pectins and a wide range of soluble sugars, alcohols, and organic acids. The basidiomycetes or wood-rotting fungi produce extracellular enzymes that degrade cellulose and hemicelluloses to soluble products.

Fungi are chiefly responsible for improving the physical structure of soil by exerting a binding effect on loose particles, and forming water-stable aggregates. The binding effect is caused by the growth of mycelia, which form fine networks that entangle the smaller particles. The soil-binding effect is enhanced by the addition of fresh organic materials whose decomposition products provide cementing substances [36,38].

YEASTS

Yeasts are simple fungi that occur in soil only to a limited extent, especially in the surface layers. Field soils contain only small numbers of yeasts; they are found most frequently in the soils of orchards, vineyards, and apiaries where special conditions favor their growth. Soil, in general, is not a favorable medium for the growth of yeasts, and they do not play a significant role in soil processes [36].

ALGAE

The algae are widely distributed in soils and are most abundant in moist, fertile soils well-supplied with nitrates and phosphates. The algae are chlorophyll-containing organisms, and in the soil surface layers where they are principally found they function as green plants, converting carbon dioxide and inorganic nitrogen into cell substance by photosynthesis. Smaller numbers of algae occur at lower depth, and in the absence of sunlight they exist heterotrophically. The soil algae are comprised of green algae (*Chlorophyceae*), the blue-green algae (*Myxophyceae*) and the diatoms (*Bacillariaceae*). Green algae predominate in acid soils, while in neutral or alkaline soils the other two groups are more prominent. Numbers of algae in soil vary widely ranging from a few hundred to several hundred thousand per gram of soil [36].

As autotrophs, the algae are important in adding to the organic matter in soils. On barren and eroded lands the algae play a fundamental ecological role by colonizing such areas and synthesizing protoplasm from inorganic substrates. Some of the blue-green algae are able to fix atmospheric nitrogen and have agricultural significance, especially in rice culture. Under the conditions required for growing rice, blue-green algae develop abundantly, and may increase the nitrogen supply as much as 20 pounds per acre (22 kilograms per hectare) [36].

PROTOZOA

The protozoa occur in all arable soils, largely confined to the surface layers. In dry, sandy soils they may penetrate more deeply. Numbers of protozoa found in soil usually range from a few hundred in dry soil to several hundred thousand per gram in moist soils rich in organic matter. Most soil protozoa are flagellates and amoebas; ciliates occur less frequently although often they are found in wet soils, and swamps. The protozoa are active in soil only when living in a film of water, and the majority are able to form inactive cysts, which can withstand desiccation.

The majority of soil protozoa, excluding the chlorophyll-containing flagellates such as *Euglena* and some saprophytic forms, feed by ingesting solid particles, mainly bacteria. Not all bacteria are suitable as food, and amoebas have decided preferences for certain bacterial species. Protozoa may thus play a role in maintaining some sort of equilibrium of the microbial flora of the soil. Overall, the effect of protozoa in soil is limited, and is not considered detrimental to the activities of the microbial population as a whole [36].

MYXOMYCETES

The slime fungi or acellular slime molds form a minor group of soil microorganisms that are intermediate in character between the flagellated protozoa and the fungi. They eventually form spores that give rise to the more common flagellated forms. Like protozoa, the myxomycetes feed on bacteria and on small particles of organic detritus [36,38].

VIRUSES AND PHAGES

Little is known of the role that these ultramicroscopic organisms play in soil processes. Viruses that attack plants and animals can in some cases be transmitted from the soil; phages (bacteriophage) having bacterial or actinomycete hosts, limit susceptible microorganisms and thus may affect the microbial balance. Those that attack the various species of symbiotic nitrogen-fixing bacteria may prevent the effective inoculation of legumes, especially where crop rotation is not practiced. This deleterious effect of phage on the nitrogen-fixing bacteria was originally attributed to their direct lytic action and subsequent dissolution of the bacterial cell. The action has been found to be more complex and may result from the development of phage-resistant auxotrophs that are less effective in fixing nitrogen than are the wild type parent strains [36].

MICROBIAL CONTENT OF SLUDGE-AMENDED SOILS

SOURCES OF SLUDGE USED AS AMENDMENT

Primary and secondary sludges are normally subjected to various treatments before they are disposed of or utilized. Treatment processes have been designed to reduce the potential of sludge as a food source for microorganisms and disease vectors; to reduce its nuisance properties, e.g., odor; or to reduce its volume. Most of these sludge stabilization processes also cause significant inactivation of microorganisms, especially viruses. Other processes are also used which have been designed specifically for pathogen reduction.

A significant percentage of the bacteria, viruses, protozoa, and eggs of parasitic worms found in wastewater become concentrated in sludge during treatment. A small proportion of these organisms may be pathogenic and capable of causing disease. The direct measurement of pathogen numbers is difficult, but pathogen reduction can be estimated by monitoring the reduction in the concentrations of indicator organisms such as total and fecal coliforms, fecal streptococci, and others. Pathogen levels can be substantially reduced by sludge treatment processes such as anaerobic digestion; although parasites such as *Ascaris* are not affected (Table 4.17).

TABLE 4.17. Typical Pathogen Levels in Unstabilized and Anaerobically Digested Sludges.

Pathogen	Typical Concentration in Unstabilized Sludge (No./100 milliliters)	Typical Concentration in Anaerobically Digested Sludge (No./100 milliliters)
Virus	2500–70,000	100–1000
Fecal coliform bacteria[a]	1,000,000,000	30,000–6,000,000
Salmonella	8000	3–62
Ascaris lumbricoides	200–1000	0–1000

[a]Although not pathogenic, they are frequently used as indicators.
Source: U.S.EPA 625/10-84-003 [47].

Both liquid and dewatered sludge can be applied beneficially to land. The sludge can be spread onto the surface or incorporated into the soil. Each choice has advantages and disadvantages. Dewatering reduces both the water and the soluble nitrogen content of sludge. The decrease in water content reduces the handling costs; while reducing the nitrogen content increases the quantity that can be applied where nitrogen loading is the limiting factor.

Sludges may also be applied to land in composted, air-dried or heat-dried forms. These dried forms are easier to store, and composted sludge is more stabilized than liquid or dewatered sludges.

Chemical stabilization (lime stabilization) of sludge is a simple process. Lime is added to sludge to produce a pH of 12 after two hours of contact. The effectiveness of lime stabilization in controlling pathogens depends on maintaining the pH at levels that kill microorganisms and inhibit growth should the finished product become contaminated. This process reduces pathogenic bacteria and viruses by over 90% (i.e., tenfold). Some helminth ova, but not ova of all species, will be substantially reduced by the lime stabilization process.

Heat treatment processes can be used both to stabilize and condition sludge. The processes involve heating sludge under pressure for a short period of time. The sludge becomes sterilized and bacterial slime layers are solubilized, making dewatering easier. Liquid sludge must be heated to 180°C (350°F) for 30 minutes, which effectively destroys pathogenic viruses, bacteria, and helminth ova. Since no reduction in organic matter is involved, proper storage is essential to avoid the problem of contamination and subsequent regrowth of pathogens.

It is possible to stabilize and sterilize sludge by incineration. This involves the burning of volatile materials in sludge solids in the presence of oxygen. Incineration is not strictly speaking a sludge disposal or use method, but a treatment that converts sludge into ash, which is either disposed of or used. Stabilization is not necessary, since incineration destroys all pathogens and other organisms. The major potential problem with this treatment method is operational reliability.

SURVIVAL OF MICROORGANISMS IN SLUDGE-AMENDED SOILS

Bacteria

Detectable numbers of bacterial and viral pathogens are low in properly treated, e.g., digested, pasteurized, composted sludges. Due to the relatively low numbers of pathogens present in properly treated sludges, studies on the die-off of bacterial and viral pathogens take one of three general forms: (1) die-off of indicator organisms in sludge-amended soil; (2) measurement of survival time of pathogens in sludge-amended soil; (3) measurement of die-off rates of pathogens seeded into sludge-soil systems and extrapolation to the low numbers detectable in sludge or the sludge-soil system. The pathogens seeded in the sludge are not subjected to digestion prior to sludge incorporation in the soil. Table 4.18 presents indicator organism die-off data from various investigations. The data, while varied, indicate that the soil is a hostile environment for most indicator organisms. In some cases, a reduction of several orders of magnitude was seen within a week, in other cases within several months. It is important when reviewing these data to realize that in some cases due to infrequent sampling an initial rapid die-off may not have been detected. Thus, a die-off to background levels may take place within a few weeks and not be detected until the next sampling period, which may be some months later.

The results of *Salmonella* die-off on soils for various investigators are also shown in Table 4.18. Results are given in terms of survival time or decline. Survival times of from one to more than 280 days are presented, though most studies indicate a survival time of less than sixty days. When interpreting this type of data for *Salmonella* it must be remembered that *Salmonella* occurs naturally in wild and domestic animals, and soils are not necessarily *Salmonella*-free before sludge application. Minimum infectious doses of *Salmonella* range from the tens to hundreds of thousands, whereas the concentrations in properly treated sludge are generally less than 100/ 100 mL. O'Brien et al. 44 found that soil levels of TC, FC, FS, SPC, and *Pseudomonas aeruginosa* were at background levels within several months after digested sludge application to land.

The Metropolitan Water Reclamation District of Greater Chicago has conducted extensive bacterial analysis of soil at its Fulton County site where sludge has been applied. Table 4.19 presents the results of soil sampling during 1975–1977. Annual duplicate samples from the control and test fields were composted and analyzed for TC, FC, FS, SPC, *Pseudomonas aeruginosa, Salmonella,* and *Staphylococcus aureus.* Where organisms were recovered, the control field counts were within the range of counts found in the test fields, excepting SPC and TC which were slightly lower in the control field. Specific pathogens *Staphylococcus aureus* and *Salmonella* sp.) were not recovered from the control or test fields.

Thus, the application of anaerobically digested sludge to agricultural soils does not increase the measurable *Staphylococcus aureus* and *Salmonella* sp. content and poses no threat to public health with respect to these organisms.

Viruses

The virus content of soils treated with wastewater sludge has been examined by several workers. In a field study in Houston, Texas, Hurst

TABLE 4.18. Survival of Indicator Organisms on Soil and Vegetables [55].

Material	Time	Decline Logs	Comments	Investigators
Soil[a]	Longer in winter than in summer			Van Donsel et al. (1967)
Alfalfa[b] canary	50 hours	"Eliminated"	Fecal coliform	Bell and Bole (1976)
Grave soil[c]	1 year	Reduced 1.3–1.8 3	Fecal coliform	Edmonds (1975)
Lettuce[d]	3–31 days	Detectable	Most remained in upper cm of soil fecal coliforms Decline after 3–7 days but still detectable after 21 days.	Nichols et al. (1971)
Irrigation water[e] on vegetables 60% raw sewage			No survival on vegetables but present in water	Dunlop et al. (1951)
Ottokee sand	6 months 6 months 6 months	2–3 2–>3 >3	TC FC 90-224 FS ton appl.	Miller (1974)
Celina silt[c] loam	6 months 6 months 6 months	3 >3 2–3	TC FC 90-224 FS ton appl.	Miller (1974)
Paulding[c] clay	6 months 6 months	>3 1–2	TC FC 90-224 FS ton appl.	Miller (1974)

[a]Seeded with brain heart infusion broth inoculated with indicator organisms.
[b]Sewage lagoon effluent—unseeded.
[c]Digested sludge—unseeded
[d]Sewage effluent—unseeded
[e]Irrigation water 60% raw sewage—unseeded.

TABLE 4.19. Cumulative Sludge Application to Test and Control Fields 1972–1977 (MT/ha) and Geometric Means of Microbial Populations per Dry Gram of Soil in Test and Control Fields Tested Annually 1975–1977 [44].

	Cumulative Sludge Appl. (MT/ha)	TC	FC	FS	SPC	P. aeruginosa	S. aureus	Salmonella
Control Field								
6	0.2	1.3×10^3	2.6×10^1	2.4×10^3	1.2×10^8	36×10^3	N.R.[a]	N.R.
Test Field								
3	174	3.1×10^3	2.1×10^2	2.5×10^3	2.1×10^3	1.5×10^3	N R	N.R.
9	47.5	2.0×10^4	1.5×10^2	3.3×10^4	4.0×10^8	8.5×10^3	N.R.	N.R.
20	143	1.7×10^3	2.3×10^1	1.5×10^1	6.8×10^8	2.2×10^4	N.R.	N.R.
21	147	3.1×10^3	8.9×10^0	3.2×10^2	1.5×10^8	2.5×10^4	N.R.	N.R.

[a]Not recovered

et al. [19] treated a land disposal site with unseeded, undigested activated sludge and detected 1–37 PFU/g (dry wt) after seven days of sludge application. All virus isolates were reported to be confirmed by a serum neutralization test. In a study conducted in Copenhagen, Denmark, Nielson, and Lydholm [45] detected 0–0.1 $TCID_{50}$/mg TSS after one day of application of unseeded digested sludge onto a field; and 0–0.02 $TCID_{50}$/mg TSS after four months of digested sludge application. The CPEs observed were not confirmed. These studies indicate a die-away of viruses after application, indicating that the soil environment is hostile to viruses.

In a study conducted at a land reclamation site in Fulton County, Illinois by Metropolitan Water Reclamation District of Greater Chicago [46] anaerobically digested lagooned sludge was incorporated into a field. Over a two-month period, samples of 200 g each were collected and processed by a polyethylene glycol (PEG) hydroextraction technique. None of the samples produced confirmed viruses.

Parasites

There are numerous reports in the literature that indicate that parasitic forms survive in sludge applied to land for from several weeks to several years (see reviews by Hays [42], Bryan [40] and Carrington [41]). The majority of the studies cited, however, are experimental and include a step in which cysts or ova are added in large numbers to raw sewage or digested sludge. Very few of these studies are field investigations wherein the survival of indigenous parasitic forms is determined. It is difficult, therefore, to extrapolate the information from these experiments to actual sludge application sites.

MICROBIAL MOVEMENT TO GROUNDWATER AND SURFACE WATER FROM SLUDGE-AMENDED SOILS

Infectious agents possibly contained in sludge-amended soils, and sludge-containing landfills may give rise to human infection through various pathways. One pathway is the runoff and leaching of pathogens from land application sites or landfills to surface water (horizontal transport) and groundwater (vertical transport), and subsequent ingestion. Viruses, bacteria, fungi, or higher parasites that survive the adverse conditions of typical sludge treatment and storage, as well as bacteria and fungi that might regrow in, or colonize the treated sludge could possibly contaminate ground or surface water supplies and therefore present a human hazard. Sorber and Moore [48] have extensively reviewed the literature dealing with the survival and transport of pathogens in sludge-amended soils. Sobsey and Shields [49] detailed the factors influencing virus survival in soils and their transport or retention in soils. Most recently, the United States Environmental Protection Agency [50,51], published a computer model and documentation for assessing the human risk from pathogens in municipal sludge applied to land. This assessment considered in detail the movement of pathogens from sludge-amended sites to ground and surface waters. Much of the following discussion will draw from the above publications.

It is generally thought that properly designed sludge application sites and landfill may pose little threat of bacterial, viral, or parasite contamination to groundwater [32]. However, although the pathogen level in treated sludges is low, organisms contained in sludges applied to land may contaminate surface waters if the application site is poorly managed and runoff is uncontrolled. For both vertical and horizontal movement of pathogens, much of the available field data in the literature provides information on the presence or absence of microorganisms in sludge-amended soils, ground, and/or surface waters, but does not provide an assessment of organism inactivation. Knowledge of the kinetics and mechanisms of inactivation would allow modeling of this process, accurate assessment of the hazard in a given situation, and provide a tool for optimizing pathogen inactivation. As a practical matter, however, microorganism contamination of ground or surface waters has not been found to be a significant problem in some large field studies and field study-based reviews of the subject [46,52–55]. Even though a practical problem of contamination of water sources has not presented itself, a consideration of the nature of the sludge-contained pathogens and a review of controlled studies dealing with their survival and movement under various conditions allow us to make certain generalizations regarding their possible movement to ground or surface waters, and aid in the prudent siting and design of sludge application and landfill operations.

Most of the infectious agents of concern here are poorly adapted to survive in soil through or over which they must move in order to reach ground or surface waters. They are killed or become inactivated as a function of agent type, predation by indigenous soil organisms, sunlight exposure, moisture, and pH. Their physical passage through soil is influenced by factors including porosity, ionic composition, and pH. Bacteria, fungi, and higher parasites do not move great distances through unfissured soil, but are physically retained by the soil. Viruses adsorb readily to clay and organic soil particles, but may pass through sandy soil. Bacteria and viruses may move more readily through saturated soils than unsaturated soils.

Ground and surface water contamination from sludge application and landfill operations has not been a significant problem at well-managed sites. One would expect any possible contamination of ground and surface waters at sludge application or landfill sites to be very limited because of the low numbers of pathogenic agents that survive most common sludge treatment processes. Also, organisms that regrow or colonize the sludge would not be a large threat to water sources because of their relatively low numbers. Bacteria, fungi, and higher parasites would not be expected to reach groundwater through unfissured soil because the intervening soil would filter these agents out of the percolating water. Viruses would likely adsorb to most soils, but might penetrate sandy soils to the groundwater, especially in cases where the groundwater was close to the surface. The impact on the groundwater, however, would be minimal because of the low levels of viruses in treated sludge.

Runoff from fields to which sludge had been applied or landfills containing sludge might be expected to contain infectious agents that could be leached out of the sludge. Again, however, the levels of such agents would be low because of the initial levels in the sludge. However, the movement of infectious agents into surface water by means of runoff is more likely than movement to groundwater because filtration or adsorption by soil does not occur in the case of horizontal movement to the extent that either does in the case of vertical movement. Sludge application to land and landfill sites must therefore be carefully managed to prevent contamination of surface waters by runoff from these sites. Prudent practice dictates that the type of soil, whether it is adsorptive to pathogens or not, whether the soil column contains fissures, and the level of the water table, should be carefully considered for siting sludge application or landfill operations in order to protect groundwater. Further, the installation and monitoring of test wells and runoff capture basins would provide additional security to ground and surface waters.

HEALTH IMPLICATIONS OF MICROORGANISMS IN SLUDGE

In spite of the extensive processing that is required or recommended for sludge prior to its ultimate use or disposal, the question of the actual disease causation in humans by pathogenic microorganisms resulting from its disposition has generated heated debates in the scientific and lay communities. The risk to humans from pathogenic microorganisms in sludge applied to land or otherwise disposed must, however, be studied in a rational manner. A fact not well appreciated is that infectious agents originally present in sludge, unlike most toxic chemicals, may be very well managed with proper treatment. Unlike the situation for many conservative toxic substances in sludge, which do not significantly degrade with treatment, pathogenic microbial agents may be drastically reduced by various modes of sludge treatment. Protection of human populations may be achieved by a careful consideration of the specific processes that act as barriers between the pathogens in untreated sludge and human populations. The temptation to express risks in terms of nonquantitative "worst case scenarios," and to use anecdotal examples of possibly hazardous situations must be resisted.

The extent to which one can quantify the movement of pathogens in sludge through the various engineered and natural barriers to human receptors determines the usefulness of any study for the evaluation of risk. Risk assessment, therefore, offers an appropriate method for studying and quantifying the risk to humans from pathogens in sludge as a function of treatment processes and environmental factors. Mathematical modeling is the basis for most quantitative risk assessments developed in the field of environmental health. If such an assessment indicates excess human risk, treatments or treatment parameters can be modified to control the situation.

The nature of the necessary changes in treatment or treatment parameters may themselves be indicated by the risk assessment model. Risk assessment has proved to be a valuable tool for the study of the hazards to human health associated with exposure to environmental and occupational toxicants [56]. It has allowed for the quantification of risks attributable to human exposure to specific toxicants through various pathways. Quantification of risk is a major goal of such studies because it facilitates the balanced, objective evaluation of potentially hazardous situations. Mathematical modeling of such situations may also facilitate the development of the most cost-effective solutions to often difficult problems. Risk assessment has proved indispensable to the establishment of rational regulations for risk management.

Accuracy of the numerical terms in a risk assessment ultimately determine the legitimacy of the assessment and hence its usefulness. The accuracy and precision of a risk assessment must be known and heavily weighed when using such an assessment for regulatory development. Although not as well developed for infectious agents as for toxic substances, risk assessment may be a very useful tool in managing the risks from infectious agents associated with the use or ultimate disposal of treated sludge. At the very least, risk assessment would provide the framework for the rational assessment of risk from infectious agents in sludges.

The United States Environmental Protection Agency (U.S.EPA) has been attempting to develop pathogen risk assessments for different sludge utilization or disposal schemes. Their first effort was the development of a pathogen risk assessment for the ocean disposal of municipal sludge [57]. The resulting risk assessment, however, was honestly termed "qualitative" rather than "quantitative" because the information available allowed only speculation as to the occurrence of human health risks from pathogens in municipal sludge disposed in the ocean. The report further concluded that more research was needed in order to develop a definitive risk assessment for the subject of ocean disposal of sludge. A major significance of the report, however, is that it was the first published effort to develop a microbial risk assessment for a potential environmental hazard and it suggested a rational means of studying the overall problem of pathogenic microorganisms in sludge.

A later effort by the U.S.EPA resulted in a pathogen risk assessment for land application of municipal sludge [50,51]. This work resulted in a computer model that could be used to evaluate agricultural utilization or distribution and marketing schemes with respect to microbial pathogens in liquid, dried, or composted municipal sewage sludge. Several municipal sewage sludge management practices are addressed by this model. They are:

- application of liquid treated sludge (1) for production of commercial crops for human consumption, (2) to grazed pastures, (3) for production of crops processed before animal consumption
- application of dried or composted sludge (1) to residential vegetables garden, (2) to residential lawns

Salmonella spp., *Ascaris lumbricoides,* and enteric viruses represent the pathogenic agents of interest in sludge. Exposure pathways include:

(1) Inhalation or ingestion of emissions from application of sludge or tilling of sludge/soil

(2) Inhalation or ingestion of windblown or generated particulates

(3) Swimming in a pond fed by surface water runoff

(4) Direct contact with sludge-contaminated soil or crops

(5) Drinking water from an offsite well

(6) Inhalation and subsequent ingestion of aerosols from irrigation

(7) Consumption of vegetables grown in sludge-amended soil

(8) Consumption of meat or milk from cattle grazing on or consuming forage from sludge-amended fields

Minimal infective doses for the modeled agents are factored into the model as well as some clinical consequences of exposure to the agents.

This latest risk assessment effort represents an excellent conceptual basis for studying the human health implications of microorganisms in sludge. However, it suffers from the lack of quantitative data describing the initial concentrations of microbial pathogens in wastewater and sludge, as well as qualitative information on the processes of microbial transport and inactivation. For practical use, the model also requires additional data on exposure levels, dose-response relationships, and receptor suscepti-bility.

Based upon their work on this risk assessment and the previous one for ocean disposal of municipal sludge, it is likely that the U.S.EPA will continue their development and will support research that will provide the data necessary for the completion of their model. For the present, however, these assessments do not allow a realistic evaluation of the health implications of microorganisms in sludge.

As an alternative to modeling and risk assessment, epidemiological studies may provide some insight into the actual health implications of sludge utilization. This approach utilizes epidemiological techniques to study a possible link between human infection and disease and exposure to sludge. Very few epidemiological studies have been designed to study human infection and disease as a consequence of sludge utilization [58,59]. The epidemiological data from the few studies conducted do not indicate a microbial health hazard from the utilization of municipal sludge.

The lack of information suggesting a microbial health hazard associated with the utilization of sludge does not alternatively prove that no hazard exists. It is possible that despite the negative studies that have been completed, and the extensive experience in sludge utilization to date, there is some unassessed increased risk of infection associated with sludge utilization practices. It seems likely, since the risk has proven so difficult to measure, that any increased risk is small. Prudence dictates, however, that the possibility of hazards continues to be studied, and that current sludge processing and handling practices for the reduction of infectious agents continue.

SAMPLING METHODOLOGY AND PROTOCOLS FOR THE ESTIMATION OF PATHOGENS AND INDICATOR ORGANISMS IN SLUDGE

The U.S.EPA has traditionally specified technology-based standards for pathogen reduction in municipal sludges. These technologies were classified into two broad categories: processes that significantly reduce pathogens (PSRPs) and processes that further reduce pathogens (PFRPs). These treatment technologies were included in 40 CFR 257.3–6.

The U.S.EPA now specifies the densities of indicator organisms and pathogens that must be attained for a particular sludge use rather than the technologies that must be employed. Implementation of this new approach is mainly due to the difficulty in assessing the equivalency of new sludge treatment technologies to the documented PSRP or PFRP processes.

In 1993, the U.S.EPA published regulations for the use and disposal of sludge, known familiarly as the 503 Regulations [60]. Included in the Preamble and Subpart D of these regulations are the requirements for pathogen reductions for sludges that are applied to land. Sludges intended

TABLE 4.20. Classifications for Sludges under the U.S.EPA's 503 Regulations [60].

Class A	
Salmonella	<3 MPN^a/4 gdw^b TS^c, or <1,000 fecal coliform MPN/gdw TS, or treated by PFRP or equivalent process
Enteric viruses	<1 PFU^d/4 gdw TS
Helminth ova	<1 viable ovum/4 gdw TS
Class B	
Fecal coliforms	<2,000,000 MPN or CFU^e/gdw TS, or treated by PSRP or equivalent process

^{a}Most probable number.
^{b}Grams dry weight
^{c}Total solids
^{d}Plaque forming units.
^{e}Colony forming units.

for land application have been subdivided into two categories: Class A and Class B, based upon the stringency of the pathogen requirements (Table 4.20). There are both human and animal contact restrictions on the use of Class B sludges. There are, however, no restrictions on the use of Class A sludges for land application, sale or giveaway programs.

The sampling and analytical techniques which must be used to demonstrate that sludges meet the Class A or Class B pathogen requirements were specified in the 503 Regulations in Subpart A, 503.8 [60], and are listed for convenience in Table 4.21. The processing methods which can be utilized to ensure that a specific batch of sludge or sludge product can meet the pathogen requirements were also specified in these regulations.

TABLE 4.21. Sampling Procedures and Analytical Protocols Specified for the Analysis of Sewage Sludges for Pathogens and Indicator Organisms under the U.S.EPA's 503 Regulations.

Procedure or Organism of Interest	Method Reference
Sampling procedures	[63,64]
Microbial populations	
Indicator organisms.	
Fecal coliforms	[65]
Pathogens:	
Salmonella species	[66,67]
Total enteroviruses	[68,69,70]
Helminth ova	[71,72,73]

EFFICACY OF METHODS

It is impractical to monitor sludges, directly, for all of the pathogenic agents of interest due to the great diversity of microorganisms present. Moreover, the methods used for the detection of pathogenic agents in sludge and other complex mixtures were not developed with these matrices in mind. Most such methods are merely adaptations of procedures originally developed for the clinical microbiology laboratory where the organisms of interest are usually present in large numbers, and do not have to compete for survival [61].

The analytical methods listed in the 503 Regulations for fecal coliforms, salmonellae, enteric viruses, and helminth ova determinations have limitations which make them less than ideal for assessing the status of sludges with respect to pathogens. They are, in addition, time consuming, tedious and costly. Of the four methods listed in Table 4.21, only the fecal coliform most probable number (MPN) method using A-1 medium, and two-stage incubation (three hours at 35°C followed by 21 hours at 44.5°C) meets the requirements of a monitoring tool which is easy to perform, relatively inexpensive, and provides quantitative information within 24 hours [62].

CERTIFICATION FOR EQUIVALENCY AS PSRP OR PFRP: A CASE STUDY

One of the more attractive features of the 503 Regulations is the provision that allows public or private wastewater treatment facilities to meet the Part 503 pathogen requirements by demonstrating that their sludge treatment process(es) are equivalent to existing, documented PSRP or PFRP processes.

Under the Class A pathogen requirements, sludges treated in a process which is equivalent to a PFRP, and which meet the Class A regrowth specifications are considered to be Class A products with respect to pathogens. The potential for regrowth of pathogenic bacteria in Class A sludges makes it important to insure that substantial regrowth has not occurred prior to disposal. Therefore, all of the Class A pathogen requirement alternatives demand that sludges meet the fecal coliform, or *Salmonella* density limits shown in Table 4.20 at the time of use or disposal, or at the time that it or a derived product is prepared for sale or giveaway.

Similarly, under Class B, Alternative 3, sludges that are treated in any PSRP or equivalent process are considered to be Class B with respect to pathogens as specified in Table 4.20.

Unlike the previous U.S.EPA regulations, equivalency under the new 503 Regulations pertains to pathogen content only. Sludges treated under equivalent processes must also meet a separate vector attraction reduction requirement as specified in §503.33(b) [60].

For consideration as equivalent to a PSRP or PFRP, a sludge treatment process must be able consistently to reduce pathogens to levels comparable to that achieved by the documented PSRP or PFRP processes listed in Tables 4.22 and 4.23, respectively. Then, as long as the process is operated under the same conditions that produce these levels of pathogens, it can continue to be recognized as equivalent.

Equivalency is, unfortunately, site specific and applies only to a particular operation, in a given location, under specified conditions. Equivalency cannot be assumed for similar processes at different locations or for processes which have been modified to meet the exigencies of a situation. On the other hand, a process that can consistently demonstrate the desired pathogen levels under conditions that are likely to be encountered anywhere, may qualify for a *recommendation of national equivalency,* which means that the process will probably be equivalent no matter where it is operated.

It is the responsibility of the permitting authority to determine equivalency under the new 503 Regulations. This is done under the guidance of the U.S.EPA's Pathogen Equivalency Committee (PEC), which is also responsible for recommendations of national equivalency. There are certain benefits that derive from a determination of equivalency, not the least of which is the reduction in the costly microbial monitoring burden (pathogen analyses) which it allows.

TABLE 4.22. Processes to Significantly Reduce Pathogens (PSRPs) Listed in 40 CFR Part 503, Appendix B.

Process	Conditions
Aerobic digestion	Liquid sludge aerated with air or O_2; solids retention time and temperature specified: 40 days at 20°C, or 60 days at 15°C
Air drying	Sludge dried on sand beds or in basins for ≥3 months; for two of three months temperature must be ≥0°C
Anaerobic digestion	Sludge treated w/o air; solids retention time and temperature specified: 15 days at 35–55°C, or 60 days at 20°C
Composting	Within-vessel, static pile or windrow methods: temperature at ≥40°C for 5 days, and must exceed 55°C for at least 4 hours during this time
Lime stabilization	Add enough lime to sludge to raise to pH 12 after 2 hours contact time

TABLE 4.23.   Processes to Further Reduce Pathogens (PFRPs) Listed in 40 CFR Part 503, Appendix B.

Process	Conditions
Composting	Within-vessel or static pile: ≥55°C for three days; windrow method: ≥55°C for ≥15 days, at least three turnings
Heat drying	Direct or indirect contact with hot gases· temperature of sludge or of wet bulk gas >80°C
Heat treatment	Liquid sludge heated to >180°C for 30 minutes
Thermophilic aerobic digestion	Liquid sludge aerated with air or O_2; 10 day solids retention time at 55 to 60°C
Beta ray irradiation	≥1.0 megarad irradiation with electron accelerator at room temperature (~20°C)
Gamma ray irradiation	≥1.0 megarad irradiation with cobalt 60 or cesium 137 at room temperature (~20°C)
Pasteurization	Sludge maintained at ≥70°C for ≥30 minutes

In August 1994, the Metropolitan Water Reclamation District of Greater Chicago (District) developed a proposal with the objective of securing from the U.S.EPA a determination of PFRP equivalency for its sludge process trains (SPTs). The District proposed to demonstrate that both its High Solids Sludge Process Train (HSSPT), and its Low Solids Sludge Process Train (LSSPT) as shown in Figure 4.1, are capable of consistently producing a Class A sludge.

The operating parameters of each of the unit processes which comprise the two SPTs (Figure 4.1) were carefully defined and controlled to effect the destruction of pathogens, and were closely monitored during their operation.

In the HSSPT, sludge with a four to five percent solids content is anaerobically digested at 35°C for 14 to 21 days. It is then conditioned with a cationic polymer, and dewatered by centrifugation, producing a cake with about a 25 percent solids content. The cake is aged in lagoons for a minimum of 18 months, and then air-dried in batches, on paved drying cells, to a total solids content of approximately 60 percent. Drying requires four to eight weeks depending on weather conditions.

In the LSSPT, sludge with a four to five percent solids content is anaerobically digested as above. The digested sludge is then lagooned for a minimum of 18 months to dewater and stabilize it. The result is a sludge with a total solids content of approximately 15 percent. The lagooned, dewatered sludge is then air-dried, on paved drying cells as described

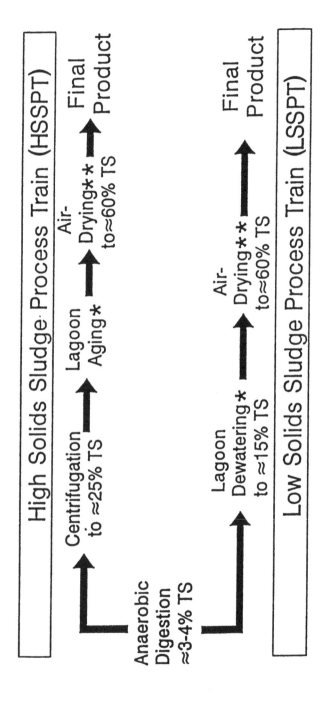

Figure 4.1 Sludge process trains proposed as equivalent to PFRP.

High Solids Sludge Process Train (HSSPT)

Anaerobic Digestion ≈3-4% TS → Centrifugation to ≈25% TS → Lagoon Aging* → Air-Drying** to≈60% TS → Final Product

Low Solids Sludge Process Train (LSSPT)

Lagoon Dewatering* to ≈15% TS → Air-Drying** to≈60% TS → Final Product

*Dewatering, Stabilization and Inactivation
**Dewatering and Inactivation

above, to a total solids content of about 60 percent. Drying requires six to ten weeks depending on the weather.

As of August 1996, more than 132 samples of sludge have been taken from selected points in each SPT and analyzed for fecal coliforms, *Salmonella* sp., enteric viruses and viable helminth ova. The sludges analyzed included samples of digester feed and draw, HSSPT and LSSPT draw, and air-dried products from both process trains.

Standard analyses were sufficient to monitor for fecal coliform and *Salmonella* sp. densities through the SPTs. However, enteric virus and helminth ova densities were so low at the end of the two process trains that the quantities of sample used for their analysis had to be greatly increased to optimize the probability of detection. The 503 Regulations specify that Class A sludges must contain fewer than one virus or viable helminth ovum per four grams, dry weight, of sludge (Table 4.20). All U.S.EPA-approved analytical procedures prescribe using samples which contain ten grams or less total solids. The District has examined individual samples for viruses containing as much as 790 grams dry weight total solids, and for helminth ova samples containing 391 grams dry weight. Table 4.24 and 4.25 contain summaries of the typical microbiological results for 72 samples of digester feed, digester draw, and the lagoon draws from LSSPT and HSSPT. Table 4.26 summarizes the results of 60 typical microbiological analyses of the final products from the LSSPT and HSSPT.

The summary results shown in Tables 4.24 and 4.25 indicate that fecal coliforms, *Salmonella*, enteric viruses and viable helminth ova all were detectable in anaerobic digester feed at the start of both SPTs. Fecal coliforms are reduced by at least one and as much as three orders of magnitude at each step in the two SPTs. *Salmonella* are reduced at least two orders of magnitude from the digester feed through the two SPTs, although they were not present in very large numbers to begin with. Both viruses and viable helminth ova were reduced from an average of 1 per four grams dry weight in the digester feed to virtually undetectable levels in lagoon draws from LSSPT and HSSPT.

The summary results shown in Table 4.26 indicated that in 40 final product samples from the LSSPT and 20 final product samples from the HSSPT fecal coliforms and salmonella were substantially reduced, while viruses and helminth ova were virtually nondetectable. In these latter analyses, more than 7,400 grams dry weight of sludge were analyzed for enteric virues, and more than 8,000 grams dry weight of sludge were analyzed for helminth ova. Moreover, not only did the final product samples consistently meet all of the criteria for Class A sludge (Table 4.20), but so did the lagoon draw samples and 29 percent of the digester draw samples (Tables 4.24 and 4.25).

TABLE 4.24. Pathogen Densities in Digester Feed and Draw in Metropolitant Water Reclamation District of Greater Chicago's PFRP Equivalency Project.

Sludge Type	Solids Content (%)	Fecal Coliform (MPN/gdw)	Salmonella (MPN/4 gdw)	Enteric Viruses (PFU/4 gdw)	Helminths (#/4 gdw)	Class A % (+/Total)
Digester feed						
G		12,300,000				
X	4.2		101	2.9	1.4	
min.	2.6	224,000	2	0.1	0.2	
max.	6.9	71,200,000	640	15	11	
n	22	22	22	22	22	0 (0/22)
Digester draw						
G		150,000				
X	3.2		17	0.9	0.46	
min.	1.7	18,000	2	0.1	0.1	
max.	4.6	109,000,000	132	4.7	1.3	
n	24	24	24	24	24	29.2 (7/24)

Notes MPN = most probable number; gdw = grams dry weight total solids; PFU = plaque forming units, +/Total = number of samples that meet the pathogen requirements for Class A sludge out of total analyzed, X = arithmetic average, min. = minimum value, max = maximum value; n = number of samples.

TABLE 4.25. Pathogen Densities in Low and High Solids Lagoon Draws from the Metropolitan Water Reclamation District of Greater Chicago's PFRP Equivalency Project.

Sludge Type	Solids Content (%)	Fecal Coliform (MPN/gm dw)	Salmonella MPN/4 gm dw	Enteric Viruses PFU/4 gm dw	Helminths #/4 gm dw	Class A % (+/Total)
Low solids lagoon draw (LSSPT)						
G		220	0.96	<0.52	<0.25	
X	14.1					
min.	10.3	16.7	<0.44	<0.02	<0.03	
max.	18.7	3,000	5.36	<1.00	<0.69	
n	17	17	17	17	17	100 (17/17)
High solids lagoon draw (HSSPT)						
G		140	<0.71	<0.40	0.20	
X	30.0					
min.	24.0	15.8	<0.24	<0.02	<0.03	
max.	43.0	850	<2.20	<1.00	0.42	
n	9	9	9	9	9	100 (9/9)

Notes: MPN = most probable number; gm = gram; dw = dry weight; PPU = plaque forming units; +/Total = number of samples which meet pathogen requirements for Class A sludge out of total analyzed; G = geometric mean; X = arithmetic average; min. = minimum value; max. = maximum value; n = number of samples

TABLE 4.26. Pathogen Densities in Low and High Solids Final Products from the Metropolitan Water Reclamation District of Greater Chicago's PFRP Equivalency Project.

Sludge Type	Solids Content (%)	Fecal Coliform (MPN/gm dw)	Salmonella MPN/4 gm dw	Enteric Viruses PFU/4 gm dw	Helminths #/4 gm dw	Class A % (+/Total)
Low solids final product (LSSPT)						
G		15				
X	64.3		<1.28	<0.57	<0.07	
min.	36.3	<0.27	<0.08	<0.01	<0.01	
max.	85.2	3,000	2.20	<1.00	0.43	
n	40	40	40	40	40	100 (40/40)
High solids final product (HSSPT)						
G		6				
X	69.2		<0.38	<0.64	<0.07	
min	56.4	<1	<0.11	<0.05	<0.02	
max.	82.8	530	<2.20	<1.00	<0.16	
n	20	20	20	20	20	100 (20/20)

Notes: MPN = most probable number, gdw = grams dry weight total solids, PFU = plaque forming units; +/Total = number of samples which meet the pathogen requirements for Class A sludge out of total analyzed; X = arithmetic average, min = minimum value, max = maximum value; n = number of samples

The District's study demonstrates that its SPTs can consistently produce a Class A final product. The low pathogen content of the intermediate products indicates an inherently large safety factor in the District SPTs.

REFERENCES

1 Metcalf & Eddy, Inc. 1979. *Wastewater Engineering: Treatment, Disposal, Reuse,* Second Edition. New York, NY: McGraw-Hill

2 Jones, W. J., D. P. Nagle, Jr. and W. R. Whitman. 1987. "Methanogens and the Diversity of Archaebacteria," *Microbiological Reviews,* 51, (1):135–177

3 Doelle, H. W. 1975. *Bacterial Metabolism,* Second Edition. New York, San Francisco, London: Academic Press

4 Moat, A. G. 1979. *Microbial Physiology,* John Wiley and Sons, Inc.

5 Curds, C. R. 1982. "The Ecology and Role of Protozoa in Aerobic Sewage Treatment Processes," *Annual Review of Microbiology,* 36:27–46

6 Taber, W. A. 1976. "Wastewater Microbiology," *Annual Review of Microbiology,* 30:263–277

7 U.S.EPA. 1989. *Environmental Regulations and Technology: Control of Pathogens in Municipal Wastewater Sludge,* EPA/625/10-89/006

8 Pike, E. B. and C. R. Curds. 1971. In "Microbial Aspects of Pollution." *Society for Applied Bacteriology, Symposium, Serial No. 2.* (Eds. G. Sykes and F. A. Skinner), pp. 123–147. London and New York: Academic Press

9 U.S.EPA. 1986. *Municipal Wastewater Disinfection,* EPA/625/1-86/021

10 Berg, G. 1966. "Virus Transmission by the Water Vehicle II. Virus Removal by Sewage Treatment Procedures," *Health Laboratory Science,* 3(2):90–100

11 Clarke, N. A., G. Berg, P. W. Kabler and S. L. Chang. 1964. "Human Enteric Viruses in Water: Source, Survival and Removability," *Proceedings of the First International Conference on Advances in Water Pollution Research, Volume 2,* W. W. Eckenfelder, Editor

12 Buras, N. 1974. "Recovery of Viruses from Waste-Water and Effluent by the Direct Inoculation Method," *Water Research,* 8:1

13 Riggs, J. L. and D. R. Spath. 1983. *Viruses in Water and Reclaimed Wastewater.* EPA/600-1-83-018

14 Curds, C. R. 1973. "The Role of Protozoa in the Activated Sludge Process," *American Zoologist* 13:161–169

15 Pedersen, D. C. 1981. *Density Levels of Pathogenic Organisms in Municipal Wastewater Sludge: A Literature Review.* Boston, MA: Camp Dresser and McKee, Inc.

16 Cliver, D. O. 1987. "Fate of Viruses during Sludge Processing," Chapter 9 in *Human Viruses in Sediments, Sludges and Soils,* V. C. Rao and J. L. Melnick, editors, Boca Raton, FL: CRC Press, Inc.

17 Hurst, C. J. 1989. "Fate of Viruses during Wastewater Sludge Treatment Processes" in *CRC Critical Reviews in Environmental Control.* Boca Raton, FL: CRC Press, Inc. 18(4):317–343.

18 Anonymous. 1988. "Development of a Qualitative Pathogen Risk Assessment Methodology for Municipal Sludge Landfilling," Project Summary, EPA/600/56-88/006.

19 Hurst, C. J., S. R. Farrah, C. P. Gerba and J. L. Melnick. 1978. "Development of a Quantitative Method for the Detection of Enteroviruses in Sewage Sludges during Activation and Following Land Disposal," *Applied and Environmental Microbiology,* 36:81–89.

20 Bertucci, J. J., S. J. Sedita and C. Lue-Hing. 1987. "Viral Aspects of Applying Wastes and Sludges to Land," Chapter 13 in *Human Viruses in Sediments, Sludges and Soils*, V. C. Rao and J. L. Melnick, eds., Boca Raton, FL: CRC Press, pp. 179–196.

21 Subrahmanyan, T. P. 1983. "Virological Hazards Associated with Sewage and Sludge: Investigations in Southern Ontario, 1972–1979," in *Biological Health Risks of Sludge Disposal to Land in Cold Climates*, P. M. Wallis and D. L. Lehman, eds., Calgary, Ontario, Canada: University of Calgary Press, pp. 259–270.

22 Sattar, S. A. 1983. "Viruses and Land Disposal of Sewage Sludge: A Literature Review," in *Biological Health Risks of Sludge Disposal to Land in Cold Climates*, P. M. Wallis and D. L. Lehman, eds., Calgary, Ontario, Canada: University of Calgary Press, pp. 271–292.

23 Graham, H. J. 1983. "Parasites and the Land Application of Sewage Sludge in Ontario," in *Biological Health Risks of Sludge Disposal to Land in Cold Climates*, P. M. Wallis and D. L. Lehman, eds., Calgary, Ontario, Canada: University of Calgary Press, pp. 153–178.

24 Reimers, R. S., M. D. Little, A. J. Englande, D. B. Leftwich, D. D. Bowman and R. F. Wilkinson. 1981. *Parasites in Southern Sludges and Disinfection by Standard Sludge Treatment*, EPA-600/2-81-166.

25 Martin, J. H., H. E. Bostian and G. Stern. 1990. "Reductions of Enteric Microorganisms during Aerobic Sludge Digestion," *Water Research*, 24:1377–1385.

26 Ward, R. L., G. A. McFeters and J. G. Yeager. 1984. "Pathogens in Sludge: Occurrence, Inactivation, and Potential for Regrowth," Sandia Report, SAND830557, TTC-0428, UC-71.

27 Jawetz, E., J. L. Melnick and E. A. Adelberg. 1987. *Review of Medical Microbiology*, 17th Edition. Norwalk, CT: Appleton and Lange.

28 Yeager, J. G. and R. L. Ward. 1981. "Effects of Moisture Content on Long-Term Survival and Regrowth of Bacteria in Wastewater Sludge," *Applied and Environmental Microbiology*, 41:1117–1122.

29 Russ, C. F. and W. A. Yanko. 1981. "Factors Affecting Salmonellae Repopulation in Composted Sludges," *Applied and Environmental Microbiology*, 41:597–602.

30 Burge, W. D., P. D. Millner, N. K. Enkiri and D. Hussong. 1986. *Regrowth of Salmonellae in Composted Sewage Sludge*, EPA/600/2-86/106.

31 Brandon, J. R., W. D. Burge and N. K. Enkiri. 1977. "Inactivation by Ionizing Radiation of *Salmonella enteritidis* Serotype Montevideo Growth in Composted Sewage Sludge," *Applied and Environmental Microbiology*, 33:1011–1012.

32 Kowal, N. E. 1985. "Health Effects of Land Application of Municipal Sludge," EPA/600/1-85/015.

33 Pelczar, M. J. Jr. and R. D. Reid. 1972. Chapter 34, "Soil Microbiology," in *Microbiology*. New York, NY: McGraw-Hill.

34 Alexander, M. 1967. *Introduction to Soil Microbiology*. New York, NY: John Wiley and Sons, Inc.

35 Richards, B. N. 1974. *Introduction to the Soil Ecosystem*. London: Longman Group Limited.

36 Lochhead, A. G. 1978. *McGraw-Hill Encyclopedia of Environmental Science*, Second Edition, S. B. Parker, ed., New York, NY: McGraw-Hill, pp. 709–713.

37 Moat, A. G. 1979. Chapters 3 and 5, *Microbial Physiology*. New York, NY: John Wiley and Sons, Inc.

38 Singleton, P. and D. Sainsbury. 1978. *Dictionary of Microbiology*. New York, NY: John Wiley and Sons, Inc.

39 Waksman, S. A. 1927. *Principles of Soil Microbiology.* Baltimore, MD: Williams and Wilkins Co., p. 38.

40 Bryan, F. L. 1977. "Disease Transmitted by Foods Contaminated by Wastewater," *Journal of Food Protection,* 20(1):45–56.

41 Carrington, E. G. 1978. "The Contribution of Sewage Sludges to the Dissemination of Pathogenic Micro-organisms in the Environment," Water Research Center Technical Report TR71, Stevenage Laboratory, Elder Way, Stevanage Herts. SG1 1th. England.

42 Hays, B. D. 1977. "Review Paper: Potential for Parasitic Disease Transmission with Land Application of Sewage Plant Effluents and Sludges," *Water Research,* 11:583–595.

43 Lue-Hing, C., D. R. Zenz, S. J. Sedita, P. O'Brien, J. Bertucci and S. H. Abid. 1980. "Microbial Content of Sludge, Soil and Water at a Municipal Sludge Application Site," Metropolitan Water Reclamation District of Greater Chicago, R&D Department Report No. 80-27.

44 O'Brien, P., S. J. Sedita, D. R. Zenz and C. Lue-Hing. 1978. "Bacterial Levels Resulting from the Land Application of Digested Sludge," *Proceedings of the First Annual Conference of Applied Research and Practice on Municipal and Industrial Waste,* Madison, Wisconsin, pp. 255–268.

45 Nielsen, A. L. and B. Lydholm. 1980. "Methods for the Isolation of Viruses from Raw and Digested Wastewater Sludge," *Water Research,* 14:175.

46 U.S.EPA. 1979. *Viral and Bacterial Levels Resulting from the Land Application of Digested Sludge,* EPA 600/1-79-015.

47 U.S.EPA. 1984. *Use and Disposal of Municipal Wastewater Sludge,* EPA 625/10-84-003.

48 Sorber, C. A. and B. E. Moore. 1987. *Survival and Transport of Pathogens in Sludge-Amended Soil: A Critical Literature Review,* EPA/600/2-87/028.

49 Sobsey, M. D. and P. A. Shields. 1987. "Survival and Transport of Viruses in Soils: Model Studies," in *Human Viruses in Sediments, Sludges, and Soils,* Rao, V. C. and Melnick, J. L., eds. Boca Raton, FL: CRC Press, Inc.

50 U.S.EPA. 1989. *Pathogen Risk Assessment for Land Application of Municipal Sludge, Vol. I, Methodology and Computer Model,* EPA/600/6-90/002A.

51 U.S.EPA. 1989. *Pathogen Risk Assessment for Land Application of Municipal Sludge, Vol. II, User's Manual,* EPA/600/6-90/002B.

52 Sedita, S. J., P. O'Brien, J. Bertucci, C. Lue-Hing and D. R. Zenz. 1977. "Public Health Aspects of Digested Sludge Utilization," in *Land as a Waste Management Alternative,* R. C. Loehr, ed., Ann Arbor MI: Ann Arbor Science.

53 Bertucci, J. J., J. S. Sedita and C. Lue-Hing. 1987. "Viral Aspects of Applying Wastes and Sludges to Land," in *Human Viruses in Sediments, Sludges, and Soils,* V. C. Rao and J. L. Melnick, eds., Boca Raton, FL: CRC Press, Inc.

54 Liu, D. 1982. "The Effect of Sewage Sludge Land Disposal on the Microbiological Quality of Groundwater," *Water Research,* 16:957–961.

55 Lue-Hing, C., D. R. Zenz, S. J. Sedita, P. O'Brien, J. Bertucci and S. H. Abid. 1980. "Microbial Content of Sludge Soil and Water at a Municipal Sludge Application Site," Report No. 80–27, Research and Development Department, The Metropolitan Water Reclamation District of Greater Chicago.

56 Hallenbeck, W. H. and K. M. Cunningham. 1987. *Quantitative Risk Assessment for Environmental and Occupational Health,* Chelsea, MI: Lewis Publishers, Inc.

57 U.S.EPA. 1988. *Qualitative Pathogen Risk Assessment for Ocean Disposal of Municipal Sludge,* EPA/600/6-88/010.

58 Hamparian, V. V., A. C. Ottolenghi and J. H. Hughes. 1982. "Viral Infections in Farmers Exposed to Sewage Sludge," paper presented at the *Annual Meeting of the American Society for Microbiology*, Atlanta, GA.

59 Brown, R. E. 1985. *A Demonstration of Acceptable Systems of Land Disposal of Sewage Sludge*, EPA 600/1-85-015.

60 U.S.EPA. 1993. 40 CFR Parts 257, 403 and 503, Final Rule. "Standards for the Use or Disposal of Sewage Sludge," *Federal Register*, Volume 58, No. 32, Friday, February 19, 1993/Rules and Regulations, pp. 9248–9415.

61 Cooper, Robert C., and John L. Riggs. 1994. "Status of Pathogen Analytical Techniques Applicable to Biosolids Analysis," In *Proceedings, Management of Water and Wastewater Solids for the 21st Century: A Global Perspective*, Washington, D.C., Water Environment Federation, Stock No. C3403, June.

62 Sedita, Salvador J. and Cecil Lue-Hing. 1994. "Pathogen Monitoring and Assessment of Biosolids: Municipal Concerns," In *Proceedings, Pathogen Assessment and Monitoring of Biosolids and Wastewater Effluents*, 67th Annual Water Environment Federation Conference and Exposition, Chicago, Illinois, October.

63 APHA (American Public Health Association). 1992. *Standard Methods for the Examination of Water and Wastewater*, 18th Edition. Washington, D.C. APHA. Parts: 9060, 9221A, 9510F.

64 U.S.EPA. 1992. "Control of Pathogens and Vector Attraction in Sewage Sludge," *Environmental Regulations and Technology*, Office of Research and Development, Washington, D.C. EPA 625/R-92/013. December.

65 APHA (American Public Health Association). 1992. *Standard Methods for the Examination of Water and Wastewater*, 18th Edition. Washington, D.C. APHA. Parts: 9221E and 9222D.

66 APHA (American Public Health Association). 1992. *Standard Methods for the Examination of Water and Wastewater*, 18th Edition. Washington, D.C. APHA. Part: 9260D.

67 Kenner, B. A. and H. P. Clark. 1974. "Detection and Enumeration of *Salmonella* and *Pseudomonas aeruginosa*," *Journal of the Water Pollution Control Federation*, 46(9):2163–2171.

68 ASTM (American Society for Testing and Materials). 1992. "Standard Practice for Recovery of Viruses from Wastewater Sludges," D4994-89, *Annual Book of ASTM Standards, Volume 11.01*, Water and Environment Technology, ASTM, Philadelphia, PA.

69 U.S.EPA. 1984. *The Manual of Methods for Virology*, EPA 600/4-84/013, as revised. Washington, D.C.

70 Goyal, S. M. et al. 1984. "Round Robin Investigation of Methods for Recovering Human Enteric Viruses from Sludge," *Appl. Environ. Microbiol.*, 48(3):531–538.

71 Fox, J. C., P. R. Fitzgerald and C. Lue-Hing. 1986. *Sewage Organisms: A Color Atlas*. Chelsea, MI: Lewis Publishers.

72 Yanko, W. A. 1987. *Occurrence of Pathogens in Distribution and Marketing Municipal Sludges*, EPA 600/1-87-014. NTIS No. PB 88-154273/AS, National Technical Information Service, Springfield, VA.

73 Tulane University. 1981. *Parasites in Southern Sludges and Disinfection by Standard Sludge Treatment*, EPA 800/2-81-166. NTIS No. PB 82-102344, National Technical Information Service, Springfield, VA.

74 AMSA. 1997. *A Sludge Analysis Survey of Member Agencies in 1996*, Association of Municipal Sewerage Agencies, Washington, DC.

CHAPTER 5

Sources and Control of Odor Emissions from Sludge Processing and Treatment

INTRODUCTION

ODORS at sewage treatment facilities may originate from wastewater treatment and sludge processing operations. Many odorous compounds, which are either discharged directly into a sewerage system or produced during anaerobic degradation of complex organic compounds, are known to cause odors in sewage and sludge treatment operations. This chapter will describe the various techniques that are available for the identification and quantitation of odors. In addition, methods for odor control in sewage and sludge treatment facilities by various physical, chemical, and biological techniques will be described.

SOURCES OF ODOR

Sewage treatment and sludge processing unit operations may cause nuisance odor problems in nearby communities. Public awareness of odors is on the increase as the expansion of urban development occurs in the vicinity of the once isolated sewage treatment works.

Odors from a sewage treatment plant may originate from such diverse sources as septic raw wastewater, overloaded secondary treatment facilities, grit, screenings, skimmings, primary and secondary clarifiers, digest-

Joseph Calvano, Prakasam Tata, and Bernard Sawyer, Metropolitan Water Reclamation District of Greater Chicago, Chicago, IL; Thomas E. Wilson, Greeley & Hansen, Chicago, IL.

ers that are not operating properly, and sludge handling and processing facilities. Sludge handling and processing facilities include unit operations, such as incineration, heat drying, composting, sand beds, lagoons, vacuum filters, belt presses, centrifuges, thickeners, storage areas, etc. [1]. Odors can be produced in sludge thickening, storage, digestion, composting, dewatering, and drying facilities.

APPROACHES FOR ODOR CONTROL

There are three general approaches for the control and treatment of odors originating from sludge processing and treatment operations. Odorous emissions can be prevented, collected and treated, or modified by masking with chemicals. Claims have also been made for the prevention and treatment of odors by bioaugmentation, where proprietary bacterial cultures are added to odor-generating units.

In many instances, odors can be reduced or prevented through improved operation and maintenance procedures. When kept clean, sludge transfer systems, such as bucket conveyors, screw pumps, belt conveyors, and conduits, will not generate odors.

Where odors are generated in enclosed spaces, such as sludge-processing buildings, covered holding tanks, and wet wells, the odorous air can be effectively treated by collecting and treating it with wet scrubbers, activated carbon, chemical absorbers, and soil or compost filters.

Masking of odorous gases such as hydrogen sulfide, methyl mercaptans, and methyl amines with fragment chemicals has been found to be a less than satisfactory method of control. Masking is not acceptable to many people because the fragrance of masking agents may be objectionable. In addition, masking of odors is not always a solution to an odor problem because it often results in an intolerable combination of sludge and masking odors. Spraying of masking agents on windy days can be ineffective even at small sludge thickening tanks and may be useless at large lagoon areas. For these reasons, the use of masking agents is the least preferred method of sludge odor control. Its use is usually limited to emergency situations to achieve temporary solutions that serve to respond to public pressure and to improve community relations.

All previously noted methods of sludge odor control, except masking, require initial confinement and collection of the odorous air. Covering of sludge blending, storage, and thickening tanks, as well as grit chambers, provides effective confinement of odors. The odorous air from enclosed unit operations, such as vacuum filters, can be kept from escaping into the atmosphere by keeping the pressure in the confined area slightly below that of atmospheric pressure. The confined odorous gases from different locations may then be collected at a central area and relevant odor treatment processes applied.

MINIMIZING SLUDGE RELATED ODORS

PRIMARY SEDIMENTATION TANKS

Primary sedimentation units can be the source of odor problems if the units are improperly designed or maintained. Hence, attention should be given to designing these units properly. If scum removal techniques are inadequate, scum will accumulate, putrefy, and produce odors. Infrequent or incomplete removal of settled solids can result in septic conditions, and the trapped gases can cause sludge to rise in clumps to the surface. When the sludge clumps break up, they release odorous gases. Gases trapped in primary effluent will also be released when flowing over the effluent weirs.

Few odor problems occur in final clarifiers if the upstream aerobic stabilization processes are properly designed and operated. The frequent removal of settled sludge from clarifiers and maintenance of a low sludge blanket level at the bottom of the clarifiers will prevent sludge from becoming septic and generating objectionable odors.

Co-thickening processes are used at some treatment plants, where the waste activated sludge is returned to the primary clarifiers to achieve a greater solids concentration along with the primary sludge. When detention times are excessive, sludge storage in the clarifiers or co-thickening of solids may cause the formation of hydrogen sulfide and mercaptans.

BIOLOGICAL WASTE TREATMENT UNITS

Odors resulting from aerobic biological waste treatment units may emit a characteristically "musty" odor. These odors do not usually constitute a major problem and are not as serious as the one caused by anaerobic processes.

If aerobic treatment processes do not receive adequate aeration, they may experience intermittent anaerobic conditions during maximum loading periods. Trickling filters and rotating biological contactors can be sources of odors when they become anaerobic. This often occurs during hydraulic and organic overload conditions. Plugging and ponding in trickling filters reduces air circulation and promote anaerobic conditions [3]. The maintenance of a thin active slime layer and aerobic conditions in these units ensures good treatment and minimum odor generation.

Activated sludge tanks do not normally emit an objectionable odor when a dissolved oxygen concentration of about 2 mg/L or greater is maintained in the mixed liquor. When poor mixing characteristics exist in these tanks, deposition of organic solids may take place on the bottom of the tanks. These solids may go anaerobic and generate odorous gases at a rate faster than the rate at which they can be oxidized by the overlying aerated mixed liquor.

The clogging of diffusers also causes uneven distribution of air. This may result in anoxic zones and solids deposition in the aeration tanks. If aeration tank walls are intermittently wetted by wastewater spray, they may develop putrescible slimes and hence odors. A regular program of maintenance to prevent the clogging of diffuser plates will help in the maintenance of an adequate dissolved oxygen concentration in the mixed liquor, which in turn minimizes the chances for the production of odorous compounds.

Some biological waste treatment systems, such as phosphorus removal and denitrification systems, are designed to operate in an anoxic mode during certain periods of the process regimen. There is a potential for the generation of odors, if the anoxic periods become excessively long or the unit processes are exposed to higher than specified organic loading rates. It is important that these types of processes be monitored closely to prevent septic conditions from occurring and to minimize odors.

SLUDGE PROCESSING UNITS

Thickening

Sludges resulting from primary and secondary clarifiers are sent to thickening units for further concentration of the solids and subsequent treatment in other unit processes, such as stabilization, dewatering, and ultimate disposal. The mixing and thickening of primary and secondary sludges may increase sulfide production in sludge handling processes because of the increased availability of food per unit mass of organisms and prevailing anaerobic conditions. As dissolved oxygen is rarely present in raw sludge, facultative bacteria will thrive and produce sulfides at the expense of the naturally occurring sulfate, which is used as an electron acceptor. Generally, the sulfate concentration in sludges is high enough to sustain sulfide production.

Sludge holding tanks and thickeners can generate odors with characteristics ranging from mildly offensive to nauseating. The odor of fresh sludges is usually less intense than septic sludges. However, when sludges become septic, they emit highly offensive and persistent odors. Generally, odors from thickeners are the result of excessive detention times and low thickened sludge pumping rates. Sludge thickeners are often the cause of odor complaints from communities residing near municipal wastewater treatment plants, due to exposure of raw sludge to the atmosphere [2].

Conditioning chemicals, such as polymers and iron salts, are used in thickening and dewatering operations. The polymers do not generally affect the release of odors from sludge, but they may be affected by other chemicals added for odor treatment. Iron salts added for improved dewaterability will be consumed by dissolved sulfide if present in the sludge. Sulfide reacts with iron salts and forms a precipitate, thereby reducing the potential for hydrogen sulfide odors.

Logical solutions to reduce the odor from thickeners include increasing the pumping rate of the thickened sludge, monitoring a low sludge blanket level, increasing the influent flow rate to the thickener without losing thickening, and chlorinating the influent sludge. Dilution water used for the elutriation of sludge for subsequent gravity thickening should contain a high dissolved oxygen concentration to aid in odor control. Sludge blankets in flotation thickeners should be removed continuously and at frequent intervals.

Stabilization

Sludge stabilization processes include anaerobic digestion, aerobic digestion, lime stabilization, composting, and chlorine oxidation. In most cases, well-operated sludge stabilization processes do not produce offensive odors. For example, digested sludge from a properly operated anaerobic digester has a characteristic tarry odor.

"Sour" anaerobic digesters, thermophilic digesters, overloaded aerobic digesters, and other poorly operated sludge stabilization unit processes will often generate offensive odors. In anaerobic digesters, the odors are usually associated with ammonia, hydrogen sulfide, mercaptans, and volatile fatty acids. Thermophilic sludges usually have a higher concentration of volatile acids than mesophilic sludges. Odorous compounds volatilize more easily at the higher temperature of the thermophilic sludge (50° to 55°C) than the mesophilic sludge (30° to 35°C). Hence, thermophilic sludge is more odorous than mesophilic sludge, particularly when it contains a high concentration of volatile acids and other sulfur-containing compounds that are easily volatilized. In aerobic digesters, the occurrence of an objectionable odor is usually the result of a low solids retention time, high organic loading rate, and inadequate aeration. The maintenance of a dissolved oxygen concentration of about 2 mg per liter, which can be achieved by increasing the aeration rate, reducing the feed rate, and increasing the solids retention time, will minimize odor generation from aerobic digesters.

Lime stabilization processes may generate large quantities of ammonia gas resulting from the high pH (~11 to 12) of the stabilized sludge. While ammonia release occurs at high pH values, sulfides are held in solution. Because of this, the odor of hydrogen sulfide is not usually perceived in lime stabilization processes. However, if for any reason the pH changes to the acidic side of neutrality, hydrogen sulfide generation may occur.

Composting operations can produce a mix of exhaust compounds, which may include sulfides, mercaptans, and ammonia, when there is inadequate aeration and poor mixing of the sludge with the bulking agent. Generally the odors develop from anaerobic conditions, which can occur in compost piles due to inadequate aeration.

When odors occur in static pile and windrow composting operations, they are considerably stronger during the first five days of aeration than in the later period. In static pile systems, odors can be reduced by passing the exhaust air stream through filters made with finished and stabilized compost. Windrow operations may produce significant odors during the turning of the windrows. This is particularly true if an insufficient quantity of bulking agent is used.

In-vessel processes generally have a lower level of odor emission because aeration can be closely controlled through various stages within the composting vessel. One additional advantage of in-vessel processes is that they are confined. Odor, if generated, is more easily controlled because it is contained in a confined exhaust air stream, which can be collected and treated.

The chlorine stabilization process will generate medicinal or chlorine odors, if operated improperly. Insufficient chlorine dosages or long-term storage of the chlorinated sludge may result in putrefaction and release of objectionable odors.

Dewatering

Belt presses, filter presses, and centrifuge operations are usually enclosed. When conditioned sludge is processed through these units, odorous gas releases can occur and cause hazardous conditions if adequate ventilation is not provided. Belt filters in particular are prone to release of odors because of the pressure differences that exist among the various regions of the belt presses. Centrifuges have only a limited release of odor to the ambient room atmosphere because they are enclosed, but they also have a small vent gas stream that may require treatment for odors. Vacuum filters allow substantial release of entrapped gases to the atmosphere, and when this occurs, the vacuum pump exhaust gases may require odor treatment.

Sludge drying beds are generally located in the open, but some of them are also covered. Fresh application of sludge to these beds may release intense odor for the first two or three days before a surface crust develops. These odors may be caused by the release of odorous gases trapped in the sludge, due to application of sludge that is only partially digested.

Any odor problem associated with sludge drying beds can be minimized by ensuring complete digestion of the sludge before it is applied or by adding chemicals to the sludge while it is being applied to the bed. Chemicals such as permanganate or hydrogen peroxide can be added to the sludge as it is being placed on the beds. Odors may also be controlled effectively by adding lime to the sludge as it is discharged to the drying beds, provided that the lime does not cause a large release of ammonia. Hydrated lime is sometimes used for odor control, but it tends to clog the sand of the drying beds.

Other Sludge Treatment Processes

Wet oxidation processes operated at the high temperatures and pressures required for conditioning sludges are often a major source of odors. Special precautions can be taken to contain and treat the odorous discharges. One of the major sources of odor from this process is decant tanks that are generally located outside of the housed sludge handling facilities, which are often uncovered, thus allowing the emission of odor.

Thermal sludge conditioning methods, such as the Porteous process, are usually associated with odors. Thermally conditioned sludges usually release odors when they are dewatered. Such dewatering systems should be enclosed, and when odorous exhaust air streams occur, they can be easily collected and treated.

Odor problems in sludge incineration processes can result from incomplete thermal oxidation of odorous stock gases, or from spillages during the transfer of sludge into the incinerator. If the volume of odorous discharges is small, it is prudent to discharge the gases into an existing incinerator, thus saving the cost of additional fuel. However, if the volume of such emissions is large, the use of a separate combustion system may be justified. Both high and low temperature combustion systems have been used for the control and treatment of odors [1]. In many cases, gas scrubbers or direct flame oxidation systems employed to meet air pollution emission control requirements are effective in reducing odorous discharges [3].

Sludge storage tanks, basins, and lagoons can be a source of odor at wastewater treatment plants. The problem is difficult to control since storage vessels are often uncovered, and large surface areas provide high exposure of the sludge to the atmosphere. Wind action on the surface of storage lagoons can add to the problem [2]. Small floating balls can be used to cover the open surfaces in holding tanks or basins to reduce the evaporation and therefore control the odors, but cost can be high. Siting lagoon facilities away from residential communities and planting trees as a wind barrier will help to reduce the impact of odors on a community.

MEASUREMENT OF ODORS

ODOROUS COMPOUNDS

The human olfactory system is extremely sensitive to a wide range of odorous chemical compounds. An individual's response to any particular compound is determined by the intensity, detectability, and character of the perceived odor. Odorous compounds in wastewater and sludges include both inorganic and organic compounds, the majority of which are produced by biological activity. With the exception of hydrogen sulfide (H_2S), the compounds of most concern are typically formed through the anaerobic or anoxic decomposition of proteins and carbohydrates that are abundant in sludges.

Table 5.1 presents a list of odorous compounds that have been reported to be found in sewage and sludge. Many of the compounds have odor thresholds in air in the parts per billion concentration range. The odor threshold refers to the minimum concentration required for an individual to perceive the odor, although the exact type of odor may not be identifiable. Due to the safety hazard and familiar "rotten egg" odor associated with H_2S, this gas often gets the most attention in odor studies and odor control work. However, it is often other compounds or combinations of compounds in Table 5.1 which cause the odors that generate community complaints. Thus, any study or field survey of wastewater and sludge odors must include the entire spectrum of odor-producing compounds in order to be successful.

SAMPLE COLLECTION

The proper collection of an odorous air sample is of the highest importance in attaining an accurate analysis of the odor. This is true for both qualitative and quantitative methods of odor analysis. The composition of an odor can range from one chemical compound to a complicated mixture of many compounds. The components that make up the odor will often dictate the method of sampling. Therefore, an insight as to which compounds or type of compounds comprises the odor is desirable. Without this insight, it is necessary to utilize a sampling method that collects a

TABLE 5.1. Odorous Compounds Found in Sewage and Sludge [4].

Compound	Characteristic Odor	Odor Threshold (ppm)
Acetaldehyde	Pungent fruity	0.004
Allyl mercaptan	Strong garlic, coffee	0.00005
Ammonia	Sharp pungent	0.037
Amyl mercaptan	Unpleasant, putrid	0.0003
Benzyl mercaptan	Unpleasant, strong	0.00019
Butylamine	Sour, ammonia-like	—
Cadaverine	Putrid, decaying flesh	—
Chlorine	Pungent, suffocating	0.01
Chlorophenol	Medicinal, phenolic	0 00018
Crotyl mercaptan	Skunk-like	0.000029
Dibutylamine	Fishy	0.016
Diisopropylamine	Fishy	0.0035
Dimethylamine	Putrid, fishy	0.047
Dimethyl sulfide	Decayed vegetables	0.001
Diphenyl sulfide	Unpleasant	0.000048
Ethylamine	Ammoniacal	0 83
Ethyl mercaptan	Decayed cabbage	0.00019
Hydrogen sulfide	Rotten eggs	0 00047
Indole	Fecal, nauseating	—
Methylamine	Putrid, fish	0.021
Methyl mercaptan	Decayed cabbage	0.0011
Propyl mercaptan	Unpleasant	0.000075
Putrescine	Putrid, nauseating	—
Pyridine	Disagreeable, irritating	0.0037
Skatole	Fecal, nauseating	0.0012
Tert-butyl mercaptan	Skunk, unpleasant	0 00008
Thiocresol	Skunk, rancid	0.0001
Thiophenol	Putrid, garlic-like	0 000062
Triethylamine	Ammoniacal, fishy	0.08

broad range of compounds. After finding out what the composition of the odor is, the implementation of the most appropriate sampling method can follow.

There are many considerations in choosing an appropriate sampling method. The physical and chemical properties of the odor will often determine which sampling method is desirable. Some of these properties are the polarity, volatility, and stability of the chemical compounds comprising the odor. In order to analyze the sample accurately, the composition of the odor must remain intact during sample collection. Therefore, condensation, adsorption, or permeation of the odorous compounds through the walls of the collection system can cause error.

Stack emissions and ambient air are the two most common sources from which odorous samples are collected. Stacks emit odors with little opportunity for dilution. Their emissions may be well above ambient temperature so that the odorants are in the vapor state at that temperature. Ambient air odor emissions usually are well diluted. The odors may still be strong even though the odorants are less concentrated.

There are two ways to collect an odor sample. In the first method the odorous air is captured in a container, as is. This is called whole air sampling. In the second method, the odorous air is passed through a trap, and the odorous compounds are removed from the air and concentrated as the sample is collected. Stainless steel canisters and Tedlar bags are examples of containers for whole air sampling of odorous air. Adsorption tubes filled with Tenax packing and/or activated carbon are the most common types of traps used for ambient air sampling.

Tedlar Bags

Sample bags made of Tedlar film have been used for collecting samples of many types of odors. The bags range in size from 1 to 50 liters of volume. Some compounds, however, can diffuse through the film or adsorb onto it. Trimethylamine in air, for example, has been shown to be unsuitable for quantitative sampling using Tedlar bags [5]. Bag sampling works well for odors that are caused by low molecular weight aldehydes or organosulfur compounds, and highly volatile compounds like hydrogen sulfide. The bags are especially useful for collecting samples for olfactometer of Butanol Wheel analyses because a large volume of air can be collected.

Canisters

Passivated stainless steel canisters are a relatively new device for collecting whole air samples. They are spherical containers, approximately one liter in volume, which are fabricated by a patented process to make the inner canister wall inert to most classes of gaseous chemical compounds. The canisters are evacuated in the laboratory, and then opened in the field to collect an ambient air sample. The canisters appear to maintain the chemical integrity of the sample better than Tedlar bags. In samples that are high in relative humidity there can be a problem with condensation on the interior walls of the canister.

Tenax

Tenax is the brand name of a class of adsorbent resins typically used as packings in gas chromatograph columns. This same material can also be used for collecting air samples for odor analysis. The Tenax is placed inside small glass or stainless steel tubes, and air is drawn through the tubes using a pump. The odorous compounds in the air are adsorbed on the Tenax. The Tenax-filled tubes are then taken back to the laboratory, where they are processed for chemical analysis typically through the use of a thermal desorption process. The desorbed odors are then passed through a gas chromatograph or gas chromatograph/mass spectrometer system for identification and quantification.

The use of Tenax is advantageous for very dilute odors and when instrumental analysis is done. Large amounts of air can pass through the Tenax, thus concentrating the odorous components. Knowing the amount of air that passed through the Tenax and the amount of the odorants trapped on the Tenax, the concentration of the odorants in the odorous air can be found. A disadvantage of sampling with Tenax is that there is incomplete recovery for many compounds. This is because they are not stable enough for this process or they do not completely desorb off the Tenax. Overloading of the adsorbent can also occur. After a certain point the odorants just pass through the Tenax without getting collected. Care should be taken in devising a method that will not allow this to happen. Using either a shorter sampling period or a longer column of the Tenax will accomplish this.

ANALYTICAL METHODS

The ability to detect and quantify odorants in sludge is a tool in the study and treatment of odors. If there is some correlation between the concentration of odorants found by an analytical method and the odor itself, then this tool is most useful. Since some odorants have low odor thresholds, the detection limit of an analytical method must be low, or else the odorants must be concentrated prior to analysis. The odorants and their concentrations in a sample will influence the choice of a method of analysis.

There are many instrumental methods that can accurately measure odorant concentrations in sludge. There are single compound analyzers, such as a hydrogen sulfide (H_2S) meter, that measures one analyte. Multiple compound analyzers, like a gas chromatograph (GC), can measure more than one analyte. There are specific detectors for a GC that are sensitive to certain types of compounds. If these types of compounds are unknown or their mixture is complicated, then a mass spectrometer detector and an electronic library of compounds can be useful.

H_2S Meters

An H_2S meter is an important tool for odor measurement, in addition to its uses as a safety instrument. It is a portable instrument that can be used for field measurements. It can be used at various locations in a plant as well as at the property lines. Although H_2S is often not the primary odor-causing compound emanating from sludge processing areas, it can often be used as an indicator that odor problems exist.

There are two types of H_2S meters. The first type has a detector that is sensitive to H_2S molecules. It can detect a range of 0–100 ppm H_2S in air. The second type is a more sensitive instrument which uses the reaction of H_2S with a gold film to detect the presence of H_2S. It can detect a range of 1–500 ppb H_2S in air.

Draeger Tubes

Detector tubes like those made by Draeger provide simple, real time analysis of selected odorants. Only one odorant can be measured at a time, and the variety of compounds that can be measured by this method is limited. A detector tube measures the amount of a compound present in air that passes through it. There are different tubes for different compounds. Inside a sealed tube is a packing that changes color when exposed to the compound. The tube is graduated to measure the progression of the color change in the packing. To use a detector tube, the sealed ends are broken off and one end is inserted into a special hand pump. Air is brought into the tube with each full squeeze or stroke of the pump. The concentration is determined from the reading of the color change on the tube and the number of full squeezes or strokes of the pump.

Ammonia, hydrogen sulfide, and dimethyl sulfide ranging from 2–30, 0.5–75, and 1–15 ppm, respectively, are examples of some detector tubes that are used. However, the limit of detection for some compounds may be too high to detect the presence of these compounds even if they are causing the odor. Also, there may be interfering compounds that cause measurement error. This method is more for determining the approximate level of odorants rather than their exact concentration.

Gas Chromatographs

Gas chromatographs can analyze many different compounds. A flame ionization detector (FID) is sensitive to hydrocarbons, like butane or benzene. It is also sensitive to organic nitrogen compounds , like pyridine or trimethylamine. Sulfur compounds, often the strongest odorants in sludge, can be detected using a flame photometric detector (FPD). Detection limits for some typical sulfur compounds using an FPD are:

Carbonyl sulfide—90 ppb
Hydrogen sulfide—130 ppb
Carbon disulfide—40 ppb
Sulfur dioxide—430 ppb

Methyl mercaptan—240 ppb
Dimethyl sulfide—120 ppb
Dimethyl disulfide—180 ppb

Although portable gas chromatographs are available, a more typical application is to collect air samples in the field and bring the samples back to the laboratory for analyses. In addition to sulfur compounds, gas chromatographs are also useful in analyzing odorous nitrogen-containing compounds and volatile organic acids.

Gas Chromatographs/Mass Spectrometer (GC/MS)

The use of a mass spectrometer (MS) as a detector for a gas chromatograph (GC) makes analysis of complicated mixtures of odorants easier. These types of complex mixtures typically occur in environmental samples. Structure and molecular weight for each compound in a sample can be determined from its mass spectrum. This spectrum is virtually unique to a compound and is the same every time it is analyzed. This allows a computer to store spectra for different compounds in a library. Unknown compounds can be easily identified if matching spectra exist in the library.

Headspace Analysis

Headspace analysis refers to a variation of GC or GC/MS analysis in which the sample to be analyzed is the air above a liquid or solid sample in a closed container. This air or vapor is referred to as headspace gas. The odorous compounds in this gas are theoretically at equilibrium with the same compounds contained in the liquid or solid phase of the sample. Headspace gases are typically more concentrated than the corresponding compounds would be in ambient air above a sample, and therefore this is usually a more sensitive method of analysis. The drawback of the method is that it doesn't tell you the actual ambient air concentration of an odorous compound that individuals may be sensing with their noses.

Olfactometer and Odor Panel

The characterization of the sensation experienced by the inhalation of an odorous sample is the objective of a sensory odor measuring program. Sensory analysis provides this because the human body can experience those sensations, process them, and then react. Of course the nose is the part of the body that is used to experience an odor. Sensory analysis is most effective for samples containing complex mixtures of odorants or odorants at concentration levels below detection of an instrumental technique. It also produces simple, useful results that are meaningful to all that are concerned. The use of an olfactometer with an odor panel is a way to perform sensory analysis. An olfactometer is an apparatus that presents an air sample containing the odorous component to an individual at varying dilutions with odor-free air. The object is to determine what level of dilution is necessary for each panelist to begin to detect an odor. From a series of these exposures, results for the odor panel can be calculated. These results can be in the form of an odor to threshold ratio, or the dilution level required for a percentage of the panel to detect the odor. One such type of measurement is the effective dosage at the 50% level (ED_{50}) value. For example, an ED_{50} of 100 means that at a 100 to 1 dilution of an odorous sample with odor-free air, 50% of the panel were able to detect the odor and 50% were not. The higher the ED_{50} value, the stronger the odor.

There are many different types of olfactometers available. A static olfactometer is one that presents different concentrations of the odor sample at an ambient rate. Examples of this can be a syringe test or the use of a scentometer. A syringe test is simply presenting samples of odorous air at different dilutions into the nostril of each panel member. A scentometer is a device that allows different amounts of odorous air to enter it and pass through an activated charcoal filter before reaching the nose.

A dynamic olfactometer is one in which a dilution of the odorous air is presented at a predetermined rate higher than ambient rate. This allows for more control of the conditions and has been shown to be better than a static method. An example of this is the Model 103 olfactometer manufactured by Illinois Institute of Technology Research Institute (Figure 5.1). It has the added feature of providing each panel member with three ports, with only one of three containing the odor for a given dilution level.

The measurement of odor using an olfactometer is fairly straightforward. Consider using an olfactometer such as the Model 103. The odorous air is introduced into a manifold (B) at a certain rate, where an increasing quantity of this air is split off by means of coils of different lengths located on the manifold (coils 1–6). From here, the odorous air travels to one of three sniffing ports at each dilution level. All three ports, however, receive nonodorous air at a certain rate so that the panelists are subjected to moving air at the same rate in all of the ports. Manifold H provides the delivery of the nonodorous air to each port. The nonodorous air can simply be compressed air passing through an activated carbon filter. So what now exists is a stepwise increasing concentration of odorous air from station 1 to station 6. The operator of the olfactometer knows which ports contain the odor. A panelist will approach the olfactometer and begins to sniff each of the three ports at station 1. Then the panelist chooses the port at which an odor is perceived. This is done by pressing a button corresponding

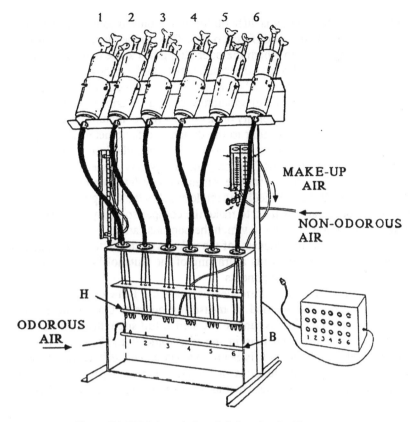

Figure 5.1 IITRI dynamic forced-choice triangle olfactometer.

to that port. A light illuminates on the control panel and the operator records that choice. This is repeated for each station. The point at which the panelists' choices begin to become consistently correct is called the detection threshold for each panelist.

The calculation of an ED_{50} value for a sample can be calculated from the detection thresholds of the panelists. There should be nine to ten panel members for an accurate calculation. A frequency tally of the detection thresholds for all of the panel members is determined. Then a series of further calculations, as described by Dravnieks et al. [6], allows the determination of the ED_{50} value for the sample.

Butanol Wheel

The intensity of an odor is an important parameter to be considered when measuring odors. However, since the characteristic odors of various compounds are so different, it is difficult for individuals to compare the relative strengths or intensities of different odors. This can be overcome by using a reference compound to which the odor strengths can be compared. In this way, odors can be analyzed so that results can be understood by other individuals not subjected to the actual odors. The reference compound that is most widely used is n-butanol. An apparatus called a Butanol Wheel is used to measure the strength or intensity of an odor by this comparative method [7]. The Butanol Wheel is similar to the dynamic olfactometer that was previously described, because it delivers the odorant and dilution air into ports to make different dilutions. The odorant in this case is the butanol vapor. The intensity of an odorous sample is measured by determining at what dilution level of the Butanol Wheel the sample matches the strength of the butanol vapor. An odor panel is used to make the comparisons. By calculating the dilution of n-butanol vapor to which the odorous sample is equivalent, it is possible to express the intensity of the unknown odor in terms of a known intensity. Figure 5.2 presents a drawing of the Butanol Wheel apparatus, highlighting its major components.

One of the principal differences between the Butanol Wheel method and the olfactometer method of measuring odors is that in the Butanol Wheel method the odorous sample is tested at full strength against a series of diluted standards, whereas in the olfactometer method, the odorous sample itself is diluted as it is being evaluated. This difference results in different sensory characteristics of the odor expressing themselves, and makes these two sensory test methods complementary to each other.

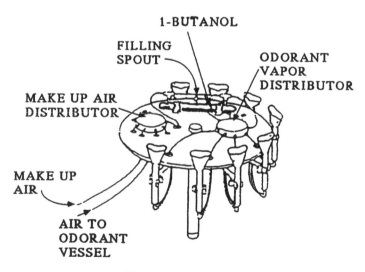

Figure 5.2 Butanol Wheel.

Prior to the widespread acceptance of the Butanol Wheel method, the standard way of expressing the strength of an odor was through the use of "category estimated scales." This method also uses an odor panel, but is more subjective than the Butanol Wheel because it uses intensity terms such as "very weak" or "very strong." An example of one such scale is:

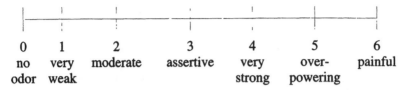

0	1	2	3	4	5	6
no odor	very weak	moderate	assertive	very strong	over-powering	painful

Each panelist estimates where the strength of the odorous sample falls on the scale, and the average of all of the panelists' responses is the odor strength of the sample. This method is still used when a Butanol Wheel apparatus is not available.

Odor Descriptors

In addition to the intensity of an odor, what an odor smells like is a big factor in determining whether it is objectionable. What an odor smells like is called the odor character. The character of an odor can be described through the use of odor descriptors. These are words or phrases that most accurately represent the quality of the particular odor of concern. Each panelist is asked to describe the odor that was sensed. A list of some odor character descriptors used in the wastewater field is shown in Table 5.2. These descriptors can be varied depending on the type of application for which they are intended.

Hedonic Scale

The degree of pleasantness or unpleasantness of an odor is called its hedonic tone. This can be determined by using a scale estimating the magnitudes of the aesthetic qualities found in odors. An example of one such scale is shown below:

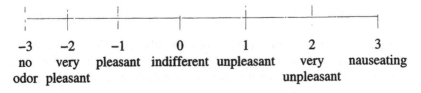

−3	−2	−1	0	1	2	3
no odor	very pleasant	pleasant	indifferent	unpleasant	very unpleasant	nauseating

TABLE 5.2. **Odor Character Descriptors.**

Sweet	Irritating
Floral	Septic
Peppermint-like	Biting
Wintergreen	Gasoline
Pine	Exhaust—diesel
Evergreen	Cresol-like
Pleasant	Rotten
Apple cider	Ammonia
Soil smell	Putrid
Charcoal	Choking
Burnt-charred	Ester-like
Smoke	Hydrogen sulfide
Stale	Butyric acid
Oil	Valeric acid
Kerosene	Mercaptans
Paint-like	Dead animal
Glue-like	Indistinct
Swampy	Corn products
Moldy	Burnt sugar
Musty	
Sour	
Manure—cattle	
Dusty, earthy	
Grass, cut	
Tarry or asphalt	

Each panelist describes the hedonic tone of a sample by judging where it falls on the scale. The average of all of the panelists' choices is the hedonic tone for the sample.

Effectiveness of Measurement Methodologies

An effective odor measurement program depends on many factors that may be interrelated. Among these are the ability to collect a representative sample, the limit of detection for the odorants, the accuracy of the results, and the cost to implement a program. These are considerations that need to be dealt with according to one's needs.

Collection Efficiency

A good sampling method must have high collection efficiencies for the odorants of interest. Samples must be collected in a way that will capture and preserve the odorants until analysis. A sampling method with a high collection efficiency for an odorant will accurately represent its field concentration at the time of sample collection.

In ambient air sampling, whole air samples collected in either Tedlar bags or stainless steel canisters will theoretically capture a representative sample. However, reactive compounds such as H_2S and other reduced sulfur gases can chemically react with other constituents in the sample or condense or adsorb on exposed surfaces of the collection device. This can result in a change in the chemical composition of the sample between the time of collection and the time of sampling. When Tenax columns or other adsorbent columns are used for sample collection the main consideration becomes the affinity of the adsorbent for the various classes of chemical compounds in the odorous sample. Tenax is currently considered to be the best all-purpose adsorbent, but it will not trap some low molecular weight sulfur compounds or many amines. Therefore, columns filled with other types of adsorbents must sometimes be coupled with the Tenax to insure a complete characterization of an air sample.

Limits of Detection

The limits of detection of individual analytical instruments have been discussed previously. Currently, ppb levels of detection are attainable for many compounds. As technology improves, it is inevitable that the limits of detection for most odorous compounds will improve, although the costs of doing the analyses may increase.

Accuracy

There are currently no standardized, widely accepted procedures for collecting odorous air samples. U.S.EPA researchers, the academic community, and private contractors are continually testing and modifying their sample collection procedures based upon the compounds of interest, detection limits, and costs. For this reason, the accuracy of any individual sampling and analysis scheme for a broad range of compounds is difficult to determine. At present, this inability to clearly gauge accuracy is the most limiting factor in conducting odor research.

Accurate odor measurements are vital to an effective odor measurement program. Correct decisions and actions hinge on them. Inaccurate results can lead to wasted time, money, and effort. Getting accurate results in odor measurement is a challenge. There is more to it than just collecting the best sample and using the most sensitive instrument. It takes a dedicated effort to be careful and thorough in performing the analysis. Proper handling of samples and good laboratory techniques are absolutely necessary in getting the best results. In most cases, samples should be analyzed within hours of collection to be sure that what is analyzed is the same as it was in the field. Results depend on the calibration of the instrument, so using accurate standards is critical. It is difficult to make these standards for odorants, especially when they are gases and at low concentrations.

Cost

In any odor-monitoring or measurement program, cost is a factor that cannot be ignored. In general, it is obvious that the lower the detection limit desired and the broader the range of compounds being analyzed, the higher the cost. A sophisticated GC/MS analysis of an air sample, not including sample collection expense, will typically cost $400–$700. If only a GC analysis is required, costs can be reduced to the $200–$300/ sample range. When odor panel work is required in order to evaluate overall odor intensity, the cost of analysis is approximately $1000/day for the odor panel. Four to six samples can typically be run for this cost.

It is clear from the above numbers that a comprehensive odor survey can cost thousands of dollars. Therefore, care must be taken to design an experimental program that will give the maximum amount of useful data within the limits of a reasonable budget.

ODOR CONTROL SYSTEMS

Odors associated with the treatment and processing of sewage sludges can be reduced or eliminated with various techniques. Process selection is dependent upon the nature of the odorous substances present, the type of sewage sludge generated, and the available resources, including cost. Among the techniques described below, some controls are aimed at the removal and/or elimination of odors from liquid or solid sludges. These are called source control. Others are for the removal of odors from the atmosphere in a confined area where the sludge is treated and processed; those discussed include:

- atmospheric dilution
- masking, modification, and counteraction
- absorption
- adsorption
- biological methods
- incineration and afterburners

SOURCE CONTROL

Process Description

Instead of trying to clean the odor-contaminated atmosphere, it is often more desirable and cost-effective to start with the control of the sludge itself, i.e., the source from which the odors are emitted, so that the emission of the odorous substances can be kept to a minimum. The most commonly used odor control techniques in sludge treatment include:

- pH adjustment
- addition of metallic salts
- aeration
- addition of nitrates
- addition of chemical oxidants

Some odors can be controlled by *pH adjustment.* Along with various organic compounds, ammonia and hydrogen sulfide are the major odorous substances associated with the treatment of sewage sludge. The chemistry of ammonia is such that at normal temperatures, it will emit an odor only at pHs higher than 7.0, with the emission becoming substantial at a pH higher than 9.5. Hydrogen sulfide is, on the other hand, not present significantly as an odorous gas at pHs above 9.0, but is the predominant form of sulfide at pHs below 6.5. Lime is used in many wastewater plants to assist sludge dewatering. A side benefit may therefore be the reduction of hydrogen sulfide emissions through the increase of pH with the lime additions. However, care should be taken to avoid overdosing, otherwise ammonia may replace hydrogen sulfide as the major odor-contributing substance. In addition, it has been reported that for certain sludges, some even worse organic odors could be given off with the addition of lime to a pH of 12 upon standing for 24 hours [8]. It has been claimed that these "high pH odors" could be eliminated by using cement kiln dust (CKD) to replace lime in the sludge treatment process.

Addition of metallic salts is primarily aimed at the reduction of hydrogen sulfide. Hydrogen sulfide can be removed through the formation of precipitates with various metal ions such as iron (ferric or ferrous) and zinc. The dissolved sulfide can normally be reduced to about 1 mg/L. It has been reported that for odor control purposes, a mixture of one part ferrous and two parts ferric by dry weight of iron is optimum [4]. The reaction is:

$$Fe^{+2} + 2Fe^{+3} + 4H_2S \rightarrow Fe_3S_4 + 8H^+$$

The end product, Fe_3S_4 (smythite), is much more insoluble than Fe_2S_3 (ferric sulfide). The use of zinc salts also may be effective. The stoichiometric dose requirements are 1.32 mg Fe or 2.04 mg Zn per mg of sulfide to be removed. This method of introducing precipitates with chemical additions is also known as the "construction process."

Aeration can be used to keep sludges from becoming anaerobic. Since many odorous substances, including hydrogen sulfide and various organics, are generated when there is a lack of oxygen supply, adequate aeration can discourage the development of these odors. Aeration may also provide mixing, which is generally needed for keeping the contents in a sludge storage tank uniform. In general, it is more suitable to use aeration for storage of the aerobically digested sludges. High oxygen demands and the potential for releasing odorous substances already generated deter the application of aeration to anaerobically digested sludges.

Aeration, along with the adequate mixing, is probably the most feasible and fundamental way to reduce odors from the sludge composting piles. Except for those originally present in the raw sewage, most of the odorous compounds associated with sludge composting are from the clumps of anaerobic zones [9]. Effective aeration is necessary, not only to support the process, but also to minimize the formation and release of the odorous substances from the compost piles. It should be noted, however, that aeration alone may not be adequate since ammonia and many odorous organic compounds can be formed via aerobic as well as anaerobic pathways.

Addition of nitrate is a chemical method of keeping liquid sludges from becoming anaerobic. In anaerobic processes, a certain group of organic compounds are obligate electron acceptors, whereby many of the odorous substances are generated. Sulfate becomes the electron acceptor with the production of hydrogen sulfide under septic conditions. Nitrate, if present, takes the place of sulfate and these organics and acts as the electron acceptor, thereby preventing the development of the odorous substances. Addition of nitrates is also used to reduce odors in sewers and sewage pumping stations.

Sludges may also be kept from becoming septic or anaerobic with the *addition of chemical oxidants* such as chlorine, hydrogen peroxide, potassium permanganate, ozone, etc. Some odorous substances may also be destroyed by those chemicals through oxidation. Potassium permanganate dosages for odor control have been reported to be from 1 to 7.5 lb/dry ton (0.5 to 3.8 g/kg) [10], or an average of 37 mg/L [2], for various sludge dewatering processes. Doses for hydrogen peroxide may typically range from 20 to 50 mg/L [2]. More on the use of chemical oxidants is given in the following discussion of absorption.

Advantages and Disadvantages

Source control should probably be the first technique considered. Even if not completely effective, a certain degree of source control will help to reduce the cost and effort needed to remove odors from the surrounding atmosphere. Methods developed for source control are straightforward and the installation costs are relatively low. These methods are, however, rather specific to the type of odorous substances to be removed. High pH, for example, favors the control of hydrogen sulfide, but enhances the emission of ammonia. Moreover, source control by itself may not be enough and may have to be supplemented with other techniques such as absorption or adsorption for reliable control and elimination of odor problems.

ATMOSPHERIC DILUTION

Process Description

This is probably the most common and oldest method of dealing with odor problems. The odorous gas stream is mixed and diluted with fresh air to such an extent that sub-odor-threshold levels have been reached. This is basically a physical phenomenon and there is no intent to remove, or even to change the characteristics, of the odorous materials. The odorous substances are no longer a nuisance if their concentrations are below their odor detection limits.

Techniques available to increase dilution include the use of booster fans, venturi nozzles, or afterburners—all with or without a high exhaust stack. The higher the exhaust stack, the more fresh air that is available for dilution and dispersion. The use of the booster fans and venturi nozzles are mechanical means to increase the speed of airflow so that the odorous gases can be released higher into the atmosphere. Afterburners (to be discussed more in detail in the following) increase the speed of exhaust gases through heating. A high fuel consumption is needed if the malodorous stream is nothing but odor-contaminated air streams.

Performance

Performance of the dilution technique is dependent upon not only the available equipment, but also very much upon the weather conditions. It is expected that the efficiency would be lower during temperature inversions, where the rate of temperature change with altitude (i.e., the lapse rate) is too low to allow the exhaust odorous gases to rise and disperse. Temperature inversions happen normally in calm, clear nights and may persist longer in the fall.

Advantages and Disadvantages

The dilution technique is simple and straightforward. Operational costs are much lower than those associated with the use of chemicals and/or sorbents. This technique, is, however, very much dependent upon atmospheric conditions. It may be difficult, if not impossible, to provide a guaranteed elimination of the odor problems by using this technique alone.

MASKING, MODIFICATION, AND COUNTERACTION

Process Description

The olfactory reaction of human beings is such that odors from different substances can interfere with each other and blends can smell quite different from individual compounds. Odors can therefore be modified by using chemicals with different odors and the resulting odor may become less objectionable, or even pleasant to the human nose. The use of perfumes is an excellent example of this type of application.

Depending upon the type of chemicals used and the intended purpose, this odor control technique has been referred to as masking, modification, interference, neutralization, counteraction, disguising, deodorizing, reodorization, perfuming, etc. It should be noted that no matter what types of chemicals are used, these techniques will not remove or reduce the odorous substances, nor the materials emitting the odors, originally present. It is merely a physiological technique to alleviate stress to the human olfactory senses. Most of the commonly used chemicals contain fragrances from flowers and fruits such as wild rose, lemon, vanilla, and pine tree. Careful experiments are normally required to select the right type, as well as the right dose, of those agents. An incorrect application could be synergistic instead of antagonistic and the result may be even more offensive. Moreover, since ''chemical perfumes'' are generally costly, this type of odor control may be justifiable only in relatively enclosed and localized areas under emergency situations. The odorants can be applied through direct addition, spraying, or metered feeding. Applications to open-air areas are almost impossible to control due to the variability of odor concentrations and the ever-changing weather conditions; this is therefore not encouraged without careful studies. It should also be noted that, for safety reasons, any gases or vapors that may be toxic (such as hydrogen sulfide) should not be disguised.

Advantages and Disadvantages

This technique generally has the lowest installation cost of any of the odor control techniques and can be used almost anywhere on a short-term basis. It should be noted, however, that this technique rarely is a long-term solution. Moreover, as mentioned above, the type of chemical and the application dose must be carefully selected or the odor problem could become even worse. Another difficulty may be the different olfactory preferences of each individual; a universally acceptable formula is difficult to find. Cost of these chemicals may also prevent using this method in large-scale applications.

ABSORPTION

Process Description

In this process, the odorous gas is brought into contact with a liquid absorbent whereby the odor-producing substances are removed through dissolution and chemical reactions. Various designs and devices are available, with all aimed at increasing the intimate contact between gas and liquid. Virtually all the so-called scrubbers and washers developed for air cleaning can be used for the removal of odorous substances in gas streams drawn from enclosed areas. The liquid phase can be water or an aqueous solution through which the contaminated gas stream is bubbled or diffused. The liquid can also be introduced into a chamber in the form of a spray, mist, or droplets, where it is allowed to pass through the flowing gas stream in a counter-current or cross-current fashion and then collected at the bottom of the chamber. The chamber can be equipped with various packing materials to increase the exposure area and contact time between the gas and liquid. In some other applications, the liquid is sprayed directly toward the area or spot where the odorous substances are emitted without using scrubbing chambers [11]. The spent liquid can simply be wasted in once-through applications, or recycled to reuse before the final exhaustion of its absorbing capacity. Figure 5.3 shows typical designs of various absorption systems.

Chemicals Used

Depending upon the nature of the odorous substances to be removed, the liquid phase can be water only, a chemical solution containing an oxidizing or a reducing agent, or an alkaline or acidic solution. For deodorizing air streams containing mixed vapors or gases of odorous substances, a series of scrubbers such as the multiple-stage scrubbing facility shown in Figure 5.3, each with a different chemical, can be used. Among the commonly used oxidants are chlorine, calcium or sodium hypochlorite, hydrogen peroxide, potassium permanganate, and ozone. The alkaline agents may include sodium or potassium hydroxide, lime, soda ash and sodium metasilicate. Sodium bisulfite is one of the few reducing agents used. Some wetting agents, such as soaps and surfactants, may be added to decrease the size of the gas bubble or the liquid droplets, to accentuate frothing, and to ''solubilize'' organic compounds at the air/water interface, thereby increasing entrainment [12]. Oil or organic solvents may also be used to replace the aqueous solution if the malodorous substances are hydrophobic.

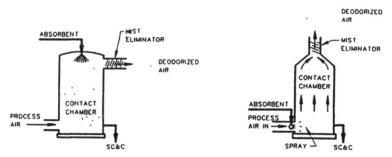

SCRUBBER WITHOUT PACKINGS

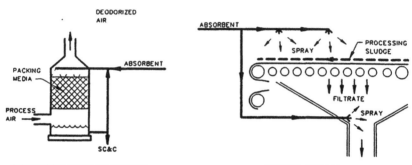

SCRUBBER WITH PACKINGS   DIRECT SPRAY

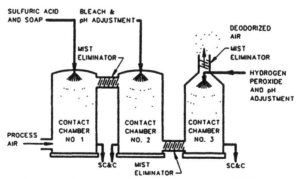

A TYPICAL MULTIPLE-STAGE SCRUBBING FACILITY

SC&C SPENT CHEMICAL AND CONDENSATION

Figure 5.3 Schematic diagrams for odor adsorption facilities.

238

Some typical chemical reactions are as follows:

- the absorption of ammonia in sulfuric acid solution

$$2NH_3 + H_2SO_4 \rightarrow (NH_4)_2SO_4$$

- the absorption and neutralization of sulfur dioxide with sodium hydroxide solution

$$SO_2 + 2NaOH \rightarrow Na_2SO_3 + H_2O$$

- the oxidation of hydrogen sulfide with chlorine

$$H_2S + 4Cl_2 + 4H_2O \rightarrow H_2SO_4 + 8HCl$$

- the oxidation of formaldehyde with potassium permanganate

$$3HCHO + 4KMnO_4 \rightarrow 4MnO_2 + 2K_2CO_3 + 3H_2O + CO_2$$

Among these, chlorine (or hypochlorite solution) and potassium permanganate are about the most popular oxidizing agents used in wastewater treatment plants for odor control purposes. It should, however, be noted that alkaline solutions of chlorine do not remove ammonia effectively, and this process can release chloramines that are just as odorous as ammonia. For this reason, a two-stage scrubbing system is often used to treat the odorous air streams containing both ammonia and other odorous compounds (such as those from composting operations). The first stage is usually designed for the removal of ammonia with sulfuric acid solution and the second stage is for oxidizing hydrogen sulfide and organics with a chemical oxidant.

Not all odorous substances can be oxidized by permanganate. Table 5.3 shows those compounds that do and do not react with permanganate [4]. It may, therefore, be desirable to identify the type of odor-producing substances before selecting chemical agents to be used in an absorption process.

The type of chemicals used, and thus the nature of the chemical reactions involved, also determine the design of the absorbing chambers. In the absorption of alkaline gases, such as ammonia, with an acid mist, such as sulfuric acid, the reaction (neutralization) is almost instantaneous, and thus a very short detention time, about 2 to 5 seconds, is adequate, provided that sufficient turbulence is present to bring about good contact between the gas and the absorbent. In an oxidation process where the odorous substances are oxidized by an absorbent such as the hypochlorite solution, a longer detention time, about 10 seconds minimum, may be needed, and the longer the detention, the more complete the reaction would be. Pilot tests are generally required to develop the design criteria, which may include the selection of chemicals, the depth of absorbing chamber, air flow

TABLE 5.3. Reactivity of Odorous Compounds with
Potassium Permanganate [4].

Compounds That Do React with Potassium Permanganate	Compounds That Do Not React with Potassium Permanganate
Aliphatic compounds·	
Formaldehyde (pungent, suffocating)	Acetone (fruity, mint-like)
Acetaldehyde (fruity, pungent)	Dipropyl ketone (fruity)
Allyl acetate (?)	Methylisobutyl ketone (sweet, camphor)
Acrolein (unbearably irritating)	Isobutyl ketone (sweet, ester)
	n-Butanol (rancid)
	Methylethyl ketone (acetone-like)
	Methylene chloride (ether-like)
Aromatic compounds.	
Benzaldehyde (nutty)	Benzene (coal tar)
Styrene (sweet)	Toluene (sweat, benzene-like)
Phenol (characteristic)	Penta-chlorophenol (phenol)
o-Cresol (phenolic, burnt)	
o-Chlorophenol (unpleasant, penetrating)	
m-Chlorophenol (phenolic)	
p-Chlorophenol (phenolic, unpleasant)	
Nitrogen-containing compounds:	
Diethylamine (putrid, ammoniacal)	Nitrobenzine (bitter almond)
Trimethylamine (fishy, ammoniacal)	Pyridine (burnt, rank)
Monoethanolamine (ammoniacal)	
Triethanolamine (ammoniacal)	
Putrescine (putrid)	
Cadaverine (putrid, decaying flesh)	
Indole (intense, fecal, gassy)	
Skatole (fecal, putrid)	
Sulfur compounds:	
Dimethyl sulfide (putrid)	Carbon disulfide (almost odorless if pure;
Dimethyl disulfide (garlic)	foul, disagreeable if impure)
Thiophene (slight aromatic, resembling benzene)	
Diethyl sulfide (garlic)	
Diethyl disulfide (garlic)	

TABLE 5.3. (continued).

Compounds That Do React with Potassium Permanganate	Compounds That Do Not React with Potassium Permanganate
Inorganic compounds:	
Hydrogen cyanide (faint bitter almonds)	Ammonia (pungent, urine-like)
Hydrogen sulfide (rotten eggs)	Carbon monoxide (faint metallic)
Nitrous oxide (slightly sweetish)	
Sulfur dioxide (suffocating, extremely pungent)	

rate, and the ratio of liquid to gas flow rates, etc. for each specific application [2].

Performance

With properly designed equipment and properly selected chemicals, absorption using scrubbers and washers has been a very effective odor control practice [13,14]. Ninety-six percent of the odor was removed by using a sodium hypochlorite/sodium hydroxide solution in a single-stage scrubber, for example [15]. The odorous substances were found to be mainly aldehydes, ketones and organic acids. It has also been reported that 97–99% of the odor intensity was reduced in a two-stage system using sulfuric acid in the first stage and sodium hypochlorite in the second stage [16]. A similar type of process also has been successful for the removal of odorous substances from a sludge composting facility. Since start-up, the measured impacts, including the odor complaints, have been reduced by more than 90% [9].

Besides chlorine and hypochlorite solutions, potassium permanganate also has been successful. In one full-scale plant application, more than 98% of the hydrogen sulfide was removed; it was, however, not very effective for the removal of hydrocarbons [13]. The effectiveness of using scrubbers and washers may vary depending upon the type of odorous compounds present, the device used, the ratio of liquid to gas flow rates, and the concentration of the chemical solution. Table 5.4 shows the typical performance of a 1% potassium permanganate solution. The percent odor reductions were determined through the measurements of ED_{50} of the contaminated air stream [17].

When removing odors from compost piles, unpacked scrubbers (or towers) could be more desirable than packed ones. It has been found that the packed towers can get clogged by dust (such as the compost fines) and biological growth (even at extreme pHs) and provisions must be made for periodically flushing the packing media.

TABLE 5.4. Performance of Odor Control Scrubbers [17].

Odorous Compound	% Odor Reduction
Acetaldehyde	80–100
Acetone	0
Acrolein	100
Allylisocyanate	80–100
Benzaldehyde	100
Benzene	0
Cadaverine	30–60
Carbon disulfide	2–8
Carbon monoxide	0
p-Chlorophenol	100
o-Cresol	100
Diethylamine	10–20
Dipropylketone	0
Formaldehyde	90
Hydrogen sulfide	100
Indole	30–50
Monoethanolamine	100
Nitrobenzene	0
Phenol	100
Putrescine	30–50
Pyridine	0
Skatole	10–30
Sulfur dioxide	100

By using 1% potassium permanganate in a 15,000 cfm low pressure scrubber
From Reference [17].

Advantages and Disadvantages

Absorption of air impurities by various scrubbers and washers is a very well developed technique. The type of chemical solution can be changed or placed in series corresponding to the nature of the odorous substances present. The major disadvantages potentially include the high moisture content of the exhaust gas, which may cause a fog, and the disposal of the spent absorbent. Care also must be taken to provide safe handling of the chemicals, which can be corrosive and irritating.

For scrubbers using water or aqueous solution, the gas leaves nearly saturated with moisture and, on some occasions, as a chemical mist. This chemical mist from hypochlorite scrubbers may contribute medicinal and chlorine odors, which may also be objectionable. Dehumidification or de-misting may be needed to prevent the chemicals from being released into the surrounding environment and also to eliminate the possible vapor plumes, which may be un-aesthetic or a health hazard. The moisture can normally be removed to a certain extent by using the impinging-type mist eliminators or by pre-condensation with a large amount of cold water.

Another disadvantage associated with the absorption process is the disposal of the spent liquid absorbent. The aqueous solution can generally be treated in the wastewater treatment plant along with the bulk of the regular sewage, but care should be taken to determine if the spent liquid is hazardous. Special procedures for handling and disposal may have to be followed if it is hazardous as defined by the current U.S.EPA rules [18].

ADSORPTION

Process Description

Differing from absorption, adsorption for odor control and removal is a "dry process" and is basically a physical phenomenon. The facilities are enclosed and the odorous air stream is passed through an adsorbent whereby the odorous constituents are removed. Figure 5.4 is a schematic diagram for an adsorption chamber. To achieve an effective removal, the adsorbent should possess a very high surface area per unit volume or unit weight. Among the major adsorbents used in odor control listed below, the various types of carbon, especially the activated carbon, are by far the most commonly used media:

Common odor absorbents

- activated alumina
- activated bauxite
- aluminosilicate
- carbons (activated carbon, caustic impregnated carbon, charcoal, etc.)
- iron oxide
- silica gel

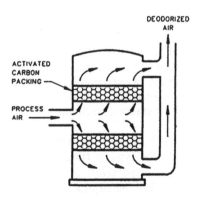

Figure 5.4 A schematic diagram for odor adsorption chamber.

The surface area of the commercial-grade granular activated carbon may vary from 1000 to 1400 m²/g [4]. The adsorption capacity may also vary, depending upon the operating conditions. In general, the capacity increases with pressure but decreases with increasing temperature and moisture content. The capacity is also higher for high molecular weight odorous substances at higher concentrations. For activated carbon, the adsorption capacity can typically reach 5 to 40% of the weight of the carbon [19].

In addition to physical adsorption, activated carbon can also act as a catalyst by providing sites for chemical oxidation. Hydrogen sulfide and mercaptans can be more easily oxidized to elemental sulfur or the less odorous, more adsorbable disulfides in the presence of activated carbon. Activated alumina can also be used as the catalyst for certain chemical reactions. One example is to pass the contaminated air through a dry filter containing potassium permanganate-impregnated activated alumina pellets [13]. The odorous substances will first be adsorbed on the pellets and then oxidized by potassium permanganate.

Depending upon the initial concentration of the odorous substance and the required removal, the carbon media can be placed either in "thin panels" or in deep bed filters. A typical thin panel measures 60 × 60 × 22 cm and contains about 20.4 kg of carbon. A deep bed filter can have 1 m or deeper media. The amount of carbon required for a particular application can be determined by using the following equation [4,20,21]:

$$W = \frac{TEQC}{(10)^6 PS} = \frac{TEQMC'}{24.04(10)^6 PS}$$

where
W = weight of carbon, kg
T = duration of service before breakthrough, hr
E = filter efficiency, fractional
Q = air flow rate, m³/hr
M = average molecular weight of the sorbed odorous substances
C = entering concentration of the odorous substances, ug/M³
C' = entering concentration of the odorous substances, ppm by volume
P = proportionate saturation of carbon before breakthrough, fractional
S = average maximum adsorption capacity (retentivity or adsorptivity) of the sorbed odorous substances, fractional

Both the filter efficiency, E, and the fraction saturated before breakthrough, P, are related to the type of filter employed. These two values are normally 0.9 and 0.165 respectively for thin panel filters, and 1.0 and 0.5 for deep bed filters. The adsorption capacity, S, varies from one substance to another; Table 5.5 is a list of those given in the cited references. The duration of service before breakthrough is typically two to six months.

Upon breakthrough, activated carbon can be regenerated and reused. The spent carbon can be regenerated with either a hot inert gas or steam, or with an alkaline solution if the carbon was originally impregnated with caustic. Each regeneration typically restores 80 to 85% of the previous capacity [14]. On-site regeneration is more cost-effective for large plants while off-site regeneration or even once-through use could be more suitable for small plants.

Performance

Performance of activated carbon filters is, in general, very reliable for the removal of odorous substances, both organic and inorganic. Among those reported, odors (largely hydrogen sulfide) in a 30,000 m^3/day (8 MGD) wastewater treatment plant were successfully removed with a 3200 kg activated carbon filter at an air flow rate of 110 m^3/min [4]. Some other reported data for adsorption capacities are given in Table 5.5.

Advantages and Disadvantages

The use of activated carbon or other adsorbing media is a well-proven technique for the removal of organics and inorganics, including odorous substances. The process is simple to install and operate, and about 80 to 100% removal of odors can usually be achieved without using chemicals, fuels, or water. There are, however, also certain disadvantages, which may include the unexpected fouling of the media due to some unknown constituents in the processing air stream. There may also be chances for unexpected reactions resulting in possibly toxic materials. The costs of carbon and its regeneration may also be a concern.

Other Installations

In addition to those listed above, there are other kinds of adsorption media more specifically used in the treatment of sewage and sewage sludge—the iron sponge for hydrogen sulfide removal [2,13], the use of scrubber piles [22], and biofilters [23]. The iron sponge is a mixture of ferric oxide and wood chips. The removal of hydrogen sulfide is accomplished by first adsorption and then converting to ferric sulfide. This is a relatively inexpensive and effective way normally used to remove hydrogen sulfide from sludge gas prior to using the gas for fuel. The same technique can be applied to the treatment of H_2S contaminated air streams.

TABLE 5.5. Adsorption Capacity of Activated Carbon [17,20].

Substance	Adsorption Capacity, %	Remarks
Acetaldehyde	7	Reagent
Acetic acid	30	Reagent, sour vinegar
Acetone	15	Solvent
Acetylene	2	Welding and cutting
Acryaldehyde	15	Acrolein, burning fats
Acrylic acid	20	
Ammonia	1–2	
Amyl acetate	34	Lacquer solvent
Amyl alcohol	35	Fuel oil
Benzene	24	Benzol, paint solvent and remover
Body odors	High	
Bromine	40 (dry)	
Butane	8	Heating gas
Butyl acetate	28	Lacquer solvent
Butyl alcohol	30	Solvent
Butyl chloride	25	Solvent
Butyl ether	20	Solvent
Butylene	8	
Butyne	8	
Butyraldehyde	21	Present in internal combustion exhaust, i.e., diesel
Butyric acid	35	Sweat, body odor
Camphor	20	
Caprylic acid	35	Animal odor
Carbon disulfide	15	
Carbon tetrachloride	45	Solvent, cleaning fluid
Chlorine	15 (dry)	
Chloroform	40	Solvent, anesthetic
Cooking odors	High	
Cresol	30	Wood preservative
Crotonaldehyde	30	Solvent, tear gas
Decane	25	Ingredient of kerosene
Diethyl ketone	30	Solvent
Essential oils	High	
Ethyl acetate	19	Lacquer solvent
Ethyl alcohol	21	Grain alcohol
Ethyl chloride	12	Refrigerant, anesthetic
Ethyl ether	15	Medical ether, reagent
Ethyl mercaptan	23	Garlic, onion, sewer
Ethylene	3	Higher adsorption capacity by reaction
Eucalyptole	20	
Food (raw) odors	High	

TABLE 5.5. (continued).

Substance	Adsorption Capacity, %	Remarks
Formaldehyde	3	Disinfectant, plastic ingredient
Formic acid	7	Reagent
Heptane	23	Ingredient of gasoline
Hexane	16	Ingredient of gasoline
Hydrogen bromide	12	
Hydrogen chloride	12	
Hydrogen fluoride	10	
Hydrogen iodine	15	
Hydrogen sulfide	3	Oxidizes to increase adsorption capacity considerably
Indole	25	In excreta
Iodine	40	
Iodoform	30	Antiseptic
Isopropyl acetate	23	Lacquer solvent
Isopropyl alcohol	26	Solvent
Isopropyl chloride	20	
Isopropyl ether	18	Solvent
Menthol	20	
Methyl acetate	16	Solvent
Methyl alcohol	10	Wood alcohol
Methyl chloride	5	Refrigerant
Methyl ether	10	
Methyl ethyl ketone	25	Solvent
Methyl isobutyl ketone	30	Solvent
Methyl mercaptan	20	
Methylene chloride	25	
Naphthalene	30	Reagent, moth balls
Nicotine	25	Tobacco
Nitric acid	20	
Nitrobenzene	20	Oil of bitter almonds; oil of mirbane
Nitrogen dioxide	10	Hydrolyzes to increase adsorption capacity
Nonane	25	Ingredient of kerosene
Octane	25	Ingredient of gasoline
Ozone	Decomposes to oxygen	Generated by electrical discharge
Packing-house odors	Good	
Palmitic acid	35	Palm oil
Pentane	12	Light naphtha
Pentylene	12	

(continued)

TABLE 5.5. (continued).

Substance	Adsorption Capacity, %	Remarks
Phenol	30	Carbolic acid, plastic ingredient
Propane	5	Heating gas
Propionic acid	30	
Propylene	5	Coal gas
Propyl mercaptan	25	
Propyne	5	
Putrescine	25	Decaying flesh
Pyridine	25	Burning tobacco
Sewer odors	High	
Skatole	25	In excreta
Sulfur dioxide (dry)	10	Oxidizes to sulfur trioxide, common in city atmospheres
Sulfur trioxide	15	Hydrolyzes to sulfuric acid
Sulfuric acid	30	
Toilet odors	High	
Toluene	29	Manufacture of TNT
Turpentine	32	Solvent
Valeric acid	35	Sweat, body odor, cheese
Water	None	
Xylene	34	Solvent

From References [17] and [20].
The adsorption capacity is also known as adsorptivity or retentivity, and is expressed in terms of percent by weight at 20°C, 760 mm Hg

Both scrubber piles and biofilter are biological methods for odor control, and will be addressed in detail in the following paragraphs.

BIOLOGICAL METHODS

Process Description

Instead of using chemicals or synthetic adsorption media, the biological methods are, as the name implies, to use microorganisms to remove odorous substances from a contaminated air or gas stream. Depending upon the nature of the wastewater treatment plant and the type of application, the commonly used biological methods may include:

- biological stabilization
- scrubber piles
- biofilters

The *biological stabilization* process consists of delivering the odorous air stream drawn from a confined area to an activated sludge aeration tank, or to pass it through a trickling filter (bio-tower), as a part of air supply to those processes. Mixed liquor in the activated sludge aeration tank and the attached biological film on the trickling filter media are therefore acting as the sorbing agents. The odorous substances are first sorbed and then decomposed along with the treatment of the regular plant influent.

Scrubber piles are exclusively used in the sludge composting process. The exhaust air stream drawn from the sludge undergoing composting is passed through a pile of well cured and screened compost whereby the odorous substances are adsorbed and then removed through biological reactions. The compost is normally placed on a layer of wood chips on the top of the exhaust pipes.

Biofilters are different from compost scrubber piles in that the adsorption media is a layer of well-ventilated and biologically cultivated soil. They can be used for the removal of odors from air streams drawn from any confined areas in a wastewater treatment plant. The "organic soil," ranging from three to ten feet in depth is placed on the top of a gravel or crushed stone bed through which runs a network of perforated pipes carrying the contaminated air. The air flow rate may range from 0.35 to 3.3 cf/min/sf (0.11 to 1.0 m^3/m^2-min) [2,23]. A layer of fabric material can be used to separate soil from the gravel. The odor removal is accomplished first by adsorption on soil particles and then by biological reactions that convert the organics to carbon dioxide and water, and hydrogen sulfide to the odorless sulfates. Various soil bacteria are the main type of microorganisms responsible for those reactions [24]. Figure 5.5 is a schematic of a biofilter and a compost scrubber pile. Pilot tests are normally required before designing the full-scale facilities [25].

Performance

Biological reactions are usually much slower than chemical and physical reactions, and the performance of biological systems is more dependent upon the properties of the odorous substances as well as the environmental conditions. In the biological stabilization process, removal is determined by both the solubility and biodegradability of the odorous substances. Odorous gases such as hydrogen sulfide may not be removed as efficiently as with chemical absorbents. An obvious restraint to this may be that the pH of the biological mixed liquor or the attached film cannot be adjusted to favor the absorption of these gases. In trickling filter (bio-tower) plants already equipped with odor removal facilities, odors generated in the sludge processing area may be directed to pass through the trickling filter

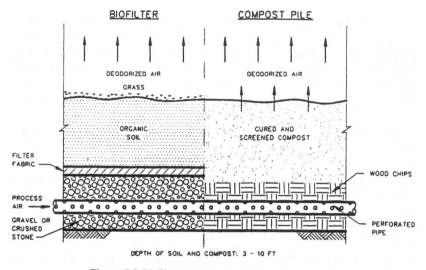

Figure 5.5 Biofilter and compost pile for odor removal.

media. The odorous substances, if not adsorbed by the biological film, could then be removed by the existing odor removal facilities. The applicability of biological stabilization for in-plant odor control is therefore very site-specific and preferably determined through pilot studies.

Both scrubber piles and biofilters are usually very effective for removal of odors [26,27]. The piles and soil bed should be carefully maintained to ensure an optimum biological activity. Lime may be used to maintain the pH suitable for biological growth and also for a better adsorption of odorous gases such as hydrogen sulfide. For compost piles where no grasses are planted, the piling material should be frequently replaced to prevent saturation with odorous substances [22].

Advantage and Disadvantages

Biological methods are usually easy to operate and require no chemicals (except some lime as addressed above) and often require minimal plant additions and modifications. Their efficiency in odor removal may not, however, be as predictable as the absorption (chemical) and adsorption (physical) processes. Very often, a supplemental process such as scrubbers may still be needed for a more reliable control. In the biological stabilization process, care should be taken that the odorous air streams do not contain toxic substances that may interfere with the primary function of the biological treatment process, i.e., the removal of BOD and nitrification. High levels of hydrogen sulfide are not desirable either because of the potential for excessive growth of the filamentous sulfur bacteria, *Thiothrix,* which may hinder the settling of the activated sludge.

The use of scrubber piles is very applicable to sludge composting facilities. No additional material or equipment is needed, and the operation is straightforward. The only precaution may be, as mentioned above, that the pile should be replaced frequently enough to prevent the compost from being saturated with the odorous substances.

Biofilters are very easy to operate, and require minimum attendance except for regular watering and cutting of the grass. The only disadvantage is probably their large land requirements. This restriction limits the use of biofilters to only small isolated plants serving rural communities.

INCINERATION AND AFTERBURNERS

Process Description

This refers to the removal of odors through thermal oxidation at high temperatures. Burning is an effective way to destroy organic substances, including both the volatile solids in sludges that contribute the odors and the vapor of the malodorous organic compounds themselves. Adequate fuel and air supplies are needed to support the process. Although the odorous organic compounds in the contaminated air streams are burnable, the heat value of the air streams is generally low and consequently a very high fuel consumption rate will be needed. Therefore, even though incineration is effective in removing the odorous organics found in sludges, it is seldom used solely for the purpose of removing odors from a contaminated atmosphere. Typically it is used along with the removal of combustible or toxic organic substances. To remove odors by burning, the commonly employed procedure is to use the odorous air stream as the combustion air for the existing sludge incinerator or heating boilers, and the odorous organic compounds are removed along with the burning of sludge and/or fuels [26].

The most important factor in sludge incineration is the burning temperature. Sludge will normally burn quite well at a temperature of 800 to 900°F (430 to 480°C), but up to 1400 and 1500°F (760 and 820°C) is required to eliminate odors and smoke [27]. It has been reported that vapors generated from the incineration of most anaerobically digested sludges requires a temperature of 1300 to 1350°F (700 to 750°C) to deodorize, and a few require 1400°F (760°C). A temperature of 1370 to 1425°F (740 to 770°C) would be needed to destroy odors from volatile fatty acids [28]. It has also been reported that a minimum temperature of 1250°F (680°C) is needed to deodorize the dewatered sludge cakes [29]. Extra fuel consumption is required to increase the burner temperature.

It is important to note that if the incineration temperature is not high enough, various odorous compounds may actually be generated upon the destruction of the volatile organics in the sewage sludge. An alternative technique for increasing the temperature, for odor control purposes, is to use afterburners. An afterburner is a separate burning chamber designed to burn the exhaust gas at elevated temperatures. It can be installed in the exhaust gas duct as a part of the sludge incinerator. For multiple-hearth incinerators, the upper one or two hearts may be modified to become the afterburner treating the exhaust gas while the incoming sludge is bypassed and fed directly onto the lower hearths [27]. Combustion temperature in the upper hearths is then increased by placing extra burners around the upper shell of the incinerator.

The operating temperature of the afterburners may normally be as high as 2000°F (1090°C). To increase the efficiency of thermal oxidation, wire screens, or plates of various metals such as silver, nickel, nichrom, platinum, and platinum-rhodium alloy have been used as the catalysts. Care should be taken to avoid excessive amounts of dust, metal fumes, and some types of halogenated hydrocarbons that can be harmful to, and would rapidly impair, the activity of the catalysts.

Performance

A well-operated incinerator can normally provide almost complete removal of the organic components of a sewage sludge while reducing both the volume and weight of the wet sludge cake by approximately 95% [30]. Including the use of afterburners, odor removing performance however varies depending upon the operating temperature and the odorous substances to be removed. According to Cox [17], in order to achieve a complete thermal oxidation of the odorous substances through burning, nascent oxygen must be available. Nascent oxygen, however, cannot be produced in sufficient quantity unless the temperature reaches 15,000°F (8300°C). This is about eight to ten times the operation temperature of an afterburner. In addition to provide a second burning, another important function of the afterburners is, as claimed by Cox, therefore simply to increase the temperature of the exhaust gas so that they can be expelled from the stack at a higher speed.

Both the height of the exhaust stack and the weather conditions will then influence the efficiency of odor removal through incineration. The odor removal performance of incineration and afterburners is therefore difficult to predict, but the experience is that, the higher the temperature, i.e., the higher the fuel consumption, the higher the removal would be [2]. Better performance is also achievable with the use of metal catalysts.

Advantages and Disadvantages

Incineration processes, along with the use of afterburners, is normally an effective way to remove odorous substances in sludges generated in the wastewater treatment process. They are relatively easy to operate with less down time but the overall power costs may be high due to high fuel consumption rates. Special attention may be placed on the exhaust gases, which could be more odorous than the sludge and the processing air streams. Very often, separate scrubbers are still needed for a more reliable control of the odors.

REFERENCES

1 United States Department of the Interior, Federal Water Pollution Control Administration. "Study of Sludge Handling and Disposal," Report No. WP-20-40.

2 United States Environmental Protection Agency. 1985. *Odor and Corrosion Control in Sanitary Sewerage Systems and Treatment Plants, Design Manual,* EPA-625/1-85-018.

3 United States Environmental Protection Agency. 1978. *Performance Evaluation and Trouble Shooting at Municipal Wastewater Treatment Facilities,* EPA-430/9-78-002.

4 WPCF. 1979. *Odor Control for Wastewater Facilities, MOP 22.* Water Pollution Control Federation, Washington, DC.

5 Leonardos, G., F. Sullivan, D. Schuetzle, S. P. Levine, R. T. Stordeur and T. M. Harvey. 1980. "A Comparison of Polymer Adsorbent and Bag Sampling Techniques for Paint Bake Oven Odorous Emissions," *J. Air Pollution Control Assoc.,* 30:22.

6 Dravnieks A, and W. H. Prokop. 1975. "Source Emission Odor Measurement by a Dynamic Forced-Choice Triangle Olfactometer," *J. Air Pollution Control Assoc.,* 25:28.

7 Markowitz, H., A. Dravnieks, W. Cain and A. Turk. 1974. "Standardized Procedure for Expressing Odor Intensity," *Chemical Senses and Flavor,* 1:235.

8 Burnham, J. C. 1990. "Reduction of Odors in Cement Kiln Dust Stabilized/Pasteurized Municipal Wastewater Sludge Cake," paper presented in the *WPCF Specialty Conference, New Orleans, LA, December 5, 1990.*

9 Wilber, C. and Murray, C. 1990. "Odor Source Evaluation," *BioCycle,* 31:3, 68.

10 WPCF. 1987. "Operation and Maintenance of Sludge Dewatering Systems," WPCF MOP OM-8.

11 1990. "Odor Control Solution." NuTech Environmental Corp, Denver, CO.

12 Hentz, L. H., Jr. et al. 1990. "Odor Control Research at the Montgomery County Regional Compost Facility," paper presented in the *63rd WPCF Annual Conference, Washington, DC, October 7–11.*

13 Cormack, J. W. et al. 1974. "Odor Control Facilities at the Clavey Road Sewage Treatment Plant," paper presented in the *47th WPCF Annual Conference, Denver, Colorado.*

14 Pope, K. J. and Federici, N. J. 1989. "Effective Odor Control Technology at Wastewater Treatment Plants," *Waterworld News*, September/October, p. 12.

15 Ewing, W. C. and Gabrieise, R. S. 1983. "Flint Fumes No Longer Nuisance," *Water/Engineering & Management*, 130:9, 59.

16 Duall Industries. 1989. "Treatment Techniques for Controlling Odors," *Waterworld News*, September/October, p. 18.

17 Cox, J. P. 1975. *Odor Control and Olfaction*. Pollution Sciences Publishing Company.

18 U.S.EPA Hazardous Waste Regulations. 40 CFR 261 Identification and Listing of Hazardous Wastes; 40 CFR 263 Transporter Standards; 40 CFR 264 and 265 Facility Standards.

19 Lovett, W. D. and Poltorak, R. L. 1975. "Activated Carbon and the Control of Odorous Air Pollutants," in *Industrial Odor Technology Assessment*, P. N. Cheremisinoff and R. A. Young, eds., Ann Arbor Science.

20 Leatherdale, J. W. 1978. "Air Pollution Control by Adsorption," in *Carbon Adsorption Handbook*, P. N. Cheremisinoff and F. Ellerbusch, eds., Ann Arbor Science.

21 Turk, A. 1986. in *Air Pollution Handbook*, A. C. Stern, ed., Academic Press.

22 WPCF. 1985. "Sludge Stabilization," WPCF MOP-9.

23 Metcalf and Eddy, Inc. 1991. *Wastewater Engineering, Treatment, Disposal, Reuse, Third Edition*. New York: McGraw Hill, Inc.

24 Carlson, D. A. and Leiser, C. P. 1966. "Soil Beds for the Control of Sewage Odors," *J. WPCF*, 38:829.

25 ASCE. 1989. "Sulfide in Wastewater Collection and Treatment Systems," ASCE Manuals and Reports on Engineering Practice No. 69.

26 WPCF. 1988. "Sludge Conditioning," WPCF MOP FD-14.

27 WPCF. 1988. "Incineration," WPCF MOP OM-11.

28 Sawyer, C. N. and Kahn, P. A. 1960. "Temperature Requirements for Odor Destruction in Sludge Incineration," *J. WPCF*, 32:1274.

29 Laboon, J. F. 1961. "Construction and Operation of the Pittsburgh Project," *J.WPCF*, 33:758.

30 U.S.EPA. 1979. "Sludge Treatment and Disposal," *U.S.EPA Process Design Manual*, EPA 625/1-79-011, September.

CHAPTER 6

Sludge Processing Technology

INTRODUCTION

SLUDGE stabilization processes are key to the reliable performance of any wastewater treatment plant. The primary objectives of sludge stabilization are reduction of odors, pathogens, and putrescibility of the sludge. Commonly used sludge stabilization processes include anaerobic digestion, aerobic digestion, composting, and chemical stabilization.

AEROBIC SLUDGE DIGESTION

Sludge digestion involves the microbiological and biochemical transformation of organic solids in wastewater sludges to innocuous end products. The principal objective of sludge digestion is the destruction of volatile solids i.e. the biodegradable degradable solids in the sludge, resulting in a reduction in the volume of sludge that requires disposal. The change in biodegradable volatile solids can be represented by a first order biochemical reaction: During aerobic digestion excess biomass (activated sludge), primary solids and trickling filter humus, alone or as mixtures are aerated in open or covered tanks. A portion of the pathogenic organisms in the

Roger T. Haug, Solids Technology Division of the City of Los Angeles, Los Angeles, CA; Richard Kuchenrither, Black and Veatch, Kansas City, MO; Joseph F. Malina, Jr., The University of Texas, Austin, TX; David Oerke, Black and Veatch, Aurora, CO; Prakasam Tata, Bernard Sawyer, Stanley Soszynski, and David Zenz, Metropolitan Water Reclamation District of Greater Chicago, Chicago, IL.

sludge also are destroyed during digestion. The aerobic digestion unit processes are separate from the liquid process system.

Aerobic sludge digestion does have some distinct advantages and disadvantages [1,2]. Advantages claimed for aerobic digestion include:

- Low capital costs result for plants under 5 MGD (220 L/s).
- It is relatively easy to operate process.
- Nuisance odors are minimal.
- Volatile solids destruction is approximately equal to that observed in anaerobic digestion as long as the ratio of primary solids to biological solids is less than 0.50.
- Supernatant exerts a low BOD_5 and contains low concentrations of suspended solids and ammonia nitrogen.
- End product is humus-like, odorless, and biologically stable.
- Pathogen reductions are high under normal design.
- Autothermal thermophilic digestion results in 100 percent pathogen destruction.

The disadvantages reported for aerobic digestion processes are:

- There are poor mechanical dewatering characteristics of the aerobically digested sludge.
- High power costs are involved to supply oxygen, even for very small plants.
- Performance is affected by type of sludge, temperature, location, and type of tank material.

The continued use of aerobic digestion has been limited to small [<5 MGD (220 L/s)] wastewater treatment facilities, because of high power costs required for aeration and mixing. The actual cost of energy will define the size of the treatment facility at which aerobic digestion will be more cost effective than anaerobic digestion.

MECHANISM

The mechanism of microbial degradation of volatile solids is different for the various individual mixtures of sludges and under different environment conditions. The aerobic digestion of excess activated sludge may be considered to be a continuation of the activated sludge process [3]. This phenomenon is illustrated in Figure 6.1.

When the soluble substrate is completely oxidized by the microbial population in the activated sludge process, the bacteria in the biomass rely on endogenous respiration of substances stored within the cells for maintenance and utilize their protoplasm for energy. The major reaction is oxidation of the constituents of the biomass. Therefore, the principal

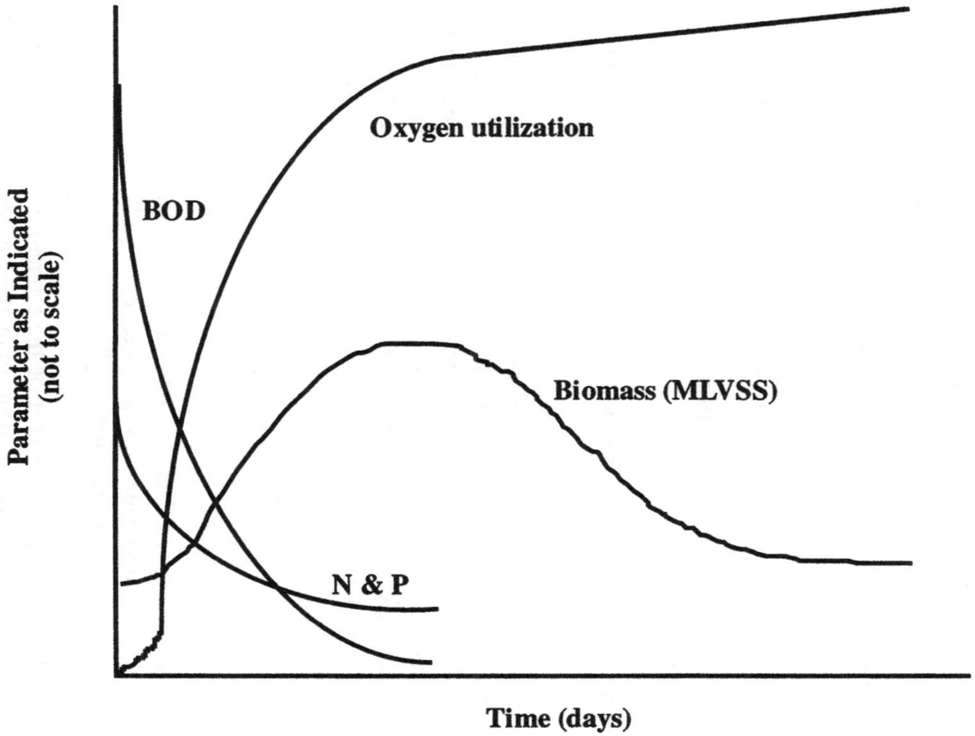
Oxygen utilization
BOD
Parameter as Indicated
(not to scale)
Biomass (MLVSS)
N & P
Time (days)

Figure 6.1 Relationship among BOD, MLVSS, oxygen utilization and nutrients.

reaction in aerobic digestion is the destruction of degradable solids (volatile solids) or the biological stabilization of the biosolids. This biooxidation of biomass results in the reduction of the volume of residual solids requiring disposal. However, this objective of volume reduction has not been completely realized in many facilities because of problems with effective dewatering of the residual solids.

The phenomenon of endogenous respiration can be described in terms of chemical reactions. The chemical composition of biomass may be written as $C_{60}H_{87}O_{23}N_{12}P$ or approximated by the expression $C_5H_7O_2N$. Biooxidation (auto oxidation) (endogenous respiration) of biomass, including the complete nitrification of the ammonia released, can be written as:

$$C_{60}H_{87}O_{23}N_{12}P + 86.5\ O_2 \rightarrow 60\ CO_2 + 12\ H^+ + 12\ NO_3^- + H_3PO_4$$
$$+ 36\ H_2O + \text{Non-Degradable Residuals}$$

This reaction may be written using the simpler representation of the cell composition ($C_5H_7O_2N$) as:

$$C_5H_7O_2N + 8.25\ O_2 \rightarrow 5\ CO_2 + H^+ + NO_3 + H_3PO_4 + 1.5\ H_2O$$
$$+ \text{Non-Degradable Residuals}$$

The biomass is made up of biodegradable and refractory (resistant to biodegradation) components as well as inorganic material. The readily biodegradable constituents are hydrolyzed and metabolized to carbon dioxide and water. However, a portion of the cell mass is composed of a polysaccharide-like material which is not readily decomposed. The overall phenomenon of degradation of biomass can be described by a first-order reaction:

$$r_{Xd} = -k_d X_v$$

in which:
r_{Xd} = rate of endogenous decay or auto oxidation (mg/L-d)
k_d = endogenous (autooxidation) rate coefficient (d^{-1})
X_v = Volatile Solids concentration (mg/L)

Approximately 75 to 80 percent of the biomass is oxidizable and 20 to 25 percent of the biomass is made up of inert material and non-degradable organic solids. Some investigators suggest that a retardant first order reaction better describes the aerobic sludge digestion of municipal wastewater [4]. This approach is based on the phenomenon that as the readily degradable solids are oxidized, the relative fraction of non degradable (refractory) solids increases.

Limited data are available on the magnitude of the endogenous decay coefficients that were observed in full-scale aerobic digestion systems. Reported values of endogenous rate coefficients observed for bench-scale aerobic digestion of wastewater solids ranged from $k_d = 0.028$ to 0.056 d^{-1} (based on degradable solids) and $k_d = 0.016$ to 0.426 d^{-1} (based on degradable volatile solids) [4]. The volatile solids fraction of the biodegradable solids was between 44 and 67%. Other data indicate that the endogenous decay coefficient increased as the concentration of volatile solids increased [4–6].

The bio-oxidation of the cellular nitrogen to ammonia and finally to nitrate could result in a decrease in the pH unless sufficient alkalinity is available to buffer the system. Theoretically, approximately 7.1 kg of alkalinity as $CaCO_3$, are required for the complete nitrification of 1 kg of ammonia [1,7].

APPLICATIONS

The successful application of aerobic digestion of wastewater sludges depends on the type of sludge and the temperature. The detention time is the primary controlling design parameter. Detention times of 15 to 30 days may be required. The organic loading is dependent on the detention time and the concentration of solids in the feed sludge. Organic loadings range from 0.67 to 1.8 kg volatile solids/m^3-day at detention times of 15 to 30 days when the feed sludge is a mixture of activated sludge and primary sludge [8]. The efficiency of volatile solids destruction is related to sludge age as illustrated in Figure 6.2.

The data represented by the curve in Figure 6.2 were reported for pilot- and full-scale aerobic digesters [5–7,9–11]. Therefore, for these sludges and at typical hydraulic detention times (i.e. 15 to 30 days) and operation temperatures of 68°F to 95°F (20°C to 35°C), volatile solids destruction of 40 to 50 percent may be achieved at sludge ages of 20 to 70 days at operating temperatures of 20°C and between 12 and 40 days at $T = 35$°C.

The theoretical quantity of oxygen required to oxidize the biomass ($C_5H_7NO_2$) and completely oxidize the ammonia released to nitrates is 1.98 pounds of oxygen per pound of biomass. However the results observed in pilot-plant and full-scale operations indicate the oxygen requirements ranged from 1.74 to 2.07 pounds of oxygen per pound of volatile solids destroyed at temperatures less than 35°C (mesophilic range). The actual specific oxygen utilization rate (pounds oxygen per 1,000 pounds of volatile solids per hour) also is a function of the sludge age and temperature [11–14]. Data observed in one study are presented in Figure 6.3. and indicate that the specific oxygen utilization decreased as the sludge age was increased and also decreased as the operating temperature decreased.

Aerobic stabilization of a combination of biomass (excess activated sludge or trickling filter humus) and primary sludges includes the sequential

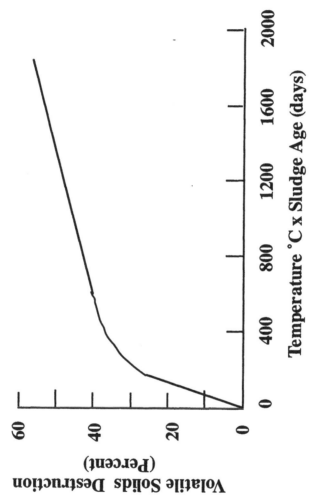

Figure 6.2 Relationship among volatile solids destruction, temperature and sludge age (adapted from USEPA [8]).

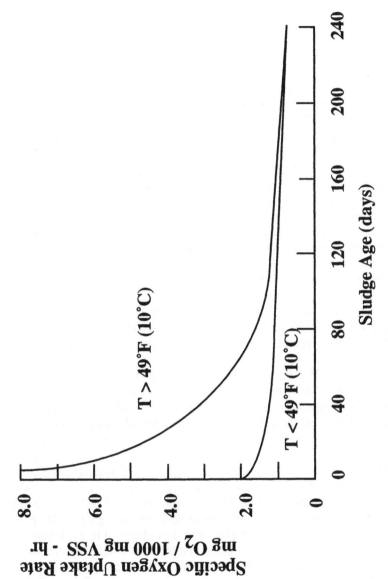

Figure 6.3 Effects of sludge age and temperature on specific oxygen utilization rate [13].

261

degradation of organic solids in the primary sludge similar to the processes in anaerobic sludge digestion along with endogenous degradation of the biomass in the biosolids. The particulate organic material must be converted to soluble compounds which can be subsequently used by the microbial population as a source of nutrients and energy. The bacterial utilization of the soluble compounds formed results in carbon dioxide, water and cell material. However, the amount of new cell material synthesized is overshadowed by the mass of volatile solids in the digesting sludge at low solids loading rates [<1.6 kg volatile solids per cubic meter of tank volume per day (kg VS/m^3-d)] (<100 pounds/ 1000 ft^3-d) [14].

At organic loadings >2 kg VS/m^3-d (160) (>125 pounds/1000 ft^3-day) aerobic digestion of primary sludge solids results in an environment in which the organic solids in the sludge are hydrolyzed and the resulting soluble organics compounds are converted to bacterial cell mass. In effect the initial degradable organic solids are converted almost quantitatively to bacterial cell material and the change in the volatile solids concentration in the digested sludge is minimal [15]. A summary of data observed during a laboratory evaluation and comparison of aerobic and anaerobic digestion of primary wastewater sludge is presented in Table 6.1. These data are averages of about ten individual analyses of samples collected after the laboratory units were operated at the desired organic loading for at least 15 days and the daily variation in each parameter was minimum and within the limits of precision of the analytical procedure.

At a volatile solids loading of 2.27 kg/m^3/day there was no difference in the degree of volatile solids destruction achieved under aerobic and anaerobic conditions [16]. However, at the higher volatile solids loadings, viz., 2.76 and 3.25 kg/m^3/day, the destruction of volatile solids anaerobically was better (46%) than that observed for aerobic digestion (26%). At the higher loadings, more than 50% of the carbon was lost during anaerobic digestion; however, the amount of carbon removed during aerobic treatment was about 23 percent. The organic nitrogen content of the digested sludges and of the feed sludge remained unchanged. However, the organic nitrogen was converted to ammonia nitrogen during anaerobic digestion.

The amount of volatile solids destroyed during anaerobic treatment increased with organic loading. The data indicate that 5.2, 7.1, and 8.7 g/ day of volatile solids were destroyed respectively at organic loadings of 2.27, 2.76, and 3.25 kg/m^3/day [16]. The data observed in the laboratory scale aerobic units, however, indicate that the total overall volatile solids removal was approximately 5 g/day for each of the three loadings.

Design criteria for aerobic digestion are presented in Table 6.2. The hydraulic detention time is the principal design criteria. Oxidation and stabilization of the volatile solids require sufficient time for the reaction to reach completion.

TABLE 6.1. Comparison of the Performance of Aerobic and Anaerobic Digestion of Municipal Wastewater Sludge [16] [T = 35°C (95°F) and Hydraulic Retention Time = 15 days].[a]

	Organic Loading (kg VS/m³-d)	pH	Total Solids (kg/m³)	Volatile Solids (kg/m³)	Total Carbon (kg/m³)	Total Nitrogen as N (mg/L)	Organic Nitrogen as N (mg/L)	NH₄⁺ as N (mg/L)
	2.27							
Feed		6.1	45.6	34.3	21.8	1,300	1,110	190
Aerobic digestion		7.8	31.2	21.6 (37%)[b]	14.8 (32%)[b]	1,275	1,115	160
Anaerobic digestion		7.0	32.5	21.2 (38%)[b]	13.4 (38%)[b]	1,280	895	385
	2.75							
Feed		5.8	56.3	41.3	26.4	1,580	1,360	220
Aerobic digestion		7.5	42.9	30.1 (27%)[b]	17.0 (36%)[b]	1,530	1,260	270
Anaerobic digestion		7.1	38.8	23.6 (43%)[b]	13.2 (50%)[b]	1,525	960	565
	3.25							
Feed		6.0	63.9	48.0	31.4	1,690	1,430	260
Aerobic digestion		7.6	50.0	35.4 (26%)[b]	24.3 (23%)[b]	1,650	1,395	255
Anaerobic digestion		7.1	42.9	26.1 (46%)[b]	14.7 (53%)[b]	1,620	1,050	570

[a]Operating conditions: air flow rate = 9.9 L/min = (350 cfm); gas was recirculated in the anaerobic digestion unit for 15 minutes each hour
[b]Numbers in parentheses are the percent removals for volatile solids and carbon, respectively.

TABLE 6.2. Summary of Aerobic Sludge Digestion Design Criteria [8] [Operating Temperature = 20°C (68°F)].

Hydraulic detention time required		
Excess activated sludge only following primary sedimentation	10–15	days
Activated sludge without primary sedimentation	12–18	days
Excess activated sludge with primary sludge	15–20	days
Trickling filter humus plus primary sludge	15–20	days
Solids loading	0 1 to 0.3	lb Volatile solids/ft³-d
	1.6 to 4.8	kg Volatile solids/m³-d
Influent solids concentration		
Excess activated sludge only	0.8 to 1 2	%
Primary + excess activated sludge	2 to 3	%
Oxygen requirements		
Includes nitrification	2.0	lb oxygen/lb VS destroyed (T <113°F) (<45°C)
	1.45	lb oxygen/lb VS destroyed (T > 113°F) (>45°C)
Nitrification is inhibited at high temperatures		
Minimum dissolved oxygen in digester	1.0 to 2.0	mg/L
Energy requirements for mixing is controlled by the geometry of the tank		
Mechanical aerators	0.75 to 1 50	hp/1000 ft³
	20 to 30	kW/1000 m³
Diffused Aeration	20 to 40	ft³/1000 ft³-min
	0.02 to 0.04	m³/m³-min
Expected maximum concentration of solids after decanting solids for degritted sludge or sludge without chemicals added	2.5 to 3.5	%

The minimum aeration time (hydraulic detention time) for excess activated sludge is 10 days and the minimum retention time increases to about 20 days when solids in the raw sewage are blended with the excess biomass. Oxygen requirements range from about 2.0 lb O_2/lb VS destroyed in the mesophilic range, and decreases to 1.45 lb O_2/lb VS destroyed at temperatures >45°C, since nitrification is inhibited at these high temperatures.

The recommended loadings for aerobically treated mixtures of primary and activated sludges or primary and trickling filter sludges is less than 1.6 kg total solids/m^3-day, and the minimum suggested detention time is 20 days. The recommended detention time for waste activated sludge is about 10 days, however 15 days is preferred [8,16]. The air requirements are 15 to 20 m^3/1000 m^3 (15–20 cfm/1000 ft^3) of tank capacity. However, when primary sludge is treated or blended with activated sludge or when the waste activated sludge is withdrawn from a final clarifier, the air supply should be increased to 25 to 30 m^3/1000 m^3 (25–30 cfm/1000 ft^3) of tank capacity.

PROCESS DESCRIPTION

The aerobic digestion of excess activated sludge resulting from the treatment of municipal and/or industrial wastewaters is an extension of the activated sludge process. "Total oxidation" and "extended aeration" are names applied to the activated sludge process to denote partial stabilization of the biological solids. However, in many treatment plants separate aerobic digestion is used for the stabilization of waste activated sludge alone and mixtures of excess activated and primary sludges. A schematic diagram of a typical aerobic digestion tank is illustrated in Figure 6.4.

In this system air is introduced through spargers that are located in a draft tube. The air provides the oxygen required to maintain an aerobic environment and the combination of diffused air and draft tube comprise an airlift pump which mixes the contents of the digester. Mechanical aeration equipment also may be used to aerate and mix the digesters. Supernatant is drawn from a sludge concentration compartment to minimize the carry over of solids into the liquid effluent. The concentrated sludge is returned to the digesting sludge through an airlift pump.

Originally, aerobic digestion was designed as a semi-batch process. Aerobic digesters still are operated in this mode at many small wastewater treatment facilities. Solids are pumped directly from the secondary clarifiers into the aerobic digester. The time required for filling the digester depends on the digester volume, volumetric flow rate of excess waste-activated sludge, precipitation (rainfall, etc.), and evaporation. During the filling operation, sludge undergoing digestion is continually aerated. When

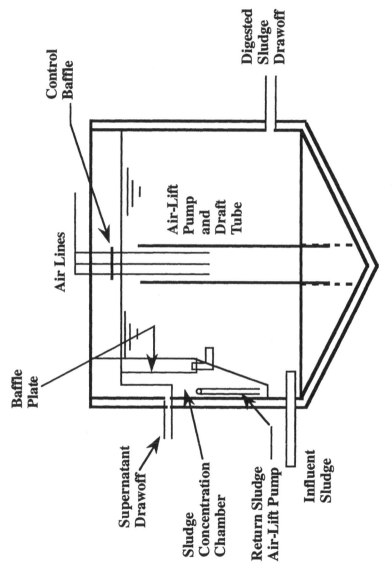

Control
Baffle

Digested
Sludge
Drawoff

Air Lines

Air-Lift
Pump
and
Draft
Tube

Baffle
Plate

Supernatant
Drawoff

Sludge
Concentration
Chamber

Return Sludge
Air-Lift Pump

Influent
Sludge

Figure 6.4 Circular aerobic digester.

the tank is full, aeration continues for two to three weeks to assure that the solids are thoroughly stabilized. At the end of the prescribed digestion time, the aeration is stopped and the stabilized solids are allowed to settle. After the sedimentation is complete, the clarified supernatant is decanted, and the thickened solids are removed at a concentration of between two and four percent. Some of the stabilized solids and supernatant are kept in the digestion tank to provide the necessary microbial population for degrading the influent wastewater solids. The cycle is repeated. The aeration device is started when excess activated sludge and or primary solids are added to the digester.

The conventional continuous aerobic digestion is operated in a similar fashion as the semi-batch process. Untreated solids (excess activated sludge or a combination of biosolids and primary solids) are pumped directly from clarifiers into the aerobic digester. The aerator operates at a fixed level. The effluent overflows into a solids-liquid separator where the stabilized solids are thickened. The thickened stabilized solids may be recycled to the aerobic digestion tank or removed for further processing.

OXYGEN REQUIREMENTS

Activated sludge biomass is most often represented by the empirical equation $C_5H_7NO_2$. Under the prolonged periods of aeration typical of the aerobic digestion process ammonium nitrogen released upon biodegradation of the biomass undergoes nitrification. Hypothetically, this equation indicates that 1.98 pounds (0.898 kg) of oxygen are required to oxidize one pound (0.45 kg) of cell mass. The results of pilot- and full-scale studies indicate that the observed oxygen requirements were 1.74 to 2.07 pounds per pound of volatile solids degraded [16]. For mesophilic systems, a design value of 2.0 lb O_2/lb VS destroyed is recommended. For autothermal systems, which have temperatures above 113°F (45°C), and where nitrification does not occur, a value of 1.45 lb O_2/lb VS is recommended.

The actual specific oxygen utilization rate, pounds oxygen per 1,000 pounds volatile solids per hour, is a function of total sludge age and liquid temperature. Specific oxygen utilization rate is seen to decrease with increase in sludge age and decrease in digestion temperature. Field studies have also indicated that a minimum value of 1.0 mg of oxygen per liter should be maintained in the digester at all times [13].

MIXING

Mixing is required in an aerobic digester to keep solids in suspension and to bring deoxygenated liquid continuously to the aeration device. Whichever of these two requirements needs the most mixing energy controls the design. Power levels of 0.5 to 4.0 hp per 1,000 ft^3 of tank volume (13 to 106 Kw/1,000 m^3) have been reported to be satisfactory. Designers should consult an experienced aeration equipment manufacturer for assistance in design.

Tank geometry affects mixing [21]. The optimum energy requirements to meet oxygen needs of the process for a particular tank geometry can be calculated for specific aeration equipment and systems. Low speed mechanical aerators have been used in noncircular basins.

pH REDUCTION

The decreases in pH and alkalinity observed in aerobic digesters under mesophilic temperature range of operation at increasing hydraulic detention times are caused by acid formation that occurs during nitrification. Although at one time low pH (<6.5) was considered inhibitory to the process, it has been shown that the system will acclimate and perform just as well at the lower pH values. It should be noted that, if nitrification does not take place, a slight decrease in pH may be observed. This phenomenon could happen at low liquid temperatures and short sludge ages or in thermophilic operation [18]. Nitrifying bacteria are sensitive to heat and do not survive in temperatures over 113°F (45°C).

DEWATERING

Mechanical dewatering of aerobically digested sludge is difficult. Dewatering properties of aerobically digested sludge deteriorate with increasing sludge age [8]. It is recommended that conservative criteria be used for designing mechanical sludge dewatering facilities for aerobically digested sludge. As an example, a designer would probably consider designing a vacuum filter for a production rate of 1.5 pounds of dry solids per square foot per hour (7.4 kg/m^2/hr), a cake solids concentration of 16 percent, with a FeCl$_3$ dose of 140 pounds (63.5 kg), and a lime dose (CaO) of 240 pounds (109 kg). This example assumes an aerobic solids concentration of 2.5 percent solids.

PROCESS PERFORMANCE

Volatile solids destruction is a function of both basin liquid temperature and the hydraulic detention time. Volatile solids reductions of 40 to 50 percent are obtainable under normal aeration conditions.

Destruction of pathogens is monitored in terms of dieoff of indicator organisms which serve as surrogates. Indicator organism die-off is high (several log reductions) in aerobic digestion at sludge ages in excess of 15 days [19].

The supernatant from aerobic digesters normally is returned to the head works of the treatment plant. Supernatant characteristics reported for several full-scale facilities operating in the mesophilic temperature range and the current design criteria for aerobic digesters are summarized in Table 6.3.

AUTOTHERMAL THERMOPHILIC AEROBIC DIGESTION (ATAD) MODE OF OPERATION

A new concept that is receiving considerable attention in the United States is the autothermal thermophilic aerobic digestion process [1]. A typical Autothermal Thermophilic Aerobic Digestion system is illustrated in Figure 6.5.

In this process, sludge from the clarifiers is thickened to a minimum of 3% before being fed to the autothermal digester; however, solids concentrations of 4 to 6% are preferred. The heat liberated in the biological degradation of the organic solids is sufficient to raise the liquid temperature in the digester to as high as 140°F (60°C). Advantages are higher rates of volatile solids destruction; production of a pasteurized sludge; destruction of all weed seeds; 30 to 40 percent less oxygen requirement than for the mesophilic process, since few, if any, nitrifying bacteria exist in temperatures above 40°C; and improved solids-liquid separation. Disadvantages include pre-thickening costs and high aeration requirements to mix the sludge at higher

TABLE 6.3. Characteristics of Supernatant after Mesophilic Aerobic Digestion.

Parameter	Ref. [14][a]	Ref. [13][b]	Ref. [19][c]
Turbidity (JTU)	120	na	na
TKN (mg/L)	40	2.9–1,350	na
NO$_3$-N (mg/L)	115	na	30
COD (mg/L)	700	24–25,500	na
PO$_4$-P (mg/L)	70	2 1–930	35
Filtered Phosphorus (mg/L)	na	0.4–120	na
BOD$_5$ (mg/L)	50	5–6350	2–5
Filtered BOD$_5$ (mg/L)	na	3–280	na
Suspended solids (mg/L)	300	9–41,800	6–8
Alkalinity (mg/L)l CaCO$_3$	na	na	150
SO$_4^-$ (mg/L)	na	na	70
Silica (mg/L)	na	na	26
pH	6.8	5.7–8.0	6 8

[a]Average of data observed for 7 months
[b]Data observed at seven operating facilities
[c]Average values

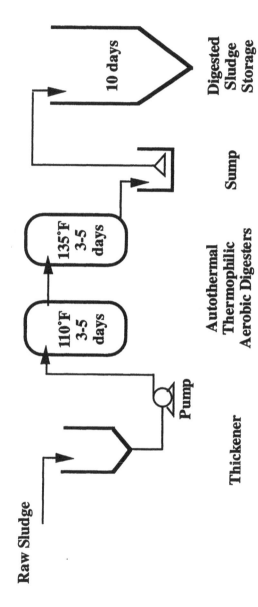

Figure 6.5 Typical autothermal thermophilic aerobic digestion system.

Raw Sludge

Thickener

Pump

Autothermal
Thermophilic
Aerobic Digesters

110°F
3-5
days

135°F
3-5
days

Sump

10 days

Digested
Sludge
Storage

solids contents. Extremely efficient aeration is required for those systems using air instead of high-purity oxygen and insulated tanks.

Design parameters for Autothermal Thermophilic Aerobic Digestion are summarized in Table 6.4. The feed sludge should contain at least 55% volatile solids. The solids concentration of the feed sludge should be at least 3% solids, which barely is sufficient to attain and maintain thermophilic conditions and a maximum of 6% which appears to be the upper concentration for efficient mixing and aeration. Fine screening (maximum bar spacing = 1.27 cm) (0.5 inch) and good grit removal from digester feed are required to remove plastic and stringy materials and to minimize abrasion on mixers and aerators.

Since the majority of aerobic digesters are open tanks, digesters liquid temperatures are dependent on weather conditions and can fluctuate extensively. As with all biological systems, lower temperatures retard the process while higher temperatures speed it up. The system should be designed to minimize heat losses by using concrete instead of steel tanks, placing the tanks below rather than above grade, and using subsurface (diffused air or oxygen) instead of surface aeration. Design should allow for the necessary degree of sludge stabilization at the lowest expected liquid operating temperature, and should meet maximum oxygen requirements at the maximum expected liquid operating temperature.

FEED SLUDGE

The feed sludge should contain at least 55% volatile solids. The solids concentration of the feed sludge should be at least 3% solids, which barely is sufficient to attain and maintain thermophilic conditions and a maximum of 6% which appears to be the upper concentration for efficient mixing and aeration. Fine screening [maximum bar spacing = 1.27 cm (0.5 inch)] and good grit removal are required to remove plastic and stringy materials and to minimize abrasion on mixers and aerators. The influent sludge can be introduced into Reactor 1 continuously, intermittently or in batches. Batch feed should be in a one-hour interval to minimize short circuiting and to ensure that the feed solids are exposed to the reactor temperature continuously for a minimum of 23 hours. This mode of operation enhances pathogen destruction. Sludge is withdrawn from Reactor 2 daily prior to introducing sludge from Reactor 1 in order to ensure that the effluent sludge has been maintained at thermophilic temperatures (55°C to 65°C) (122°F to 149°F) for at least 24 hours and to minimize the possibility of contaminating the treated sludge by contact with partially treated sludge. After the transfer of sludge from Reactor 1 to Reactor 2 is completed, raw sludge is introduced to Reactor 1.

TABLE 6.4. Design Parameters for Autothermal Thermophilic Aerobic Digestion [1,19].

Reactors	Two (2) or more stages of equal volume operating in series, depending on the size of the plant and volumetric flow rate of sludge to be treated, daily batch operations
Reactor types	Cylindrical tanks with a height to diameter ratio of 0.5 to 1.0
Sludge type	Primary solids, municipal and industrial biosolids (excess activated sludge, trickling filter humus), manure
Pretreatment requirements	Efficient grit removal, fine screening, gravity thickening, belt thickening or dissolved air flotation to concentrate solids
Influent total solids concentration	40 to 60 kg/m^3 (4% to 6%)
Required volatile solids concentration	≥25 kg/m^3
Detention time	5 to 6 days
Minimum reaction time	20 hr/stage
Temperature and pH	Reactor 1: T = 35°C to 50°C (95°F to 122°F) pH ≥ 7.2
	Reactor 2: T = 50°C to 65°C (122°F to 149°F) pH ~ 8.0
Air input	4 m^3/hr/m^3 of active reactor volume
Specific power	85 to 105 W/m^3 of active reactor volume
Energy requirement	9 to 15 kWh/m^3 of sludge
Heat potential for recovery	23 to 30 kWh/m^3 of sludge

REACTORS

Typically two (2) reactors in series are installed. Concrete and steel tanks have been used for the construction of ATAD reactors. Steel tanks are less susceptible to heat stress and are less costly to construct than concrete tanks. However, steel tanks require 14 cm (about 6 inches) of mineral insulation and are clad with ribbed aluminum sheeting to protect the insulation from the elements and for aesthetic purposes.

Reactor hydraulic detention times are from 6 to 8 days with 3 to 4 days per reactor. about 60% of the volatile solids destruction takes place in Reactor 1. Mixing and aeration systems include: aspirating aerators, venturi aeration equipment and immersible mechanical aerators. Surface foam is controlled with foam cutters. Control of the foam layer is important in the ATAD process; however, the exact role of the foam layer has not been completely elucidated. The foam seems to improve oxygen utilization, provides insulation and enhances biological activity. However, excessive foam inhibits air from entering the digesting sludge mass. Sufficient free-board [0.5 to 0.1 m (1.65 to 3.5 ft)] must be included in the design of the reactors to allow for accumulation and removal of the foam volume. The foam cutters break up the large foam bubble into small bubbles to form a compact foam layer on the liquid surface. The design and operation of the foam cutters are empirical and must consider the surface area of the reactors, concentration of solids undergoing aerobic digestion, and the type and intensity of aeration.

The digested sludge is cooled in an holding/storage tank, gravity thickener, or by passing the hot sludge through an heat exchanger. The final sludge usually is stored for about 20 days after aerobic digestion. Off gases from the ATAD system have a musty odor and contains ammonia. The off gases can be treated in a water scrubber, or a biofilter or may be diluted with ambient air, depending on the location of the treatment facility.

COST

Evaluation of the available cost data indicate that the total capital cost for the complete system including thickening, ATAD system and storage facilities ranged from $410/lb/d of sludge treatment capacity ($900/kg/d) for plants with wastewater flows of about 300,000 gallons per day (13 L/s) to approximately $100/lb/day of sludge treatment capacity ($215/kg/d) for plants with treatment capacities of 3 MGD (130L/s). The capital cost for the ATAD system alone ranged from approximately $300/lb/d of sludge capacity ($650/kg/d) to about $65/lb/d of sludge capacity ($140/kg/d), respectively. Estimated capital costs for a ATAD system to treat an annual average of 8,846 lb/d of sludge (3,848 kg/d) were reported to be about $225/lb/d sludge treatment capacity ($495/kg/d). This estimate is for a population equivalent of 100,000 or approximately a design wastewater flow rate of 10 MGD. [20]

AEROTHERM PRETREATMENT (PRE-STAGE)

The aerobic thermophilic digestion system has been applied as pretreatment prior to mesophilic anaerobic digestion. The principal objective of this system in addition to volatile solids destruction is the elimination of pathogens and indicator organisms. The final digested sludge should meet the requirements for Class A sludge. A schematic flow diagram for a typical system combining aerobic thermophilic digestion and anaerobic digestion is illustrated in Figure 6.6.

The pre-stage system includes a thickener, sludge/sludge heat exchanger, aerobic thermophilic digester, an equalization tank to adjust the effluent of the aerobic tank and the anaerobic feed cycle, and an auxiliary water/sludge heat exchanger. The exhaust gases from the aerobic thermophilic digester contains volatile acids and other reduced compound and could cause odor problems and requires pretreatment before discharge, usually into the aeration tank of the activated sludge system. The auxiliary heat exchanger is heated by burning methane produced in the anaerobic digesters. Dual-fuel boilers commonly used in anaerobic digestion systems provide startup and backup heating capacity as required. Design parameters for the pre-stage systems are summarized below in Table 6.5.

The pre-stage aerobic thermophilic system normally includes only one aerobic thermophilic reactor which is designed with an hydraulic detention time of less than one (1) day. The operating cycle is semi-continuous. The cycle starts when hot effluent from the aerobic thermophilic digester is introduced into the sludge/sludge heat exchanger and used to heat the thickened cold raw sludge. The sludges are kept in the heat exchanger until the thermophilic sludge reaches a temperature of approximately 40°C (104°F). The preheated raw sludge is pumped into the aerobic reactor where aeration and mixing continue. Additional raw sludge is not introduced into the aerobic digestion system for at least 1 to 2 hours in order to enhance sludge disinfection.

Sludge is added intermittently to the anaerobic digestion tank during the course of a day. Usually about one-tenth of the volume of the aerobic digester is transferred into the anaerobic digester in each of ten (10) transfers. After the transfer from the aerobic digester is complete, raw sludge is introduced from the thickener into the aerobic thermophilic digester. Supplemental heating (heat exchanger) is required in these systems to maintain the proper temperature in both the aerobic and anaerobic digester.

The aeration is limited in the aerobic thermophilic digester; therefore, the sludge stabilization in the aerobic stage is not complete. Anaerobic

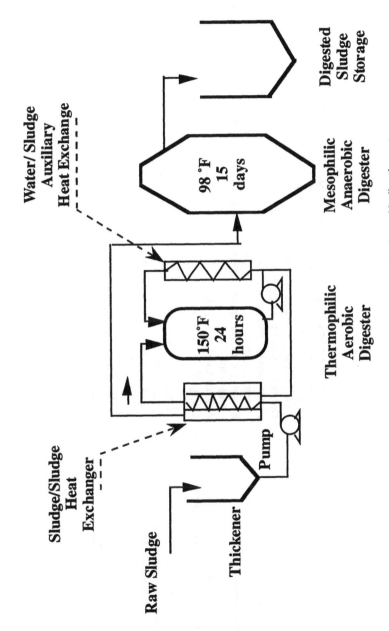

Figure 6.6 Typical pre-stage aerobic thermophilic digestion/anaerobic digestion system.

275

TABLE 6.5. Design Parameters for Pre-Stage Aerobic Thermophilic Digestion [1].

Reactors	One (1) or more stages of equal volume operating in parallel, depending on the size of the plant and volumetric flow rate of sludge to be treated, semi-continuous operations
Reactor types	Cylindrical tanks with a height to diameter ratio of 2 to 5
Sludge type	Primary solids, municipal and industrial biosolids (excess activated sludge, trickling filter humus), manure
Pretreatment requirements	Efficient grit removal, gravity thickening, belt thickening or dissolved air flotation to concentrate solids
Influent total solids concentration	40 to 60 kg/m³ (4% to 6%)
Required volatile solids concentration	≥25 kg/m³
Detention time	12 to 24 hours (average)
Minimum reaction time	1 to 2 hours
Temperature and pH	T = 60°C to 65°C (140°F to 149°F) pH ≥ 7.2 in final reactor
Air input	1 m³/hr/m³ of active reactor volume
Specific power	~ 100 W/m³ of active reactor volume
Energy requirement	Limited design data are available 3 to 5 kWh/m³ of sludge have been reported

digestion is required to complete the degradation of the volatile solids and to produce a stabilized sludge. During anaerobic digestion volatile solids are converted to methane gas. Typical methane production is approximately 8 to 10 ft^3 of methane per pound of volatile solids destroyed in the anaerobic digestion tank (0.6 to 0.6 m^3/kg) VS destroyed [23].

CONCLUSIONS

Aerobic digestion is a cost effective system for the biological treatment of municipal wastewater sludges for small treatment facilities (flow rate <5 MGD). However, aerobically digested sludge is difficult to dewater and contains indicator organisms in the final digested sludge. The recent adoption of the U.S.EPA Part 503 sludge regulation has focused more attention on other cost effective sludge digestion systems for small communities. The Autothermal Thermophilic Aerobic Digestion system and the AeroTherm Pretreatment system offer two alternatives.

ANAEROBIC SLUDGE DIGESTION

Anaerobic digestion of wastewater sludges is the transformation of the organic solids contained in sludges in the absence of oxygen to gaseous end products (e.g., methane and carbon dioxide) and to innocuous and easily dewatering substances. A net reduction in the quantity of solids and volume of sludge requiring disposal also are realized. Sludge digestion may be accomplished under aerobic or anaerobic conditions. Aerobic sludge digestion is more cost effective for small wastewater treatment facilities while anaerobic digestion is used at large treatment plants. In the anaerobic digestion process a portion of the organic solids are microbiologically converted to methane and carbon dioxide gases. Destruction of pathogenic organisms also is accomplished and the final product is a stable, innocuous sludge which can be used as a soil conditioner.

The principal advantages of anaerobic sludge digestion include:

- The produced digester gas contains between 60 and 75 percent methane by volume and is a source of usable energy. The energy produced exceeds the energy required to maintain the temperature of the digesting sludge in most locations. It can also be used to heat buildings, drive the engines for the aeration blowers, or to generate electricity which could be used to drive the pumps at the treatment facility.
- Approximately 25–45% (weight basis) of the influent sludge solids are destroyed and resulting in a reduction in the mass and volume

of the sludge requiring disposal and a decrease in the associated costs of disposal.

- The anaerobically digested sludge contains nitrogen and phosphorus and other nutrient as well as organic material which can improve the fertility and texture of soils.
- A large fraction of the pathogens and parasite ova associated with the raw sludge are inactivated during anaerobic digestion.

The principal disadvantages of anaerobic sludge digestion are:

- The capital costs are high. Large, covered tanks along with pumps for introducing raw sludge and circulating sludge, heat exchangers and gas compressor(s) or pump(s) for mixing are required.
- Long hydraulic detention times, in excess of 10 to 15 days, are required for adequate sludge stabilization, methane-production, and pathogen inactivation.
- Supernatants from anaerobic digestion contain suspended solids, oxygen consuming compounds, nitrogen, and phosphorus. These return flows require additional treatment and when returned to the treatment plant add increased loads to the facility.

MECHANISM OF ANAEROBIC SLUDGE DIGESTION

Anaerobic digestion of wastewater sludges is a controlled process in which there is an orderly degradation of volatile solids and other organic material to a stable, innocuous sludge. A consortium of acid-forming and methane-producing bacteria, in the absence of oxygen, transform the degradable fraction of the initial sludge into methane, carbon dioxide and trace amounts of hydrogen and hydrogen sulfide gases. A schematic diagram of the conversion of volatile solids to stable products and gases is presented in Figure 6.7.

This schematic is based on 100 kg of total solids in the raw sludge and a volatile solids content of 70% (fixed solids content is 30%). It should be noted that not all of the volatile solids in sludge are biologically degradable. In this schematic, Figure 6.7, the volatile content of the digested sludge is 50%; therefore, the amount of volatile solids in the digested sludge is the same as the amount of fixed solids (noncombustible solids). The water associated with the raw and digested sludges is not included in the mass balance presented in this illustration.

The microbial stabilization of wastewater sludges is a sequential process in which volatile solids (organic materials) are hydrolyzed to simpler soluble organic compounds by facultative heterotrophic organisms. These soluble organic compounds are fermented by acid-producing facultative bacteria to volatile acids, carbon dioxide and some hydrogen gas. The

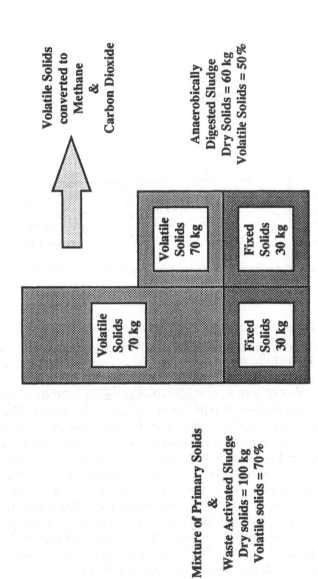

Volatile Solids converted to Methane & Carbon Dioxide

Anaerobically Digested Sludge
Dry Solids = 60 kg
Volatile Solids = 50%

Volatile Solids 70 kg

Fixed Solids 30 kg

Volatile Solids 70 kg

Fixed Solids 30 kg

Mixture of Primary Solids & Waste Activated Sludge
Dry solids = 100 kg
Volatile solids = 70%

Figure 6.7 Schematic diagram of the conversion of volatile solids (dry weight basis) to gas.

279

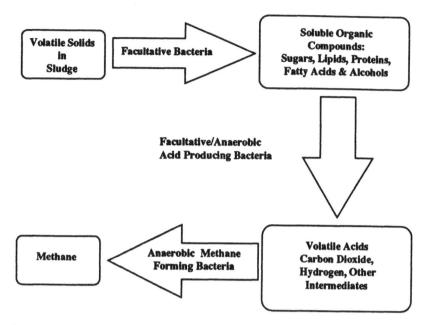

Figure 6.8 A simplified schematic diagram of the mechanism of anaerobic degradation of sludge.

volatile acids are converted primarily to methane gas by anaerobic-forming bacteria. A simplified schematic diagram of the mechanism of anaerobic sludge digestion is illustrated in Figure 6.8.

The acid forming bacteria are primarily facultative, with some obligate anaerobes, and represent a wide variety of microbial genera. The acid forming bacteria are relatively tolerant to changes in pH and temperature. The facultative consortium also can use molecular oxygen (dissolved oxygen) during metabolism; therefore, these microbes protect the methane-forming bacteria, that are strict anaerobes from dissolved oxygen that may be introduced with the feed sludge into the anaerobic digestion system. Acid-forming bacteria convert volatile solids and soluble intermediate organic compounds into volatile acids and simpler organic compounds. There is a redistribution among the organic fraction of the solids from particulate matter to simpler dissolved organic compounds and the release of carbon dioxide, hydrogen and hydrogen sulfide gases.

The principal intermediate compounds produced during acid fermentation are the volatile acids which, in turn, are used by the methane-forming bacteria. The pH may decrease, if the volatile acids accumulate and the concentration increases to a level at which the methane forming bacteria are inhibited. The primary volatile acids produced are acetic, propionic, and butyric acids; however, other volatile acids frequently found in smaller quantities in anaerobically digesting sludge include formic, valeric, isovaleric, and caproic acids.

The chemical formulations for these volatile acids are listed in Table 6.6. These volatile acids are the substrate for the methane forming bacteria which convert the acids to methane and carbon dioxide gases. The mechanism of anaerobic digestion of sludge is sequential in nature; however, acid fermentation and methane fermentation take place simultaneously and synchronously in a well-buffered, actively digesting system.

Effective anaerobic digestion of sludge requires the maintenance of a balance among the rates of the acid production and conversion of volatile acids to methane. The rate of methane fermentation controls the overall rate of sludge stabilization. If the pH decreases below pH = 6.0, methane-forming bacteria will be inhibited and volatile (organic) acids will continue to accumulate. Alkalinity also is produced during the anaerobic degradation of solids and tends to provide the buffering capacity for the system.

The methane-forming bacteria are strict anaerobic organisms and produce methane from the anaerobic fermentation of simple organic compounds. However, each species of methane-producing bacteria can ferment only a relatively restricted group of simple compounds to methane; therefore, several different species of methane forming bacteria are necessary for the anaerobic stabilization of the organic fraction of sludges. A summary of the methane forming bacteria and substrate utilized by each species is presented in Table 6.7 [24].

These bacteria have one major similar characteristic, besides being strict anaerobes, namely, they all produce methane from the anaerobic fermentation of simple organic compounds. However, each species of methane bacteria can ferment only a relatively restricted group of simple compounds to methane; therefore a number of different species of methane forming bacteria may be necessary to completely stabilize some wastes anaerobically.

Each of the longer-chained organic acids will be degraded to either

TABLE 6.6. Volatile Acids Commonly Found during Anaerobic Sludge Digestion.

Common Name	Formula
Formic acid	$HCOOH$
Acetic acid	CH_3COOH
Proprionic acid	CH_3CH_2COOH
Butyric acid	$CH_3CH_2CH_2COOH$
Valeric acid	$CH_3CH_2CH_2CH_2COOH$
Isovaleric acid	$(CH_3)_2CHCH_2COOH$
Caproic acid	$CH_3CH_2CH_2CH_2CH_2COOH$

TABLE 6.7. Classification of Methane Bacteria by Substrate.

Rod-Shaped Cells		
Non-spore forming:	*Methanobacterium*	
Methanobacterium formicicum		formic acid, CO_2, H_2
Methanobacterium propronicum		propionic acid
Methanobacterium sohngenii		acetic acid, butyric acid
Spore forming *Methanobacillus*		
Methanobacillus omelianskii		primary and secondary alcohols, H_2
Spherical Cells		
Methanococcus mazei	*Methanococcus*	acetic acid, butyric acid
Methanococcus vannielii		formic acid, H_2
	Methanosarcina	
Methanosarcina barkerii		methanol, acetic acid, carbon monoxide, H_2
Methanosarcina methanica		acetic acid, butyric acid

butyric or propionic acid and then to acetic acid and methane. One more of propionic acid is formed per mole of odd-numbered carbon long chain fatty acid during methane fermentation; however, no propionic acid is formed when long chain even-numbered carbon fatty acids are fermented.

ANAEROBIC PROCESS REQUIREMENTS

The rate limiting reaction in anaerobic digestion of sewage sludge is the conversion of volatile acids to methane gas by the methane-forming bacteria which are strict anaerobes and extremely sensitive to changes in temperature and pH. Therefore, it is essential that the environment in the anaerobic digestion tank be maintained at conditions optimum for methane forming bacteria. Several environmental conditions must be maintained for optimal operation of anaerobic digestion of sludges. Optimum conditions for maximum methane production during anaerobic digestion are listed in Table 6.8 [25,26]. The range of extreme conditions under which sludge digestion and methane production also have been reported and these are included in Table 6.8.

The methane produced during anaerobic digestion of organic solids contains much of the energy released as the amount of energy required to support the growth of the facultative and anaerobic bacteria is relatively low in comparison to the energy released through aerobic oxidation of the same quantity of organic solids and available for the synthesis of new organisms. The growth of new bacterial cells (biomass) is not evident during the anaerobic digestion of wastewater sludges because the increase

TABLE 6.8. Environmental and Operating Conditions for Maximum Methane Production during Anaerobic Digestion of Wastewater Sludges.

Variable	Optimum	Extreme
pH	6.8–7.4	6.4–7.8
Oxidation reduction potential (ORP) (mv)	−520 to −530	<490; >550
Volatile acids (mg/L as acetic acid)	50–500	>2,000
Alkalinity (mg/L as CaCO₃)	1,500–3,000	<1,000; >5,000
Temperature		
Mesophilic	86–95°F	<68; >104°F
	30–35°C	<20; >40°C
Thermophilic	122–132°F	<113; >140°F
	50–56°C	<45; >60°C
Hydraulic detention time (days)	10–15	<7; >30
Gas composition		
Methane (CH₄) (%v)	65–70	<60; >75
Carbon dioxide (CO₂) (%v)	30–35	<25; >40

in the concentration of suspended solids resulting from microbial cell synthesis is relatively low and is overshadowed by the overall reduction in the concentration of volatile solids in the sludge and the mass of undigested or partially digested sludge still remaining in the system.

Sufficient alkalinity is essential for proper pH control. Alkalinity is derived from the breakdown of organic compounds and is present primarily in the form of bicarbonates which are in equilibrium with the carbon dioxide in the gas at a given pH.

DESIGN CONSIDERATIONS FOR ANAEROBIC SLUDGE DIGESTION

The purpose of anaerobic digestion is the biological destruction of a portion of the volatile solids in the sludge to enhance the dewaterability of the digested sludge and to minimize the putrescibility of the sludge. Volatile solids degradation is time dependent. Therefore, the design criteria for anaerobic digestion systems is based on the solids detention time required to achieve a specific reduction in the volatile solids content of the digested sludge. The hydraulic detention time and solids retention time (mean cell residence time) are the same for an anaerobic digestion system with no recycle. The design parameters that must be considered and be controlled include hydraulic detention time, uniform solids loading, temperature, and mixing [27].

Hydraulic Detention Time

The generation times (i.e. the time required to double the number of bacteria or to double the microbial population) of methane-forming bacteria range from less than two days to more than 20 days at a temperature of 95°F (35°C). Therefore, typical detention times for anaerobic sludge digestion is about 15 to 20 days. However, hydraulic detention times as low as 7 days may be used at facilities where a high level of operational control of the process is maintained. The conversion of volatile solids to gaseous products is controlled by the hydraulic detention time. Therefore, the design detention time is a function of the final disposition of the digested sludge, i.e. land application or incineration. Some operating data for full-scale anaerobic sludge digestion systems are summarized in Figure 6.9 [28,29].

The data reported are for full-scale plant operations of operating anaerobic sludge digesters collected by Estrada [23] (solid squares). The solid triangles represent data published by Torpey [29] who continued to add concentrated sludge to the anaerobic digesters until impending process failure was observed. These data indicate a wide variation in the concentration of volatile solids at the same hydraulic detention time. The reported

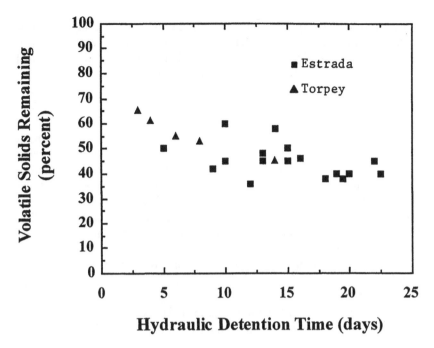

Figure 6.9 Effects of hydraulic detention time on the destruction of volatile solids.

performance data reflect the effects of variations in the composition of raw sludge on the characteristics of the digested sludge. A concentration of 50% volatile solids in the digested sludge usually is considered satisfactory. Therefore, the data presented in Figure 6.9 indicate that this level of volatile solids could be achieved at a minimum hydraulic detention time of five days; however the bulk of the data indicate that the design hydraulic detention time should be at least 10 days.

Longer hydraulic detention times may be desirable depending on the level of operator skills available at the treatment facility. The ultimate disposal of the digested sludge determines the degree of digestion required.

Solids Loading

The hydraulic detention time controls the degree of stabilization of volatile solids. The actual volatile solids loadings to the anaerobic digestion tank is controlled by the efficiency of the sedimentation and thickening processes in removing solids and in concentrating the sludge that is pumped to the digester. Therefore, the concentration of solids in the feed sludge actually controls the loading and the size of the anaerobic digester. The ability to thicken the sludge becomes an important design and operating consideration and may be a major limitation to digester loadings. Pretreatment of sludge may involve blending of primary sludge with thickened excess activated sludge or thickening the blended primary and biological sludges to maintain the organic loading to the digester. Design solids loading to anaerobic digestion systems should be between 200 and 450 pounds of volatile solids per 1,000 cubic feet per day (3.2 to 7.2 kg VS/m³-day) [27]. (Note: 100 lb/1,000 cu ft-day = 1.6 kg/m³-day.)

The hydraulic detention time affects the size of digestion tank required to achieve the desired destruction the volatile solids during anaerobic digestion. In turn the size of the digestion tank and the concentration of solids in the feed sludge dictate the solids loading possible at the required minimum hydraulic detention time. The relationship between the concentration of solids in the feed sludge and the organic loading to the digestion tank for a given hydraulic detention time is illustrated in Figure 6.10.

For example, in order to operate an anaerobic digestion system at a volatile solids loading of 200 lb/1,000 cu ft-day (3.2 kg/m³-day) and a detention time of 10 days, the concentration in the feed sludge must be about 3.2% based on volatile solids [30]. This concentration translates to a solids concentration of approximately 4.5% total solids (assuming

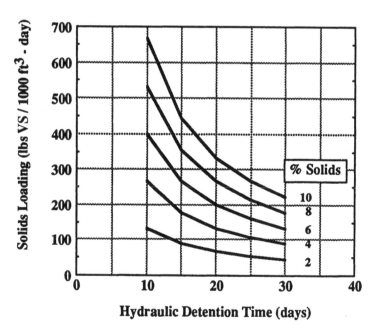

Figure 6.10 The effects of solids concentration in the feed sludge and hydraulic detention time on the organic loading to anaerobic digestion systems.

72% volatile solids). In order to achieve this loading of the combined primary and biological sludges, the biological sludges would have to be thickened to at least 2% and the solids concentration of primary sludge would have to approach 7%. If the design hydraulic detention time is 15 days, the solids concentration would have to be increased to 4.6% based on volatile solids (6.4% total solids) to maintain a solids loading of 200 lb VS/1000 cu ft-day (3.2 kg/m³-day). Separate continuously operated gravity or dissolved air flotation thickeners, belt presses, of thickening centrifuges would be required to concentrate excess activated sludges at most treatment plants to insure continuous volatile solids loading as high as possible.

Temperature

It is essential that the operating temperature be maintained as constant as possible. Sharp and frequent fluctuations in temperature affect the performance of the methane-forming bacteria. The effects of temperature on total gas and methane production are illustrated in Figure 6.11.

Methane-forming bacteria are active in two temperature zones, namely, in the mesophilic range between 85°F to 95°F (29.5°C to 35°C) and in the thermophilic range of 122°F to 140°F (50°C to 60°C) [31]. At temperatures above 104°F (40°C) and below 122°F (50°C), methane production is inhibited [32]. Therefore, anaerobic digestion systems should not be designed to operate in the temperature range of 100°F to 120°F (38°C to 49°C). Anaerobic sludge digestion also can be operated successfully at temperatures as low as 68°F (20°C) as long as sufficient residence time for the methane-producing bacteria is provided. Gas production is higher at temperatures of 95°F (35°C) and 131°F (55°C) [33,34]. However, at temperatures near 113°F (45°C) the gas production is lower, indicating that the methane forming bacteria are inhibited at this intermediate temperature range.

One important advantage of thermophilic digestion is a greater destruction of pathogenic organisms than those reported for sludges digested at mesophilic temperatures [35]. The increased cost of providing the additional heat required to maintain the thermophilic conditions is not offset by increased gas production or more complete digestion.

The principal advantages of mixing anaerobic digestion tanks are the elimination of scum layers and thermal stratification and maintenance of a uniform temperature throughout the tank. The major disadvantage of completely mixing a digestion tank, in addition to the cost of mixing, is the need for a facility which will enhance the separation of the digested solids from the liquid phase, depending on the final disposal of the digested sludge [36].

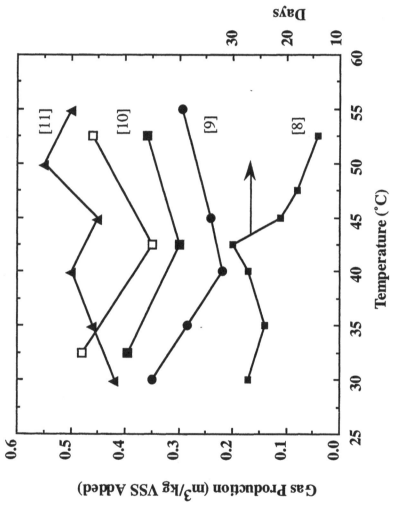

Figure 6.11 Effects of temperature on total gas and methane production.

PRODUCTS OF ANAEROBIC SLUDGE DIGESTION

The main products of anaerobic sludge digestion are gas containing methane and carbon dioxide and innocuous digested sludge solids. The underflow from mixed anaerobic digestion tanks is a liquid sludge that contains high concentrations of suspended and dissolved solids, organic and inorganic materials) that include nitrogen and phosphorus compounds.

Gas Production

Typical gas production rates range from 12 to 16 cubic feet of gas per pound of volatile solids destroyed (cu ft/lb VS) or (0.75 to 1.0 m³/kg VS destroyed). The methane content of the gas is 60 to 75% by volume [37]. The energy content of the methane produced is about 1,000 BTU per cubic foot. Therefore, assuming a gas production rate of 15 cu ft/lb VS destroyed and a methane content of 67%, the energy produced is approximately 10,000 BTU per pound of volatile solids destroyed.

The potential for recovery and using the energy from anaerobic digestion are important considerations, especially in light of the costs of energy supplied to the treatment facility. A schematic diagram of a gas recovery and reuse system is presented in Figure 6.12 [37].

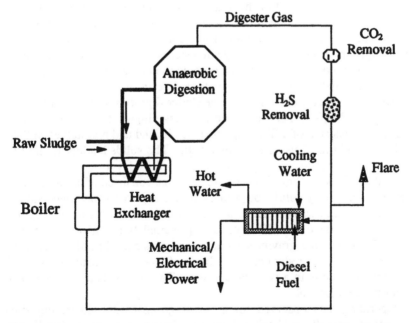

Figure 6.12 Recovery and reuse system for gas produced during anaerobic sludge digestion.

The energy in the digester gas can be used to heat the raw sludge and to maintain the digester at the desired temperature, as well as to heat the buildings. At large treatment facilities, the amount of energy available in the gas produced during the anaerobic digestion of the sludges is in excess of that required for heating and may be converted to mechanical and/or electrical energy which could be used in plant operations.

In the system illustrated in Figure 6.12, a portion of the produced digester gas is used as fuel for the dual fuel engine and is converted to mechanical and/or electrical energy. The mechanical energy generated can be used to drive the compressor for the diffused aeration system for an activated sludge system while the electrical energy can be used to pump the raw sewage and to meet other electrical energy requirements of the treatment facility. Cooling water which is discharged from the engine at a temperature of 160°F to 185°F (70°C to 82°C) is used to preheat the sludge. In addition the hot exhaust gases from the engine are passed through a heat exchanger to heat the sludge. A portion of the digester gas also is used to heat the sludge to the desired temperature at which the digester is to be operated and to heat the buildings.

Solids Separation After Digestion

The ultimate disposal of the anaerobically digested sludge dictates the need to separate and concentrate the digested sludge from the liquid phase (supernatant). Some high-rate anaerobic sludge digestion systems include two-stages in series. The first stage digester is heated and mixed and most of the active biodegradation of the organic matter and gas production take place in the first stage. The second-stage digester is similar in design and size as the first-stage unit. A two-stage sludge digestion system is illustrated in Figure 6.13 [38]. This full-scale two-stage high-rate anaerobic sludge digestion consisted of a first stage that was heated and maintained at 94°F (34.4°C) and mixed using recirculated gas. The solids loading to the first stage unit was approximately 45 lb volatile solids/day-1,000 ft^3 of digester capacity (0.72 kg VS/m^3-day). The feed sludge consisted of approximately equal amounts (dry solids) of primary sludge and excess activated sludge. The second stage was neither heated nor mixed and the hydraulic detention time in each stage was 39 days. The principal function of the second-stage digester is the gravity separation and concentration of the digested sludge solids; thus reducing the volume of sludge requiring disposal. A supernatant results in a two-stage system. The supernatant requires treatment before disposal.

Operating data observed for a two-stage anaerobic sludge digestion system at the Village Creek Wastewater Treatment Plant in Fort Worth, Texas are summarized in Table 6.9 and Table 6.10 [38]. These data

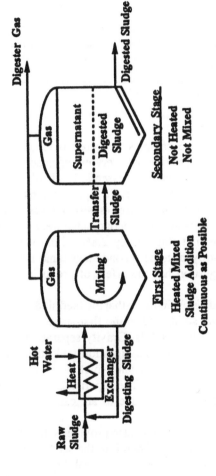

Figure 6.13 Schematic of a two-stage anaerobic sludge digestion system (Fort Worth, Texas).

291

TABLE 6.9.   Average Chemical and Physical Characteristics of Sludges from a Two-Stage Anaerobic Sludge Digestion System.

Component	Concentration (mg/L)[a]			
	Feed Sludge	Transfer Sludge	Supernatant	Stabilized Sludge
pH	5.7	7.7	7.8	7 8
Alkalinity (as CaCO₃)	758	2,318	2,630	2,760
Volatile acids (as acetic acid)	1,285	172	211	185
Total solids	35,600	18,200	12,100	32,800
Fixed solids	9,000	6,600	3,310	12,300
Volatile solids	26,600	11,600	8,790	20,500
Total nitrogen (as NH₃)	1,559	1,425	1,182	2,146
Ammonia nitrogen (as NH₃)	213	546	618	691
Organic nitrogen (as NH₃)	1,346	879	564	1,455

[a]Except pH

represent daily composite samples collected over a one month period during the summer and provide an overview of anaerobic digestion performance. These data indicate that essentially all of the stabilization of volatile solids occurred in the first-stage digestion tank in which approximately 57% of the volatile solids were converted to gas. Solids degradation and gas production observed in the second-stage were negligible compared to the activity in the first stage. Only 2.8% of the volatile solids in the raw sludge were destroyed in the second-stage tank. Destruction of solids during anaerobic digestion has the effect of producing a dilute digested sludge.

The data presented below indicate that the feed sludge contained approximately 3.56% of total solids and in the first-stage tank the concentration was reduced to 1.86% total solids. Gravity separation and concentration did not occur in the second-stage digestion tank; therefore, most of the digested solids were in the supernatant which was recycled to the primary clarifiers at this treatment facility. The supernatant solids are then returned to the digestion system in the sludge or are carried out of the treatment plant in the effluent.

The second-stage digestion tank also provides standby digester capacity and storage of digested sludge. Therefore the second-stage digestion tank should be designed and equipped with capability to be mixed and with appropriate piping connected to the heat exchange units so that feed sludge can be added directly to the second-stage tank through the heat exchanger. The second-stage digestion tank also assures against short-circuiting of raw sludges through the digestion system and provides a margin of safety for enhanced pathogen reduction.

TABLE 6.10. Materials Balance for a Two-Stage Anaerobic Sludge Digestion System.

	Quantity (tons)				Gas	
	Feed Sludge	Transfer Sludge	Supernatant	Stabilized Sludge	1st Stage	2nd Stage
Total solids	106.8	53.5	32.2	13.6	—	—
Volatile solids	79.9	34.1	23.4	8.5	—	—
Fixed solids	26.9	19.4	8.8	5.1	—	—
Carbon	46.2	20.4	11.8	4.5	22.1	2.7
Total nitrogen (as NH_3)	4.66	4.20	3.25	0.89	0.47	0.04
Ammonia nitrogen (as NH_3)	0.64	1.61	1.64	0.28	—	—
Organic Nitrogen (as NH_3)	4.02	2.58	1.50	0.60	—	—

293

Two-stage digestion systems permit the separation of the solids from the liquid carrier. The data discussed above indicate that the supernatant resulting from solids separation contains a considerable amount of volatile solids and organic constituents and cannot be disregarded in evaluating digester performance or estimating the load on the treatment plant. The basic cause of the problem is that anaerobically digested sludges do not settle well. Two factors contribute to this phenomenon: flotation of solids and high proportion of fine-sized particles.

A schematic flow diagram and a solids mass balance for a typical municipal treatment facility with anaerobic sludge digestion are presented in Figure 6.14 [39]. The treatment system illustrated in Figure 6.14 includes a thickener for the excess activated sludge and a blend tank to mix the primary and thickened excess biological solids prior to introduction into a two-stage sludge digestion system. The influent wastewater characteristics (BOD$_5$ and suspended solids) are recalculated to reflect the contributions of the return flows from the processing and digestion of the sludge solids. Summary calculated results of a solids mass balance are presented in Table 6.11. These data represent the conditions at steady-state operations for each of the units producing sludge as well as for the sludge handling and processing units.

In this example, approximately 52% of the volatile solids were destroyed and the volatile solids content of the sludge was reduced from about 68% to 50%. The return flows contribute approximately 3.6 m^3/s of flow and 558 kg/d of suspended solids

Other concentrations and quantities of sludge that require digestion can be substituted into this type of mass balance for other situations. The chemical and physical characteristics of the return flows can be estimated for the new conditions.

The digested sludge still contains relatively high concentrations of organic material and suspended solids as well as nitrogen and phosphorus. The stabilized sludge solids may be dried on sand beds or disposed of by other methods such as application to land as a soil nutrient. The drainage from the sludge drying beds must be returned to the influent of the plant.

Treatment of Digester Supernatant

The characteristics of the supernatant from two-stage anaerobic sludge digestion systems for different wastewater treatment facilities are presented in Table 6.12. These characteristics represent a summary of the wide range of data observed at operating facilities and include supernatants of digestion systems in which the feed sludge was either primary sludge alone, primary sludge plus trickling filter humus or primary and activated sludges [40,41].

All parameters were not reported for all facilities for which operating data were observed; therefore, the average, maximum and minimum con-

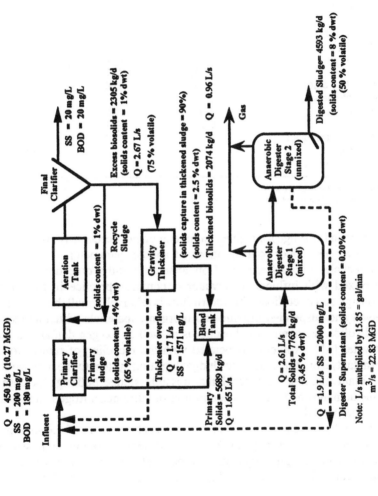

Figure 6.14 Solids mass balance—anaerobic sludge digestion system.

Q = 450 L/s (10.27 MGD)
SS = 200 mg/L
BOD = 180 mg/L

Influent

Primary Clarifier

Aeration Tank

(solids content = 1% dwt)

Recycle Sludge

Final Clarifier

SS = 20 mg/L
BOD = 20 mg/L

Excess biosolids = 2305 kg/d
(solids content = 1% dwt)
Q = 2.67 L/s
(75 % volatile)

Primary sludge
(solids content = 4% dwt)
(65 % volatile)

Gravity Thickener

(solids capture in thickened sludge = 90%)
(solids content = 2.5 % dwt)   Thickened biosolids = 2074 kg/d   Q = 0.96 L/s

Thickener overflow
Q = 1.7 L/s
SS = 1571 mg/L

Blend Tank

Primary Solids = 5689 kg/d
Q = 1.65 L/s

Q = 2.61 L/s
Total Solids = 7763 kg/d
(3.45 % dwt)

Anaerobic Digester Stage 1 (mixed)

Gas

Anaerobic Digester Stage 2 (unmixed)

Digested Sludge= 4593 kg/d
(solids content = 8 % dwt)
(50 % volatile)

Q = 1.9 L/s SS = 2000 mg/L

Digester Supernatant (solids content = 0.20% dwt)

Note: L/s multiplied by 15.85 = gal/min
m³/s = 22.83 MGD

TABLE 6.11. Mass Balance for Solids for an
Activated Sludge Treatment Plant and Effects of
Thickener Overflow and Anaerobic Digestion Supernatant.

	Total Solids (kg/d)[a]	Volatile Solids (kg/d)[a]	Flow Rate (m³/d)[b]
Primary sedimentation			
sludge	5689	3698	142
Activated sludge process			
Effluent	778	584	
Waste sludge	2305	1729	230
Thickener			
Overflow	231	173	147
Underflow	2074	1556	83
Blend tank to digester	7763	5254	225
Digester effluents			
Supernatant	327	213	164
Sludge	4593	2297	58
Volatile solid destroyed		2744	
Gas produced			2744
Methane Produced			1922

[a]Multiply kg/d by 2 206 to get lb/d.
[b]Multiply m³/d by 264.2 to get gal/d

centrations in the range for a given parameter do not necessarily correspond to the lower and upper concentrations of another parameter.

The recycle of the supernatant from anaerobic sludge digestion systems to be mixed with influent wastewater is common practice in treatment plant design and operation. This return flow is relatively small, but contains dissolved and suspended organic and inorganic materials which add suspended solids, nitrogen and phosphorus as well as oxygen demand to the wastewater [41]. Problems associated with the recycling of return flows (supernatant) from the anaerobic sludge digestion system to the headworks of the plant include: odor problems, possible sludge bulking, increased chlorine demand and higher concentrations of nutrients, nitrogen and phosphorus, in the effluent.

Separate biological, chemical and/or physical treatment of the supernatant may be considered depending on the effluent limitations imposed by the local or federal regulatory agency. Process alternatives for the treatment of supernatants from anaerobic digestion are summarized in Table 6.13.

DISPOSAL/REUSE OF DIGESTED SLUDGE

Anaerobically digested sludge may be mechanically dewatered by belt presses, screw presses, vacuum filters, and centrifuges after it is conditioned

TABLE 6.12. Characteristics of Supernatants for Two-Stage Anaerobic Sludge Digestion Systems.

Parameter	Concentration[a] (mg/L)		
	Primary Sludge[b]	Primary and Trickling Filter Sludges[c]	Primary and Waste Activated Sludge[b]
Total solids	9,400	4,545	814 983 1,475 2,160
Volatile solids	4,900	2,930	
TSS			
Average	4,277	2,205	383 143 740 1,075 4,408
Maximum	17,300	7,772 32,400	14,650
Minimum	660	100 85	100
VSS			
Average	2,654	1,660	299 118 750 3,176
Maximum	10,850	4,403 17,750	10,650
Minimum	200	135	100
BOD$_5$			
Average	713	1,238	515 667
Maximum	1,880	6,000	2,700
Minimum	200	135	100
COD		4,565 2,230	1,384 1,310 1,230
TOC		1,242	443 320
Total (PO$_4$)–P		143	63 87 100
NH$_3$–N		853	253 559 360 480
Organic–N		291	53 91
pH	8.0	7.3 7.2	7.0 7.8 7.0 7.3
Volatile acids (as acetic acid)		264	322 250
Alkalinity (as CaCO$_3$)	2,555	3,780	1,349 1,434

[a] Except for pH, all values are average for the sampling period studied
[b] Values indicated are a composite from seven treatment plants.
[c] Values indicated are a composite from six treatment plants.
Source: Adapted from References [40,41]

TABLE 6.13.   Process Alternatives for the Treatment of Supernatants.

Constituent	Removal Process Alternatives
Suspended solids	Coagulation with iron salts, filtration
BOD$_5$	Removal with suspended solids, stripping of volatile acids, aerobic biological treatment, activated carbon adsorption
Carbon dioxide	Precipitation with lime, stripping, ion exchange
Nitrogen	Removal of total Kjeldahl nitrogen (organic nitrogen) with suspended solids, chemical precipitation, ion exchange, ammonia stripping may occur at pH > 8 3
Phosphorus	Removal with suspended solids, chemical precipitation, ion exchange

with chemical conditioning agents. Anaerobically digested sludge may be dried in drying beds, or evaporation basins followed by application of the dewatered sludge on the land for non-agricultural or silvacultural purposes. Liquid anaerobically digested sludge also may be applied directly to the non-agricultural land for the production of edible crops, grasses, ornamental plants or trees. Dried sludge also may be mixed with a bulking agent, usually wood chips and composted [active aerobic digestion for 5 days at a minimum temperature of 53°C (127°F)] followed by aging for 30 days or longer. The wood chips are separated from the composted mixture. Composting sludge under these conditions results in a Class A sludge.

Alkaline stabilization of digested sludge to meet Class A requirements. Lime, kiln dust and combination of the two materials are added to sludge to raise the pH to 12 and the sludge is heated to 70°C (158°F) to affect pathogen reduction. Heat pasteurization also is used to reduce the pathogen concentration in sludge to satisfy the Class A requirements.

The addition of quicklime to digested sludge takes advantage of the heat released during hydration of the lime to raise the temperature to approximately 70°C (158°F) [42]. The lime also raises the pH to >12. Under these conditions of temperature (70°C) (158°F) and pH (pH > 12) pathogens are reduced to below detection limits. Lime disinfection of sludge approaches the conditions required to produce a Class A sludge. A Class A sludge being one which contains <1,000 fecal coliforms/g, <1 virus/4 g, <3 salmonella sp./4 g, and <1 helminth ovum/4 g.

DESIGN OF ANAEROBIC SLUDGE DIGESTION SYSTEMS

Tank Configuration

Advances in the design of anaerobic digestion tanks are related to improving performance and minimizing operating problems. In some cases aesthetic considerations were controlling in selecting digester configuration. Therefore, recent design approaches to the design of anaerobic digestion systems have considered various configurations centering on the egg-shaped digesters and different modifications. Concerns over hydrocarbon emissions have focused attention on fixed covers instead of floating covers for anaerobic digestion. Excessive foaming in anaerobic digestion tanks is linked to using gas for mixing the tank contents. Therefore, external recirculation of the digesting sludge using pumps and mixers located outside the digestion tank is becoming more popular.

Anaerobic digestion tanks may be either cylindrical or egg-shaped. The shape of the anaerobic digestion tank effects the type of mixing required. In the United States anaerobic digestion tanks historically have been circular and relatively shallow with the diameters larger than the depth [43]. The diameters range from 20 feet to 125 feet (6 m to 38 m) with side water depths between 20 feet to 40 feet (6 m to 12 m). The walls usually are reinforced concrete or post-tensioned concrete. The bottoms of the tanks slope (minimum slope = 6 horizontal: 1 vertical) towards the center forming a cone in which digested sludge solids could concentrate. A schematic diagram of a shallow cylindrical digestion tanks with external mixing is presented in Figure 6.15.

Some digestion tanks are designed with waffle bottoms in which the tank floor is subdivided into 12 pie-shaped hoppers each sloping towards a separate drawoff port along the outside wall of the digestion tank [44]. This design allows for steeper floor slopes, reduces the distance the settled solids must travel, and minimizes the amount of grit accumulation in the tank. However, the removal of the accumulated grit is difficult and expensive in addition to the higher costs of construction for this type of bottom because of the complex excavation, form work and piping required.

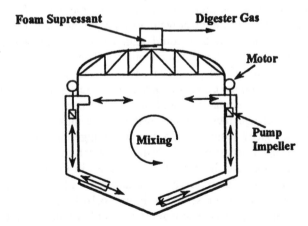

Figure 6.15 Schematic of a fixed cover shallow cylindrical anaerobic digestion tank with external mixing using recirculation pumps.

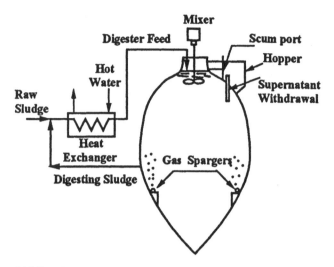

Figure 6.16 Egg-shaped anaerobic sludge digestion tank with external heat exchanger.

Floating covers were used historically for anaerobic digestion tanks in which the cover rises and falls with the depth of sludge in the tank. In some designs the floating cover also served to store the gas accumulation in the head space above the sludge; thus maintaining a relatively constant positive pressure in the tank. However digestion tanks with fixed covers are the preferred design.

External mixing is provided by recirculation pumps that ensure uniform temperature and distribution of raw and digesting sludge in the tank. These external pumps are preferred to gas recirculation systems which have been linked to excessive foaming in the digestion tanks. Foaming is controlled by completely filling the digester, which is possible with a fixed covered tank and using digested sludge or water to suppress the foam; thus minimizing the amount of foam that could gain entry into the gas collection system. A gas storage tank is essential with fixed cover anaerobic digestion facilities. A portion of the produced gas is stored under pressure and is available to flow into the digestion tank and fill the volume that may result from the transfer of sludge into and out of the digestion tank. Failure to include a gas storage facility could result in the introduction of air into the head space gases resulting in a potential dangerous situation.

The egg shaped digestion tank design in which the depth of the tank is much larger than the diameter has been used in the Federal Republic of Germany for more than 50 years. A schematic illustration of an egg-shaped digestion tank is presented in Figure 6.16 [45].

Recently this design has been gaining in popularity in the United States. The purpose of forming an egg-shaped tank is the elimination of the need for cleaning the digester sides which form a cone so steep at the bottom that grit cannot accumulate. In the egg shaped digester the sludge surface at the top is small and the gas produced causes mixing as the gas rises to the surface of the sludge contents. The mixing cause by rising gas is augmented by pumped circulation of sludge from the bottom of the egg-shaped digester through the heat exchanger to the top of the tank. Gas spargers may be located along the inside walls of the tank. The gas discharged through these spargers is effective in keeping the walls clean and in detaching any materials adhering to the walls or to increase mixing. The need for external mixing and problems with excessive grit and scum accumulation are minimal. The scum contained at the surface is kept fluid with a mixer and removed through a special scum removal port. The construction of egg-shaped tanks requires complex form work and special construction techniques; therefore, the construction costs are higher than for more conventional type tanks.

The shallow cylindrical tanks and the egg-shaped tanks represent the two extremes of tank configurations used for anaerobic digestion. Other tank configurations that were developed in the Federal Republic of Germany include the "Conventional" and the spherical/cone anaerobic sludge digestion tanks which are illustrated in Figure 6.17.

Design and operating characteristics of several anaerobic sludge digestion system in the Federal Republic of Germany are summarized in Table 6.14 and Table 6.15 [45]. Materials of construction included steel and concrete and digester configurations included: egg-shaped, spherical and conventional tanks. The hydraulic detention times ranged from 16 to 32 days. The detention times in the conventional and spherical digestion tanks were longer than that in the egg-shaped digestion tanks, namely, 25 to 32

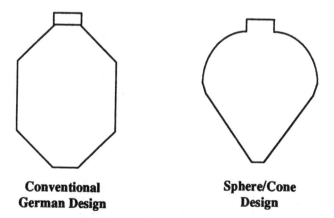

Conventional   **Sphere/Cone**
German Design   **Design**

Figure 6.17 Anaerobic sludge digestion tank configurations.

TABLE 6.14. Characteristics of Anaerobic Digestion Facilities in Germany.

Facility	Shape	Primary Digesters	Material	Volume (m³)	Loading (kg VSS/m³ d)	Solids Retention Time (days)
Bremen	Egg	4	Concrete	6,000[b]	1.66[c]	16 to 19
Hamburg	Egg	10	Concrete	8,000	1.69	18 to 19
Mannheim	Egg	2	Steel	7,500	1.83	16
Weinheim	Egg	2	Steel	4,700	2.23	—[d]
Buhl	Sphere[a]	2	Steel	3,200	0.54	32
Neckarsulm	Sphere[a]	2	Steel	2,700	1.05	25
Hanover-Gummerwald	Conventional	2	Concrete	7,500	1.03	30

[a]A modified egg shape consisting of a sphere on top of a cone bottom.
[b]Multiply by 35.3 to get ft³.
[c]Multiply by 0.0623 to get lb/ft³-d
[d]Not available.

TABLE 6.15. German Anaerobic Digester Characteristics.

Facility	Shape	Volume (m³)	Approximate Mixing Energy (kW/m³)			Primary Pumping (m³/s)	Primary Turnover per Day
			1st Stage	2nd Stage	Recirculation		
Bremen	Egg	6,000	0.0025[a,b]	0.0061[a]	0.0025[a]	0.70	3.3
Hamburg	Egg	7,500	0.0023[b]	None	0.0020[d]	1.0	11
Mannheim	Egg	3,200	0.0023[b]	0.0018[a] (Gas)	0.0029[a]	0.70	8
Weinheim	Egg	7,500	0.0032[b]	0 005[a] (Gas)	0.003[d]	1.00	18
Buhl	Sphere	8,000	Recirculation.	0.0036[a] (Gas)	0.0003[d]	0.01	0.3
Neckarsulm	Sphere	2,700	Recirculation	Gas[a]	0.0008[d]	0.042	1.3 to 1.4
Hanover-Gummerwald	Conventional	4,700	0.0020[b]	None	0.0054[d]	0.70	8

[a]Operates intermittently.
[b]Mechanical draft-tube mixer.
[c]A modified egg shape consisting of a sphere on top of a cone
[d]Estimated

303

days compared to 16 to 19 days. The design detention times may reflect the higher costs associated with the construction of egg-shaped digesters rather than a more efficient digestion system.

The data presented in Table 6.15 reflect the energy required to mix the contents of the various shaped anaerobic digestion tanks. The turnover rates appear to be affected by the energy applied, recirculation energy and the primary pumping rate. Turnover rates, which indicate the degree of mixing varied from 3.3 to 18 per day for the egg-shaped digestion tanks, 0.3 to 1.3 per day for the spherical digestion tanks, 8 for the conventional digestion tank. There is no consensus regarding the degree of mixing or pumping required and the optimum turnover rates.

Preliminary design data for a number of anaerobic digestion systems in the United States are presented in Table 6.16 and Table 6.17 [45]. This information is presented for illustrative purposes regarding dimensions, solids loadings, mixing requirements. The actual constructed facilities may vary from these proposed design parameters.

AUXILIARY MIXING

A certain amount of mixing occurs naturally in an anaerobic digestion tank as the rising gas bubbles mix the sludge and the recirculation of heated sludges causes some thermal convection currents and turnover of the digestion tank contents. However, auxiliary mixing is essential to maximize the advantages of complete mixing and to ensure stable performance of and control of the anaerobic digestion process.

Various systems for mixing anaerobic digesters include: external pumped recirculation, recirculation of compressed digester gas or mechanical mixing [43]. The first application of gas recirculation was originally employed to break the scum layer. Pumped circulation is relatively simple and effective at keeping the scum layer moist so that gas produced during digestion can escape. However, the large flow rates required for complete mixing of the volume of the digestion tank limits the use of pumped circulation as the only method of mixing.

Minimum power requirements for pumped circulation is 0.2 to 0.3 Hp/1000 ft^3 (0.005 to 0.008 kW/m^3) and may be higher, if friction losses are excessive. Cost effective use of pumped circulation is in combination with other mixing systems. Pumped circulation is used to circulate the digesting sludge through external heat exchanger where the digesting sludge is blended with the raw sludge and heated prior to return to the digestion tank.

The first gas recirculation system was used to break the scum layer using lances to introduce the gas at a point about 10 to 12 ft (3.05 to 3.66 m) below the scum surface [43]. The quantity of gas required to mix a

TABLE 6.16. Proposed Design Parameter for Egg-Shaped Anaerobic Digestion Systems in the U.S.

	Deer Island Boston, MA	Back River Baltimore, MD
Digester shape	Egg	Egg
Number of digesters	14	2
Volume per digester	11,355 m³	11,355 m³
	(401,057 ft³)	(410,057 ft³)
Height	41.5 m (136 ft)	42.4 m (139 ft)
Maximum diameter	25.9 m (85 ft)	26.0 m (85 ft)
Volatile suspended solids loading	1.41 to 2.19 kg/m³ · d	2.77 to 2.84 kg/m³ · d
	(0.09 to 0 14 lb/ft³ · d)	(0.14 to 0 18 lb/ft³ · d)
Sludge age	17 to 26 days	11.1 to 16.7 days
Primary mixing system-mechanical draft tube		
Motor	37.3 kW (50 hp)	44.7 kW (60 hp)
Draft-tube diameter	0 70m (2.3 ft)	0.9 m (3 ft)
Pumping rate	1 m³/s (35 ft³/s)	1.31 m³/s (46.3 ft³/s)
Turnover rate	7.6/d	10/d
Mixing energy level	0.003 kW/m³	0.0035 kW/m³
	(0.11 hp/1000 ft³)	(0.13 hp/1000 ft³)
Secondary mixing system unconfined gas		
Compressors		
Motor		76.8 kW² (103 hp)
Mixing energy level		0.0055 kW/m³
		(0.21 hp/1000 ft³)
Pumped recirculation		
Purpose heating	Digester	Digester
Pump capacity	0.045 m³/s (1.59 ft³/s)	0.048 m³/s (1.7 ft³/s)
Maximum pump capacity	0.109 m³/s (3.85 ft³/s)	0.096 m³/s (3.39 ft³/s)
Turnover rate	0.8/d	0.7/d

305

TABLE 6.17. Preliminary Design Parameters for Anaerobic Digestion System in Hyperion, Los Angeles, CA.

	German conventional	Sphere-cone
Digester shape		
Number	18	18
Volume	9460 m³	9460 m³
	(334,126 ft³)	(334,126 ft³)
Height (inside)	32.7 m	33.4 m
	(107 ft)	(110 ft)
Maximum diameter (inside)	25.0 m	25.6 m
	(82 ft)	(84 ft)
Volatile solids loading	1.8 to 2.3 kg/m³ · d	1.8 to 2.3 kg/m³ · d
	(0 11 to 0 14 lb/ft³ · d)	(0.11 to 0.14 lb/ft³ · d)
Sludge age	14 to 19 4 days	14 to 19.4 days

Primary mixing system-mechanical draft-tube

Purpose	Mixing
Motor	40.3 kW (54 hp)
Draft-tube diameter	0.56 m (1.84 ft)
Pumping rate	1.0 m³/s (35 ft³/s)
Turnover rate	9/d
Mixing energy	0.0038 kW/m³
	(0.14 hp/1000 ft³)

Pumped recirculation

Purpose	Digester heatmg
Number pumps per digester	1
Motor	29.8 kW (40 hp)
Pumping capacity	0.22m³/s (7 8 ft³/s)
Turnover rate	2/d
Mixing energy level	0.0024 kW/m³
	(0.09 hp/1000 ft³)
Operation	Continuous

digestion tank varies with the volume of sludge undergoing digestion, the concentration of sludge in the digester, the volatile solids content of the digesting sludge and the diameter of the tank [44]. In the Pearth process the number of gas discharge points and the gas discharge rate is a function of the diameter of the tank. For example, three or four discharge points discharging an average of between 5 to 9 cubic feet per minute (cfm) per 1000 ft^3 of digester capacity (5 to 9 L/m^3) are used for 20 to 30-foot (6 to 9 m) diameter digestion tanks. However for digestion tanks which are 100 to 110 feet in diameter, 6 to 8 discharge points are typical and the average gas discharge rate is between 0.66 to 1.0 cfm/1000 ft^3 of digester capacity (0.66 to 1.0 L/m^3). The actual horsepower requirements range from about 0.049 Hp/1000 ft^3 of digester capacity (0.0013 kW/m^3) for 100 to 110-foot (30 to 33.5 m) diameter tanks to 0.375 Hp/1000 ft^3 (0.01 kW/m^3) of digestion capacity for the smaller tanks [20 to 30 feet (6 to 9 m) in diameter]. These power requirements correspond to nameplate horsepower ratings of 0.07 to 0.73 Hp/1000 ft^3 digester capacity (0.002 to 0.19 kW/m^3), respectively.

Recirculated compressed gas may be introduced through spargers into a draft tube. The released gas acts like a gas lift pump in the digester and sludge is carried upward through the draft tube to the surface of the digesting sludge where the sludge is directed radially towards the tank periphery. The rising gas bubbles cause sludge near the bottom of the tank to be drawn into the draft tube. This recirculation pattern causes mixing of the entire tank volume. Operating data indicate that the power requirements for this type of gas recirculation system is between 0.1 to 0.12 Hp/1000 ft^3 digester capacity (0.0026 to 0.0032 m^3) [43].

Various mechanical methods of mixing the digesting sludge have been employed. Mixing can be maintained by a number of mechanical draft tube mixers located to provide maximum mixing. Sludge also may be recirculated by an external pump which draws sludge from the central portion of the digester discharges the sludge tangentially through nozzles near the surface to break the scum, and near the bottom of the tank. Surface mixers also have been used to mix the sludge in anaerobic digestion tanks. The circulation pattern is from the bottom of the central portion of the tank and radially along the surface to the periphery. One disadvantage of the use of rotary machines to mixed digesting sludge is potential problems with rags and other stringy materials in the sludge wrapping around the drive shaft which could cause eccentric torques on the shaft resulting in uneven wearing of the bearings. Accumulation of rags, hair and other stringy material on the impeller (propeller) of axial type mixers also would reduce the efficiency of the mixers. However, the development of radial flow mixers has eliminated many of the problems associated with the propeller-type mixers. These radial pumps can be operated either in the clockwise or counterclockwise direction; therefore, the mixer impellers are self cleaning. The radial-flow type of mixer also minimizes foaming in the digestion tank.

The ability of the mixing devices to circulate completely the contents of a digestion system is not without limits. The relative content of fixed and volatile solids in the feed sludge affects mixing. Mixing by gas recirculation may be hindered as the total solids content of the digesting sludge exceeds 5%. Feed sludges should not be concentrated to more than 8% total solids, if the volatile solids content is less than 70%; that is, the fixed solids exceed 30% [43]. However, in most municipal wastewater treatment plants mixing capabilities do not control the solids loading as much as the ability to concentrate the feed sludge in the sedimentation and thickening processes.

TEMPERATURE CONTROL

External heat exchange units are used to heat raw sludge and to maintain the temperature of the sludge undergoing digestion. Pumping the digesting sludge from the digestion tank through the external heat exchange unit permits the seeding of raw sludge with digesting sludge; however, this external mixing does not markedly affect the circulation pattern in the digester and does not replace the need for effective mixing of the digesting sludge.

Steam injection directly into the sludge has been used to heat sludge. However, the addition of steam only adds more water to the sludge and dilutes the concentration of the digested sludge. In addition a greater volume of supernatant is generated. The schematic diagram presented in Figure 6.18 illustrates the installation of an external heat exchanger as part of a high-rate anaerobic sludge digestion system.

The digesting sludge also may be heated with internal heat exchangers. Water jackets are placed around the periphery of draft tubes through which the sludge is pumped mechanically or by gas recirculation. The recirculation rate must be high enough to prevent a build-up of sludge cake on the surface of the interior walls to maximize heat exchange. Accumulation of sludge cake must be cleaned periodically by removing and servicing the draft tube from the digestion tank. The hot water is recirculated through and external boiler heated by burning digester gas.

The temperature of the feed sludge must be increased to the temperature of the digestion tank and the temperature of the digesting sludge must be maintained at the operating temperature. The concentration of solids in the sludge markedly affects the heat requirements (BTU/lb solids/°F) (kg-cal/kg/°C). As the percent solids content increases and the amount sludge solids remains constant, the amount of water associated with the sludge is reduced; therefore, the heating requirements per pound of solids de-

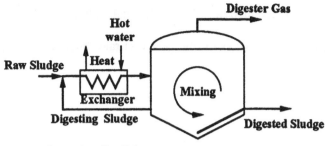

Operating Conditions:

**Heated to Constant Temperature, Mixed, Continuous
Loading and Withdrawal of Digested Sludge
Minimum Hydraulic Detention Time : 10 to 15 days**

Figure 6.18 High rate anaerobic digestion with external heat exchanger sludge heating.

creases. For example, sludge at 2% solids may have a heating requirement of approximately 60 BTU/lb solids/°F; however, the heating requirements are reduced to about 15 BTU/lb solids/°F when the sludge is concentrated to 8% solids by removing water.

The amount of heat required to raise the temperature of the incoming sludge is:

$$q_s = Q_m C_p (T_2 - T_1)$$

where:

q_s = heat required to raise the temperature of the incoming sludge from T_1 to T_2, (BTU/hr) (kcal/hr)

Q_m = mass flow rate of sludge (pounds/hour) (kg/hr)

C_p = specific heat of the sludge, (approximately 1.0 BTU/pound · °F) (1.0 kcal/kg · °C)

T_1 = temperature of feed sludge, (°F) (°C)

T_2 = temperature desired within the digestion tank, (°F) (°C)

The heat requirements must be adjusted to make up for heat losses to the air and to the soil surrounding the digestion tank. These heat losses depend on the shape of the digestion tank, materials of construction, and the temperature gradient from the temperature of the digesting sludge and the outside air and/or soil temperature. The heat transfer coefficient is affected directly by the film coefficient for the interior surface of the digestion tank, and the film coefficient for the exterior surface of the tank. The heat transfer coefficient also is affected inversely by the thickness and the thermal conductivity of the individual wall and roof materials of construction.

The general expression for the heat loss rate through the component structures is

$$q_L = U A (T_2 - T_1)$$

where:

q_L = heat-loss rate, (BTU/hr)

U = heat transfer coefficient, (BTU/hr · ft² · °F) (kcal/hr · m² · °C)

A = area of the material at right angle to the direction of the flow of heat (ft²) (m²)

T_1 = ambient temperature outside the digestion tank, (°F) (°C)

T_2 = temperature desired within the digestion tank, (°F) (°C)

Typical overall heat transfer coefficients are listed in Table 6.18.

The heat losses can be reduced by insulating the digestion tank cover and walls exposed to the ambient air. Common insulating materials used include: a dead air space, lightweight insulating concrete, glass wool, insulation board, and urethane foam. The insulation material frequently is covered to protect the insulation and for aesthetic purposes. Common facing materials include: brick, metal siding, precast concrete panels, and stucco.

INDICATORS OF REACTOR PERFORMANCE

Unbalanced anaerobic treatment operation generally is indicated by an increase in the carbon dioxide content and a corresponding decrease in the methane content of the gas produced and an increase in the concentration of volatile acids in the digesting sludge. Eventually, there will be a decrease in the total quantity of gas produced daily, and a decrease in the pH. Anaerobic digestion may be unbalanced temporarily by sudden change in temperature, organic loading, and/or composition of the sludge. This type of imbalance generally can be remedied by not feeding the digester temporarily and/or providing sufficient time to permit the microbial population to adjust to the new environment. Frequently, control of the pH near neutral may be required.

Prolonged or relatively permanent imbalances of the anaerobic system may be the result of the introduction of materials which are toxic to the methane forming bacteria and/or an extreme drop in pH. This type of imbalance frequently is encountered in starting up a new anaerobic digester. The population of methane forming bacteria require time to develop sufficient numbers to convert the volatile acids to methane and carbon dioxide.

TABLE 6.18. Overall Heat Transfer Coefficients for Materials of Construction Anaerobic Digestion Tanks.

Material of Construction	Overall Heat Transfer Coefficient (U) (BTU/hr · ft² · °F)
Fixed steel cover (0 25 inch plate)	0.91
Fixed concrete cover (9 inch thick)	0.58
Floating cover (wood composition roof)	0.33
Concrete wall (12 inch thick) exposed to air	0.86
Concrete wall (12 inch thick), 1 inch air space and 4 inch brick exposed to air	0.27
Concrete wall or floor (12 inch thick) exposed to wet earth (10 feet thick)	0.11
Concrete wall or floor (12 inch thick) exposed to dry earth (10 feet thick)	0.06

BTU/hr · ft² · °F = 4 9 (kcal/hr · m² · °C)
inches × 2 54 = cm
feet × 30 = cm

311

At times during start up, the pH drops below the minimum pH tolerated by the methane-producing bacteria. If a prolonged imbalanced condition is caused by toxic material, the toxic substance must be removed. The imbalances caused by extreme drops in pH may be remedied by providing sufficient time and control of the pH at near neutral until the treatment returns to normal efficiency.

The different chemical parameters which characterize the anaerobic environment are interrelated and one variable may directly or indirectly affect the others. Therefore, a thorough understanding of these interrelationships and an accurate estimate of the pH, alkalinity, volatile acid concentration and carbon dioxide content of the gas are essential to the control of the anaerobic system. The anaerobic digester contents are usually well-buffred and pH can not by itself be used as an indicator of a suitable environment.

There is no single parameter which can be isolated as the best indicator of the environment within the digester. In the past, the rate of gas production was considered the best indication of digester performance; however, this parameter does not describe the environment within the digester and usually when gas production drops off, it is the result of an environmental upset and remedial measures can be drastic.

The parameters which would provide an insight into the condition of the environment early enough to avert digester failure are the volatile acids concentration, alkalinity, and the carbon dioxide content of the gas along with pH [43]. Sudden large changes in these parameters will signal impending digester failure much sooner than a drop in the rate of gas production. The onset of digester failure is signaled by a sharp increase in the free volatile acid concentration and the carbon dioxide content of the gas and a decrease in the pH and possibly alkalinity. The daily gas production may also drop slightly. Digester upset may usually be controlled by maintaining the pH near neutral and/or reducing the loading on the digester.

Daily evaluation of the pH, volatile acid concentration, alkalinity of the sludge and the carbon dioxide content of the gas is imperative to good digestion and relatively simple and inexpensive analytical techniques are available. Electronic pH meters are available and relatively simple to use. A direct titration method of evaluating the volatile acid concentration has been developed. This method consistently yields results which are higher than those obtained by the distillation methods. However, since relative values are important in digester control, this shortcoming of the direct titration method is not significant. Alkalinity may easily be estimated by titrating the sample with a standard sulfuric acid solution to an end point of pH = 4.5. The color of the supernatant frequently precludes the use of any indicator solutions in this analysis.

The carbon dioxide content of the gas may be evaluated by using gas chromatographic techniques; relatively simple and inexpensive gas partitioners are available. Another simple technique for estimating the carbon dioxide content of a digester gas is to pass a known volume of gas through a carbonate free sodium hydroxide solution to absorb the CO_2 as Na_2CO_3. Barium chloride solution is added to precipitate the Na_2CO_3 as $BaCO_3$, and the excess NaOH is titrated with 1 N hydrochloric acid solution.

pH CONTROL IN ANAEROBIC DIGESTION

The relationships among the carbon dioxide content of the digester gas, and the alkalinity and pH in the digesting sludge are presented in Figure 6.19 [46] and may be represented by the following reactions:

$$CO_2 + H_2O \rightarrow H_2CO_3$$
$$H_2CO_3 \rightarrow H^+ + HCO_3^-$$

The hydrogen ion concentration [H+] and the pH of the system may be calculated from the following equilibrium equation for the ionization of H_2CO_3:

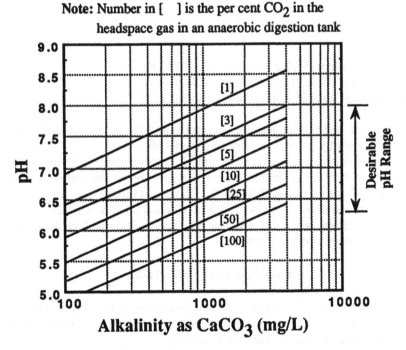

Note: Number in [ ] is the per cent CO_2 in the headspace gas in an anaerobic digestion tank

Figure 6.19 Relationship among carbon dioxide in the gas, pH, and bicarbonate alkalinity.

$$[H^+] = k_1 \frac{[H_2CO_3]}{[HCO_3^-]} \text{ , and}$$

$$pH = -\log [H^+] = \log \left[\frac{1}{[H^+]}\right]$$

At values of pH between pH = 6.6 and pH = 7.4 and at a carbon dioxide content in the gas of 30 to 40% by volume, the bicarbonate alkalinity will range between 1,000 and 5,000 mg/L as $CaCO_3$. The concentration of bicarbonate alkalinity should be approximately 3,000 mg/L as $CaCO_3$. The bicarbonate alkalinity is approximately equal to the total alkalinity of the anaerobic system.

A portion of the alkalinity appears as "volatile acid salts" alkalinity which results from the reaction of volatile acids with the bicarbonate present, yielding carbon dioxide. At low volatile acid concentrations, the bicarbonate alkalinity represents approximately the total alkalinity; however, as the volatile acids concentration increases, the bicarbonate alkalinity is much lower than the total alkalinity. About 83.3% of the volatile acids concentration contributes to the alkalinity as "volatile acids salts" alkalinity and the following equation may be used to estimate the concentration of bicarbonate alkalinity:

$$BA = TA - (0.85)(0.833)TVA$$

in which:
BA = bicarbonate alkalinity (mg/L as $CaCO_3$)
TA = total alkalinity (mg/L as $CaCO_3$)
TVA = total volatile acids (mg/L as acetic acid)

The factor (0.85) accounts for the fact that 85% of the "volatile acid salts" alkalinity is measured by titration to pH = 4.0.

Lime frequently is used for pH control; however, one should remember that the pH of the anaerobic system is a function of the carbon dioxide content of the gas as well as the bicarbonate alkalinity and the volatile acids concentrations of the environment. Control of pH is generally desired as the pH drops below pH = 6.5 or 6.6. The addition of lime at this pH results in the following reaction:

$$Ca(OH)_2 + 2CO_2 \rightarrow Ca(HCO_3)_2$$

However, as the concentration of bicarbonate alkalinity approaches 500 to 1,000 mg/L, the continued addition of lime results in the precipitation of the insoluble calcium carbonate, namely:

$$Ca(OH)_2 + CO_2 \rightarrow CaCO_3\downarrow + H_2O$$

Therefore, although lime is effective in the control of the pH of an anaerobic digester, careful control of the addition of lime is imperative because of the interrelationship among the lime added, dissolved carbon dioxide, pH, bicarbonate alkalinity, and the concentration of volatile acids. Sodium bicarbonate may be used because of the care that must be exercised when lime is used, except where gross pH correction is required.

Anhydrous ammonia also may be used to control pH in an anaerobic digestion tank. Anhydrous ammonia also has been reported to aid in dissolving the scum layer. Ammonia is closely related to the alkalinity in the control of the anaerobic process. The ammonia reacts with the carbon dioxide and water, resulting in ammonium carbonate which provides alkalinity to the system. The ammonium carbonate is thus available to react with the free volatile acids which are present in an upset anaerobic digestion system resulting in the formation of volatile acid salts. These reaction may be written as follows:

$$CO_2 + H_2O + NH_3 \rightarrow NH_4HCO_3$$
$$NH_4HCO_3 + RCOOH \rightarrow RCOONH_4 + H^+ + HCO_3^-$$

Therefore, anhydrous ammonia can improve the environment in a "stuck" digester by providing buffering alkalinity and by eliminating free volatile acids. The anhydrous ammonia must be added carefully, because the indiscriminate addition could lead to excess ammonia which is toxic to the bacteria.

The required amount of anhydrous ammonia may be calculated based on the free volatile acid and alkalinity concentrations. The bicarbonate alkalinity which is available for buffering the system can be calculated from the expression discussed previously:

$$BA = TA - (0.85)(0.833)TVA$$

If the bicarbonate alkalinity calculated from this equation is less than the alkalinity required to provide adequate buffering, namely, 500 to 1,000 mg/l as $CaCO_3$, anhydrous ammonia may be added to increase the alkalinity. The amount of anhydrous ammonia which should be added can be calculated from the following expression which is based on a commercial grade containing about 80% ammonia:

$$NH_3 \text{ (pounds)} = 3.5 \times 10^{-6}(V)(AR)$$

in which
V = volume of sludge in the digestion tank (gallons)
AR = required alkalinity in mg/L as $CaCO_3$

The required alkalinity (AR) is the algebraic sum of the desired alkalinity in the digesting sludge and the bicarbonate alkalinity (BA) available for buffering.
An example will illustrate these calculations. Assume the following data:
Total alkalinity (TA) = 2,500 mg/L as $CaCO_3$
Total volatile acids (TVA) = 4,000 mg/L as acetic acid
Volume = 100,000 gallons.
Alkalinity available for buffering = 1,000 mg/L as $CaCO_3$

The bicarbonate alkalinity is

$$(BA) = 2,500 - (0.85)(0.833)4,000 = -332 \text{ mg/L as } CaCO_3$$

Therefore the alkalinity is less than the volatile acids and free volatile acids are present. The anhydrous ammonia added must equal this deficit plus the desired alkalinity. The amount of anhydrous ammonia to be added is $1,000 - (-332) = 1,332$ mg/L as $CaCO_3$ or 3.5×10^{-6} (100,000) 1,332 = 466 pounds (211.6 kg) of commercial grade anhydrous ammonia.

DIGESTER SEEDING AND ADDITIVES

The inoculation of raw sludge with digested sludge is an important factor in the anaerobic digestion process. Seeding of the incoming sludge ranks in importance with the control of temperature and mixing. Seeding is brought about by the intimate contact of raw sludge and digesting sludge.
Various biocatalysts and enzymes have been added to sludge digesters in an attempt to improve their performance [47,48]. A biocatalyst consisting of a group of preserved living cultures of anaerobes, together with their enzyme systems was added in concentrations of 5,400 to 53,300 mg/L to mixtures of raw and digested sludge. This sludge-biocatalyst mixture was incubated at temperatures ranging from 39° to 104°F (4°–40°C). The results indicated that: (a) There was no change in the rate of gas production and the volume of gas produced; (b) there was no improvement in the volatile solids destruction; and (c) the biocatalyst did not expedite the beginning of the methane fermentation stage of digestion.
These results have been confirmed in plant-scale tests of biocatalysts. A commercial inoculum used to seed a digester reduced the time required for full gas production, but this preparation compared unfavorably with seed sludge from an active digester. Another enzyme preparation was

applied to a digester at a rate of one pound per day for a ten day period and resulted in the elimination of scum and grease problems, which had brought the digesters to a standstill.

These results indicate that sludge digestion once established is not materially affected by the addition of biocatalysts and enzymes. However, some preparations may have merit in the control of excessive formation of scum blankets, and in accelerating the recovery of an anaerobic digestion system which may have experienced upset conditions.

TOXICITY

Organic and inorganic materials which are toxic to the methane-forming bacteria must be removed from the incoming sludge or chemical compounds which alleviate or reduce the toxic effects may be added. Toxic organic materials usually are not associated with wastewater sludges in concentrations which would inhibit methane-forming bacteria.

Cations may affect the rate reaction of anaerobic organisms to a greater extent than the concentration of organic material. The evaluation of the stimulatory or inhibitory and/or toxic effects of various cations, anions or organic materials on the anaerobic digestion process is very complicated. A limited amount of published data on stimulatory and inhibitory effects of various materials on anaerobic systems is available; however, the emphasis on toxicity reduction and the disposal of hazardous substances along with the limitations imposed by the new sludge regulations directed at the control of the disposal of municipal sludges should result in the development of additional information dealing with stimulatory and inhibitory effects of various organic and inorganic compounds.

Ammonia, which is an essential source of nitrogen to bacteria is released during the anaerobic degradation of organic matter in sludges. The toxic form of nitrogen is free ammonia (NH_3) not the ammonium ion (NH_4^+) and the pH defines the form of ammonia in the aqueous system. Ammonia nitrogen is beneficial at concentrations of 50 to 200 mg/L and at neutral pH However, ammonia nitrogen concentrations in excess of 1,250 mg/L but less than 3,000 mg/L were inhibitory to anaerobic digestion systems at $7.4 < pH > 7.6$. Ammonia is toxic at concentrations of ammonia nitrogen in excess of 3,000 mg/L.

Sulfides also can be toxic to bacteria in an anaerobic system at concentrations in excess of 200 mg/L at a pH near neutral. However, at concentrations between 50 and 100 mg/L, sulfides are tolerated with little or no acclimation. Sulfides may be introduced into the system as a component of the waste or may be produced by the biological reduction of sulfates or degradation of proteins which contain sulfur. Some of the sulfide leaves the system as hydrogen sulfide gas, while a portion of the sulfide will be precipitated as heavy metal salts if these metals are present. However, a portion of the sulfide remains dissolved as a weak acid which ionizes depending on the pH of the solution.

Heavy metals may be toxic to the bacteria responsible for anaerobic digestion. Hexavalent chromium is reduced under anaerobic conditions to the trivalent state which is relatively insoluble at a pH near neutral which is typical of anaerobic digestion.

The heavy metals which are toxic to bacteria at relatively low concentrations are copper, zinc, and nickel. Iron and aluminum are insoluble at neutral pH; therefore, these metals are not toxic. Toxicity of copper, zinc, and nickel can be reduced by reacting these metals to precipitate as metal sulfides which are essentially insoluble at the neutral pH and reduced environment in anaerobic digestion tanks. Therefore, in evaluating the tolerance of anaerobic treatment systems to heavy metals, the concentration of reducible sulfur compounds must be considered. Approximately 1.8 to 2.0 mg/L of heavy metals are precipitated as metal sulfides by 1.0 mg/L of sulfide ($S^=$). Sodium sulfide (Na_2S) or ($Na_2S \cdot 9H_2O$) frequently is added to precipitate heavy metals, if sufficient reducible sulfur is not present in the digesting sludge.

SUMMARY

The mechanism of anaerobic digestion involves complex biochemical reactions by which a portion of the organic material in the sludge is converted to methane and carbon dioxide. The optimum conditions required for effective anaerobic digestion include thorough mixing of the contents of the digestion tank either mechanically or by means of gas recirculation and maintaining the temperature of the digesting sludge between 86°F and 95°F (30°C to 35°C).

The hydraulic detention time in the anaerobic sludge digestion system is one of the most important parameters that controls the quality of the digested sludge that is discharged from the digestion system. A detention time of 10 days should be sufficient although 15 days is preferable. Maintaining the solids loading and hydraulic loading as uniform as possible will improve digester performance. The recommended loading for anaerobic digestion tanks which are mixed and heated is 200 lb VS/1,000 cu ft-day (3.2 kg/m³-day), although higher loadings can be treated effectively, if a more concentrated sludge could be fed to the digester. However, the solids loading is controlled by the minimum required detention time and the concentration of the solids in the feed sludge.

Recent design approaches to the design of anaerobic digestion systems have considered various configurations centering on the egg-shaped digesters and different modifications. Concerns over hydrocarbon emissions have focused attention on fixed covers instead of floating covers for anaerobic digestion. Excessive foaming in anaerobic digestion tanks is linked to using gas for mixing the tank contents. Therefore, external recirculation of the digesting sludge using pumps and mixers located outside the digestion tank is becoming more popular.

Anaerobic digestion tanks may be either cylindrical or egg-shaped. The shape of the anaerobic digestion tank affects the type of mixing required. In the United States anaerobic digestion tanks historically have been circular and relatively shallow with the diameters larger than the depth.

Various mechanical methods of mixing the digesting sludge have been employed. Mixing can be maintained by a number of mechanical draft tube mixers located to provide maximum mixing. Sludge also may be recirculated by an external pump which draws sludge from the central portion of the digester discharges the sludge tangentially through nozzles near the surface to break the scum, and near the bottom of the tank.

The ability of the mixing devices to circulate completely the contents of a digestion system is not without limits. The relative content of fixed and volatile solids in the feed sludge affects mixing. Mixing by gas recirculation may be hindered as the total solids content of the digesting sludge exceeds 5%. Feed sludges should not be concentrated to more than 8% total solids, if the volatile solids content is less than 70%; that is, the fixed solids exceed 30%. However, in most municipal wastewater treatment plants mixing capabilities do not control the solids loading as much as the ability to concentrate the feed sludge in the sedimentation and thickening processes.

Control of anaerobic digestion is complicated since the different parameters which characterize the environment within the digestion tanks are all interrelated and one variable may directly or indirectly affect the others. Therefore, daily laboratory analysis of the volatile acids and the alkalinity concentrations as well as the pH of the digesting sludge and the carbon dioxide content of the gas are essential for efficient digestion with a minimum number of problems.

Digesters which are upset or "stuck" are characterized by an environment which is not conducive for the development and metabolism of the methane-producing bacteria. In many cases, the addition of lime, sodium bicarbonate, and/or anhydrous ammonia can be used to maintain the pH of the contents of the digestion tank to a level near neutral. The combination of maintaining a neutral pH and providing sufficient time for the system to recover is effective in improving the performance of an anaerobic digestion system which was upset.

THE ROLE OF LIME STABILIZATION PROCESSES IN WASTEWATER SLUDGE PROCESSING AND DISPOSAL

CHEMICAL STABILIZATION

Treatment of wastewater sludges with chemicals, such as chlorine and lime, can also be an effective stabilization process. Chlorine and lime are the primary chemicals that have been researched and used. Chlorine stabilization is rarely used; however, lime is widely used, and it is one of the lowest cost alkali available in the wastewater industry.

In the past, lime was used for reducing odors in privies, increasing pH in stressed digesters, and removing phosphorus in advanced wastewater treatment. Today, the most common use of lime in wastewater treatment plants is to condition sludge prior to dewatering. It is also the principal sludge-stabilizing chemical in numerous municipal treatment plants ranging from 0.1 to 32 MGD [49]. According to the latest United States Environmental Protection Agency (U.S.EPA) Needs Survey, over 250 municipal wastewater treatment plants (WWTP) use lime treatment to stabilize their sludge [50]. Lime stabilization can be part of a sludge conditioning process prior to dewatering (pre-lime stabilization) or following a dewatering step (post-lime stabilization).

Stabilization Objectives

The purpose of sludge stabilization is twofold: to substantially reduce the number and prevent regrowth of pathogenic organisms and, thereby, minimize the health hazard associated with the sludge; and to substantially reduce the number of odor-producing organisms and, thereby, minimize nuisance conditions created during sludge disposal.

Because most wastewater sludge in the United States is eventually disposed of on land, stabilization is of major importance. The sludge must be nonhazardous to humans, biologically inactive, free of offensive odors, and aesthetically acceptable.

Regulations governing the disposal or reuse of sludges on land were promulgated by the U.S.EPA in October 1979. These regulations, which are published in the Code of Federal Regulations, Title 40, Part 257 (40 CFR 257), divided sludge processes into those that significantly reduce pathogens (PSRP) and those that further reduce pathogens (PFRP) [51]. A PSRP is a process that reduces pathogenic viruses by 90% and fecal and total coliform by 99%. A PFRP reduces pathogens to below detectable limits.

Table 6.19 and Table 6.20 summarize the U.S.EPA definitions of PSRPs and PFRPs, and indicate typical operating parameters for each.

TABLE 6.19. Regulatory Definition of Process to Significantly Reduce Pathogens (PSRP).

Aerobic Digestion. The process is conducted by agitating a sludge with air or oxygen to maintain aerobic conditions at residence times ranging from sixty days at 15°C (59°F) to forty days at 20°C (68°F), with a volatile solids reduction of at least 38%.

Air Drying. Liquid sludge is allowed to drain and/or dry on underdrained sand beds, or on paved or unpaved basins in which the sludge depth is a maximum of 9 inches A minimum of three months' retention is required, two months of which must have average daily temperatures above 0°C (32°F)

Anaerobic Digestion. The process is conducted in the absence of air at residence times ranging from sixty days at 20°C (68°F) to fifteen days at 35°C (95°F) to 55°C (131°F), with a volatile solids reduction of at least 38%.

Composting. Using the in-vessel, static aerated pile, or windrow composting methods, the solid waste is maintained at minimum operating conditions of 40°C (104°F) for five days. For four hours during this period, the temperature must exceed 55°C (131°F).

Lime Stabilization. Sufficient lime is added to reach and maintain a pH of 12 for two hours.

Wastewater sludge applied to the land or mixed into the soil must meet at least the PSRP requirements. Public access to the application site must be controlled for twelve months, and grazing by animals, whose products are consumed by humans, must be prevented for at least one month. The sludge must be stabilized by a PFRP if the sludge is applied to land where crops for direct human consumption are grown less than eighteen months after application.

The *Federal Register* defines a lime treatment process for pathogen control in Section 257, Appendix II, as "sufficient lime is added to produce a pH of 12 after two hours of contact." A minimum dose of 6% lime, plus the addition of 20 to 40% cement or lime kiln dust (wet weight basis) and maintenance of a 50% total solids sludge at pH above 12 for three days or dried to 65% total solids qualifies as a PFRP. A PFRP can be considered a pasteurization process because it significantly reduces the number of pathogens (an additional one log kill) when compared to a PSRP process.

Lime kiln dust (LKD) and cement kiln dust (CKD) used in the PFRP chemical stabilization process are reclaimed by-products of the lime and cement manufacturing industries, respectively. LKD and CKD have alkaline properties similar to those of lime and extremely large surfaces areas that can provide adsorptive properties for additional odor control, etc. LKD typically has a higher concentration of calcium oxide (CaO) than CKD. Calcium oxide is the active stabilizing agent present in alkaline materials. LKD and CKD are approximately one-third the price of lime;

TABLE 6.20. Regulatory Definition of Process to
Further Reduce Pathogens (PFRP).

Composting. Using the in-vessel or static aerated pile composting methods, the solid waste is maintained at operating conditions of 55°C (131°F) or greater for three days. Using the windrow composting method, the solid waste attains at temperature of 55°C (131°F) or greater for at least fifteen days during the composting period. Also, during the high-temperature period, there must be a minimum of five turnings of the windrow.

Heat Drying. Dewatered sludge cake is dried by direct or indirect contact with hot gases; the moisture content is reduced to 10% or lower Sludge particles reach temperatures well in excess of 80°C (176°F), or the wet bulb temperature of the gas stream in contact with the sludge at the point where it leaves the dryer is in excess of 80°C (176°F).

Heat Treatment. Liquid sludge is heated to a temperature of 180°C for 30 minutes

Thermophilic Aerobic Digestion. Liquid sludge is agitated with air or oxygen to maintain aerobic conditions at residence times of ten days at 55°C (131°F) to 60°C (140°F), with a volatile solids reduction of at least 38%.

Other Methods. Other operating conditions may be acceptable if pathogens and the vector attraction of the wastes (volatile solids) are reduced to an extent equivalent to the reduction achieved by any of the above methods.

Any of the processes listed below, if added to a PSRP, will further reduce pathogens.

Beta Ray Irradiation. Sludge is irradiated with beta rays from an accelerator at dosages of at least 1.0 megarad at room temperature, approximately 20°C (60°F).

Gamma Ray Irradiation. Sludge is irradiated with gamma rays from certain isotopes, such as 60 cobalt and 137 cesium, at dosages of at least 1.0 megarad at room temperature 20°C (60°F)

Pasteurization. Sludge is maintained for at least 30 minutes at a minimum temperature of 70°C (158°F)

Other Methods. Other methods or operating conditions may be acceptable if pathogens are reduced to an extent equivalent to the reduction achieved by any of the above add-on methods

however, typical recommended dose rates are two to three times greater than lime dose rates on a dry weight basis.

The PFRP chemical stabilization process has several advantages and disadvantages when compared to PSRP chemical stabilization. The advantages include being less prone to odor production, less restricted for distribution, and more community acceptable. The primary disadvantages are additional chemicals and equipment required for maintaining a temperature elevation and for accelerated drying.

In February 1993, the U.S.EPA published new sludge regulations stating that pathogen reduction will no longer be process-based, as were the requirements of PSRP and PFRP. Laboratory analyses will be required to prove that the sludge meets pathogen and bacterial testing standards. Three new classes of pathogen/vector attraction have been established: Class A, Class B, and Class C. Sludge pathogens include pathogenic bacteria, animal viruses, protozoa, and helminth ova. Vector attraction refers to the attraction of rodents, flies, mosquitoes, or other organisms to sludge. The most restrictive pathogen reduction standards, Class A, will be required for sludges with the most public health exposure.

Basic Theory

The lime stabilization process theory is a simple one. Lime is added to raise pH, and adequate contact time is provided. At a pH of 12 or more with sufficient contact time, pathogens and microorganisms are either inactivated or destroyed. Chemical and physical sludge characteristics are also altered by the reactions. The chemistry of the process is not well understood, although it is believed some complex molecules split by reactions such as hydrolysis and saponitrification.

In properly stabilized sludge, little or no decomposition occurs, and no odors are produced from the biologically induced emission of gases that would otherwise take place [49]. The adjustment of pH alone can cause gas release; at a high pH (pH greater than 10.5) ammonia gas is given off, while at low pH (pH less than 6.0) hydrogen sulfide release is probable.

Advantages and Disadvantages

Both pre-lime and the post-lime stabilization processes are reliable, low in capital cost, and easier to operate than other stabilization processes. The major equipment items include conveyors, pug mill, or similar mixing devices, and storage and feeding systems. Many municipalities of the wastewater industry that use lime stabilization have indicated that pathogens and odors are greatly reduced [52,53]. The stabilized sludge can then be applied to agricultural land and provides a good source of nitrogen and calcium, as well as beneficial organic matter. The sludge may also partially or fully replace liming agents used on acid soils. The lime stabilization process acts to fix or immobilize specific metal ions and, therefore, restricts the possible uptake of metals by plants. Lime is safer, cheaper, and easier to handle than chlorine.

Improved filterability of the stabilized sludge occurs when the proper doses of conditioning agents, such as iron, aluminum salts, or polymers are used in conjunction with lime. Filterability also depends on the type of sludge being stabilized; primary sludge tends to be more easily dewatered than secondary sludge.

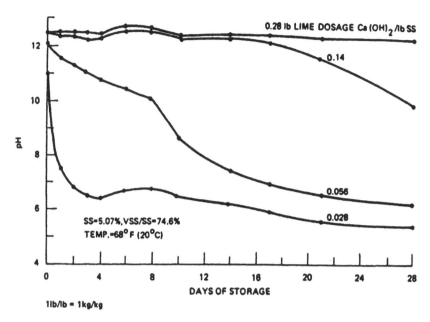

Figure 6.20 Change in pH during storage of primary sludge using different lime dosages. (Source: U.S.EPA. 1979. *Process Design Manual for Sludge Treatment and Disposal,* EPA 625/1-79-011, Cincinnati, Ohio: U.S.EPA.)

Nevertheless, there are disadvantages. Lime is less effective than chlorine in converting sludge to biologically stable material [54]. However, it is safer, cheaper, and easier to handle. One principal disadvantage is that lime does not destroy the organics that promote the growth of biological organisms [55]. Failure to dose the sludge to a pH of 12 or more for two or more hours can lead to a drop in pH during storage and subsequent odor problems, as well as possible regrowth of pathogenic organisms (Figure 6.20).

Another disadvantage of lime stabilization, as compared to digestion processes, is that sludge mass is not reduced. In fact, it is actually increased because of lime and chemical formations. Since there is an increase of solids, the amount to be disposed is essentially proportional to the dosage rate. This may result in higher transportation and ultimate disposal costs, depending on how much volume reduction can be achieved by improved dewatering. Any increase in transportation and disposal costs must be weighed against the capital savings in using lime stabilization rather than another process.

A fourth disadvantage is that agricultural utilization of lime-stabilized sludge is generally not appropriate where the soils are inherently alkaline, as in many parts of the drier western states. Also, lime-stabilized sludge has lower concentrations of soluble nitrogen and phosphorus when compared to anaerobically digested sludge.

Applicability

Lime stabilization is recognized for stabilizing wastewater sludge prior to land application and landfilling. Traditionally, lime stabilization has only seriously been considered in the following situations:

- *stabilization facilities at small WWTPs:* Lime stabilization is appropriate at small treatment plants where the associated small sludge production has access to land for agricultural utilization or ultimate landfill disposal. The process is also practical at small plants that store sludge for later transportation to larger facilities for further treatment and/or disposal.
- *backup for existing stabilization facilities:* A lime stabilization system can be started (or stopped) quickly. Therefore, it can be used to supplement existing sludge quantities, exceed design levels, or replace incineration and dryers during fuel shortages. Full sludge flows can be lime treated when existing facilities are out of service for cleaning or repair.
- *interim sludge handling:* Lime stabilization systems have a comparatively low capital cost and, therefore, may be cost-effective for plants operated a short term.
- *expansion of existing facilities or construction of new facilities to improve odor and pathogen control:* Lime stabilization is particularly applicable in small plants or when the plant will be loaded only seasonally.

Pre-Lime Stabilization versus Post-Lime Stabilization

Lime stabilization can be part of a sludge conditioning process prior to dewatering (pre-lime stabilization) or following a dewatering step (post-lime stabilization). Although pre-lime stabilization is most common, post-lime stabilization has significant advantages, particularly in terms of re-duced lime requirements. In addition, post-lime stabilization does not require special constraints on conditioners or equipment for dewatering.

The standard approach to lime stabilization is to add wet lime to liquid sludge for achieving a pH of 12 for two hours following treatment. This process has been limited to small plants with short haul distances.

Lime addition alone, without the addition of other conditioners such as aluminum or iron salts, may produce a sludge that does not dewater well. Therefore, pre-lime stabilization without additional conditioners is not economical over a long term.

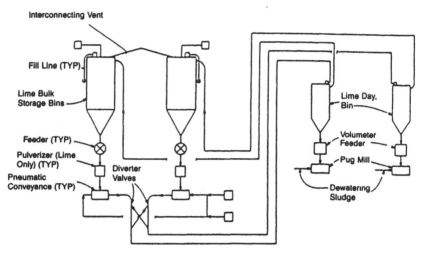

Figure 6.21 Typical post-lime stabilization feed system. (Source: Black and Veatch. May 1989. "Draft Phase II Report, Residual Solids Management Planning Study—Southwest Area," Metropolitan Waste Control Commission.)

Larger plants have developed two modifications of the lime stabilization process to improve sludge dewaterability and overall economics [51,55]. One modification increases the lime dose in a standard iron and lime conditioning mixture prior to vacuum filtration (pre-lime stabilization).

The second modification involves the addition of dry lime (quick or hydrated) to dewatered sludge cake (post-lime stabilization). Significant advantages of post-lime stabilization include a variety of conditioners used for dewatering not only those suited for high pH applications, no special requirements for sludge dewatering equipment, and avoidance of high lime-related abrasion, corrosion, and scaling problems to the mechanical dewatering equipment, a common situation when a pre-lime stabilization process is used. The operator has the option to use either hydrated lime or quicklime (without slaking). In addition, there is the possibility of using heat generated during slaking of quicklime in the lime-sludge mixture to enhance pathogen destruction. Quicklime is less costly than hydrated lime, particularly when slaking is eliminated or at large-scale facilities using lime in excess of 3 to 4 tons per day [49].

A post-lime stabilization system requires adequate mixing to avoid pockets of putrescible material. An effective system is a pug mill with double paddle mixers. Figure 6.21 shows a post-lime stabilization process.

Existing Post-Lime Stabilization Facilities

Several wastewater treatment plants successfully use post-lime stabilization. The following list outlines post-lime stabilization projects in the United States:

- Blue Plains Wastewater Treatment Plant (WWTP), Washington, D.C: Blue Plains is a 300-MGD wastewater treatment plant that produces about 1500 wet tons of sludge per day. Approximately 70% of the sludge is applied to agricultural land after it is dewatered by vacuum filters and stabilized with hydrated lime.
- Eastside WWTP, High Point, North Carolina: Post-lime stabilization was used by Eastside temporarily while their digesters were being repaired. The Eastside WWTP is a 16-MGD facility. Raw sludge was dewatered with belt filter presses and mixed with quicklime in a pug mill.
- Boone, Iowa, WWTP: The Boone, Iowa, WWTP is a 2-MGD facility and uses cement kiln dust (N-Viro soil PSRP) to treat about 1.5 tons per day of belt filter press dewatered sludge. The stabilized sludge is stored offsite in an uncovered pit on a nearby farmer's field. The farmer mixes the sludge with cow manure and applies it to cropland.
- T. E. Maxson WWTP, Memphis, Tennessee: The T. E. Maxson plant is an 80-MGD facility that uses post-lime stabilization to treat only primary sludge. Primary sludge is dewatered with belt filter presses and mixed with about 12% by weight (dry basis) of quicklime. The stabilized sludge is land applied with a truck-mounted, beater-type spreader and is incorporated into the soil with a tractor and disk.
- Stamford, Connecticut: The Stamford 20-MGD municipal treatment plant uses land disposal as a backup for their coincineration process. When land disposal is required, digested sludge is dewatered, post-lime stabilized to control odors, and then disposed on land.
- Toledo, Ohio: The Bay View WWTP is a 100-MGD facility that has piloted the AASSAD process (post-chemical PFRP stabilization) and is scheduled to complete the construction of a full-scale facility in summer 1989. Digested primary sludge will be mixed with raw waste activated sludge, dewatered, and treated with lime and CKD. The mixtures will be aerated for at least seven days and applied to cropland.
- Dayton, Ohio: As part of a sludge management program, dewatered raw primary and secondary sludge are lime treated for pathogen control prior to land application.

The managers of these programs stated that they were satisfied with the overall post-lime stabilization process for its simple operation and good productivity. The worst problems using the lime handling systems were inadequate ventilation and plugged storage bins. Care must be taken in the design of the chemical storage and handling systems to ensure reliability and a safe working environment. Few odor problems were reported.

OTHER PROCESS ALTERNATIVES AND MODIFICATIONS

There are several other lime stabilization process alternatives or modifications to the basic approaches previously discussed. Some of these alternatives evolved when using other processes in a treatment plant. Other approaches deal with only lime stabilization.

An example of using other processes combines lime-treated primary sludge for phosphorus removal, with raw secondary sludge [56]. A second example uses existing digesters or other available plant tankage to thicken lime-stabilized sludge prior to dewatering and disposal [54].

One alternative associated strictly with the lime stabilization process uses two sludge-mixing vessels—one for increasing the initial pH to a value greater than 12, and one for providing adequate contact time and allowing excess lime addition to maintain the pH within the desired range [57].

Process Design Criteria

The primary factors used to design a pre- or post-lime stabilization system are pH, contact time, and lime dosage.

Contact Time and pH

Contact time and pH are directly related because the necessary pH must be maintained for an adequate time to destroy pathogens. The lime dosage must provide enough residual alkalinity to maintain a high pH for several days before disposal. This alkalinity will prevent the pH from dropping and permitting growth or reactivation of odor-producing and pathogenic organisms.

The drop in pH (referred to as pH decay) occurs in the following sequence. Atmospheric carbon dioxide (which forms a weak acid when dissolved in water) is absorbed, then gradually consumes the mixture's residual alkalinity, and the pH gradually decreases. Eventually, a pH is reached where bacterial action resumes, continuing the drop in pH from the production of organic acids (similar to the reactions within the anaerobic digestion process). Numerous studies have examined the pH and contact time required to stabilize sludge. It is agreed that a significant reduction in pathogens and odors occurs when pH is increased to 12.5 for 30 minutes (which keeps pH above 12 for two hours). The pH usually decreases from its initial value during the stabilization process and, therefore, should be increased to more than 12 and maintained. Sludge does not have to be contained within a contact vessel for the stated time, as long as the sludge can be monitored to ensure an adequately high pH has been maintained for the desired period.

Lime Dosage

The third major design factor is the required lime dosage. This dosage depends on a number of factors such as the type of sludge (primary, waste activated, etc.), chemical composition (including organic content) of the sludge and liquid, and solids concentration of the sludge.

Table 6.21 shows the range of pre-lime stabilization dosages required to maintain a pH of 12 for 30 minutes for various types of sludge [49]. Numerous researchers have confirmed these doses [58]. Chemical composition also determines the required lime dose. This composition is a combination of the type of sludge as well as the treatment process from which it is produced (such as whether chemical coagulation is used). The final factor affecting lime dose is solids concentration. Usually, as the percent of solids concentration increases, the required dose increases.

Figure 6.22 displays the general relationship between lime dosage and pH for a typical municipal sludge at several solids concentrations. Table 6.22, calculated from data on Figure 6.22, shows a relatively constant lime

TABLE 6.21. Lime Dosage Required for Chemical Treatment at Lebanon, Ohio.[a]

Type of Sludge	Average Solids Concentration (percent)	Average Lime Dosage [lb $Ca(OH)_2$/lb dry solids]	Average pH Initial	Average pH Final
Primary sludge[b]	4.3	0.12	6 7	12 7
Waste activated sludge	1.3	0.30	7.1	12.6
Anaerobically digested combined	5.5	0.19	7.2	12.4

[a]Dose required to attain pH for 30 minutes.
[b]Includes waste activated sludge
Source. United States Environmental Protection Agency Technology Transfer, September 1979 *Process Design Manual for Sludge Treatment and Disposal.* EPA 625/1-79-011, pp 6–105.

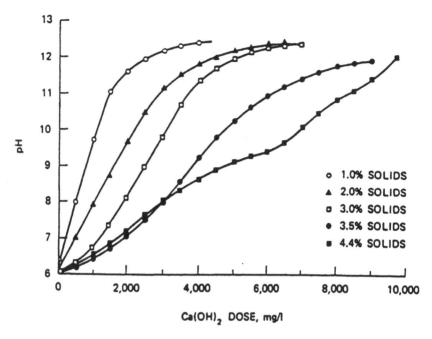

Figure 6.22 Lime doses required to raise pH of a mixture of primary sludge and trickling filter humus at different solids concentrations. (Source: U.S.EPA. April 1975. *Lime Stabilized Sludge: Its Stability and Effect on Agricultural Land,* EPA-670/2-75-012. Washington, D.C.: National Environmental Research Center.)

dose per unit mass of sludge solids required to attain a particular pH level. Therefore, lime requirements are more closely related to the total mass of sludge solids than to sludge volume. Volume reduction by thickening may have little or no effect on the lime amount required, because the sludge solids mass is not changed. The pre-lime stabilization dose for a given mass of sludge solids is relatively the same regardless of the solids concentration, over a range of solids concentrations typically found in wastewater treatment (0.5 to 4.5%). Actual required dose per unit mass of sludge solids tends to be somewhat higher for dilute sludges than for concentrated sludges, because more lime is required to raise the pH of water: about 1 g/L (0.01 lb/gal) to reach pH 12; about 5 g/L (0.04 lb/gal) to reach pH 12.5.

The data presented in Tables 6.21 and 6.22 can be used for preliminary design of a lime stabilization facility; yet, we recommend that the required dosage be determined on a case-by-case basis because of the many factors that affect this dosage. In order to prevent pH decay and the associated growth of organisms, an excess of lime over the required dose may be necessary [58]. The exact dosage required for any particular sludge can be determined through laboratory-scale testing.

TABLE 6.22. Lime Doses Required to Keep pH above 11 at Least 14 Days.

Type of Sludge	Lime Dose [lb Ca(OH)₂/lb Suspended Solids]
Primary sludge	0.10–0.15
Activated sludge	0.30–0.50
Septage	0.10–0.30
Alum-sludge[a]	0.40–0.60
Alum-sludge–plus primary sludge[b]	0.25–0.40
Iron-sludge[a]	0.35–0.60

[a]Precipitation of primary treated effluent
[b]Equal proportions by weight of each type of sludge.
1 lb/lb = 1 kg/kg.
Source: Farrell, J. B., J. E. Smith, Jr. and S. W. Hathaway. 1974. "Lime Stabilization of Primary Sludges," *Journal Water Pollution Control Federation,* 46:113.

Figures 6.23, 6.24, and 6.25 show curves indicating the characteristic pH drop that occurs when lime is added to sludge [49,59]. Notice when the dose is too low in Figures 6.23 and 6.24, the lime-sludge mixture may initially attain the target pH of 12, but a rapid pH decay may occur.

Note that minimum lime doses of 25 to 40% as Ca(OH)₂ were required for pre-lime stabilization prior to vacuum filtration (Figure 6.24). Gener-

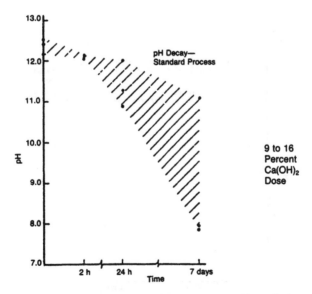

Figure 6.23 Example of pH decay following lime stabilization by the standard process (liquid-liquid contact, pre-lime stabilization). (Source: Westphal, A. and G. L. Christensen. 1983. "Lime Stabilization: Effectiveness of Two Process Modifications," *Journal Water Pollution Control Federation,* 55:1381.)

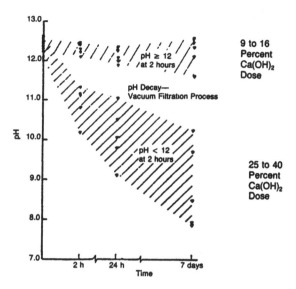

Figure 6.24 Example of pH decay following lime stabilization prior to vacuum filtration (pre-lime stabilization). (Source: Westphal, A. and G. L. Christensen. 1983. "Lime Stabilization: Effectiveness of Two Process Modifications," *Journal Water Pollution Control Federation,* 55:1381.)

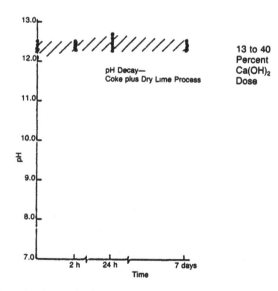

Figure 6.25 Example of pH decay following lime stabilization by dry quicklime addition of raw primary/waste activated sludge cake (post-lime stabilization). (Source: Westphal, A. and G. L. Christensen. 1983. "Lime Stabilization: Effectiveness of Two Process Modifications," *Journal Water Pollution Control Federation,* 55:1381.)

ally, lower minimum lime doses of 25 to 30% as $Ca(OH)_2$ are required for effective post-lime stabilization. Figure 6.25 shows a wide range of minimum post-lime stabilization doses of 13 to 40% as $Ca(OH)_2$.

EFFECTIVENESS AND PROCESS PERFORMANCE

Properly designed and operated pre- and post-lime stabilization systems reduce odors and odor production potential in sludge; reduce pathogen levels; and improve dewatering characteristics of the sludge. The nature and extent of the effects are described in the following paragraphs.

Odor Reduction

Lime stabilization systems substantially reduce odor. Initially, when air mixing systems are used, ammonia odors increase as a result of ammonia stripping. Once the initial ammonia odors have been emitted, odors can be reduced by a factor of 10^4. In fact, one of the major odors in sludge processing facilities, hydrogen sulfide, is essentially eliminated as the pH increases to 9 and above because it is converted to nonvolatile ionized forms (Figure 6.26).

Pathogen Reduction

Several studies have demonstrated that both pre-lime and post-lime

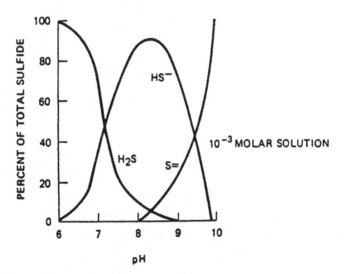

Figure 6.26 Effect of pH on equilibrium between hydrogen sulfide and nonvolatile ionized sulfides. (Source: U.S.EPA. 1979. *Process Design Manual for Sludge Treatment and Disposal,* EPA 625/1-79-011, Cincinnati, Ohio: U.S.EPA.)

stabilization achieve significant pathogen reduction, provided a sufficiently high pH is maintained for an adequate period of time [51,60].

Pathogen reductions can be achieved in sludges that have been lime-treated to pH 12.0. Table 6.23 lists bacteria levels measured during the full-scale studies at Lebanon, Ohio, WWTP and shows that lime stabilization of raw sludges reduced total coliform, fecal coliform, and fecal streptococci concentrations by more than 99.9%. The numbers of *Salmonella* and *Pseudomonas aeruginosa* were reduced below the level of detection. Table 6.23 also shows that pathogen concentrations in lime-stabilized sludges ranged from 10 to 1000 times less than those in anaerobically digested sludge from the same plant.

Christensen, a professor at Villanova University, researched the pathogen reduction performance of post-lime stabilization using dry quicklime using lime doses of 13 and 40% as $Ca(OH)_2$. His results indicated post-lime stabilization can achieve a reduction in fecal coliform and fecal streptococcus pathogens of at least two orders of magnitude. This was as good as and, in some cases, better than standard liquid sludge pre-lime stabilization and pre-lime stabilization followed by vacuum filtration (Figures 6.27 and 6.28). No growth of either organism occurred by day 7. Christensen also reported that pre-lime and post-lime stabilization processes performed as well as or better than mesophilic aerobic digestion, anaerobic digestion, and mesophilic composting in reducing densities of fecal coliform and fecal streptococcus (Table 6.24).

Additional studies on dry lime stabilization of wastewater sludge showed a 4 to 6 log reduction of fecal streptococcus at pHs around 12 [64]. Such a treatment scheme can yield PFRP sludges by using an addition of 1000 kg of lime per 500 kilograms of sludge at 20% solids, as a result of the heat for lime hydrolysis, raising the temperature of the sludge to greater than 55°C [65]. When the treatment approaches a dosage of 50% lime to sludge, the cost of this treatment process is approximately $20 to $100 a day per dry ton [64].

The U.S.EPA recently approved the use of post-chemical stabilization using 10% lime and 25% alkaline CKD (dry weight basis) followed by accelerated drying as a PFRP process. The final pH of the sludge/chemical mixture is between 12 and 12.4. The inactivation of the pathogens, excluding parasites, appeared to be a result of the high pH along with accelerated drying [66].

There is little information about the amount of virus reduction during lime stabilization, but a few studies suggest rapid destruction at a pH of 12 [49]. However, qualitative analysis by a microscopic examination has indicated substantial survival of higher organisms (such as hookworms, amoebic cysts, and *Ascaris* ova) after twenty-four hours at high pH [54]. It is not known whether long-term contact would eventually destroy these organisms.

TABLE 6.23. Reduction in Bacteria at Lebanon, Ohio.

Type of Sludge	Bacterial Density (number/100 mL)				
	Total Coliforms[a]	Fecal Coliforms[a]	Fecal Streptococci	Salmonella[a]	Ps. aeruginosa
Raw					
Primary	2.9×10^9	8.3×10^8	3.9×10^7	62	195
Waste activated	8.3×10^8	2.7×10^7	1.0×10^7	6	5.5×10^3
Anaerobically digested					
Mixed primary and waste activated	2.8×10^7	1.5×10^6	2.7×10^5	6	42
Lime-stabilized[c]					
Primary	1.2×10^5	5.9×10^3	1.6×10^4	<3	<3
Waste activated	2.2×10^5	1.6×10^4	6.8×10^3	<3	13
Anaerobically digested	18	18	8.6×10^3	<3	<3

[a] Millipore filter technique used for waste activated sludge. MPN technique used for other sludges
[b] Detection limit = 3.
[c] To pH equal to or greater than 12.0

Source. United States Environmental Protection Agency Technology Transfer. 1979. Process Design Manual for Sludge Treatment and Disposal, U.S.EPA 625/1-79-011, pp 6–109

335

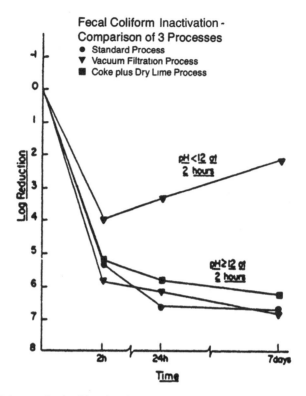

Figure 6.27 Average fecal coliform inactivation by two pre-lime stabilization processes and one post-lime stabilization process. (Source: Westphal, A. and G. L. Christensen. 1983. "Lime Stabilization: Effectiveness of Two Process Modifications," *Journal Water Pollution Control Federation,* 55:1381.)

Sludge Dewaterability

Sludge dewaterability may be improved in the pre-lime stabilization process, particularly if additional conditioners are used as required. One pilot-scale study showed that lime-treated sludge dewatered more rapidly on sand drying beds and yielded higher ultimate solids. Iron substantially improves the mechanical dewaterability of both primary sludges and mixtures of primary and waste sludge [67]. Additional lime in excess of the dose required for conditioning is typically necessary for stabilization [59] and also results in improved dewaterability over that of untreated sludge [54]. In addition to lime dose, factors such as organic and inorganic constituents affect dewaterability.

Chemical changes that occur in sludge during sludge dewatering include:

- There is an increase in the total suspended solids concentration, as a result of adding lime solids and precipitation of dissolved solids.
- There is an increase in total alkalinity.
- There is a reduction in volatile suspended solids concentration of 10 to 35% as a result of dilution with lime and some minor loss of volatile organics to the atmosphere. However, little if any volatile solids destruction occurs. In comparison, aerobic and anaerobic digestion stabilization processes can reduce volatile solids content by as much as 45 to 55%.
- There is a reduction in soluble phosphorus values because of the reaction with orthophosphate to form calcium phosphate precipitate.
- There is a reduction in nitrogen values as a result of ammonia stripping; however, this loss is usually minor unless the sludge is vigorously mixed or surface applied to land while at high pH.

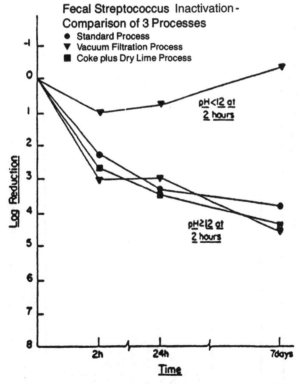

Figure 6.28 Average fecal streptococcus inactivation by two pre-lime stabilization processes and one post-lime stabilization process. (Source: Westphal, A. and G. L. Christensen. 1983. "Lime Stabilization: Effectiveness of Two Process Modifications," *Journal Water Pollution Control Federation*, 55:1381.)

TABLE 6.24. Inactivation of Bacteria by Different Sludge Stabilization Processes.

Process	Fecal Coliform[a]	Fecal Streptococcus	Reference
Anaerobic digestion (35°C)			
Mean	1.84	1.48	[52]
Range	1.44 to 2.33	1.1 to 1.94	[52]
Aerobic digestion			
20°C[a]	1	1	[52]
30°C[a]	2	1.64	[61]
Composting	≥	2 9	[62]
Lime Stabilization			
Raw primary	5.1	2.4	[52]
Waste activated	3.2	3.2	[52]
Mixed primary and trickling filter humus, 4 percent solids	2.6	1.8	[52]
Storage[b]			
10°C	—	1	[63]
20°C	—	1.5	[63]
30°C	—	2.0	[63]

[a]Laboratory study, 35-day detention time
[b]Laboratory study, 30-day detention time.
Source. 1985. "Sludge Stabilization," Manual of Practice FD-9. Water Pollution Control Federation, Washington, D.C.

One potential drawback to mechanical dewatering of lime-stabilized sludge is the likelihood of scaling problems resulting from the high lime doses.

PFRP ALKALINE STABILIZATION TECHNOLOGIES

Envessel Pasteurization

The envessel pasteurization process is a PFRP-approved, alkaline stabilization process patented by the RDP Company. In this process, dewatered sludge is preheated in an insulated electrically heated screw conveyor, called a thermofeeder, prior to being transferred to a pug mill mixer [69]. The heated sludge and quicklime are mixed in a heated and insulated pug mill mixer, called a thermoblender. Because supplemental heat is used to elevate the temperature of the sludge, alkaline material must only be added in sufficient quantities to elevate the pH. This results in a lower lime dose than if exothermic hydration reactions are used as the sole source to elevate sludge temperatures. The mixture is conveyed to a heated and insulated vessel reactor where it is retained at a minimum temperature of 70°C for 30 minutes to meet the U.S.EPA PFRP requirements. The envessel pasteurization process does not require an additional drying step to achieve PFRP stabilization, and therefore, requires less onsite land. However, it does require more building space for the thermoblenders and vessels. The end product also has a high moisture content compared to processes that use large chemical dosages or supplemental drying.

The envessel pasteurization process has been successfully demonstrated during several pilot tests. Lower chemical consumption and corresponding projected annual costs for PFRP alkaline stabilization have been shown. However, no full-scale facilities were in operation at the time of this study.

Biofix

Bio Gro Systems, Inc. has developed four separate alkaline stabilization processes to meet increasingly stringent operational and regulatory requirements. The Biofix I and II processes only meet the PFRP requirement and therefore will not be discussed.

Biofix Stage III processing meets the criteria for PFRP by utilizing the exothermic reaction of quicklime with water: $CaO + H_2O \rightarrow Ca(OH)_2 +$ heat. Each pound of 100% of quicklime produces 491 BTUs of heat and extracts free water from the sludge. This reaction achieves temperatures in excess of 70°C for more than 30 minutes, as required by federal regulations (40 CFR, Part 257) for add-on pasteurization to meet PFRP criteria.

For all states of Biofix processing, odor-control reagents and/or nutrient enhancements may also be incorporated as required.

N-Viro Process

The N-Viro process is a proprietary, patented process that satisfies PFRP requirements. Technically, the process is defined as "advanced alkaline stabilization with subsequent accelerated drying." However, the process is generally identified by its trade name, N-Viro, and will be referred to as such throughout this chapter. Two alternative methods of conducting the N-Viro PFRP process have been approved by the U.S.EPA and PFRP equivalent processes. These two alternatives are as follows.

Alternative 1

Alkaline materials, such as cement kiln dust (CKD), lime kiln dust, quicklime fines, pulverized lime, or hydrated lime, are added and mixed to the sludge in sufficient quantity to achieve a pH of 12.0 or greater for at least seven days. Following mixing, the alkaline-stabilized sludge is dried for at least thirty days and until a minimum solids concentration of at least 65% is achieved. Sludge solids shall be kept above 60% before the pH drops below 12.0. Mean ambient temperature must be above 5°C for the first seven days.

Alternative 2

Alkaline materials are added and mixed to the sludge in sufficient quantity to achieve a pH greater than 12.0 for at least seventy-two hours. Concurrent with maintaining this high pH, the sludge is heated to a temperature of at least 52°C and maintained at that temperature for at least twelve hours. Stabilized sludge is dried until a minimum solids concentration of 50% solids is achieved. This alternative is similar to Alternative 1 with the exception that it includes a thermal treatment process (or heat pulse) step and is not subject to the ambient air temperature limitations.

Although two specific methods have been approved as meeting PFRP sludge stabilization, these two methods can be implemented in a variety of ways. In actual operation, the N-Viro process is usually conducted as an extension of traditional post-lime stabilization.

SLUDGE DISINFECTION BY IONIZING RADIATION

INTRODUCTION

As indicated in Table 6.20, beta ray irradiation (which should more properly be referred to as electron-beam irradiation) and gamma ray irradiation are both listed by U.S.EPA as PFRPs for sludge. Electron-beam and gamma ray irradiation are both defined as ionizing radiations, meaning that they possess sufficient energy to remove bound electrons from the atoms in the matter through which they pass, thus creating charged ions. In an aqueous system, the main ions formed are hydrated electrons, hydrogen free radicals, and hydroxide free radicals. These free radicals exert disinfection power by damaging the DNA and RNA in bacteria, viruses, and other pathogens, as well as reacting with oxygen and other molecules in aqueous systems to produce ozone and hydroperoxides which attack organics (including pathogens) in the sludge through various oxidation and dissociation reactions.

The pathogen reducing ability of ionizing radiation is a function of the absorbed radiation dose. The unit of measure of this absorbed dose is the rad, which is defined as the absorption of 100 ergs of energy per gram of material. The absorbed dose typically depends on three factors:

(1) The type of material

(2) The type of radiation

(3) The energy absorption mechanism

The use of ionizing radiation for sludge disinfection is predicated on applying a sufficient absorbed dose in a uniform manner to a large sludge volume in an economically reasonable process. The U.S.EPA specifies in the Part 503 Regulations that this dose must be at least 1.0 megarad (Mrad) at 20°C for the process to be considered a PFRP.

As previously stated, the two processes recognized by U.S.EPA for achieving PFRP criteria are electron-beam (beta rays) and gamma rays. However, as of 1996, there were no full-scale or large pilot scale sludge irradiation facilities in operation in the United States, Canada, or Western Europe. Thus, there is a lack of current information on the performance and economics of sludge disinfection by ionizing radiation. The most comprehensive recent review of information on this subject can be found in a 1992 American Society of Civil Engineers (ASCE) report on radiation treatment of water, wastewater, and sludge [103]. The following information is principally abstracted from that report, as well as from the U.S.EPA *Process Design Manual on Sludge Treatment and Disposal* [104].

GAMMA RAYS

Gamma rays are produced during the decay of certain radioactive elements. The two radioisotopes that have been used in sludge disinfection studies are cobalt-60 and cesium-137. Cobalt-60 is a manmade isotope produced by bombarding cobalt-59 with neutrons. Cobalt-60 decays with emission of two gamma rays with energies of 1.17 and 1.33 million electron volts (MeV). The half-life of cobalt-60 is 5.26 years. Cesium-137 is a by-product of the fission of uranium-235 fuel in nuclear reactors. Cesium-137 decays with the emission of one gamma ray of 0.66 MeV. The half-life of cesium-137 is 30 years.

Gamma rays have strong penetrating power and can thus provide uniform dosages to bulk materials. The isotopes provide continuous and reproducible source strengths, with exposure time being the key design parameter. The main disadvantages of gamma ray sources are limited availability of the required isotopes, a significant regulatory compliance burden imposed by the Nuclear Regulary Commission in the United States and similar agencies in other countries, and the fact that the radiation cannot be turned off when not in use.

The world's longest operating full-scale gamma ray sludge irradiator was located in Geiselbullach, Germany. It began operation in 1973 and remained in service through the early 1990s. In this facility anaerobically digested sludge was irradiated for disinfection prior to land application. The sludge was processed in a batch mode using a cobalt-60 source.

The facility consists of an above ground sludge feed tank which holds liquid digested sludge and a below ground concrete lined sludge irradiation shaft and sludge recirculation system where the sludge is exposed to the gamma ray source in a recirculating system to provide a uniform dose to the entire sludge volume. The irradiation time depends on the sludge characteristics, and is set to provide a dose of 300 krad to the sludge mass. Typical irradiation times are 3 to 5 hours. The sludge processing capacity of the facility ranges from 30–130 m³/day.

Tests on sludge samples at the Geiselbullach facility indicate that a two-log reduction in total bacterial count, a two-log reduction in Enterococcus, and a four-log reduction in Enterobacteriaceae can be achieved at a 300 krad dosage.

The major gamma ray disinfection work conducted in the United States was done at Sandia National Laboratories in Albuquerque, New Mexico. The test facility began operation in 1978 and continued in operation until 1985. Cesium-137 was used as the gamma ray source. The system was designed to process dewatered sludge with a solids concentration of 40 percent or greater.

The impetus for much of the experimental work at the facility was the promulgation by the U.S.EPA in 1979 of the 40 CFR, Part 257 Regulations entitled "Criteria for Classification of Solid Waste Disposal Facilities and Practices." These regulations were the predecessor of the Part 503 Regulations and contained the concept of PSRP and PFRP processes for treating sludge. Gamma ray irradiation at a 1 Mrad dose was listed as a PFRP in the Part 257 Regulation.

The Sandia facility consisted of two concrete lined vaults. One vault was the Cs-137 storage pool and the other was the irradiation chamber.

During operation, the Cs-137 was moved to the irradiation chamber by a remote-controlled conveyor. Dried sludge was then transported into the irradiation chamber in a bucket conveyor system. The bucket conveyor carried the sludge in a serpentine pattern both above and below the Cs-137 source in order to achieve a uniform dose. The applied radiation dose was varied from 420–1000 krad over the years of operations. The facility had a sludge processing capacity of approximately 6000 kg/day.

Yeager and Ward [105] reported on the effectiveness of gamma ray irradiation for reducing pathogens in sludge based upon work at the Sandia facility. They conducted tests to determine the dose rate needed to achieve a 90% reduction in various pathogen populations and then calculated the effect that a 1 Mrad dose of gamma irradiation should have. They determined that a 1 Mrad dose of gamma irradiation would reduce *E. coli* bacteria by 46 orders of magnitude, viruses by three orders of magnitude, and *Ascaris* ova by 20 orders of magnitude.

ELECTRON BEAMS

High-energy electrons are produced by high voltage linear accelerators. This is a manufactured piece of electronic equipment which can produce a focused beam of 1 to 10 MeV electrons, with 1 to 2 MeV being typical. Certain radioisotopes also emit beta radiation, but the energy of these electrons is too small for practical use in sludge treatment.

The main advantage of electron beams is that the radiation source can be turned on and off with the flick of a switch. The main disadvantage of electron beams for sludge disinfection is the low penetrating power of the beam. A 1 MeV electron can only penetrate about 1 cm of water. For this reason, the design of an electron-beam sludge disinfection facility is considered more complicated than a gamma ray facility.

Two major electron-beam sludge disinfection projects have been conducted in the United States. The first was conducted at the Deer Island Treatment Plant in Boston, Massachusetts on an intermittent basis over the period of 1976–80. The Deer Island facility consisted of a constant head liquid sludge feed tank which was equipped with an underflow discharge weir. Sludge is discharged under the weir in a thin film and then flows down an inclined ramp. An electron accelerator producing an 0.85 MeV electron beam is positioned above the ramp, and directs the beam back and forth across the sludge path at a frequency of 400 times per second. The thickness of the liquid sludge film was kept at approximately 0.2 cm. The estimated applied radiation dose was 400 krad. The system had a sludge processing capacity of 400 m³/day.

Reported results for anaerobically digested sludge indicate that total bacteria were reduced by four logs at a dose of 280 krad, total coliform by five to six logs at a dose of 200 krad, and fecal streptococci by over three logs at a dose of 400 krad. At the 400 krad dose, total viruses as measured by plaque forming units were reduced by one to two logs.

The second project was located at the Virginia Key Wastewater Treatment Plant in Miami, Florida, and was operated from 1984–85. The facility consists of a liquid sludge feed system which delivers the sludge to a 1.2 m wide weir, over which the liquid sludge flows at an estimated thickness of 0.5 cm. A 1.5 MeV electron accelerator is positioned above the weir, and projects a scanning electron beam into the sludge as it flows over the weir. The applied radiation dose was calculated to be 400 krad at a sludge flow rate of 680 m^3/day.

Reported results indicate that at the 400 krad dose fecal coliform were reduced from 100,000/mL in the influent to below detection limits in the effluent, and salmonella were reduced from 2000/mL in the influent to below detection limits in the effluent.

At present, it appears that routine use of either gamma rays or electron beams for sludge disinfection is many years away, and will require further research. Cost data for a large scale installation must still be determined.

The continuing public apprehension about the use and storage of radioactive materials has led most researchers to conclude that electron-beam technology is actually the only feasible irradiation process for the future. In 1997 it is anticipated that a new sludge irradiation research project will begin at the Virginia Key electron-beam facility [106]. This project is aimed at evaluating electron-beam irradiation in the context of the Part 503 pathogen reduction requirements, and determining meaningful cost data for the process.

SLUDGE CONDITIONING

INTRODUCTION

Definition and Purpose of Conditioning

In this section, the term sludge conditioning is defined as the overall process of enhancing the aggregation of suspended sludge particles by chemical or physical means. Conditioning of sludge neutralizes or destabilizes the chemical or physical forces acting on colloidal and/ or particulate matter suspended in a liquid (e.g., water). This destabilization process brings about the growth of the otherwise small visible sized particles into larger aggregates known as flocs, which have an indefinite shape with random and noncrystalline arrangement. The purpose of such enhancement of floc growth or conditioning is usually to increase the efficiency of dewatering or thickening of the sludge through various treatment processes intended to concentrate the sludge solids. Chemical conditioning of sludge includes the use of organic polymers, inorganic salts, and a combination of both. Physical conditioning of sludge includes heat treatment, freezing followed by thawing, and elutriation.

Theories and Mechanisms of Chemical Conditioning

In general, two types of mechanisms are operative in chemical sludge conditioning. One is particle charge neutralization and the other is particle bridging. These two mechanisms are distinguished by the terms coagulation and flocculation, respectively. Such a sharp division of mechanisms of sludge conditioning is most likely an oversimplification, and a combination of both mechanisms in some proportion is probable in most cases. The conditioning of inorganic and biological sludges with inorganic chemicals is predominantly associated with the charge neutralization mechanism, whereas the conditioning of biological sludges with organic polymers is predominantly associated with the bridging of the sludge particles.

In the destabilization or neutralization of particle charges, the layers of electrical charge surrounding the suspended particles are discharged by the addition of conditioning agents with an opposite electrical charge. Since most suspended sludge particles in wastewater-related sludges are negatively charged, cationic conditioning agents are used for charge neutralization. The higher the positive charge, the more effective is the conditioning agent. Charge neutralization helps the particles to come into contact with each other, as there will be no repulsion between them. This effect results in the formation of flocs, which can settle and compact well. The charge destabilization essentially produces flocs that can closely pack together.

Depending on the net charge of a sludge suspension, conditioning of sludge may be achieved by cationic (positively charged) and anionic (negatively charged) polymers.

In addition to destabilizing surface electrical charges on dispersed sludge particles, polyelectrolytes flocculate sludge by a process called bridging. Bridging is the simultaneous attachment of polymer molecules, which disperse as a long chain in a sludge medium, to two or more sludge particles. In this way, polymers bridge the gaps between particles and draw them together in a lattice structure to form floc. The higher the molecular weight, charge density, and geometric length of a polymer molecule, the greater is its effectiveness as a conditioning agent. In this mechanism, discharge of surface electrical charges on a sludge particle may not occur in some cases and the electrostatic repulsion will often prevent close packing of the suspended particles. Thus, this mechanism tends to produce flocs that are loosely packed together.

A variety of theoretical explanations have been advanced to provide details for these mechanisms. The purpose of the following sections is not to provide a detailed presentation of the theories described in the literature. They can be found elsewhere [70].

Electrical Double Layer and Zeta Potential Theory

The electrical charge surrounding the suspended sludge particles is regarded as consisting of two regions. These are composed of an inner layer (Stern layer) located at a distance from the surface equal to a hydrated ion radius, and an outer diffuse layer located farther from the surface than the Stern layer, in which ions are distributed according to electric forces and thermal motion.

The difference in electrical potential between the surface of a sludge particle and the liquid it is suspended in, is called the zeta potential. It is essentially a measure of the electrostatic charge of the particle in the outer diffuse region or its electric double layer. The rate and direction of movement of such particles is used to determine zeta potential under an applied voltage. Wastewaters have zeta potentials of approximately 45 millivolts (MV) and effective conditioning reduces this value to approximately 5 MV. Thus, zeta potential may be used to monitor conditioning performance and to optimize conditioner dosage in some cases.

D.L.V.O. Theory

Named after the investigators Derjaguin, Landau, Verwey, and Overbeek [70,73], this concept attempts to quantify particle stability in terms of attractive and repulsive forces that are a function of interparticle distance. The attraction is due to van der Waal's interaction and the repulsion is due to electric double layer overlapping. Attraction will predominate at small and large interparticle distances and repulsion may predominate at intermediate interparticle distances. When a conditioning agent is added to sewage sludge, it attempts to reduce or eliminate the repulsive forces at intermediate distances so that the attractive forces between particles will predominate and stimulate floc formation.

Enmeshment and Sweep Floc

Some inorganic and organic long-chain-molecule conditioning agents having a high molecular weight produce a voluminous three-dimensional lattice structure or matrix, which allows the entrapment of suspended particles or flocs. As the matrix contracts and settles, the suspended material remains enmeshed in the lattice and it appears to be "swept" from the medium in which it is suspended. This is known as the sweep floc process of conditioning and may occur in conjunction with either of the two mechanisms mentioned previously [70,73].

Mechanisms of Physical Conditioning

Cell Structure Disruption

Conditioning of sludge by heat treatment or freeze/thaw cycles alters the surface properties of suspended solids and ruptures the cells of biomass. Biological sludges contain water and cellular material that exist in a gel-like structure. Much of the water is bound inside the cellular material. The gel structure breaks down when sludge is subjected to stress and lysis of the cellular material also occurs. The bound water associated with the gel structure is released, resulting in the aggregation of the suspended solids [71,72,74,75].

Elutriation

The dewaterability of a sludge generally improves by a process called elutriation. Elutriation consists of a thorough mixing of approximately equal volumes of sludge and wash liquid (water or final effluent) followed by settling in a gravity thickener. This washing process results in the dilution of dissolved solids (e.g., alkalinity) and partial removal of fine suspended solids. Essentially, a washing of the sludge lowers the concentration of components that would impede dewatering performance or hinder chemical conditioning. Elutriation or sludge washing lowers the dosage of inorganic or organic chemical conditioners and improves their effectiveness [72,74].

FACTORS AFFECTING SLUDGE CONDITIONING

Many factors affect sludge conditioning, which is an important step prior to dewatering of sludges. These factors may broadly be categorized as physical, chemical, and biological, and are discussed as follows.

Physical Factors Influencing Sludge Conditioning

Perhaps the single most significant factor that influences sludge conditioning is the particle size distribution of the sludge suspended solids. As the particle size decreases, the surface area of the suspended solids mass increases. This means greater hydration and increased resistance of the suspended solids conditioning, which is intended to aid in the release of bound water and aggregation of suspended solids.

Any process that reduces the size of the suspended sludge particles will have a negative effect on conditioning, since it will increase the degree of conditioning required and/or limit the effect produced. Pumping of sludge, for example, subjects the sludge suspended material to shear forces and thus reduces the size of sludge particles. The level of shear stress imparted on sludge suspended solids is a function of the type of pump used, flow rate, and the geometry of the piping network used for the transport of sludge. Shear stress, which causes particle size reduction, is also imparted by mixing and agitation of the sludge, and is a function of the intensity of turbulence created by the mixing equipment used. Sludge storage and aging also have a negative effect on conditioning of sludge, although it is not as severe. These issues concerning the effect of shear stress on particle size are of even more significance after sludge conditioning is achieved by chemicals, as different chemicals may produce flocs of varying floc strength or fragility. Obviously, sludges containing flocs that are more fragile will dewater poorly in comparison to sludges having flocs that are less fragile.

Sludge origins and the nature of the solids matrix from a certain unit process also influence sludge conditioning. For example, primary sludge requires less conditioning than secondary sludge (e.g., waste activated sludge). Primary and secondary sludges from biological wastewater treatment plants usually require cationic polymer conditioning, while chemical sludges often require anionic polymer conditioning.

The degree of mixing in the preparation of solutions of organic conditioning agents is an important factor because high-speed mixing tends to cut or fracture long polymer chains into shorter, less effective fragments. A proper degree of mixing is needed to achieve solution blending. Excessive mixing, however, results in polymer degradation. Another major factor is the degree of mixing required to distribute chemical conditioners in the sludge. The intensity of mixing must be such that it provides uniform dispersal of the conditioning agent throughout the sludge, which allows for good floc formation. However, it should not be excessive to destroy the flocs that are formed. The degree and/or the intensity of mixing is related to the viscosity of the chemical conditioner solution and the suspended solids concentration of the sludge. As these two factors increase, it becomes more difficult to achieve effective dispersal of the conditioner in the sludge [71–73,77,78].

Chemical Factors Influencing Sludge Conditioning

The alkalinity and pH of a sludge are perhaps the most important of the chemical factors influencing its conditioning. The charge densities on the surface of the suspended sludge particles and on a conditioning agent are functions of pH. Both of these charge densities affect the surface chemistry of particles involved in conditioning, depending on the sludge type and conditioner. The pH of sludge also governs the ionization state of other chemical species in the sludge. The charge type and geometry of these species, the degree of particle hydration, and the nature of the bound water matrix may in turn influence the dewaterability of sludges. Thus, pH determines the floc structure and affects its properties. Organic conditioners tend to be effective over a broader pH range than inorganic conditioners. However, the effect of pH may also be specific to a given situation and is difficult to generalize.

Alkalinity is an important sludge property that modifies conditioning requirements and the pH of sludge affects alkalinity. High alkalinities tend to have detrimental effects on the performance of both inorganic and organic conditioning agents and tend to increase the dosage requirements of a conditioning agent. Also, a high ionic strength of the sludge liquid phase and of the makeup water tends to have a detrimental effect on floc stability, conditioning agent performance, and dosage requirements of both inorganic and organic conditioners. Moreover, some species of dissolved inorganic salts that are balanced in anion and cation charge (e.g., NaCl, $CaSO_4$) impair conditioning by reducing zeta potential toward zero. Thus, a high concentration of such dissolved materials tends to nullify the effectiveness of conditioners that utilize charge as their principal mechanism of floc formation [73]. Some organic acids also inhibit conditioning. Fulvic acid and humic acid, for example, impair floc formation with both organic and inorganic conditioning agents [73].

The concentration of the sludge suspended solids is an important factor and increases the dosage of a conditioner in proportion to the suspended solids concentration.

The ash or fixed solids content of sludge affects the charge type of the conditioner that is effective with a given sludge. In general, sludges with fixed solids less than 50% require cationic conditioners while sludges with fixed solids greater than 50% require anionic or nonionic conditioners.

The charge density and molecular weight of a conditioning agent govern its performance. In general, as the charge density and molecular weight of a conditioning agent increase, so does its effectiveness in forming good flocs. However, it will be progressively more difficult to prepare solutions of conditioning agents as their molecular weight and charge density increase. The charge density is associated with the neutralization of repellent electrical forces, which inhibit floc formation. The molecular weight (i.e., length of molecular chain) is associated with particle bridging that results in floc formation. The charge type of the conditioner is also a factor and the demand for the conditioner is related to the effective particle charge in the sludge.

The concentration of a conditioning agent is an important factor in the dewatering of sludges and is related to the viscosity of its solution. The concentration and dose of a conditioning agent must be sufficient to meet the requirement for sludge conditioning within a given flow and pumping rate of the sludge and conditioning agent, and yet not so concentrated that the viscosity of the solution hinders its effective distribution in the sludge. Water used for diluting the conditioner may often be applied close to the mixing point of the sludge and conditioner, in order to obtain effective distribution of the conditioner.

Organic conditioning agents, in addition to being fragile and difficult to mix into solution, also require a specific waiting period (aging time) in order to activate fully in the diluted state. Initially, long-chain polymers are coiled and ineffective for conditioning. In solution, a waiting period (aging) allows the coils to unwind and expose the many charged sites along each molecule. Thus, the conditioning agents become more effective during the electrokinetic and bridging mechanisms of floc formation. Low molecular weight polymers may only require a few seconds to activate while high molecular weight polymers may require as much as an hour to activate. Dilute organic polymer solutions, however, begin to deteriorate in performance after several hours. Mannich type polymers are the most rapid in their degradation rate in comparison to emulsion or other powder type of polymers [71–73,77,78].

Biological Sludge Conditioning Factors

The biological solids content of sludges is a significant factor in influencing conditioning. In general, biological sludges have different conditioning requirements than chemical sludges. Processed biological sludges have different conditioning requirements than raw and primary sludges. Anaerobically digested sludges tend to have lower requirements for charge neutralization than the raw sludges from which they are derived, and hence can be dewatered by conditioning agents with a moderate cationic charge. Aerobically digested sludges, however, tend to require more highly charged cationic conditioning agents. Biological sludges from activated sludge processes with long solids retention times tend to have lower conditioner requirements for charge neutralization than those with short solids retention times. This affects the charge requirement on the conditioning agent to be utilized for effective floc formation. Cationic polymers, in general, are the most efficient in conditioning biological sludges.

Factors such as an increase in the concentration and geometry of biocolloids, degree of filamentous growth, concentration of pinpoint flocs, and production of exocellular biopolymers, will increase polymer dosage requirement for conditioning biological sludges. The conditioning efficiency of polymers decreases progressively with waste activated sludges as the organic loading rate (*F/M* ratio) of a treatment plant increases and the mixed liquor dissolved oxygen concentration decreases. The blending of waste activated sludge with other types of sludges usually increases the requirement of a chemical conditioner and degrades the performance of a conditioning agent [71,72].

SLUDGE CONDITIONING METHODS

Inorganic Chemical Conditioning

Chemicals Used for Conditioning Sludge

The earliest use of inorganic chemicals for aggregation of suspended sludge particles as a plant process was in the 1920s. Since that time, iron and aluminum salts, alone or with lime, have been used widely for sludge coagulation/flocculation. Recently, however, organic conditioners have begun to replace inorganic conditioners in many sludge dewatering and thickening applications.

Lime [CaO or $Ca(OH)_2$] and liquid ferric chloride are the two most commonly used inorganic conditioning agents. Liquid ferrous sulfate, anhydrous ferric chloride, aluminum sulfate, aluminum chloride, and potassium permanganate are also in use, although less frequently.

Lime is used mainly in conjunction with sludge for pH control and to some extent as a bulking agent that increases sludge porosity while resisting compression. Other ancillary benefits, such as pathogen reduction and drying due to release of heat, have also been reported [71–73,75].

Iron and aluminum salts dissociate to form divalent or trivalent metal ions that react with hydroxyl and other alkaline ions. The hydroxides of these metals are very insoluble and precipitate rapidly. Many slightly soluble intermediate species occur during the formation of the metal hydroxide precipitates. These include various partially hydrated complexes such as MOH^{+2}, $M(OH)_2^{+1}$, and $M(OH)_4^{-1}$ where M is the metal radical. In addition to these species, the intermediate hydrates can polymerize into long chains of metal hydrate units, which have many partial charge sites along the chains. These metal hydrates produce precipitates of long sticky molecules with very complex structures. The formation of these hydropoly-mers is a function of the pH of the sludge plus chemical conditioner matrix. They are solubilized with structure alteration at both high and low pH, and are stable only within specific pH ranges. The optimum pH range for trivalent iron salts is between 6 and 7, while for trivalent aluminum salts, it is between 4.5 and 5.5. Below the pH of 6 and 4.5, trivalent iron and aluminum species become predominantly positively charged, and above the pH of 7 and 5.5 these species become predominantly negatively charged, respectively. Lime is most often used to control pH within these ranges. The optimization of pH is critical for effective conditioning with these chemicals.

Although the metal ions Fe and Al are of primary importance in achiev-ing the coagulation of sludge, the associated anions (usually sulfate or chloride) and their valence state are also important as factors in condition-ing performance. For example, trivalent iron salts are more effective than divalent iron salts, and chloride salts are more effective than sulfate salts. If alkalinity is present in the sludges, iron and aluminum salts will release carbon dioxide and the alkalinity will serve as a buffer in pH control. The major technical difference of practical significance between iron and aluminum salts as conditioners, is their solubility at a pH of 7. Aluminum salts are relatively soluble at this pH, whereas iron(III) salts are relatively insoluble. Thus, lime addition for pH control may not be necessary with iron(III) salts in some applications, which can be a considerable advantage. Similar considerations hold for divalent iron salts, which are relatively soluble at a pH of 7.

Availability and Storage of Inorganic Chemicals Needed for Conditioning Sludge

Ferric chloride is sold commercially in liquid form as a 30 to 35% solution. It is very corrosive and requires proper handling and storage. Containers made of materials such as epoxy, rubber, ceramic, PVC, and vinyl are recommended for its storage. It can be stored for long periods of time without deterioration. Ferrous sulfate (also known as copperas) is similar to ferric chloride in its handling and storage characteristics. It is available in granular form and produces acidic solutions. However, it also cakes up above 20°C and oxidizes in moist air, which requires special care for handling.

Lime is available in two major dry forms: Pebble quicklime (CaO) and powdered hydrated lime [Ca(OH)$_2$]. In either form it is caustic, prone to form a precipitate when made into a slurry, and may form a calcium carbonate scale.

Aluminum sulfate (alum) is available in dry form as a lump or powder.

It does not have the same type of caustic properties as lime, but produces acidic solutions like ferric chloride.

Dosage Requirements and Cost

Dosage information for these conditioners is highly specific to a given application and must be determined from laboratory, pilot-scale, or full-scale tests.

Typical chemical costs are $0.40/kg ($0.19/lb) for ferric chloride, $0.23/kg ($0.11/lb) for alum, and $0.17/kg ($0.08/lb) for lime. These costs are a function of availability and transportation requirements [72–75].

Organic Chemical Conditioning

Nature and Advantage of Organic Polymers

Organic chemical conditioners first came into use in the 1960s. The appearance of sophisticated dewatering equipment, along with improvements in polymer effectiveness, led to increasing use of organic polymers with virtually all dewatering processes. Unlike inorganic chemical conditioners, which are only a few in number to consider, the organic chemical conditioners consist of a vast array of products that differ greatly in chemical composition, functional effectiveness, and cost effectiveness. Moreover, new products are being developed continually. In considering the advantages and disadvantages of using organic versus inorganic chemical conditioners, the following could be considered as reasons for selecting organic polymers:

(1) The dosage for conditioning is usually less with organic polymers than inorganic conditioners.
(2) Inorganic conditioners increase the sludge mass and volume substantially (adding to the disposal problem), whereas organic polymers do not.
(3) Polymers do not lower the fuel value of the sludge cake if incineration is to be used.
(4) Material handling is cleaner with polymers (with reduced operation and maintenance problems).
(5) Polymers are more effective in terms of achieving a high solids recovery and producing a cake with a high solids content in many dewatering applications.

Typical characterizing features of synthetic organic polymers used for conditioning are charge type, ''backbone'' unit structure, molecular weight, charge density, and active solids.

The charge type refers to the positive or negative electrical charge sites along the polymer chain. Polymers are called cationic (positively charged), anionic (negatively charged), nonionic (neutral), or amphoteric (having both positive and negative charges), depending on the nature of the predominant charge sites in the molecule. The "backbone" unit (or monomer) is the repeating unit from which polymers are synthesized through polymerization reactions. Various functional groups can be added (such as carboxyl or amino) in order to alter the polymer properties. The molecular weight is a measure of polymer length and is described as low (<0.1 million), medium (<0.1 to 0.5 million), high (0.5 to 6 million), and very high (6 to 18 million). Charge density is a measure of the amount of predominantly charged sites associated with the overall charge type of the polymer. Charge density is expressed as a percentage of available sites and varies from low (10 to 15%) to high (90 to 100%). Depending on the polymer charge type and physical form, the active solids can range from 2% to 95%.

A widely used backbone monomer is acrylamide. When polymerized, polyacrylamide is formed, which is a long-chain molecule with molecular weight in the millions and which is essentially nonionic. By adding carboxyl functional groups, an anionic product is obtained. By adding amino, imino, or quaternary amino functional groups, a cationic product is obtained. It is also possible to copolymerize acrylamide with an anionic monomer to obtain an anionic long-chain copolymer. Similarly, copolymerization of acrylamide with a cationic monomer yields a cationic long-chain copolymer. By such methods, entire families of polymers with varying degrees of charge and molecular weight can be produced. The cationic polymers are most widely used for sludge conditioning since most sludge solids have a negative charge.

Examples of typical cationic polymers include: polyamines, polyethylimines, polyimidamines, polybutadienes, polyamidamines, polyquaternaries, and substituted polyacrylamides.

Cationic polymers are also produced by what is known as mannich reaction. In this reaction, polyacrylamide is reacted with formaldehyde and dimethylamine to form an aminomethylated polyacrylamide. These polymers are called mannichs and are extremely cost-effective in many applications, but they tend to lose conditioning effectiveness in sludges with high alkalinity or with pH values of 7.2 or higher. Copolymers are less sensitive to pH and alkalinity but they may not be as cost-effective. Examples of typical anionic polymers include: polyacrylates, carboxylic polymers, and substituted polyacrylamide. These polymers are used less frequently in sludge conditioning but are effective with sludges high in iron or lime. Examples of typical nonionic polymers include: polyglycidyl polymers and polyacrylamides. These polymers have very limited use in sludge conditioning. Very little information is available on amphoteric polymers or their use in sludge conditioning.

Polymers for sludge conditioning are manufactured in five different forms: dry, emulsion, liquid, mannich, and gel. Each has unique storage and handling characteristics. All require specific feed equipment and special procedures for solution preparation. Advantages and disadvantages of each form must be carefully evaluated for any given conditioning application.

Dry polymers (solids) are available as powders, granules, beads, and flakes. The active solids can be as high as 95% (or higher) and the appearance is generally white. Dry (as-received) polymers can be stored for a year or more without loss in activity. Dilute water solutions begin to deteriorate within twenty-four hours. Emulsion polymers are dispersions of polymers in hydrocarbon oil. The active solids range from 25 to 50% and the appearance is generally milky white. Emulsions can be stored for several months without loss in activity. Solutions of emulsions made with water begin to deteriorate within twenty-four hours.

Mannich polymers contain 2–6% active solids and are clear. They can be stored for several weeks without loss in activity. However, diluted water solutions begin to deteriorate within several hours. Mannichs have high pH values and can cause problems with scale formation.

Liquid polymers are water solutions of non-mannich type polymers with lower viscosities than the mannichs mentioned previously (25% in value, approximately). The active solids range from 10 to 50% and the appearance is generally clear. Liquid polymers can be stored for several months without loss in activity. However, when diluted with water, they begin to deteriorate within twenty-four hours. Generally they have a neutral pH and scale formation is not a problem.

Gel polymers are water solutions of polyacrylamide type polymers that have tough, rubbery consistency and are available as 23 kg (50 lb) logs 46 cm (18 inches) long by 23 cm (9 inches) in diameter. The active solids range from 30 to 35% and their appearance is generally clear. Stability of as-received polymer and diluted water solutions is similar to the liquid form polymers mentioned previously.

Mechanism of Organic Polymer Conditioning

A schematic representation (Figure 6.29) of polymer interaction with sludge particles through charge attraction and bridging, which are the primary mechanisms in dewatering, consists of the following sequence:

(1) Initial adsorption of polymer on sludge particles
(2) Floc formation of destabilized particles
(3) Secondary adsorption of polymer on sludge particles
(4) Initial adsorption with excess polymer

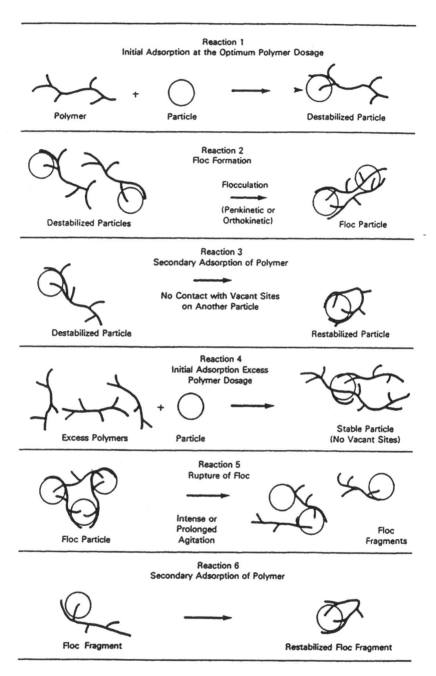

Figure 6.29 Schematic representation of the bridging model for the destabilization of colloids by polymers [71].

356

(5) Floc break-up and fragmentation

(6) Secondary readsorption of polymer on floc fragments

The entire polymer conditioning phenomenon is considered to encompass all six of these activities. These are not necessarily of equal importance and the degree of importance is a function of the unit process of conditioning.

Dosage Requirements and Cost

Dosage requirements of these conditioners for any sludge dewatering application is highly specific for the unit-process used and must be determined from laboratory-, pilot-scale, or full-scale tests.

Typical chemical costs are $0.13/kg ($0.06/lb) for mannichs and $3.90/kg ($1.80/lb) for dry solid polymers. Emulsions, liquids, and gels are intermediate in cost between mannichs and dry solid polymers, with the cost dependent on the active solids concentration. General costs are also a function of availability and transportation requirement [71–75,77,80].

Physical Conditioning

Thermal Conditioning

Originally, thermal sludge conditioning was intended to promote solid/liquid separation from sludges through the release of bound water from cells lysed during the conditioning process. Eventually, other benefits were recognized, such as enhanced waste activated sludge digestibility and elevated cake heating value. The concept of thermal conditioning of sludge for the release of bound water from sludges was developed by W. K. Porteous in 1900. The process involved heating sludge by steam injection. Even though technically effective, it was not a commercial success. The earliest successful systems for thermal conditioning of sludge were developed in Europe in the 1930s. The first of such systems to be built in the U.S.A. was at Levittown, Pennsylvania. It was built in 1967. By the 1970s, there were 120 installations in North America. Today, thermal conditioning of sludge is not as popular as other sludge conditioning methods and interest in it is diminishing due to economic considerations.

The most recent thermal processes can be classified according to the temperature and pressure used, and whether air or oxygen is added to the process. If air or oxygen is not added, the process is called nonoxidative heat treatment, and if air or oxygen is added, the process is called oxidative heat treatment. Nonoxidative heat treatment is usually conducted at temperatures of 150°C to 200°C and pressures of 1034 to 2068 kPa (150 to 300 PSI). In oxidative heat treatment, a portion of the solubilized organics is oxidized to carbon dioxide and water. The degree of oxidation that occurs is a function of the temperature, pressure, oxygen concentration, and contact time. Oxidative heat treatment is usually conducted at temperatures of 175°C to 360°C and pressures of 1279 to 11,377 kPa (200 to 1650 PSI). In both processes, the heat treatment in the reactor is applied for a time period of 15 to 40 minutes, depending on the specific characteristics of the sludge involved.

Advantages of thermal sludge conditioning include:

(1) Excellent sludge dewatering characteristics
(2) Additional chemical conditioning usually not required
(3) Sludge stabilization and pathogen destruction
(4) Increased heat value for incineration
(5) Sludge stabilization uninfluenced by toxic materials
(6) Performance stability with regard to changes in sludge characteristics

Disadvantages of thermal sludge conditioning include:

(1) There are high capital costs.
(2) It requires careful supervision, skilled operators, and preventative maintenance.
(3) It produces a malodorous gas stream that must be treated.
(4) It produces sidestreams with high concentrations of organics and ammonia that must be treated.
(5) Scale formation in heat exchangers requires cleaning by difficult and potentially hazardous procedures.
(6) Subsequent centrifugal dewatering may require polymer addition to control fine particle capture [71,72,74,75].

Elutriation

Elutriation was originally patented by Genter in 1941. The usual operation consists of two steps—a mixing step where the sludge is mixed with a washing liquid, and a settling step where the sludge suspended solids are recovered in their original volume. These two steps may be repeated and each such repetition is called a stage. Normally, elutriation systems have a mixing time of one minute and a gravity settling time of three to four hours.

A typical elutriation operation consists of washing sludge with effluent or water before subjecting it to chemical conditioning. Such a washing process removes soluble components of sludge (particularly alkalinity) that would otherwise increase the dosage and decrease the performance of a chemical conditioner. Commonly, anaerobically digested sludge is washed with plant effluent to yield lower alkalinity and fine solids. Ferric chloride or other chemical agents are used in a subsequent sludge conditioning step. Elutriation of sludge to remove alkalinity is also helpful in applications where mannich type polymers are employed with anaerobically digested sludge.

A potential problem with elutriation is that fine solid particles are returned to the plant with the washwater, and these fine solids can be discharged via the plant effluent causing a deterioration in effluent quality and creating operational problems throughout the treatment plant. Currently, elutriation is not used extensively in the U.S.A. for sludge conditioning, due to the widespread acceptance of polymers as chemical conditioners [72,74].

Freezing and Thawing

It has been observed that sludge dewatering characteristics improve when sludge is allowed to freeze and then allowed to thaw. At least one facility using natural freezing is operating in Canada, but no full-scale artificial freeze/thaw systems exist due to unfavorable economics. The capital costs, operating costs, and space requirements are all considerably higher than for conventional chemical conditioning of sludge [72,74].

Irradiation of Sludge

It has also been observed that sludge dewatering characteristics improve if the sludge is irradiated with gamma radiation. Irradiation of waste activated or digested sludge reduces the dosage of ferric chloride for conditioning and enhances the dewatering of sludge. The irradiation procedure ruptures the bacterial cell walls and releases the bound water. Currently, no full-scale sludge irradiation facilities exist for the purposes of dewatering. However, small facilities exist for the purposes of disinfection [72].

Other Conditioning Methods

In some cases, the use of an inorganic conditioner in combination with an organic polymer is more beneficial than using either conditioning agent alone. It is observed that low organic polymer concentrations can substantially reduce the dosage of the inorganic coagulant required for sludge conditioning [73]. In other cases, when a single organic polymer cannot be found to produce a complete coagulation, it may be possible to use a dual polymer treatment method. For example, an anionic polymer can be added first and then a cationic polymer can be added to complete the conditioning process. Each conditioning agent, in these cases, functions to destabilize different portions of the suspended sludge solids. The best sequence of addition and the appropriate conditioners to use must be determined experimentally. Although such combination processes are known to be effective, few actual applications have been reported in the technical literature [72,73].

Fly ash, newspaper pulp, and crushed coal have been used to condition sludge and to enhance the dewatering of sludge by mechanical means. Although such treatment is effective, sludge volumes may not be decreased significantly due to the large quantities of bulking agent that need to be added. Also, high inorganic content of the resulting cake can produce problems in the ultimate utilization or disposal of the sludge product [72,73].

Sludge acidification, permanganate addition, and solvent extraction are three processes of sludge conditioning that deserve only a passing mention as far as dewatering is concerned. Acidification is effective, but not nearly as much as chemical conditioning or thermal conditioning.

Permanganate can help lower subsequent chemical conditioning costs as well as help control odors. However, permanganate is expensive and may not be cost-effective to use on a continuous basis.

Solvent extraction to improve sludge dewatering has been tried and found to be effective. However, it is not cost effective when compared to conventional sludge conditioning methods [72].

Conditioning for Various Unit Processes

Thickening Applications

In gravity thickening applications, organic polymers are usually the preferred choice for conditioning. However, inorganic conditioners, such as alum and ferric salts, are also used with or without lime. Normally, cationic polymers with moderate charge and high molecular weight are effective. Although conditioning increases the solids loading rates to thickeners by a factor of two to four and improves solids capture, it has only a small effect on the underflow solids concentration. In general, the greatest value of conditioning in gravity thickening is to prevent operational problems caused by solids carryover.

The efficiency of the dissolved air flotation (DAF) thickening process can be improved by cationic polymers with moderate charge and high molecular weight. Some sludges, however, do not require polymer conditioning when DAF units are operated at low hydraulic and solids loading rates. The present solids capture rate in DAF units without conditioning is normally 95%. Polymer addition may, however, increase the solids capture up to 98%. It may also be possible to double the solids loading rate and increase the float solids concentration with polymer addition.

Aerobically stabilized biological sludges do not generally require conditioning for centrifugal thickening, unlike anaerobically digested sludges, which usually require polymer addition to achieve an effective solids capture. Many centrifuge installations use high molecular weight cationic polymers for dewatering anaerobically digested sludges. If alum, ferric salts, or lime are used for thickening, an anionic polymer may be required.

In rotary and belt thickeners, the polymer addition point is very important. It will vary with the type of unit, polymer, and sludge type. Sludge lagoons are also used for sludge thickening, and the use of polymers can improve solids capture in them.

Dewatering Applications

In the past, conditioning of sludge with inorganic chemicals has been the most widely used procedure in vacuum filtration dewatering. Ferric chloride and lime were the most-used chemicals. In recent years, however, organic polymers have received more attention and many wastewater treatment facilities prefer them as conditioning agents.

When compared to inorganic conditioning agents, the advantages of polymers are: more convenient materials handling, lower equipment requirements at about the same chemical cost, less equipment corrosion, and lower mass of sludge solids for disposal. The disadvantages associated with the use of polymers for dewatering are: significantly lower cake dryness and a greater degree of attention needed by the operator during the vacuum filtration process, inferior cake release in some cases in contrast to inorganic chemical conditioning, especially if the sludge has a high grease and alkalinity content.

Polymers have traditionally been used for conditioning sludges in centrifugal dewatering in order to improve the percent solids capture and percent cake solids concentration. Cationic polymers with moderate to high charge and high molecular weight are ordinarily used. Due to the high shear forces found in centrifuges, polymer requirements are often site-specific and very selective. Polymers with a high floc strength are required for effective performance. In many cases it is advantageous to provide additional polymer/sludge contact time to improve mixing. The correct conditioning agent and dosage can be determined by pilot or full-scale tests.

Even though sludge drying beds are commonly used for dewatering municipal sludges, the practice of using conditioning agents for improving their performance is not widespread. Nevertheless, polymer addition has been used effectively on both gravity and vacuum-assisted drying beds to achieve significantly reduced drying times, lower bed area requirements, and increased loading rates. Relatively small dosages are required (as low as 50 mg/L) to considerably improve drainage properties of digested sludges on sand beds.

Belt filter presses have traditionally used polymers for sludge conditioning, with solids recovery from about 95% to 98%. The higher the proportion of the biological solids in a sludge is, the higher the polymer dosage requirement is. The conditioning of sludge excessively with polymers can cause the blinding of media. In general, a strong, medium-to-large sized floc is desirable for effective performance in the gravity drainage section of the belt filter press.

Inorganic chemicals have traditionally been used for dewatering sludges with plate filter presses. Ferric chloride and lime, either alone or in conjunction with fly ash are commonly used. Polymers have gained some acceptance with the plate filter press operators with the development of high cationic, high molecular weight products. Generally, plate filter presses yield slightly lower cake solids with polymer conditioning than with ferric chloride and lime [71,72,75].

SELECTION OF CONDITIONING AGENTS

Tests for the Selection of Conditioning Agents

The selection of organic and inorganic conditioning agents can be done by means of laboratory, pilot-scale, and full-scale tests. Sludge conditioning tests must be performed with fresh sludge samples within several hours (or within a day at the most) of their collection, because storage of organic sludges affects their dewatering properties and will produce erroneous conditioning results. Time considerations are also relevant with regard to the conditioning agent solutions. Diluted polymer solutions should be used within a day, as their activity would decrease otherwise.

Many laboratory tests have been described in the literature for selecting conditioning chemicals. The most popular ones and their methodology have been summarized by Vesilind [74]. Buchner funnel filtration and capillary suction time (CST) test are the most commonly used tests.

Buchner Funnel Filtration Test

The Buchner funnel filtration test is one of the simplest ways to determine the dewatering characteristics of sludges and the dosage of chemicals to achieve the optimum dewatering. It can provide information for all dewatering applications, but it is best known for its use in vacuum filtration. The test consists of measuring the volume of filtrate collected from sludge aliquots treated with varying doses of conditioner over a constant time period. The optimum dosage of the conditioner can be obtained by plotting the filtrate volume versus conditioner dosage. Conditioning agents may be ranked according to the filtrate volume collected at their optimum dosage. With a given sludge, the chemcial that gives the highest filtrate volume among a group of conditioning chemicals tested at a constant filtration time is the best conditioner, and vice versa. The test may also be run by measuring the time it takes to filter a constant volume with varying doses of conditioner until an optimum is found. In this case, filtering times versus conditioner dosage may be plotted to find the optimum dosage. The conditioner ranking is according to filtering time at the optimum dosage (minimum time is the best).

Capillary Suction Time Test

The measurement of capillary suction time (CST) is also a very simple way to determine the sludge dewatering characteristics and the dosage effects of conditioning agents. CST tests are much quicker to perform than Buchner funnel tests. The instrument that mesuresd the CST consists of a hollow stainless steel well (1 cm in diameter and about 4 cm in height) which serves as a sludge reservoir resting on a filter paper. The filter paper and the sludge holding well are made to rest on a plastic base, which is equipped with two electrodes (Figure 6.30). When a sludge sample

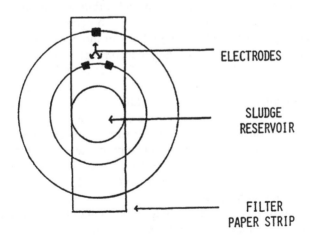

ELECTRODES

SLUDGE
RESERVOIR

FILTER
PAPER STRIP

Figure 6.30 Filter paper arrangement [79].

is poured into the reservoir, filtrate is drawn out of the sludge and moves outward as it saturates the filter paper.

When the filtrate reaches the first electrode mounted on the plastic base touching the filter paper, a timer is started. When the filtrate reaches the second electrode touching the paper, the timer stops. The time interval that the filtrate takes to travel the distance (10 mm) between the two electrodes is called CST. It is evident that the filter paper serves as both a filter medium and a filtrate volume container.

In effect, the CST apparatus is a type of miniaturized Buchner funnel filtration apparatus, where the time to filter a constant volume (into a graduated cylinder) is measured. In the case of the CST apparatus, the time to filter a constant volume (in the pores of the filter paper between the electrodes) is measured.

The CST measurement is a very versatile tool in providing information for all dewatering applications. However, it is best known for vacuum filtration, filter belt press filtration, and plate/frame press filtration applications.

As in the Buchner funnel tests, the conditioner dosage may be varied and the CST measured at each dose. The CST versus the conditioner dosage may then be plotted to find the optimum dosage, and the conditioners may be ranked according to the CST noted at their optimum dosages. A conditioner that yields the lowest CST values is the best among the conditioners tested for ranking their performance (Figure 6.31).

In the Buchner funnel filtration and CST tests, the mixing/stirring protocol (mixing rate, mixing time, sample size, etc.) used with the sludge/conditioner mixture for a specific dewatering application is very important. This protocol is a function of the sludge matrix and the degree of floc destruction that occurs from agitation and turbulence during a specific dewatering application. Floc deterioration also occurs as a result of the shear force imposed upon the sludge flocs in the conditioning chamber, piping network, and the pumping equipment through which the conditioned sludge passes. When flocs having a low floc strength deteriorate, the free water released from the flocs by the conditioning agent is reabsorbed into a bound condition in the sludge matrix, although not necessarily as strongly as in the initial state. If this happens, the dewatering performance deteriorates. Failure to consider the actual shear stress imparted to a conditioned sludge in the CST test will result in an inappropriate dosage for conditioning the sludge on a plant scale. This is because the floc shearing condition that occurs in the CST test does not simulate the floc shear condition of the full-scale dewatering unit.

The degree of destruction of sludge flocs, conditioned by a chemical in any dewatering application, may be determined empirically by noting the change in CST (or Buchner funnel filtration time) that occurs at

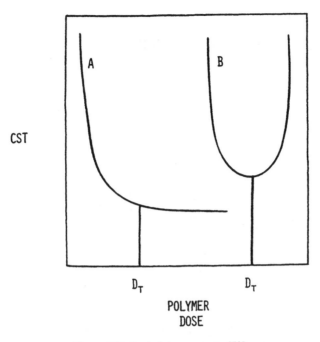

CST

D_T D_T

POLYMER
DOSE

Figure 6.31 Typical dosage curves [79].

the full-scale dewatering unit. In the laboratory, this change in CST (or Buchner funnel filtration time) can be simulated by appropriate mixing rate and mixing time. It can range from a simple swirling by hand to stirring for several minutes, at as much as 1000 rpm, with a variable speed mixer.

Conditioning agents that exhibit similar CST values, may be compared and ranked further for their efficacy from a knowledge of the results obtained in floc strength tests. The floc strength is an expression of the ratios of CST values at different mixing speeds and/or times. It can be logically extended into the concept of a slope obtained from a graph of CST values versus cumulative stirring time of the sludge/conditioner mixture (at a specific stirring rate). This slope may be viewed intuitively as a measure of floc destruction per unit of applied shear stress. In a floc strength test, sludge flocs that resist destruction per unit of applied shear stress, will exhibit small slopes since CST will not tend to change much. The smaller the value of the slope is, the greater the floc strength is. Thus, polymers that are used to condition a specific sludge can be ranked by the values of slopes (or floc strength) from plots of CST vs. cumulative stirring time (Figure 6.32).

The cumulative stirring time and stirring rate are a function of the shear

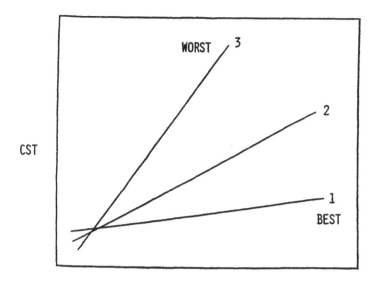

STIRRING TIME

Figure 6.32 Typical floc strength curves [79].

destruction exhibited in a specific dewatering application and must be determined empirically in the laboratory. These parameters are also a function of sludge total solids concentration and conditioning agent concentration.

Table 6.25 presents the stirring rates of a laboratory variable speed mixer that were found optimal to determine floc strength characteristics in centrifuge dewatering of anaerobically digested sludges having different total solids concentrations. When 200 mL of a polymer conditioned sludge in a 600 mL beaker was stirred at the stirring rate indicated, over the cumulative stirring time interval of 100 to 400 seconds, the floc strengths obtained correlated well with the efficacy of polymers tested in centrifuges [79]. The polymer dosages for these floc strength determinations were obtained from the minimum CST on a plot of CST vs. polymer dosage.

The floc strength concept may be used with the Buchner funnel test

TABLE 6.25. Stirring Rates Employed to Determine Floc Strength of Anaerobically Digested Sludge in Centrifuge Dewatering [79].

Sludge (% total solids)	2–3	3–4	4–5
Stirring rate (RPM)	1000	750	500

(using filtration time at a constant volume) in an analogous fashion, but it is substantially more tedious to do so and the sample size requirement may make it unrealistic.

Although Buchner funnel filtration time and CST are closely related, several distinctions can be made:

(1) The area of filtration in the CST apparatus is very small.
(2) Only a very small amount of filtrate is withdrawn from the CST sludge well.
(3) For highly conditioned sludges, the filtration time is so short that very little media blinding and compaction of sludge solids can occur in the CST test.

In some dewatering applications, such as vacuum filtration, these distinctions may be important. Three general principles that provide practical guidance in interpretation of results from both tests are:

(1) A single CST or Buchner funnel filtration time value attributed to a sludge provides very little information about its dewatering potential.
(2) Neither of these measurements is a comprehensive dewatering indicator, and attempts to use them as such can result in an erroneous conclusion.
(3) The effective evaluation of sludge dewatering performance with a particular conditioning agent and a specific dewatering application may require additional tests that simulate important characteristics of the application.

One example of such an additional test is the filter leaf test, as neither the CST nor Buchner funnel test can provide an indication of all relevant properties that influence dewatering characteristics. The filter leaf test consists of a fine circular mesh device over which a cloth filter medium is stretched in order to simulate a vacuum filter application. Vacuum is applied to the filter leaf apparatus. As a consequence, a sludge cake is formed on the filter medium. In this way, the pickup and release properties of the sludge cake can be studied with different conditioning agents.

Another example is the strobe light technique for measuring the settling rate of conditioned sludges (or the rate of cake formation) and for studying the solids concentration, texture, and compaction properties of the cake produced in a bench-top centrifuge. This is an attempt to simulate highly relevant characteristics in a centrifuge dewatering application. Details of these examples are provided in the literature [74,79].

Specific Resistance Test

Another test that is useful in evaluating sludge dewaterability is the specific resistance test. Specific resistance represents the relative resistance a sludge offers to drainage of its liquid component. Typical specific resistance values for municipal sludges are from 3 to 40 × 10^{11} m/kg for conditioned digested wastewater sludge and 1.5 to 5 × 10^{14} m/kg for primary sludge [72].

Specific resistance of a sludge can be determined by performing the Buchner funnel filtration test. The data on volume of filtrate collected at different times is obtained and a plot of the time/volume (t/V) vs. volume (V) of the filtrate collected is made. The specific resistance of the sludge is derived using the slope (b) of this plot with the aid of the following equation:

$$R = \frac{2bA^2P}{\mu c}$$

where

R = specific resistance (m/kg)
b = slope of t/V versus V (s/m^6)
A = area of filter (m^2)
P = test pressure (N/m^2)
μ = dynamic viscosity (n-s/m^2)
V = volume of filtrate collected (m^3)
t = time from start (s)
c = mass of solids per unit volume of filtrate (kg/M^3)

The specific resistance is a function of the vacuum applied in a Buchner funnel filtration. The determination of the specific resistance of a sludge is tedious. Specific resistance can be correlated with CST measurements, which are easy and simple to produce.

Pilot and Full-Scale Tests for Selection of Polymers

Both pilot and full-scale tests are also performed for the judicious selection of polymers for dewatering applications. In general, full-scale tests are preferable to test conditioning agents for all the dewatering applications. Pilot-scale tests and laboratory tests are useful for guidance and preliminary screening of polymers and determining test conditions to develop the full-scale test protocols.

Usually, the test protocol consists of ranking the conditioning agents according to the performance specification and the dosage requirement to achieve that performance specification. Tests are conducted with each conditioning agent under consideration. In these tests, characteristic curves of performance, such as conditioner dosage vs. percent cake solids and percent capture, should be developed. Once the characteristic performance curves are developed for each conditioning agent, the dosage required to achieve the performance specification can be determined. If other input variables, such as machine settings, influence the percent cake solids and solids capture, then criteria must be specified to optimally set the level of those variables. Algorithms may be used for the estimation of parameters of the models developed and for determining the optimum dosage in those cases where it cannot be ascertained by the performance specifications alone.

In the case of simple horizontal bowl centrifuges, if the feed sludge flow is kept constant for the duration of the tests, the only input that influences the performance of both cake solids and solids capture is the sludge conditioner dosage (assuming no internal machine characteristics are changed during the tests). By setting a performance specification of 95% solids capture, the dosage requirement can be determined from a graph of solids capture versus conditioner dosage obtained from the test results. Once the dosage requirement is determined, the cake solids performance at that dosage can be determined from a graph of cake solids versus conditioner dosage, which is also obtained from the test.

The conditioners may be ranked according to cake solids at 95% solids capture. A minimum percent cake solids specification may be set (if desired) to eliminate those conditioners with poor cake solids performance. A conditioner may be purchased, which is the least expensive among those that meet the minimum percent cake solids specification.

Many variations on the above are possible once the characteristic performance curves are obtained, by taking sufficient samples to reliably draw the graphs or estimate the parameters of the models that represent the curves.

In more complex centrifuges, where one or more internal machine characteristics can be continuously changed (backdrive speed for example), the situation becomes complicated. The models representing the cake solids and the solids capture data are no longer a function of the polymer dosage alone, but are related to both the dosage and the machine variables. In this case, specifying a solids capture of 95% is not sufficient to determine the dosage requirement from a graph of solids capture versus conditioner dosage. This is because there is a family of performance curves corresponding to various settings of the backdrive speed. There is also a family of cake solids performance curves corresponding to various settings of the backdrive speed. Some criteria must be specified, which will optimally set the level of the backdrive speed and allow a unique determination of conditioner dosage and cake solids performance at that dosage.

However, in order to select a polymer and determine its optimum dosage, many samples of cake and centrate have to be taken during a test run, at several polymer dosage rates and backdrive speeds, using a factorial or fractional factorial statistical design. Analysis of the data collected using a multivariate analysis approach would achieve the desired objective of selecting a polymer.

PROCEDURE FOR THE SELECTION AND PROCUREMENT OF POLYMERS FOR CENTRIFUGAL DEWATERING

Many POTWs use polymers for sludge conditioning. For example, The Metropolitan Water Reclamation District of Greater Chicago (District) spends over three million dollars annually to purchase polymers to condition anaerobically digested sludge prior to centrifuge dewatering at three of its water reclamation plants (WRPs). Due to the large sums of money involved, it is important to the District to procure an effective polymer for the least cost for use at its centrifuge facilities. Hence, it has developed a procedure, which includes a testing protocol for the selection and procurement of polymers. This procedure has been in use for the last seven years to the satisfaction of the District's Purchasing and Maintenance and Operations Departments as well as the vendors, who compete in the District's polymer bidding process.

The availability of a wide array of polymer products makes selection of a polymer for optimum performance with a given dewatering process or a particular sludge very difficult. Polymer characteristics alone are not adequate to allow for such a selection. As a result, empirical test procedures involving bench-, pilot-, or full-scale tests must be used to determine which polymer works best for a given dewatering device and sludge. Bench- and pilot scale tests although convenient are less reliable than full-scale tests. However, full-scale tests are cumbersome, time consuming and resource intensive. Obviously, the bench- and pilot-scale tests allow for more controlled conditions than full-scale tests. But, ultimately, the greater reliability of full-scale test offers the best opportunity for providing guidance in the determination of the best polymer for the least cost. Although bench- and pilot-scale tests followed by full-scale tests were originally used at the District, currently full-scale tests alone are used to select the best polymer for dewatering from the standpoint of cost and performance.

The District dewaters its anaerobically digested sludge with high performance rotating bowl centrifuges These machines required the development of a sophisticated full-scale test procedure for the selection of polymers. Due to the complexity of the machine control variables for these centrifuges, the polymer selection procedure for these high performance machines requires the use of performance models developed by optimization techniques for the selection of the best polymer at the least cost. The software used to implement the polymer selection procedure and to determine the polymer dosage rate has been developed by the District.

UNDERLYING PRINCIPLES AND RATIONALE OF THE POLYMER SELECTION PROCEDURE

The objective of the polymer selection procedure is to rank polymers according to the dosages required to achieve a specified performance criterion and to select a polymer that has the lowest cost. In the actual polymer testing protocol, performance characteristic curves describing polymer dosage vs., percent cake solids and polymer dosage vs., percent solids capture are developed for each of the polymers tested. From these curves, the actual dosage required to achieve a performance specification is determined using optimization techniques.

The polymer performance characteristic curves are obtained from two mathematical models. For a given sludge, these models include three input variables, which have the most influence on the percent cake solids and percent solids capture. These variables are the centrifuge bowl speed, pinion speed, and polymer dosage. The inclusion of sludge characteristics as inputs into these models is not necessary. It is assumed that the sludge used in different tests conducted within a short span of time, usually a week to ten days, will not change significantly in its characteristics.

Nonlinear algorithms describing the relationships between the percent solids capture, cake solids, and bowl and pinion speed of the centrifuge are developed by using the least squares technique and are used to estimate the model parameters from the data collected during the full-scale polymer evaluation tests. One polymer is tested per day at different dosage rates and at different pinion speeds. A family of characteristic performance curves (polymer dose vs., percent cake solids and percent solids capture) corresponding to various pinion speed settings are developed. Optimization techniques are used with the data collected to determine the optimum pinion speed, because the optimum polymer dosage rate occurs at this pinion speed. Unlike the bowl speed, which is held constant (a constant machine parameter), the optimum pinion speed varies with the polymer tested. Hence, optimization procedures are developed to estimate the optimum pinion speed from test data, which can then be used to obtain the optimum polymer dosage rate. Criteria are provided to optimally determine and set the pinion speed to condition the sludge at an optimal polymer dosage. As the optimum pinion speeds vary for different polymers, they can not be arbitrarily set to a constant value to determine the optimum polymer dosage for all the polymers tested.

For centrifuges that do not have a variable pinion speed control, the rationale for polymer selection is the same. The testing procedure is simplified since optimization of the pinion speed is not required. The performance models are correspondingly simplified as well.

MODELS AND DETERMINATION OF OPTIMUM PINION SPEED, POLYMER DOSAGE RATE, OPTIMUM CAKE SOLIDS, AND SOLIDS CAPTURE

Based on the results obtained from the full-scale tests conducted for polymer evaluation and selection, the performance characteristic curves have the that can be characterized as Type 1 and Type 2, respectively (Figure 6.33). Type 1 describes an asymptotic linear or quadratic relationship, whereas Type 2 describes a maximum quadratic relationship. In Type 1 cases (performance characteristic curves which are asymptotic linear and asymptotic quadratic forms), the percent cake solids reaches a maximum and stays constant [Ck (max)] under any polymer dose and pinion speed. In the Type 2 case (quadratic maximum), however, the percent cake solids drops off after the maximum is achieved as the polymer dose is increased beyond the optimum polymer dosage. In most polymer tests, the performance characteristic curves fall in the Type 1 category (asymptotic linear or quadratic) and the Type 2 cases (quadratic maximum) are very rare. The shape of a typical performance characteristics curve for percent solids capture is shown in Figure 6.34.

Algorithms were developed for the above two types of cases and software was written for deriving the model fitting paratmeters and for the optimization of pinion speed. The estimation of model parameters is done with a combination of the Nelder-Mead Simplex algorithm (1965) and the Golub-Pereyra algorithm (1973). A Fortran IV coding of the Nelder-Mead Simplex algorithm is provided by Olsson (1974) and a Fortran IV coding of the Golub-Pereyra algorithm is provided by Ottoy and vanSteenkiste (1980). A commercial software package called "Scientist" can implement a variation of both of these algorithms and also provides an exceptional graphics capability. Many difficulties with convergence can be avoided by using these algorithms for parameter estimation. It is also highly beneficial to implement the optimization procedure on a commercially available software packages such as "TK Solver," which is an equation solver that allows for automated and convenient solutions to nonlinear equations without the need for sophisticated programming skills. At the District, the optimization procedure is carried out using both the "Scientist" and "TK Solver" packages.

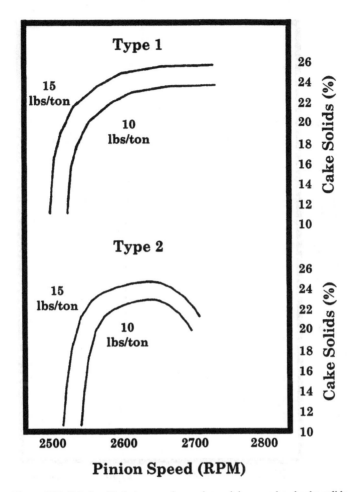

Figure 6.33 Relationship between polymer dose, pinion speed and cake solids.

From the fitted performance curves using the optimization procedures, the optimum pinion speed is read. The optimum polymer dosage occurs at the optimum pinion speed. and it is read from the polymer dosage scale for the optimum pinion speed determined. In the District's polymer testing protocol, the performance specification is chosen to be 95 percent solids capture. The percent solids capture can be obtained by the following equation.

$$\%CP = \left[\frac{FD - CN}{FD}\right] \times \left[\frac{CK}{CK - CN}\right]100$$

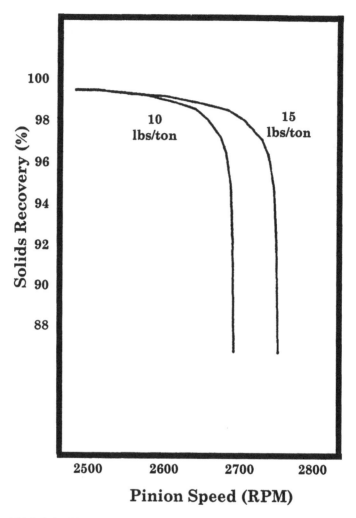

Figure 6.34 Relationship between polymer dose, pinion speed and percent solids capture.

where

%CP = percent solids capture
FD = feed solids (%)
CN = centrate solids (%)
CK = cake solids (%)

Some polymers may not yield the specified percent solids capture and such polymers are rejected. Those polymers, which satisfy the performance criterion are then ranked according to the optimum dosage rates determined. A polymer which meets all the specified performance criteria and has the lowest cost can then be purchased at a quantity that is needed to condition a specified number of dry tons of sludge at the optimum polymer dosage (lbs per dry ton) determined in the test protocol.

TOTAL PROCESSING COST

In its polymer selection procedure, the District has developed a cost function that not only considers polymer cost, but also other relevant costs such as transportation of the cake to a given location and agitation drying of the centrifuge cake to a fixed percentage dry solids, i.e., 65 percent. This cost function has been incorporated in an equation that determines the overall cost associated with a particular polymer. The total sludge processing for the polymers tested is calculated according to the relationships given below:

$$\text{Processing Cost(\$)} = (A)(B) + C_1 + C_2(1/D)$$

where

Processing Cost (\$) = dollars per dry ton of sludge

A = pounds of polymer per dry ton of sludge (optimum dose) as determined in the polymer evaluation test

B = cost (\$) of polymer per pound of polymer

C_1 = a value specific to a processing site and reflects interfacility transportation cost

C_2 = a value to reflect agitation drying costs specific to a particular drying site

D = percent cake solids at optimum polymer dose

The costs associated with all polymers for processing a dry ton of solids are then ranked. The polymer with the lowest total processing cost is selected for purchase.

POLYMER SELECTION PROCEDURE

The District's Purchasing Department has adopted the following polymer testing and selection protocol for open bidding, which consists of the following steps:

(1) Sending of advertisements to polymer manufacturers and receipt of responses from manufacturers (four weeks)

(2) Laboratory tests to determine full-scale test variables (one week)
(3) Full-scale tests (three weeks)
(4) Data analysis (two weeks)
(5) Bidding process and contract award

The Purchasing Department issues bid documents to various polymer vendor and also advertises for the procurement of polymers. After the responses to the bid documents are received within a specified time (usually four weeks from the date of advertisement), full-scale testing of the polymers submitted by vendors (a maximum of two per vendor) is scheduled and the vendors are informed of the dates on which their respective polymers will be tested. Sometimes, it may be necessary to conduct lab tests to determine the full-scale test variables for the polymers submitted by the manufacturers prior to such full-scale testing. Usually, a full-scale test to evaluate one polymer takes one full day.

In the full-scale test, the sludge flow rate to the centrifuge is kept constant during the test. The same centrifuge is used with all polymers. Cake solids, centrate, and centrifuge feed samples are taken at various dosages (lbs/dry ton of solids) and pinion speeds using a factorial or fractional factorial sampling design over the operating range of minimum and maximum torque, for each polymer to be tested. The full-scale testing protocol starts at high pinion speeds (high torque, low capture, high cake condition) and follows with progressively lower pinion speeds (low torque, high capture, low cake condition) in equally spaced intervals. If sampling were to begin at significantly lower pinion speeds, there is a high risk of the cake liquefying and spilling over from the conveyor belts of the centrifuges, thereby causing downtime for cleanup. All samples are taken after at least 15 minutes of centrifuge operation at a given polymer dose. These samples are then analyzed for percent total solids and the percent solids capture calculated as previously described.

Polymers that do not produce a percent capture value greater than 95 percent at some point on the performance characteristic curve (polymer dose vs., percent capture) are rejected.

After the full-scale scale tests are conducted, the pinion speed, polymer dose, percent capture, and cake solids data are tabulated. These data are then subjected to model selection and optimization procedures using software programs to determine the optimum pinion speed and the optimum dosage. These programs are available from the District's Research and Development Department for model selection and optimization. After determining the optimum dose and the percent optimum cake solids, the total processing cost is determined according to the equations presented above. The polymers are then ranked according to the cost of processing per dry ton of sludge and the polymer that has the lowest processing cost is selected.

COMPOSTING

INTRODUCTION

Aerobic composting is a method of sludge stabilization in which sludge organics are decomposed by microorganisms in the presence of oxygen. The result of sludge composting is the production of a stabilized, humus-like product that can be used, for example, as a soil amendment, for erosion control, as a mulch, or other soil-like products. Aerobic composting is dependent on several operational parameters, which include oxygen availability within the compost, moisture content, temperature, and biodegradable volatile solids content of the compost. Each of these parameters affects the growth of the aerobic organisms responsible for decomposition of the compost.

An adequate oxygen supply is necessary for the growth of aerobic organisms. If the oxygen supply to the composting organisms is not sufficient, anaerobic organisms become predominant in the compost, resulting in the production of nuisance odors. Numerous methods are available for ensuring an adequate oxygen supply to the aerobic organisms. Some composting systems depend on natural convection to supply oxygen; other systems use forced aeration. Further enhancement of aerobic conditions is provided by the addition of a bulking agent to improve the porosity of the compost and mechanical mixing of the compost to prevent air channelization. Most composting systems use a combination of two or more of these measures to ensure an adequate oxygen supply.

Another important parameter of composting is the moisture content of the compost. The moisture content must be sufficient to support biological activity but not so high as to eliminate the void spaces within the compost that are required for oxygen transfer. Sludge is usually dewatered prior to composting, but most dewatered sludge contains too much moisture to be composted alone. A bulking agent must be added to the sludge to increase the total solids content. For most composting systems, an initial compost total solids concentration of 40% is desired. The addition of a bulking agent not only increases total solids content, it also improves the porosity of the compost.

During composting, biological decomposition of volatile organics produces temperatures within the compost ranging from 50°C to 70°C. These high temperatures evaporate water from and destroy pathogenic organisms in the compost. The desired operating temperature ranges from 55°C to 60°C; temperatures in excess of 60°C can reduce biological activity, and temperatures below 55°C are not sufficient for adequate pathogen destruction or compost drying. These temperatures must be maintained for a minimum duration of time in order to be effective in destroying pathogens. Compost temperatures can be controlled to some extent by forced aeration:

if temperatures become too high, air can be drawn through the compost to remove hot air, thereby reducing the temperature.

The amount of water that can be removed from the compost is related to the quantity of volatile organics available for decomposition. Decomposition of volatile organics produces heat, which evaporates water from the compost. The water vapor is then removed from the compost by aeration. If compost drying is insufficient, additional volatile organics may be required to produce additional heat. One method for increasing the volatile organics is to add an amendment to the compost, such as wood chips, sawdust, tree trimmings, recycled compost, or a mixture of these materials. An amendment can also be used to increase the total solids concentration and the porosity of the compost.

The amendment quality can impact the quantity of sludge that can be composted. If amendment quality is poor, i.e., low total solids or low volatile solids concentrations, then more amendment is required. This reduces the quantity of sludge that can be composted. If amendment of high quality is used, less amendment is required; therefore, more sludge can be composted within a given volume.

There are a variety of systems available for municipal sludge composting. These systems are often divided into two categories—reactor or in-vessel composting systems, and nonreactor composting systems.

Reactor systems physically contain the sludge in a vessel during composting. Horizontal and vertical flow are the two basic types of reactor systems. The primary difference between these two types is the method used for moving the compost through the reactor. Reactor systems can be further classified by the method used for maintaining aerobic conditions within the compost. All reactor systems use forced aeration; some systems use mechanical mixing as well as forced aeration to maintain aerobic conditions.

Nonreactor composting systems include windrow and aerated static pile systems. Nonreactor systems are often open to the atmosphere, but they can be covered to reduce the detrimental effects of precipitation and aid in odor control. Aerated static pile systems mix sludge with a bulking agent and place the mixture in large piles. Air is drawn or forced through the piles by a blower system. These piles are not moved or mixed throughout the composting period. Windrow systems mix sludge with a bulking agent or amendment and place the mixture in long piles called windrows. Machines are used to regularly mix the windrows to ensure adequate aeration. Some windrow systems use forced aeration in addition to mechanical mixing to aerate the compost.

The types of sludge that can be composted include both digested and undigested primary and secondary sludges. Although digestion of a sludge prior to composting reduces the potential for odor production, it also reduces the volatile organics available for decomposition.

In addition to producing a stabilized soil amendment product, composting also produces a liquid sidestream that consists of condensate and leachate. Condensate and leachate can be collected and, in most cases, are returned to the wastewater treatment plant headworks.

DESIGN CRITERIA

Detention Time

For wastewater systems, two detention times can be defined, one based on single-pass liquid retention time and one based on solids residence time. Detention time based on liquid retention time is usually termed "hydraulic retention time" (HRT). Detention time based on the average residence time of solids in the system is usually termed "solids residence time" (SRT). For a reactor without recycle of solids, the HRT and SRT are equivalent. If solids are recycled, however, the SRT will be greater than the HRT.

The distinction between SRT and HRT is an important concept. For homogeneous, liquid phase systems, the efficiency of biodegradable volatile solids (BVS) decomposition is determined by the system SRT. The minimum HRT is determined primarily by time constraints imposed by oxygen transfer and the ability to maintain the required microbial concentrations. The same is somewhat true for composting systems. The extent of microbial decomposition should be determined by the system SRT, whereas reactor or process stability is determined by HRT. The single-pass hydraulic residence time is defined as:

$$HRT = V/(Q + q)$$

For liquid phase systems, the volumetric flow rate remains essentially constant across the reactor, simplifying the calculation of average residence time.

HRT for a composting system can be defined as the single-pass, mean residence time of the mixed materials, including recycle. The volume of mixed materials entering a composting reactor does not remain constant with time. Moisture is lost by evaporation, and BVS solids are lost by microbial decomposition. As a result, mixture volume usually decreases across the system. Recall that mass is conservative but volume is not. To account for the volumetric changes that occur during composting, the basic equation for HRT can be modified to:

$$HRT = V/(Q + q) \text{ in } + (Q + q) \text{ out}/2$$

HRT is based on the average of infeed mixture volume and outfeed mixture volume.

HRT is an important parameter that affects composting temperature, output solids content, and the stabilization of BVS. Process stability is favored by longer detention times. A minimum design HRT of twelve to twenty days is a reasonable choice for most reactor systems because the process is stable over a wide range of conditions and high output solids contents and BVS reductions are possible. Longer detention times produce smaller improvements in these parameters for a comparable increase in reactor volume. Many reactor systems are designed for an HRT of about fourteen days, which is consistent with the previous discussion. Batch processes may require longer detention times because of the lag phase encountered at the start of composting.

SRT for a liquid phase system is defined as:

$$SRT = V/Q \quad (16)$$

SRT is the residence time of the feed components excluding the volume of any recycled material. By analogy, SRT for a composting system can be defined as: the minimum system SRT for design is likely a function of at least the following:

(1) The extent of process control incorporated into the design and the processes used for the high-rate and curing phases
(2) The types of amendments used and whether the feed sludge is raw or digested
(3) The extent to which kinetic rate limitations can be avoided
(4) The end use of the product

Design parameters used for successfully operating systems were also reviewed. These data, together with the data referenced above, suggest that a minimum system SRT of about sixty days is required to produce a compost with sufficient stability and maturity to avoid reheat and phytotoxic effects. This assumes that the feed is properly conditioned with typical amendments and recycled product to close the energy balance and reduce kinetic limitations. It further assumes a reasonable level of process control to prevent excessive rate limitations from developing during the process.

An SRT of sixty days should be adequate for most composting systems and end product uses. However, longer SRTs may be required where a very stable end product is desired, for example, with high end uses such as in landscaping and horticulture.

Aeration Requirements

Air must be supplied to a composting material for three basic purposes: (1) to satisfy oxygen demands imposed by organic decomposition; (2) to remove moisture from the composting material to provide drying; and (3) to remove heat generated by organic decomposition to control process temperatures. The ability to control aeration is one of the key points of process control and is an important consideration in process selection.

The stoichiometric oxygen requirement can be determined from the chemical composition of the feed solids and the extent of degradation during composting. Assuming an average composition of sludge organics of $C_{10}H_{19}O_3N$,

$$\overset{201}{C_{10}H_{19}O_3N} + 12.5O_2 = \overset{400}{10CO_2} + 8H_2O + NH_3$$

the oxygen demand is estimated to be 2.0 lb O_2 per pound of BVS.

Assuming all ammonia is oxidized, the maximum nitrification demand can be estimated to be 0.32 lb O_2/lb sludge BVS.

The nitrification demand is significantly less than that required for organic oxidation. Also, most of the ammonia will not be oxidized. Some will be used directly for cell synthesis, but the largest fraction will probably be lost from the compost by volatilization. Therefore, oxygen demands for nitrification generally are not considered.

Amendments such as sawdust are largely cellulosic in nature, with a decomposition reaction that can require 1.2 lb O_2 per pound BVS.

The stoichiometric demand for oxygen varies from a low of about 1.0 lb O_2/lb organic for highly oxygenated wastes such as cellulose to a high of about 4.0 lb O_2/lb organic for saturated hydrocarbons. On a practical basis, the values for domestic sludge and cellulosic material should not vary significantly from those stated previously.

The quantity of moisture in saturated air increases with increasing air temperature. Gas that leaves a wet composting material will be near saturation and about the same temperature as the composting material. If thermophilic temperatures are maintained, considerable moisture will be removed with the exit area.

If the temperature difference between inlet and outlet air is greater than about 45°F (25°C), relative humidity of the inlet air will have a minor effect on moisture removal. This means that drying can occur even in climates with high ambient humidity in the feed air.

Saturation water vapor pressure is a function of temperature and can be approximated by an equation of the form:

$$\log_{10} PVS = (a/T, + b)$$

Knowing the relative humidity, the actual water vapor pressure can then be estimated as:

$$PV = (RH)PVS$$

The specific humidity, w, is defined as the ratio of the mass of water vapor to the mass of dry air in a given volume of gas.

The specific humidity at 10°C, 100% relative humidity, and 760 mm Hg total air pressure, is

$$0.0081 \text{ lb water/lb dry air}$$

Similarly, at 55°C and 100% relative humidity, the specific humidity is 0.1146 lb water/lb dry air. Therefore, the net removal of water vapor from the composting mixture in this example 0.1146 − 0.0081 = 0.1065 lb water/lb dry air.

If the required moisture removal is 3.587 lb water, the air required to remove this water is

$$33.7 \text{ lb dry air/lb dry sludge}$$

Again, this value represents a *quantity* of air that must be supplied during the composting cycle to remove a given quantity of moisture from the mixture. The air required for moisture removal is normally significantly greater than the stoichiometric demand for biological oxidation. The drying demand is influenced largely by cake solids content and exit air temperature. Only at cake solids approaching 30 to 40% and exit gas temperatures of 70°C do the air requirements for drying and biological oxidation become reasonably equivalent. At cake solids of 20%, the air requirement for drying can be ten to thirty times that for biological oxidation. Indeed, this is the case for the example presented here.

Rates of biochemical reactions generally increase exponentially with temperature. However, process temperatures can elevate to the point of thermal inactivation of the microbial population. Temperature then becomes rate-limiting. To maintain process temperatures in an optimum range for the microbial population, air supply can be increased, thus increasing the rate of heat removal.

The heat of combustion per electron transferred to a methane-type bond is essentially constant at about 26.05 kcal per electron transferred, or since O_2 accepts four electrons, 104.2 kcal/mole O_2. This in turn equals 3.256 kcal/gm O_2 or 5866 BTU/lb O_2.

If the total oxygen demand was estimated to be 1.003 lb O_2/lb dry sludge, heat release is then 9943 BTU released/lb feed BVS. This heat will be removed primarily by the hot, moist exhaust gases leaving the process. Some heat will be removed by the outfeed solids and some lost to the surroundings. However, these are relatively small by comparison and will be neglected here. Again, assume inlet air is saturated at 10°C (50°F) and exit gas saturated at 55°C (131°F).

Heat released by organic decomposition will heat the incoming dry air and water vapor to the exit temperature, supply the heat of vaporization, and heat the evaporated water vapor to the exit gas temperature. Let x be the lb dry air required to maintain the compost temperature at 55°C. Then:

$$\text{Heat of vapor, at } 50°F = (x)(01.065)(1065.6)$$

$$\text{Heat moisture to } 131°F = (x)(0.1065) = (0.0081)(0.42)(131 - 50)$$

$$\text{Heat dry air to } 131°F = (x)(0.24)(131 - 50)$$

where 1065.6 is the heat of vaporization (BTU/lb) at 50°F; 0.42 and 0.24 are the specific heats of water vapor and air (BTU/lb-°F), respectively.

$$(x)(113.5 + 3.90 + 19.4) = 5883$$

where x = 43.0 lb air/lb dry sludge, or 72.7 lb air/lb feed BVS.

The air demands for moisture and heat removal represent total demands that must be supplied over the composting cycle. It should be noted that the weight of air moved through the compost process can be thirty to fifty times that of the dry weight of sludge. Composting has often been described as a problem of materials handling. The designer and operation should be aware that input air and exhaust gas are usually the largest components of the mass balance.

For a batch process, such as the windrow and aerated pile, the total air *quantity* can be converted to a *rate* of aeration by considering the time duration of the process. Assume a batch process with a total air demand based on heat removal of 43.0 lb air/lb dry sludge. Assume that this quantity is supplied at a constant rate over the twenty-five-day composting cycle. Then the *average* rate of aeration will be 1920 scfh/dt sludge.

A number of factors can cause the *peak* rate of aeration to exceed the average rate calculated above. First, the rate of organic oxidation, and therefore the rate of heat release, will vary throughout the composting cycle. Second, the type of aeration control may increase the peaking factor. An on-off aeration control logic is used in some systems. Obviously, the total air demand can be supplied only during the "on" sequence. Temperature feedback logic is popular with many composting systems and can also result in high peaking factors.

The peak rate of air supply can be estimated from data developed by Wiley [80], Jeris et al. [81], and Schultz [82]. These researchers observed that the rate of oxygen consumption is a function of temperature. Rates of 4 to 5 mg O_2/gm Vs-hr were observed in the temperature range of 50 to 65°C.

Using a theoretical model of the aerated pile process operated with temperature feedback control logic, Haug [83] predicted peak aeration rates of about 3800 to 4800 scfh/dt of sludge for raw sludge blended with wood chips. Murray et al. [82] reported that peak aeration rates of 4000 to 5000 cfh/dt were sometimes insufficient to keep process temperatures below 60°C when composting a raw sludge/wood chip mixture. Both the theoretical and measured values compare favorably with the estimate based on oxygen consumption data.

It should be noted that the peak aeration rate may be maintained for only a short period until the peak demand has passed. In the model presented by Haug [83] the required aeration rate exceeded 4000 scfh/dt for about two days, 3000 scfh/dt about four days, and 2000 scfh/dt about eight days.

If the aeration system cannot meet the peak demand, process temperatures will exceed the desired set point. The designer must make a tradeoff between aeration system capacity, and hence capital cost, and the needs for process temperature control. It may be more cost-effective to size the system for less than the peak demand and accept process temperatures above set point for a short period. Dynamic process models are useful to develop the statistical information, such as that presented by Haug [83] previously, on which to base such a decision.

Aeration systems for batch processes take many forms. Some aerated pile and windrow systems use a single blower connected to a single pile. In this case, the peak demand is carried by the blower. Other systems use a central blower facility connected to many individual piles. In this case, the peaking factor can be "shaved" because it is unlikely that all piles simultaneously would be at the point of maximum oxygen consumption.

Minimum aeration rates should also be considered because the system must be capable of operating over the turndown range from peak to minimum rates. For an individual pile or windrow the minimum average rate usually occurs at the very beginning of the batch cycle and may be as low as 200 to 500 scfh/dt sludge [85].

Many reactor systems operate on a continuous or semi-continuous feeding schedule. Typically, sludge is conditioned and loaded into the reactor on a daily basis. Reactor HRT is usually greater than ten days, so that a daily feeding schedule can be modeled as a continuous process.

For a continuous feed process, the quantity of required air can be converted to an average rate of aeration by considering 1 lb/day of feed sludge. Using the air quantity for temperature control determined above, the aeration rates become 43.0 lbs/day of air per lb/day of dry sludge. This is equivalent to 47,900 scfh/dtpd of sludge (standard cubic feet per hour/dry ton per day).

Using dynamic process models, Haug [85] predicted maximum sustainable aeration rates between 40,000 to 60,000 scfh/dtpd of sludge for an autogenous raw sludge/recycle mixture composted in a continuous feed, complete mix reactor.

With continuous feeding, the *peak* aeration demand should not exceed the average demand over the composting cycle (i.e., peaking factor = 1). With semi-continuous feeding, some peaking might occur between feed periods. However, in practice it has not been necessary to consider this effect. The 15 to 20% contingency normally applied to blower design capacity appears to be sufficient to handle any above average aeration demands.

The designer should consider that feed quantity and composition may vary throughout the life of the project. The process should be examined under all expected conditions to determine whether seasonal or periodic peaking factors should be applied. Similarly, minimum flow conditions should be considered. This is particularly important if the process will operate below design loadings in early years. There is a tendency to focus attention only on average or peak design loadings. However, the full range of operation should be considered to avoid control problems during minimum turndown conditions.

LAGOONING OF SLUDGE

INTRODUCTION

Process Definition

One of the definitions given for the word "lagoon" by the *Webster's Third New International Dictionary* is, "a shallow artificial pond for the natural oxidation of sewage and ultimate drying of the sludge" [86]. Lagoons have been used as a cost-effective means of processing residential, municipal (residential and industrial), industrial, and agricultural wastes for many decades, because of their ability to stabilize the organic matter contained in these wastes and to store the sludge solids that settle to the bottoms. Lagoons are also used for the processing and disposal of water and wastewater treatment plant sludges. Natural depressions in the ground are sometimes used as lagoons for the processing of wastewaters and sludges. They can be built, however, according to rationally developed design criteria to achieve the desired results of wastewater treatment, such as stabilization of organic matter, by-product recovery (single cell protein harvesting from oxidation ponds growing algae), or biogas generation and fertilizer/soil conditioner (anaerobic lagoons with covers).

Lagoons are known by different names. These include ponds, sewage lagoon, waste treatment lagoon, stabilization lagoon, waste pond, oxidation pond, stabilization pond, maturation pond, and waste stabilization pond. Qualifiers such as aerobic, facultative, anaerobic, aerated, sludge drying, sludge storage, permanent, and manure, are also used to describe the nature of the lagooning process. Based on the work done in developing and developed countries, extensive literature exists on lagoons used for the processing of municipal wastewater. Because of the cost-effectiveness, lagoons provide a low-cost method for treating domestic and industrial wastewater in developing countries. However, the emphasis on lagoons as presented in this section will be on the application of lagoons to the processing of sewage sludge.

Application and Role of Lagoons in Sludge Processing and Disposal

According to a classification by the Ministry of Housing and Local Government of Great Britain, sludge lagoons can be characterized as follows, based on their application: (1) thickening, storage, and digesting, (2) drying, and (3) permanent [87].

Lagoons used for Thickening, Storage, and Digestion

Sometimes when mechanical units fail or become overloaded, lagoons may be used as a substitute or a temporary remedy for storage until the situation is corrected. Lagoons, when used for the storage of raw or undigested sludge solids, have the potential for odor generation; hence they are rarely used for this purpose. However, certain industrial sludges, which are inert and not biodegradable, can be readily stored with minimal potential for odor generation. When sludge is stored for lengthy periods, i.e., extending to a number of years, thickening of sludge occurs due to compression of solids. During such long periods of storage, digestion of sludge does take place in lagoons, although it progresses at a slow rate. This is due to the fact that lagoons are rarely heated like the anaerobic digesters used for the stabilization of raw sludge. In cold climates, the cycles of winter freezing and spring thawing provide an opportunity for the natural physical conditioning of sludge, which enhances its dewatering and consolidation.

Lagoons Used for Drying

Drying lagoons are used in place of sand drying beds. Where conditions are conducive to dewatering, such as in sandy soils, sludge is periodically added to the lagoon, allowed to dry, and the dried solids removed by scrapers or other heavy material handling equipment.

Permanent Lagoons

Where land area is plentiful, the construction and use of these types of lagoons is justified for sludge disposal because sludge is not usually removed from them for many years. Decantation of supernatant liquor into another lagoon operated in series will enhance the performance and effectiveness of these types of lagoons in terms of achieving a sludge with a high solids content.

Lagoons processing sludge may also be classified as (1) aerobic, (2) facultative, or (3) anaerobic, depending on their dissolved oxygen regimen.

Aerobic Lagoons (Aerated Sludge Storage Basins)

Lagoons are rarely operated under totally aerobic conditions for treating raw sludge (or even digested sludge) because of the high oxygen demand that is exerted by the sludge. Nevertheless, aerobic digestion is an accepted way of stabilizing waste activated sludge and primary sludge. Stabilization of these sludges is achieved in carefully designed aerobic lagoons (digesters). The difficulty in transferring oxygen economically and maintaining a dissolved oxygen concentration through the entire mass of a sludge column renders aerobic lagoons not generally cost-effective for sludge storage, treatment, and disposal. However, as indicated earlier, aerobic waste stabilization ponds are used for the treatment of domestic sewage, where oxygen is supplied by mechanical aeration, photosynthetic activity, and surface reaeration.

Facultative Sludge Lagoons

Facultative lagoons are those where an aerobic layer containing a lower solids concentration is maintained in the top one-half to one meter depth of a sludge lagoon. The top layer of a facultative lagoon is kept aerobic by mechanical aeration, algal photosynthesis, or by direct natural surface reaeration. The top layer should be maintained without the buildup of a scum layer, which will otherwise reduce the efficiency of oxygen transfer. The middle layer and the bottom layers of facultative sludge lagoons are anaerobic. Facultative lagoons are built and operated successfully without odor problems at the Manukau sewage treatment plant in Auckland, New Zealand and in Sacramento, California, U.S.A. [88].

Anaerobic Sludge Lagoons

An anaerobic lagoon is usually an open structure like the facultative lagoon. Sometimes, the depth of an anaerobic lagoon can be even greater than that of a facultative lagoon in relation to its surface area. It is analogous to an anaerobic digester except that it is not heated. The main objective of anaerobic lagoons is to settle those solids that have a higher specific gravity than water and provide storage for sludge at the bottom. Hence, there is a need for greater depth for such lagoons. No deliberate attempt is made to settle or remove any floatable material. As a consequence a thick scum layer can occur in these types of lagoons. Digestion of sludge solids occurs at a lower rate than in conventional anaerobic digesters, as anaerobic lagoons are not heated. In contrast to the facultative lagoons, no attempt is made to maintain an aerobic zone on the top of the anaerobic lagoons. Sludge loading rates to anaerobic lagoons are higher than those used in facultative lagoons.

Advantages and Limitations

Table 6.26 lists the advantages and limitations of the aerobic, facultative, and anaerobic sludge storage lagoons.

DESIGN CONSIDERATIONS

When lagoons are constructed according to rationally developed design criteria and operated under careful supervision, they are very cost-effective for the storage, stabilization, utilization, and disposal of sewage sludge. However, many lagoons built in the past were emergency situations for the handling of sludge, rather than as products of a strategic master planning of sewage sludge management and disposal schemes.

The factors that influence the siting and design of lagoons are (1) proximity to residential areas, (2) available land area, (3) climate, (4) subsoil permeability, (5) depth, (6) sludge loading rate, (7) sludge characteristics, (8) groundwater table, and (9) supernatant withdrawal facilities [89].

If the intended use of a lagoon is for emergency purposes, it can be designed with a smaller volume than the volume that would be required for achieving evaporation and consolidation of sludge over a long period of time.

Sludge Drying Lagoons

According to the *Ten State Standards* published in the U.S.A. [90], the following are the design criteria listed for drying lagoons: (1) the soil must be reasonably porous and the groundwater table must be at least 45 cm (18 inches) below the bottom of the lagoon; (2) areas

TABLE 6.26. Advantages and Limitations of Lagoons.

Advantages	Limitations
Aerobic Sludge Lagoons	
1. Odor problems are minimal if operated properly.	1. Energy requirements are high.
2. Yields a highly stabilized sludge in a relatively short time unlike with facultative sludge lagoons (FSLs) and anaerobic lagoons.	2. Maintenance and operational costs are high.
3. Drier sludge can be obtained in a relatively short time due to shallow depths.	3. Cannot be operated economically and effectively at higher depths
4. Can be used with raw sludge.	4. Not practical and economical for long-term storage like facultative sludge lagoons (FSLs) and anaerobic lagoons.
5. Oxygen requirements are lower for anaerobically digested sludge than for raw sludge	
6. Nitrification is possible.	
Facultative Sludge Lagoons (FSLs)	
1. Provides long-term storage with acceptable environmental impacts (odor and groundwater contamination risks are minimized).	1. Can only be used following anaerobic stabilization. If acid phase of digestion takes place in lagoons they will stink.
2. Continued anaerobic stabilization, with up to 45 percent vs reduction in first year.	2. Large acreages require special odor mitigation measures.
3. Decanting ability assures minimum solids recycle with supernatant (usually less than 500 mg/L) and maximum concentration for storage and efficient harvesting (> 6 percent solids) starting with digested sludge of < 2 percent solids.	3. Requires areas of land, for example, 15 to 20 gross acres (60 to 8 ha) for 10 MGD, (438 L/s) 200 gross acres (80 ha) for 136 MGD (6000 L/s) carbonaceous activated sludge plants.
4 Long-term liquid storage is one of few natural (no external energy input) means of reducing pathogen content of sludges.	4. Must be protected from flooding.

(continued)

389

390

TABLE 6.26. (continued).

Advantages	Limitations
Facultative Sludge Lagoons (FSLs)	
5. Energy and operational effort requirements very minimum.	5. Supernatant will contain 300–600 mg/L of TKN, mostly ammonia.
6. Once established, buffering capacity is almost impossible to upset.	6. Magnesium ammonium phosphate (struvite) deposition requires special supernatant design.
7. Allows for all tributary digesters to operate as primary complete-mix units (one blending unit may be required for large installations).	7. May have higher energy requirements than anaerobic lagoons, when mechanical mixers are used to keep the top layer aerobic.
8. Provides an environmentally acceptable place for disposal of digester contents during periodic cleaning operations.	
9. Sludge harvesting is completely independent from sludge production.	
Anaerobic Lagoons	
1. Same as for facultative sludge lagoons.	1. Same as for facultative sludge lagoons (1 through 6).

surrounding lagoons must be graded to prevent discharge of surface water runoff into the lagoon; (3) lagoon depth should not be more than 60 cm (24 inches); and (4) at least two lagoons should be provided. Ettlich et al. [91], and Burd [89] summarized the following design criteria for sludge drying lagoons based on the available literature (Table 6.27). Solids loading rates of 112.1 kg/m^2 (500 tons/acre) and 8.3 and 8.6 g/m^2 (1.7 and 1.84 lb/ft^2) per thirty days of bed use were also reported in the literature [89].

Facultative Sludge Lagoons (FSLs)

Design considerations for FSLs include area loading rate, surface agitation requirements, physical layout, and other site specific features.

Operational experience at the Manukau Sewage Purification Works, Auckland, New Zealand, and Sacramento Regional Wastewater Treatment Plant, Sacramento, California, U.S.A., showed that facultative sludge lagoons can be operated successfully at a volatile solids loading rate of 0.0975 kg/m^2/d (20 lbs/1000 ft^2/d) [88,92,93].

Studies conducted at Sacramento also showed that facultative sludge lagoons can be operated successfully without any odor problems at a volatile solids loading rate of up to 0.195 kg/m^2/d (40 lbs/ 1000 ft^2/d) during summer months. The design criteria for these sludge storage basins are reported by Schafer [92] and are given in Table 6.28.

Surface agitation is required for facultative sludge lagoons to prevent the formation of a thick scum layer, which will inhibit oxygen transfer to the

TABLE 6.27. Design Criteria for Drying Lagoons [81,91].

	Design Parameter
a. Solids loading rate	
Primary sludge	96.1 kg/m^3/year
—(lagoon as a digester)	(6 lbs/ft^3/year)
Digested sludge	35–38 kg/m^3/year
—(lagoon for dewatering)	(2 2–2.4 lbs/ft^3/d)
b. Area required	
Primary sludge	0 0929 m^2/capita
(dry climate)	(1 ft^2/capita)
Activated sludge	0 31586 m^2/capita
(wet climate)	(3.4 ft^2/capita)
c. Dike height	60 cm (2 ft)
d. Sludge depth after decanting—depths of 60 cm to 1.2 m (2–4 ft) have been used in very warm climates	38 cm (15 in)
e. Drying time for depth of 38 cm (15 in) or less	3 to 5 months

TABLE 6.28. Design Criteria for Sludge Storage Basins—Sacramento
Regional Wastewater Treatment Plant, Sacramento, California, U.S.A. [92].

Total number of sludge storage basins	20
Surface area—hectares (acres)	50.6 (125)
Depth at normal operation—m (ft)	4.57 (15)
Solids loading rate—	
kg/m²/d (lbs/1000 ft²/d)	0.0975 (20)
Stored solids concentration, %	>6
Surface mixers for aeration	40
Barrier wall height, m (ft)	3.64 (12)
Supernatant return flow metering	3 154–17.0
90° V-notch weir, L/s (gpm)	(50–270)
30.5 cm (12 in) Parshall flume	8.77–175.3
L/s (MGD)	(0.2–4.0)

top layers of the lagoon. Facultative sludge lagoons require the operation of two brush-type floating surface mixers [2.4 m (8 feet) long rotors driven by 11.2 kw motors at a speed of 70 rpm] for about six to twelve hours per day to maintain a satisfactory operation of lagoons having a surface area of four to seven acres [88].

The depth of facultative sludge lagoons can be established by the practical limitation of the movement of dredges in lagoons to remove sludge from them. Depths can range from about 3.5 and 4.7 m (11.5 to 15 feet) [88].

In order to minimize the dissemination of odor, the long and short sides of the lagoons should be arranged such that the side having the shortest dimension is oriented parallel to the maximum prevailing winds. Also, lagoon area and siting should be such that the impact of odors when generated is at a minimum on any nearby communities. Experience at Sacramento Regional Wastewater Treatment Plant showed that the maximum size of a battery of facultative lagoons is 50 to 60 acres for controlling the transport of odors.

Bubbis [94] reported on the utility of underdrains in the lagoons for enhancing dewatering and drying. However, such installations have not been widely used. He also reported that by providing underdrains in lagoons, sludge was able to be dried and removed in eighteen months; in contrast it took about twenty-four months in lagoons which are not equipped with underdrains [95]. Lagoons with pea gravel substratum and underdrains are in operation in Peoria, Illinois, U.S.A., but data are not available on the efficiency of these lagoons in comparison to those without such a substratum and underdrains.

Aerobic Sludge Lagoons

Aerobic sludge lagoons are also known as aerated sludge storage basins.

The design considerations for aerated sludge storage basins include the following: (1) assurance of uniform solids concentration in the lagoon and transfer of oxygen, (2) availability of sufficient air or oxygen to ensure aerobic conditions in the storage basin, and (3) ability to accommodate rainfall and hence facility to handle liquid level variability.

The aeration requirement of aerated sludge storage basins can be determined as in the case of aerobic sludge digesters except that digested sludge storage basins have a lower oxygen requirement per unit mass of sludge solids because the sludge is already stabilized. A dissolved oxygen level of about 0.5 mg/L is quite adequate.

In situ Enhancement of Solids Concentration in Lagoons

Laboratory-, pilot- and full-scale lagoon experiments were conducted at the Metropolitan Water Reclamation District of Greater Chicago to determine whether it would be possible to accelerate the dewatering and drying of anaerobically digested sludge in lagoons by in-site chemical conditioning of the sludge. Krup, Prakasam, and Zenz [96] reported on the results of a laboratory lagoon study using plexiglas columns of 15 cm (6 inches) diameter and 18.3 m (6 feet) length. They reported that anaerobically digested sludge containing 4 to 4.5% solids was concentrated to about 7.5% in about seven days by polymer conditioning. Allied Colloids polymer Percol 728 was used at a rate five kg per MT (10 lbs/sh·t) of solids at sludge depths of 51 cm and 76.2 cm (20 and 30 inches). The rapid rate of thickening was achieved by removing the bound water released from the lagoons. It was also found that digested sludge was dried completely without polymer conditioning in fourteen, twenty-five, and thirty-six days when it was spread to depths of 3.75, 7.5, and 12.5 cm (1.5, 3, and 4.5 inches), respectively. However, the corresponding days for drying were fourteen, thirty-six, and fifty-four days when the sludge was conditioned with polymer and the released bound water was not removed from the experimental lagoons. No bound water was released by polymer conditioning of a previously lagooned sludge, which had a solids content of 20%.

Jeffrey [97,98] conducted laboratory scale experiments to determine the effect of sludge depth on the dewatering rates of sludge by drainage and transpiration and evaporation. He reported that drainage was independent of sludge depth during the first twenty days of operation and was proportional to depth after this period. During the first twenty days the lagoons did not produce a cake that could be easily removed. Evaporation by natural processes, and transpiration accomplished by growing a plant cover on the sludge were considered to be important factors in achieving rapid drying rates in lagoons. From other experiments he conducted, it was found that dewatering rates were affected initially by the permeability of the supporting media. However, the rate of drainage decreased after a short time and the effect due to supporting media was not significant.

Pilot-scale experiments were conducted at the Metropolitan Water Reclamation District of Greater Chicago with lagoons (4.8 m × 4.8 m) equipped with a 45 cm deep (18 inches) sand bed at the bottom of the lagoons and a 10 cm (4 inches) diameter underdrain laid at the bottom of the sand bed. Anaerobically digested sludge containing about 4% solids was conditioned with 150 kg of ferric chloride/mt (300 lbs/sh·t) of solids and applied to a depth of 120 cm (4 feet) and 240 cm (8 feet) in the lagoons. Initially, a very small volume of drainage resulted from the lagoon. Subsequently, i.e., after a few hours, no drainage resulted although excellent coagulation of the sludge was achieved.

However, when the same sludge was applied to a depth of 71 cm (28 inches) onto an open lagoon of 46.45 m^2 (500 ft^2) having a sand bed of 45 cm (18 inches) depth at its bottom, clear drainage resulted overnight. A sludge cake containing about 10 to 12% solids was achieved. The sludge surface cracked within two to three days. Cake solids concentration reached 23% within three to four weeks. Two repeated dosings of the lagoon with the same type of ferric chloride conditioned sludge onto previously dewatered sludge still left in place resulted in progressively poor drainage, although excellent coagulation of the sludge was achieved and bound water was released. These experiments suggest the following: (1) accelerated dewatering and drying of sludge can be achieved in lagoons by proper chemical conditioning and drainage conditions; (2) application of ferric chloride conditioned sludge onto lagoons equipped with sand bed bottoms can yield cakes containing a high solids content when sludge is dosed up to a depth of 71 cm (28 inches); and (3) the permeability of the sludge cake resulting from ferric chloride conditioning limits the process of repeated applications of chemically conditioned sludge for achieving *in situ* dewatering and drying of sludge.

Studies on a full-scale lagoon were conducted also, which has a clay bottom and a surface area of 10 hectares (~25 acres). During one trial in Chicago in the winter of 1987, approximately 37,850 m^3 (10 million gallons) of anaerobically digested sludge conditioned with 150 kg ferric chloride/mt (300 lbs/sh·t) was applied to this lagoon. A clear discharge containing less than 10 mg of suspended solids/L resulted immediately and escaped from the drawoff box of the lagoon. Before the beginning of the following spring the solids concentration in the lagoon averaged about 25%. During the summer of 1988, the solids content of the sludge was over 90%. A follow-up application of another 49,205 m^3 (13 million gallons) during the winter of 1988 resulted also in the discharge of a clear supernatant from the lagoon. But a clear lake of the supernatant accumulated in the lagoon extending to about 100 m (~300 ft) from the drawoff box. This clear liquid could not be removed by gravity, as the hydraulic profile of the lagoon was not conducive for it. If the lagoon was properly designed to drain this water completely or the water was removed by pumping, a significant degree of dewatering of the sludge could have been achieved in a matter of days. Further studies are required to develop design criteria for optimizing *in situ* dewatering and drying of chemically conditioned sludge in lagoons as a significant potential cost savings exists in such a process.

PROCESS MANAGEMENT AND PERFORMANCE

Results of Lagoon Stabilization and Dewatering

When lagoons are designed and operated properly, they offer a very cost-effective means of treatment, utilization, and disposal of sludge. An efficient management of lagoon operations not only achieves good sludge dewatering, resulting in a very high solids concentration sludge, but also yields a well-stabilized sludge with a low volatile solids content. Like the aerobic lagoons, facultative lagoons, when operated with an aerobic layer on the top, provide an effective means of controlling odors. The following case studies also illustrate the performance results achieved by lagooning processes.

Case Studies

A brief history of sludge lagoons and early experiences of operating sludge lagoons has been reported by a Committee on Sewage Disposal, Engineering Section, American Public Health Association chaired by Pearse [99] of the Metropolitan Sanitary District of Greater Chicago. In this report, several case studies of bygone years have been given. A few major contemporary sludge lagoon case histories are given below.

Sludge Storage and Drying Lagoons, Metropolitan Water Reclamation District of Greater Chicago (MWRDGC)

Dalton and Neil [100] and U.S.EPA [40] described the lagoon operations of the MWRDGC. Lagoons are an important part of the overall sludge disposal practice of the MWRDGC. Currently about one-half of the volume of the anaerobically digested sludge (~250 dt/day of solids) generated by the Stickney Water Reclamation Plant of the MWRDGC is pumped into twenty-two drying storage lagoons located at its Lawndale Solids Management Area. Each lagoon has an average volume of 153,000 m³ (230,000 y³) at an average depth of about 5 m (16 feet). They are rectangular in shape. At the northern end of each lagoon, a sludge influent pipe delivers anaerobically digested sludge brought through a pipeline from the Stickney Water Reclamation Plant. A drawoff box is located at the southern end of each of the lagoons, from which supernatant is drawn off and recycled to the head end of the Stickney Water Reclamation Plant.

Sludge containing 4–5% solids (~55% volatile solids) is pumped from the anaerobic digesters to fill a layer of approximately 15 cm (6 inches) each time into a lagoon. Lagoons, which are available, are dosed with sludge in rotation until they are filled. Supernatant, which usually appears within a week after the application of a dosing of digested sludge, is drained by adjusting the drawoff boxes or weirs to stop sludge from escaping from the lagoons. An advantage associated with pumping sludge at 4–5% solids concentration is that it spreads evenly on the sludge deposited previously in the lagoons and has undergone some thickening. Spreading of sludge evenly promotes evaporation, and hence the drying of sludge.

As the supernatant is removed from the lagoons, the sludge attains a solids concentration of about 8–10%. As the sludge is further concentrated by evaporation, a crust develops. The evaporation rate decelerates as the crust is formed. The process of concentrating the sludge is repeated after each application of sludge into the lagoon.

During spring, a slackline operation is executed to further concentrate the sludge. The purpose of this operation is to maintain and improve the surface drainage by scraping the sludge crust on the top of the lagoon to the side of the lagoon. The following benefits are derived from this operation:

(1) Scraping the drier material in the crust to the side and blending it with wetter sludge will result in a higher concentration of solids than in the wet sludge, which can then be removed.

(2) The drainage pattern created will improve the drainage of released water, and any water deposited by precipitation will run off rather than be absorbed by sludge.

(3) Scraping the sludge from the surface exposes wetter sludge for evaporation.

(4) Drier sludge scraped to the sides can easily be removed by a clam bucket and hauled away in trucks.

The slackline system is comprised of a 36.3 mt (40 tons) rated 1.53 m³ (2 y³) dragline crane, a 27.2 mt (30 tons) dozer which is used as a tail anchor, and a crescent-shaped scraper and carrier for material transfer. The crane has two cables. The first one acts as a track cable, extends from a hoist drum over to a boom point and across the lagoon to the bulldozer, which acts as the tail anchor. The crane also supports the crescent scraper. The second cable (inhaul or drag cable) extends from another hoist drum through the fairlead to the crescent scraper and carrier. The crescent scraper rides on the track cable and is used for scraping the sludge surface to the side of the lagoon. The crescent scraper and carrier is pulled to the side of the lagon with the drag cable while the track cable maintains the scraper at the appropriate height, which is generally one foot of sludge depth. Enough tension is applied to the track cable in order to return the scraper to the middle of the lagoon.

In the past, the lagoons of the MWRDGC stored and processed sludge for many years and yielded sludge having a solids content in the range of 25 to 50% with an average of 30 to 35%. The current mode of practice, however, consists of filling each lagoon to capacity with sludge and subjecting it to the slackline operation for achieving proper conditioning and a product containing a total solids content of about 15% within five years. At the end of five years, sludge taken out of the lagoons is dried in drying cells by mechanical agitation using heavy equipment (Brown Bears) to a solids content above 50%.

The MWRDGC also operates lagoons at its Calumet and Hanover Park Water Reclamation Plants very effectively in managing its sludge utilization and disposal schemes.

Anaerobic Liquid Sludge Storage Lagoons, Fulton County, Illinois, U.S.A. [88]

For roughly two decades, the Metropolitan Water Reclamation District of Greater Chicago (MWRDGC) has owned and operated the world's largest sludge reclamation and recycle program in Fulton County, Illinois, which is popularly known as the Prairie Plan. Fulton County is located about 320 km (200 miles) from Chicago. A land area of over 15,000 acres, which is basically stripmined land, is large enough to recycle all the sludge produced by MWRDGC. There are four lagoons with a surface area of 89.1 hectares (220.2 acres). These are built to store anaerobically digested sludge hauled by barge at a solids content of about 4–6% from the Stickney Water Reclamation Plant of the MWRDGC. Lagoons 1 and 2 have an average depth of 10.7 m (35 feet). Lagoons 3a and 3b have an average depth of 5.5 m (18 feet). Lagoon 3b is primarily used for supernatant storage; other lagoons are used for sludge storage. Sludge removed from the lagoons had a 17% lower volatile solids content than the sludge put into the lagoons. These lagoons are not currently being used for receiving sludge from the Stickney Water Reclamation Plant of the MWRDGC because of the high sludge transportation costs. The Fulton

County Site is currently receiving shipment of air-dried sludge solids (65% total solids) processed in the lagoons at the Lawndale Solids Management Area mentioned previously.

Facultative Sludge Lagoons (Sludge Storage Basins, SSBs), Sacramento Regional Wastewater Treatment Plant, Sacramento, California, U.S.A.

The Sacramento Regional Wastewater Treatment Plant (SRWTP) is a pure oxygen activated sludge plant having a design capacity of 567,750 m³ per day (150 MGD). Over the past seven years, anaerobically digested sludge solids generated at this plant are disposed of on a carefully managed site of approximately 75 hectares (185 acres) [92].

Twenty facultative lagoons having a surface area of 50.5 hectares (125 acres) and an average depth of 4.56 m (15 feet) receive anaerobically digested sludge from nine digesters via a sludge blending unit, which has a detention time of about three days. The storage capacity of these lagoons is five years. The design criteria of these lagoons are given in the "Design Considerations" section.

An aerobic surface layer of about one to three feet in depth with a high pH is maintained by mechanical aeration, natural surface reaeration, and photosynthetic activity. The high pH and dissolved oxygen concentration in the top layer minimizes the production and emission of odors. In the middle layer, facultative organisms thrive and break down organic matter. The bottom layer of the lagoons is anaerobic and the accumulated solids in this layer are decomposed by anaerobic bacteria. These lagoons are dosed at an average solids loading rate of 0.0975 kg/m²/d (20 lbs/1000 ft²/d). During warm weather periods, when sunlight is in abundance, these lagoons can be fed at a solids loading rate of 0.195 kg/m²/d (40 lbs/1000 ft²/d). This is possible due to higher reaction rates of organic matter degradation and photosynthetic activity during summer months.

The anaerobically digested sludge fed into the lagoons usually has a solids concentration of less than 2% and has a volatile solids content of 68% of the total solids. By the time the sludge is removed from the lagoons for dedicated land disposal, it is consolidated to a total solids concentration of about 6% and has a volatile solids content of 54% of the total solids. The volatile solids reduction achieved due to lagoon storage and treatment averages about 43%.

Schafer [92] has summarized the experiences on the performance of facultative sludge lagoons at different sites including the lagoons of Sacramento Regional Water Reclamation Plant. These are given in Table 6.29. These lagoons have been operating satisfactorily for several years. Some concern has been expressed about the applicability of facultative sludge

TABLE 6.29. Experiences with Facultative Sludge Lagoons [92].

	Plant Location				
	Auckland, New Zealand Manukau Sewage Works	Sacramento, CA Central W.T.P.	Corvallis, OR	Red Bluff, CA	Salinas, CA
Number of lagoons	3	Batteries I, II: 8 each	2	2	3
Total water surface, hectares (acres)	23.47 (58)	Battery I: 16.2 (40) Battery II: 23.1 (57)	1.82 (4.5)	0.38 (0.93)	2 42 (6)
Time system in operation as of 1982	2 basins, 21 years 3rd basin, 11 years	Battery I: 4 basins, 8 years 4 basins, 6 years Battery II: 8 basins, 1 year	3 years	3 years	4 years
Type of sludge to FSLs[a]	Digested (mainly P)	Digested (P + S)	Digested (P + S)	Digested (P) Raw S	Digested (P + S)
Use surface mixers?	Yes		Yes	Yes[b]	Yes[b,c]
Algae present?	Yes		Yes	Yes	Yes
Loading rate kg/m²/d (lbs/1000 ft²/d)	0.09 (18.4)	0 097 for most Battery I basins. Testing has been completed on a broad range of loading rates (20).	0.049–0.073 (10–15 estimate)	0.097–0.122 (20–25)	0.073–0.122 (12–15 estimate)

(continued)

399

TABLE 6.29. (continued).

	Plant Location				
	Auckland, New Zealand Manukau Sewage Works	Sacramento, CA Central W.T.P.	Corvallis, OR	Red Bluff, CA	Salinas, CA
Immediately adjacent to urban development	Yes	Not at present time. Future development will be about 610 m (1000 ft) distant	Yes, within few hundred feet	Yes, within few hundred feet.	Yes, less than 305 m (1000 ft) distant.
Comments	No odor problem.	Odor problems experienced with higher loading rates and also when surface mixers not used.	Operations is very good. No odor.	Raw waste-activated sludge is added direct to the FSLs. Occasional odors are controlled with additional surface mixing.	Significant odor problems prior to FSL operation. None, after.

[a]FSL = facultative lagoon; P = primary sludge, S = secondary sludge.
[b]Only used when necessary
[c]Windy area provides natural surface agitation.

lagoons in colder climates. Schafer [92] has reported that Flagstaff, Arizona and Colorado Springs, Colorado in the U.S.A., which have colder climates are implementing these types of lagoons currently. Ice does not seem to accumulate on these lagoons during winter periods due to the operation of surface mixers at night, the high solids content of sludge acting as a heat source, and the high salt content of sludge depressing the freezing point of water.

AGITATION DRYING OF SLUDGE

THEORY OF AGITATION DRYING

Agitation drying is a fairly new technique used to dry municipal sewage sludge. This technique has great appeal since it relies mainly on the forces of nature and is therefore attractive from both an operating and capital cost viewpoint.

The agitation drying process consists of placing a layer of sludge on an open surface and then periodically agitating it with mechanical equipment until it dries to a desired solids content. Unlike sand drying beds, which rely mainly upon drainage to remove water from sludge, evaporation is the chief means by which the solids content of agitation-dried sludge is increased. Agitation of sludge, especially previously dewatered sludge, tends to expose a greater surface area to the effects of sunlight and wind. Also, agitation tends to break up the dried surface crust on sludge exposed to air, thereby also serving to speed the evaporation of water.

Obviously climate conditions play an important role in agitation drying. Factors such as the amount and rate of precipitation, percent of sunshine, air temperature, relative humidity, and wind velocity play an important role. Weather, being uncontrollable, prevents the establishment of a standard operating procedure for agitation drying.

Sludge exposed to air dries by evaporation to a moisture content that depends upon temperature, wind velocity, and other climate conditions. Rain lengthens the drying time, but its effect is less important if the sludge has dried to a point where the rainfall does not make it liquid.

The effect of temperature on the rate of sludge drying is well established. Quan and Ward [101] reported that the evaporation rate doubled when converting from low temperature and low humidity to one of high temperature and high humidity. Fleming [102] also found that sludge dries in six weeks in the summer but required twelve weeks to dry in the winter.

The nature and moisture content of sludge affects drying. Sludge with high initial moisture content takes longer to dry. Greasy sludges take longer to dry than sludges with considerable grit. Primary sludge dries faster than secondary sludge and stored sludge dries slower than sludges that are not.

PROCEDURES PRIOR TO AIR DRYING

Agitation drying proceeds best with dewatered sludges with a solids content of at least 12%. Otherwise the sludge is too liquid and it is difficult for the mechanical agitation operation to expose a large surface area. Dewatered sludge tends to form a dried crust that is broken up by the agitation equipment, thereby enhancing evaporation.

Agitation drying by its very nature results in a large surface area of sludge exposed to the open air. It is imperative that sludge placed in agitation drying beds be stabilized to reduce its volatile solids content and thereby reduce odors.

AGITATION DRYING AT THE METROPOLITAN WATER RECLAMATION DISTRICT OF GREATER CHICAGO (MWRDGC)

The MWRDGC has the largest agitation drying operation for municipal sewage sludge in the United States. Currently the District has 415 acres of agitation drying in operation.

The final processing step in the MWRDGC's sludge management plan is agitation drying in the open at ambient temperatures. In this process, dewatered digested sludge from centrifuges or lagoons is placed onto a mildly sloped paved drying surface. This sludge is then spread out over the surface with a rubber-tired bulldozer to a depth of about 45.7 cm (18 inches). The sludge is then agitated periodically to break up the dried surface material and expose the underlying wet sludge to the atmosphere. Agitation is performed with either a bulldozer or a rubber-tired "Brown Bear" tractor equipped with a front mounted auger.

When the sludge reaches about 30% solids, it is windrowed and the windrows are periodically agitated with the Brown Bear. When the windrow reaches 50% solids or higher, it is removed from the drying surface and placed into trucks for transportation.

The MWRDGC's experience with the agitation drying system is that, once dried, sludge is not significantly wetted by rainfall, provided that the accumulated water is allowed to drain freely from the drying surface. The agitation drying sites are designed such that runoff from rain events is directed to drawoff boxes at the low end of the site and returned to the plant for treatment.

Annual precipitation in the Greater Chicago area averages about 83.9 cm (33.0 inches) per year and nearly equals the annual evaporation rate of 83.0 cm (32.7 inches) per year. The effectiveness and economics of agitation drying in areas that have significantly higher precipitation would be much different than that experienced by the MWRDGC. In addition, the Chicago winter, during which no agitation operations take place, provides the opportunity for several freeze-thaw cycles to help release bound water from sludge placed upon the drying areas before spring operations begin. This phenomenon is significant and greatly reduces the amount of water to be evaporated from the sludge.

The operations and maintenance costs associated with the MWRDGC's agitation drying operation for 1987 are $22.7 per mt ($20.64 per ton). However, these costs do not include costs for land or capital costs for mechanical equipment. The MWRDGC uses outside contracting services for most of its agitation operations and the land space for agitation drying was made available from the existing real estate inventory owned by the MWRDGC.

REFERENCES

1 U.S. Environmental Protection Agency. 1990. *Autothermal Thermophilic Aerobic Digestion of Municipal Wastewater Sludge,* EPA625/10-90/007.

2 Dreier, D. E., 1965. "Aerobic Digestion of Sludge," Paper Presented at Sanitary Engineering Institute, The University of Wisconsin, Madison, Wis.

3 Eckenfelder, W. W., Jr., 1956. "Studies on the Oxidation Kinetics of Biological Sludges," *Sewage and Industrial Wastes,* 28:983–989.

4 Krishnamoorthy, R., R. C. Loehr, 1989. "Aerobic Sludge Stabilization—Factors Affecting Kinetics," *Journal of Environmental Engineering, ASCE,* 115:283–301.

5 Ganczarczyk, J., M. F. Hamoda, and H. L. Wong, 1980. "Performance of Aerobic Digestion at Different Sludge Solids Levels and Operating Patterns," *Water Research,* 14(11):627–633.

6 Reynolds, T. D., 1973. "Aerobic Digestion of Thickened Waste Activated Sludge," *Proc. 28th Industrial Waste Conference,* Purdue University, W. Lafayette, IN, pp. 12–37.

7 Droste, R. L. and W. A. Sanchez, 1986. "Modeling Active Mass in Aerobic Sludge Digestion," *Biotechnology and Bioengineering,* 28(11):465–573.

8 Lawton, G. W. and J. D. Norman, 1964. "Aerobic Sludge Digestion Studies," *Journal Water Pollution Control Federation,* 36:495–504.

9 Koers, D. A. and D. S. Mavinic, 1978. "Aerobic Digestion of Waste Activated Sludge at Low Temperatures," *Journal Water Pollution Control Federation,* 50(3):460.

10 Murphy, K. L., 1959. "Sludge Conditioning by Aeration," Unpublished Master's Thesis, University of Wisconsin, Madison.

11 U.S. Environmental Protection Agency. 1979. *Process Design Manual for Sludge Treatment and Disposal,* EPA625/1-79/011.

12 Kelly, H. G., H. Melcer, and D. S. Mavinic, 1993. "Autothermal Thermophilic Aerobic Digestion of Municipal Sludges: A One-Year, Full-Scale Demonstration Project," *Water Environment Research,* 65(7):849–861.

13 Ahlberg, N. R. and B. I. Boyko, 1972. "Evaluation and Design of Aerobic Digesters," *Journal Water Pollution Control Federation,* 44:634.

14 Jaworski, N., G. W. Lawton, and G. A. Rohlich, 1961. "Aerobic Sludge Digestion," *International Journal of Air and Water Pollution,* 4:1/2.

15 Burton, H. N. and J. F. Malina, Jr., 1965. "Aerobic Stabilization of Primary Wastewater Sludge," *Proceedings of the Nineteenth Industrial Waste Conference, Part Two, Engineering Bulletin,* 49(2):716–723, Purdue Univ., W. Lafayette, IN.

16 Hostetler, J. B., and J. F. Malina, Jr., 1964. *A Comparison of Aerobic and Anaerobic Sludge Stabilization Studies,* Technical Report, EHE 05-6502, 91 pp., Environmental Health Engineering Laboratories, The University of Texas, Austin, TX.

17 Malina, J. F., Jr., 1967. "Aerobic Sludge Stabilization—Discussion," *Advances in Water Pollution Research, Volume II,* Water Pollution Control Federation, Washington, D.C.

18 Matsch, L. C. and D. F. Drnevich, 1977. "Autothermal Aerobic Digestion," *Journal Water Pollution Control Federation,* 49:296.

19 Stankiewicz, M. J., Jr., 1972. "Biological Nitrification with the High Purity Oxygen Process," *Proceedings 27th Purdue Industrial Waste Conference,* Purdue University, Lafayette, IN.

20 Martin, J. H., H. E. Bostian, and G. Stern, 1990. "Reductions in Enteric Microorganism during Aerobic Sludge Digestion," *Water Research,* 24(11):1377–1385.

21 Water Pollution Control Federation, 1985. "Sludge Stabilization," *Manual of Practice FD-9,* Alexandria, VA.

22 Vik, T. E. and J. R. Kirk, 1993. "Evaluation of the Cost Effectiveness of the Autothermal Aerobic Digestion Process for a Medium Sized Wastewater Treatment Facility," Preprint, Water Environment Federation, Anaheim, CA.

23 Malina, J. F., Jr., 1964. "Thermal Effects on Completely Mixed Anaerobic Digestion," *Water and Sewage Works,* 95:52–56.

24 Barker, H. A., 1956. *Bacterial Fermentations,* John Wiley & Sons, New York, NY.

25 Malina, J. F., Jr., 1962. "Variables Affecting Anaerobic Digestion." *Public Works,* 93(9):113–116.

26 McCarty, P. L., 1964. "Anaerobic Waste Treatment Fundamentals," *Public Works,* 95(9):107–112; 10:123–126; 11:91–94; 12:95–99.

27 Anonymous, 1987. "Anaerobic Sludge Digestion," *Manual of Practice No. 16,* 2nd Edition, Water Pollution Control Federation.

28 Estrada, A. A., 1960. "Design and Cost Consideration in High Rate Digestion." *Proceedings, Sanitary Engineering Division, American Society of Civil Engineers,* 86, SA3, 111.

29 Torpey, W. N., 1955. "Loading to Failure of Pilot High Rate Digester," *Sewage and Industrial Waste Journal,* 27:121–133.

30 Zablatzky, A. R. and G. T. Baer, 1971. "High-Rate Digester Loadings," *Journal Water Pollution Control Federation,* 40:2021.

31 Fair, G. M. and E. W. Moore, 1934. "Time and Rate of Sludge Digestion and Their Variation with Temperature." *Sewage Works Journal,* 6:3.

32 Malina, J. F., Jr., 1962. "The Effect of Temperature on High-Rate Digestion of Activated Sludge." *Proceedings of the 16th Purdue Industrial Waste Conference,* Purdue University, Lafayette, Indiana, Engineering Bulletin, 46(2):232.

33 Malina, J. F., Jr., 1964. "Thermal Effects on Completely Mixed Anaerobic Digestion." *Water and Sewage Works,* 95:52.

34 Golueke, C. G., 1958. "Temperature Effects on Anaerobic Digestion of Raw Sewage Sludge." *Sewage and Industrial Wastes Journal,* 30:1225.

35 Keefer, C. E., 1959. "Effects of Premixing Raw and Digested Sludge on High Rate Digestion." *Sewage and Industrial Wastes Journal,* 31:388.

36 Walker, J. D., 1979. "Successful Anaerobic Digestion—A Review," Unpublished paper.

37 Galwardi, E. F., V. Behn, M. J. Humenick, J. F. Malina, Jr., and E. F. Gloyna, 1974. "Recovery of Useable Energy form Treatment of Municipal Wastewaters," Technical Report EHE-74-06, CRWR-116, Center for Research in Water Resources, The University of Texas at Austin.

38 Woods, C. E., and J. F. Malina, Jr., 1965. "Stage Digestion of Wastewater Sludge," *Journal Water Pollution Control Federation,* 37:1495.

39 Malina, J. F., Jr., 1992. "Anaerobic Sludge Digestion," in *Design of Anaerobic Processes for the Treatment of Industrial and Municipal Wastes,* Edited by, J. F. Malina, Jr. and F. G. Pohland, Water Quality Management Library, Technomic Publishing Co. Inc., Lancaster, PA.

40 Anonymous, 1979. *Process Design Manual for Sludge Treatment and Disposal,* Chapter 6, Municipal Environmental Research Laboratory, U.S. Environmental Protection Agency, Washington, D.C.

41 Malina, J. F., Jr. and J. Difilippo, 1971. "Treatment of Supernatants and Liquids Associated with Sludge Treatment," *Water and Sewage Works—Reference Number.*

42 Malina, J. F., Jr., F. Coba-Para, and H. M. Wu, 1993. "Effectiveness of Municipal Sludge Treatment Processes in Eliminating Indicator Organisms," *Proceedings, Joint Residuals Management Conference,* American Water Works Association and Water Environment Federation, Phoenix, AZ.

43 Malina, J. F., Jr. and E. M. Miholits, 1968. "New Developments in the Anaerobic Digestion Sludges," in *Advances in Water Quality Management,* The University of Texas Press, Austin, TX.

44 Boarer, J., 1978. "Unusual Waffle Bottom Digester Design," Paper presented at the California Water Pollution Control Association, Northern Regional conference, Redding, California.

45 Stukenberg, J. R., J. H. Clark, J. Sandino, and W. R. Nayda, 1992. "Egg-Shaped Digesters: from Germany to the U.S.," *Water Environment & Technology,* April.

46 Speece, R. E., 1996. *Anaerobic Biotechnology for Industrial Wastewaters,* Archae Press, Nashville, TN.

47 Malina, J., 1961. "The Effects of Temperature on Anaerobic Digestion of Activated Sludge," Doctoral Dissertation, Univ. of Wisconsin.

48 Rudolfs, W., 1932. "Enzymes and Sludge Digestion," *Sewage Works Journal,* 4:782–789.

49 U.S.EPA. 1979. *Process Design Manual for Sludge Treatment and Disposal,* U.S.EPA 625/1-79-011, U.S.EPA, Cincinnati, Ohio.

50 U.S. Environmental Protection Agency. 1989. *1988 Needs Survey of Municipal Wastewater Treatment Facilities,* U.S.EPA 430/09-89-001. Cincinnati, Ohio: U.S.EPA.

51 Christensen, G. L., 1982. "Dealing with the Never-Ending Sludge Output," *Water Engineering Management,* 129:25.

52 Westphal, A. and G. L. Christensen, 1983. "Lime Stabilization: Effectiveness of Two Process Modifications," *Journal Water Pollution Control Federation,* 55:1381.

53 Kamplemacher, E. H. and L. M. Van Noorle Jansen, 1972. "Reduction of Bacteria in Sludge Treatment," *Journal Water Pollution Control Federation,* 44:309.

54 Farrell et al. 1974. "Lime Stabilization of Primary Sludges," *Journal Water Pollution Control Federation,* 46:113.

55 Metcalf and Eddy, Inc. 1979. *Wastewater Engineering: Treatment, Disposal, Reuse.* New York: McGraw-Hill Book Company.

56 Paulsrud, B. and A. S. Eikum, 1975. "Lime Stabilization of Sewage Sludges," *Water Resources* (G.B.), 9:297.

57 Counts, C. A. and A. J. Shuckrow, 1975. *Lime Stabilized Sludge: Its Stability and Effect on Agricultural Land,* U.S.EPA 670/2-75-012, Battelle Memorial Institute, Richland, Washington.

58 Ramirez, A. and J. Malina, 1980. "Chemicals Disinfect Sludge," *Water and Sewage Works,* 127(4):52.

59 Otoski, R. M., 1981. *Lime Stabilization and Ultimate Disposal of Municipal Wastewater Sludges,* U.S.EPA-600/S2-81-076, Cincinnati, Ohio: U.S.EPA.

60 Bitton, G. et al. 1980. *Sludge—Health Risks of Land Application.* Ann Arbor, Michigan: Ann Arbor Science Publishers, Inc.

61 Novak, J. T. et al. 1982. "Stabilization of Sludge from an Oxidation Ditch," presented at the *55th Water Pollution Control Federation Conference,* St. Louis, Missouri.

62 Pederson, D. C., 1980. "Density Levels of Pathogenic Organisms in Municipal Wastewater Sludge—A Literature Review," EPA Project Report.

63 Lodderhose, J. R. et al. 1982. "The Stability of Oxidation Ditch Sludge for Land Application," paper presented at the *55th Water Pollution Control Federation Conference,* St. Louis, Missouri.

64 Anderson, J., 1980. "Lime Treatment of Sewage Sludge," unpublished paper, Strabenken AB, Stockholm, Sweden.

65 Bick, K. June 1986. "Applications of Cement Kiln Dust in Wastewater Sludge Management," Report by Dallas Water Utilities, Dallas, Texas, p. 43.

66 Burnham, J. C., 1986. "A Report: The Effect of Cement Kiln Dust and Lime on Microorganism Survival in Toledo Municipal Wastewater Sludges," Report for Medical College of Ohio to City of Toledo, Ohio, p. 31.

67 Christensen, G. L. and D. A. Stule, 1979. "Chemical Reactions Affecting Filterability in Iron-Line Sludge Conditioning," *Journal Water Pollution Control Federation,* 51:2499.

68 1985. Sludge Stabilization. *Water Pollution Control Federation Manual of Practice No. FD-9.*

69 U.S.EPA 1989b. September 1989. *Environmental Regulations and Technology—Control of Pathogens in Municipal Wastewater Sludge,* United States Environmental Protection Agency, EPA/625/10-89/096.

70 Weber, W., 1972. *Physiochemical Processes.* New York: John Wiley & Sons, Inc.

71 U.S.EPA. 1987. *Design Manual for Dewatering Municipal Wastewater Sludges.* U.S.EPA 625/1-87/014. Center for Environmental Research Information, Cincinnati, OH.

72 WPCF/MOP FD-14. 1988. *Sludge Conditioning Manual of Practice.* Alexandria, VA: Water Pollution Control Federation.

73 Schwoyer, W., 1981. *Polyelectrolytes for Water and Wastewater Treatment.* Boca Raton, FL: CRC Press, Inc.

74 Vesilind, P. A., 1979. *Treatment and Disposal of Wastewater Sludges,* Revised Edition. Ann Arbor, MI: Ann Arbor Science Publishers.

75 WPCF/MOP OM-8. 1987. *Operation and Maintenance of Sludge Dewatering Systems Manual of Practice,* Alexandria, VA: Water Pollution Control Federation.

76 Baskerville, R. and R. Gayles, 1968. "A Simple Automatic Instrument for Determining the Filterability of Sewage Sludges," *Journal of the Institute of Water Pollution Control,* No. 2, Great Britain.

77 Mace, G., 1990. "Specifiers Guide to Polymer Feed Systems," *Pollution Engineering,* 22:75.

78 Mishra, S., 1989. "Polymer Flocculation of Fine Particles: Theoretical Developments," *Pollution Engineering,* 21:102.

79 Soszynski, S., 1991. *Capillary Suction Time—Applications Beyond Filtration.* Chicago, IL: Metropolitan Water Reclamation District of Greater Chicago.

80 Wiley, J. S. "A Preliminary Study of High-Rate Composting," *Trans. ASCE,* Paper No. 2895.

81 Jeris, J. S. and R. W. Regan, 1973. "Controlling Environmental Parameters for Optimum Composting," *Compost Science,* Jan.–Feb.

82 Schultz, K. L., 1962. "Continuous Thermophilic Composting," *Compost Science,* Spring.

83 Haug, R. T., 1982. "Modeling of Compost Process Dynamics," *Proceedings of the Nat. Conf. on Composting of Municipal and Industrial Sludges,* Wash. D.C., published by Hazardous Materials Control Research Institute, Silver Spring, Maryland.

84 Murray, C. M. and J. L. Thompson, 1986. "Strategies for Aerated Pile Systems," *BioCycle, Journal of Waste Recycling,* 27(6).

85 Haug, R. T., 1980. *Compost Engineering—Principles and Practice.* Lancaster, PA: Technomic Publishing Co., Inc.

86 Gove, P. B., ed. 1967. *Webster's Third New International Dictionary.* Springfield, Massachusetts, G. & C. Merriam Company.

87 Ministry of Housing and Local Government. 1954. *Treatment and Disposal of Sewage Sludge.* Her Majesty's Stationery Office, London, England.

88 United States Environmental Protection Agency. 1979. *Process Design Manual, Sludge Treatment and Disposal.* U.S.EPA Report No. 625/1-79-001, Municipal Environmental Research Laboratory, Cincinnati, Ohio, U.S.A.

89 Burd, R. S., 1968. *A Study of Sludge Handling and Disposal.* Publication No. WP-20-4, *Water Pollution Control Research Series,* Office of Research and Development, U.S. Department of the Interior, p. 326.

90 1971. *Recommended Standards for Sewage Works, Great Lakes-Upper Mississippi River Board of Sanitary Engineers.* Albany, New York: Health Educational Service.

91 Ettlich, W. F., D. J. Hinrich and T. S. Lineck, 1978. *Operations Manual—Sludge Handling and Conditioning.* Publication No. U.S.EPA-430-9-78-002, U.S. Environmental Protection Agency, Office of Water Program Operations, Washington, D.C. pp. XIII-1 to 5.

92 Schafer, P., 1982. "Odor Control Features Make Lagoons an Acceptable Sludge Process," paper presented at the *Sixth Mid-America Conference on Environmental Engineering Design,* Kansas City, Missouri, June 14–15, 1982.

93 Kido, W. H. and W. R. Uhte, 1990. "Seven Years of Low-Cost Sludge Disposal at a 150-MGD Plant." paper presented at the *International Association for Water Pollution*

Research and Control Conference on Sludge Management, Los Angeles, California, U.S.A., January.

94 Bubbis, N. S., 1953. "Sludge Drying Tests at Winnipeg," *Sew. and Ind. Wastes,* 25(11):1361–1362.

95 Bubbis, N. S., 1962. "Sludge Drying Lagoons at Winnipeg," *J. Water Poll. Contr. Fed.,* 34(8):830–832.

96 Krup, M., T. B. S. Prakasam and D. R. Zenz, 1982. "The Effect of Polymer Addition on the Drying Rate Digested and Lagooned Sludge," The Metropolitan Sanitary District of Greater Chicago, Research and Development Department Report No. 82-19, p. 24.

97 Jeffrey, E. A., 1959. "Laboratory Study of Dewatering Rates for Digested Sludge in Lagoons," *Proceedings of the 14th Purdue Industrial Waste Conference,* Purdue University, West Lafayette, Indiana, U.S.A., pp. 359–384.

98 Jeffrey, E. A., 1960. "Dewatering Rates for Digested Sludge in Lagoons," *J. Water Poll. Contr. Fed.,* 32(11):1153–1160.

99 Pearse, L. et al. 1948. "Sludge Lagoons," Report of the Committee on Sewage Disposal, Engineering Section, American Public Health Association, *Sew. Works. J.,* 20(5):817–831.

100 Dalton, F. E. and F. C. Neil, 1983. "Sludge Program at the Metropolitan Sanitary District of Greater Chicago," report of the Metropolitan Sanitary District of Greater Chicago, Chicago, Illinois, U.S.A.

101 Quan, J. E. and G. B. Ward, 1955. "Corrective Drying of Sewage Sludge," *International Journal of Air and Water Pollution,* 9:311.

102 Fleming, J. R., 1950. "Sludge Utilization and Disposal," *Public Works,* 90(8):12–122.

103 ASCE Air and Radiation Management Committee, 1992. *Radiation Energy Treatment of Water, Wastewater, and Sludge—A State-of-the-Art Report,* Environmental Engineering Division, ASCE, Publication No. JFBNO-87262-901-S, New York, NY.

104 USEPA, 1979. *Process Design Manual for Sludge Treatment and Disposal,* U.S.EPA 625/1-79-011, USEPA, Cincinnati, Ohio.

105 Yeager, J. G. and R. L. Ward, 1980. "Effectiveness of Irradiation in Killing Pathogens," In *Use of Cesium-137 to Process Sludge for Further Reduction of Pathogens,* SAND80-2744, pp. 69–90, Sandia National Laboratories, Albequerque, New Mexico.

106 *Workshop on Enhancement of Wastewater and Sludge Treatment by Ionizing Radiation,* National Science Foundation/University of Miami, Key West, Florida, January 13–15, 1997.

CHAPTER 7

Municipal Sewage Sludge Management at Dedicated Land Disposal Sites and Landfills

FOR many years, municipal sewage sludge has been disposed of in landfills. Normally such sludge is placed into landfills where municipal solid waste is also disposed. However, some municipalities have elected to construct their own landfills for disposal of only municipal sewage sludge.

Although the utilization of the plant fertilizer value and soil conditioning properties of municipal sewage sludge has become widespread, some municipalities have found that disposal of their sludge at dedicated sites is advantageous. At dedicated land disposal (DLD) sites, the sludge is regularly applied at high rates to the surface soil. No crops are grown and the site is used only for the application of municipal sewage sludge.

Municipal sewage sludge can be used for two beneficial purposes at landfills. First, municipal sewage sludge can be substituted for, or mixed with, topsoil and used to grow a vegetative cover on closed landfills. Secondly, municipal sewage sludge can often be substituted for soil used for daily cover at a municipal solid waste landfill. Daily covering of municipal solid waste is required by both state and federal regulations to reduce odors and blowing litter as well as for other reasons. Municipal sewage sludge can often be used instead of soil for this purpose.

This chapter will present information on the planning, design, and operation of sludge-only landfills and DLD sites. The use of municipal sewage sludge as daily and final cover at landfills is also addressed.

Robert A. Griffin, University of Alabama, Tuscaloosa, AL; Cecil Lue-Hing and David Zenz, Metropolitan Water Reclamation District of Greater Chicago, Chicago, IL; Ronald B. Sieger, CH2M Hill, Dallas, TX; Warren Uhte, Brown and Caldwell, Seattle, WA.

409

The last section of this chapter discusses and describes the mathematical models that can be used to determine the groundwater transport (if any) of municipal sewage sludge constituents. Mathematical models are being increasingly used for this purpose, and this section should be invaluable to those who plan and design landfills for municipal sewage sludge disposal.

DESIGN AND OPERATION OF SLUDGE-ONLY LANDFILLS AND DEDICATED LAND DISPOSAL SITES

DEFINITION: SLUDGE-ONLY LANDFILLS AND DEDICATED LAND DISPOSAL SITES

Sludge-only landfills are defined as sludge disposal sites that are used exclusively for the disposal of wastewater sludge. The application of the sludge can either be in the solid (dewatered) or liquid state. However, the sludge is most often combined with a processing, dewatering, or fixing material and is usually applied in the dewatered state.

Dedicated land disposal (DLD) sites are those sites where the sludge is applied to the surface of the land on a routing basis where the objective is sludge disposal rather than sludge utilization. At sludge utilization sites the fertilizer and soil conditioning properties of the sludge are utilized to grow crops. At DLD sites no crops are grown. Such DLD sites normally employ annual application rates of 11.2 to 22.4 dry kg/m² (50 to 100 dry tons per acre) per year. The sludge is usually applied as a liquid.

As can be seen by the case studies listed herein, the full bottom and side containment of sludge-only landfills is of extreme importance. The most cost-effective sludge-only landfills are those where the natural soils at the site support such a development. Environmental issues involving surface and groundwater probably the most sensitive and important to the successful operation of sludge-only landfills.

DEWATERED SLUDGE APPLICATION

Dewatered sludge application is usually used when the disposal site is located some distance from the treatment plant or when climate conditions make it necessary to store the sludge in a confined space for up to six months. Thirty-two to forty kilometers (twenty to twenty-five miles) is usually the distance where liquid transport becomes more expensive than dewatered sludge transport. For very small plants this distance may increase, because it is often more practical to transport thickened liquid sludge than to operate a dewatering system.

Some plants have incorporated dewatered sludge application in nearby facilities. This application is usually governed by the need to store sludge during inclement weather or to achieve exceptionally large applications per hectare (acre). Another reason for dewatered sludge application involves the process used to stabilize or fix the sludge. Cement kiln dust, lime, and fly ash are some of the materials often combined with sludge in this type of operation. Usually when these materials are used, the sludge is not stabilized.

Dewatered sludge application involves the placement of the dewatered sludge in the soil in a manner that allows the maximum disposal of sludge in a given area. This has been accomplished in narrow trenches, wide trenches involving side dumping of sludge, and mounding of sludge combined with fixing material. Almost all dewatering applications involve the mixing of the soil with the sludge, either continuously or at least on a daily basis. Dewatered sludge is covered daily to maintain the vector and odor control necessary to meet sludge disposal regulations.

A proposed design for the Village Creek plant in Fort Worth, Texas calls for the application of the dewatered sludge to 1.52-meter (5-foot) diameter circular holes 6.10 to 9.14 meters (20 to 30 feet) deep. The holes would be drilled just days prior to their use. The entire site is underlined with an impervious hardpan condition that assures the complete containment of the sludge. An impervious cut-off wall was to be installed to confine all contaminated water to the landfill site.

LIQUID SLUDGE APPLICATION

If the sludge disposal site is located near the treatment plant or in some cases the separate sludge treatment facility, it is often cheaper to apply the sludge in its liquid state. This type of operation eliminates the need for the operation of an expensive dewatering facility. Liquid sludge operation does depend on either the plant's ability to store liquid stabilized sludge for sufficient periods of time to assure adequate application time, or the plant's location being in a weather pattern that provides adequate application time on a routine basis.

Environmentally the most acceptable means of applying the liquid sludge to the DLD site is by direct injection. This can be accomplished by using specially equipped tank trucks that two injection equipment or by umbilical systems involving a tractor with injection equipment connected by hose to a pressurized liquid sludge pipeline. The former is usually used by smaller plants, while the latter is most often applied to larger plants or separate sludge treatment and disposal operations.

Other methods of sludge application to DLD sites involve the liquid surface spreading of the sludge by tank truck to the soil surface with its subsequent incorporation into the soil by plowing or by spraying the sludge over a large surface area with high-pressure nozzles connected to liquid sludge pipelines. When these systems are used, the soil surface must be plowed periodically to assure maximum application rates and to maintain the vector and odor control necessary to meet sludge disposal regulations.

Surface runoff from DLD sites must be controlled so that rainfall is not allowed to collect and transport the sludge contents down into the application soil. This control involves sloping the application site to assure rapid runoff, collecting the runoff in a manner that minimizes transport of soil, and return of the runoff to either the treatment plant or a special evaporation holding pond. When the runoff is returned to the treatment plant, the field runoff collection system is often provided with a small storage reservoir designed to assure that large storms can be processed without jeopardizing the goal of protecting the application area.

CASE STUDIES

The following case studies provide some insight into the design and operation of successful sludge-only landfills and DLD sites in the United States. Sludge-only landfills include those operated by the North Shore Sanitary District in Lake County, Illinois; and the Trinity River Authority of the Texas Central Regional Wastewater Treatment Plant in the city of Grand Prairie, Texas. The Dallas Water Utilities Southside Wastewater Treatment Plant in Dallas County, Texas operates both a sludge-only landfill and a DLD site. The DLD sites include the Dublin San Ramon Services District, San Ramon, California; the city of Colorado Springs Hannah Ranch Sludge Treatment Plant, El Paso County, Colorado; and the County of Sacramento Regional Wastewater Treatment Plant, Sacramento, California.

CASE STUDY A—NORTH SHORE SANITARY DISTRICT— LANDFILLING OF FLY ASH STABILIZED DEWATERED SLUDGE

Background and History

The Newport Township sludge-only landfill, located in Lake County near Waukegan, Illinois, receives sludge generated by four North Shore Sanitary District treatment plants serving over a quarter million people. The design capacity of the four plants is 2.50 cubic meters per second [57 million gallons per day (MGD)]. The wide trench, individual cell landfilling operation started in July 1974.

Prior to January 1985, the District operated the landfill by burying vacuum filter dewatered sludge cake using a trench-and-fill method. The cake consisted of 22% solids which had been conditioned with lime and ferric chloride. The design and operation of this 113.72-hectare (281-acre) sludge-only landfill is described in detail in the Environmental Research Center Information Center Publication EPA-625/4-78-012, dated October 1978 and entitled *Sludge Treatment and Disposal—Sludge Disposal—Volume 2,* and Publication EPA-625/10-84-003, dated September 1984 and entitled *Use and Disposal of Municipal Wastewater Sludge.*

By 1980 the District recognized that in order to continue sludge land-filling, additional land would have to be acquired. Based on the present Illinois regulations and laws, the District estimated it would take at least six years of litigation to permit an additional 186.16-hectare (460-acre) site of a sludge-only landfill. In addition, they recognized the "not in my backyard" syndrome that would pertain to even the best designed and operated facility.

In 1983 the District held preliminary conversations with The American Fly Ash Company, which operated a fly landfill immediately north of the District's landfill. Preliminary evaluations by both the District and American Fly Ash laboratories indicated that a pozzolanic reaction could be obtained in a fairly wide range of sludge and fly ash mixtures. As a result of these favorable indications, the District and American Fly Ash entered into an agreement to prepare a preliminary design report for the continuing disposal of sludge and fly ash on the Newport Township Land-fill site.

The fly ash involved is a by-product from the burning of coal and recovered from power plant flue gases by mechanical and/or electrical precipitation. This fly ash is comprised of very fine particles, the majority of which are glassy spheres. The fly ash primarily contains silica (SiO_2), alumina (Al_2O_3), ferric oxide (Fe_2O_3), calcium oxides (CaO), and smaller quantities of other oxides and alkalies. The fly ash is an artificial pozzolan and is not cement-like in itself, but with the presence of water, it forms a cement-like product. The moisture for this application will be supplied by the water within the sludge.

Fludge is the term used to describe the resultant material formed by blending the fly ash and sludge. Depending on the ratio of the constituents and the type of ash, the fludge is a soil-like material with behavioral properties ranging from those similar to a granular material, deriving its strength through internal friction, to a rock-like material, deriving its strength through cementation of the particles resulting from hydration of the ash.

Tests indicated that all of the mixes tested produced a material suitable for use in a self-supporting fill. Wet densities of the fludge, when compacted to 90% of ASTM D698, ranged from 44.8 kg/m³ (75.5 pounds per cubic foot) to as high as 62.2 kg/m³ (104.8 pounds per cubic foot). Strength values ranged as high as in excess of 24,413.35 kg/m² (5000 pounds per square foot) and angles of internal friction in excess of 35 degrees. Permeability tests showed values ranging from 5×10^{-6} to 1×10^{-7} centimeters per second. The fludge was found not to be classified as a hazardous waste.

Landfill Design

The original wide trench landfill operations used up the south 60.71 hectares (150 acres) of the landfill site. The new fludge landfill site was to be located on the north 53.02 hectares (131 acres) in an "I" configuration. The northern, southern, and central portions of the "I" are approximately 22.66, 25.09, and 5.26 hectares (56, 62, and 13 acres), respectively. The cells are designed to provide the District with twenty years of landfill capability. The central portion of the site is the location of the processing facilities. Figure 7.1 provides a view of this site layout.

The site elevations ranged from 222.50 to 224.64 meters (730 to 737 feet) mean sea level along the northeast edge of the 53.02 hectares (131 acres) to 211.23 meters (693 feet) along Ninth Street at the southwest corner of the property. The landfill will top out at elevation 237.13 meters (778 feet) and, depending on whether it is in Cell A or Cell B, will have its bottom located at elevation 212.14 meters (696 feet) or 210.31 meters (690 feet). The bottom elevations have been set to minimize fill height and maintain sufficient clay embedment.

The general design criteria for the landfill are as follows:

(1) Mix ratio 1:1 to 3:1, fly ash to sludge. A ratio of 2:1 is preferred.
(2) A 60.96-meter (200-foot) buffer zone between property lines and fludge boundary, except American Fly Ash property line, which is 27.43 meters (90 feet) is needed.
(3) Initially bury fludge in the area north of processing center (pug mill).
(4) Side slopes of internal 2:1, external 3:1, except for east side on Green Bay Road, which must be 4:1, are needed.
(5) Berms are keyed into clay a minimum of 0.91 meter (3 feet).
(6) The clay on the bottom and sides must have a permeability of 1×10^{-7} centimeters per second and be at least 3.05 meters (10 feet) thick.
(7) All surface water is in contact with the active subcells and will be pumped to the discharge holding pond.
(8) Clay cover should be 0.15 meter (6 inches) of clay after each subcell phase construction.
(9) No daily cover is required.
(10) Top cover should be 0.61 meter (2 feet) of clay plus topsoil.

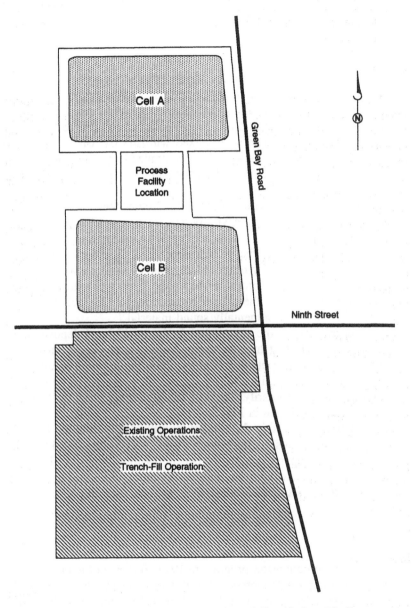

Figure 7.1 Layout—sludge landfill site—North Shore Sanitary District.

415

(11) Slopes must be seeded during and after construction.

(12) Berms shall be a minimum of 3.05 meters (10 feet) perpendicular to the face of the fill, and are constructed of 1×10^{-7} centimeters per second permeability or greater.

Operation

Originally the process facilities were designed to operate on sludge from the vacuum filter dewatering operation at the Waukegan plant. Since startup of the fly ash operation, the District has eliminated the central dewatering operation and is now accepting dewatered sludge from each of its four plants. The material is mixed with the fly ash as soon as it arrives at the processing area. This decentralized operation resulted in the saving of considerable costs by eliminating over a dozen operating positions.

The dewatered sludge is dumped into 38.23-cubic meter (50-cubic yard) hoppers that feed progressive cavity pumps for transporting the sludge to the pug mill. The dry fly ash is delivered in tractor trailer tanks and pneumatically unloaded in the fly ash silos. An airslide system is used to transfer the fly ash to the pug mill. A water makeup system is also provided to assure proper moisture content in the fludge. The three loading systems are controlled to provide the proper mixture at the pug mill to form a homogeneous and environmentally sound material.

The pug mill mixes the two materials thoroughly and deposits them onto conveyor belts that transfer the mixture to a stockpile outside the building. After initial reaction, this stockpile is loaded onto dump trucks for disposal to the landfill. During inclement weather this thoroughly mixed material is temporarily stored in a large storage shed. A "fludge" process diagram is shown in Figure 7.2.

The material is placed in the various landfill lifts by trucking from the processing center with six-wheel dump trucks. A D-5 Cat is used to strike the material off to approximately 0.91-meter (3-foot) depth. A plain steel drum roller is used to roll the material in place. As existing subcells are excavated, the excavated material is used to raise the berms of the excavation.

Monitoring

The original groundwater program in 1974 consisted of ten monitoring wells approximately 9.75 meters (32 feet) deep in the clay till. These wells have always tested negative. In 1987 the Illinois Environmental Protection Agency requested further studies of the geology of the area and asked the District to monitor the uppermost aquifer. The uppermost aquifer ranges from 12.19 meters (40 feet) deep on the southwest corner of the property

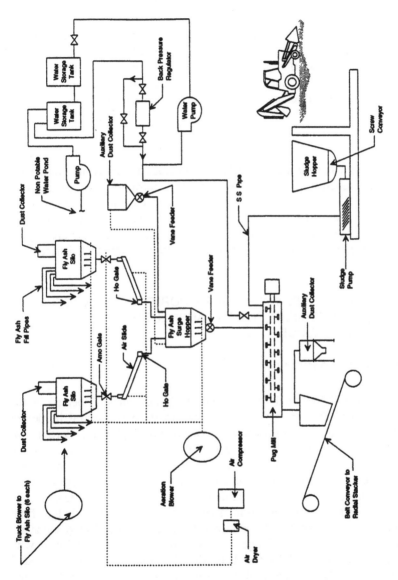

Figure 7.2 Fludge process diagram.

417

to more than 30.48 meters (100 feet) deep on the northeast corner of the property. The new groundwater monitoring system consists of seven deep wells in the uppermost sand aquifer and five shallow wells in the superficial till. These wells are still testing negative. Figure 7.3 provides a site plan showing these groundwater monitoring well locations.

The proposed U.S.EPA Part 503 regulations for sludge disposal put maximum limits for monofill (sludge-only) landfills at 6 log 10 organisms per gram VSS (1×10^6 organisms) for fecal coliform and fecal streptococcus. The District has checked its sludges and found that the sludges in any form—either thickened cake or fludge—exceed this limit.

Costs

The original wide trenching landfill operation carried out from 1974 to 1988 resulted in a cost of $68.34 per dry metric ton ($62 per dry ton) of solids. This cost includes both annualized capital costs and annual operating costs. The new fludge operation is costing $121.25 per dry metric ton ($110 per dry ton) of solids. Although these costs are higher, they are well below any other alternative disposal system the District could implement at this time.

CASE STUDY B—TRINITY RIVER AUTHORITY OF TEXAS CENTRAL REGIONAL TREATMENT PLANT—LANDFILLING OF DEWATERED SLUDGE

Background and History

The Central Regional Wastewater Treatment Plant is located in the city of Grand Prairie, which is adjacent to the west-central border of Dallas. It is at the confluence of two major highways. Portions of Dallas, Fort Worth, Arlington, and eighteen other customer cities are served by this 5.91 cubic meter per second (135 MGD) treatment plant. The influent flow is largely domestic and the sewerage system contains no combined sewers. The treatment process includes screening, grit removal, primary sedimentation, fine-bubble air-activated sludge, final sedimentation, automatic backwash filters, and chlorination/dechlorination with SO_2. Primary sludge is gravity thickened and combined with gravity belt thickened/dissolved air flotation thickened waste-activated sludge. Combined sludge at 4–6% solids is conditioned with 20% lime and 5% ferric chloride (by dry sewage solids weight). Conditioned sludge is dewatered on five 1300 mm by 1900 mm and three 1500 mm by 2000 mm 689.5-kilopascal (100-psi) recessed chamber filter presses to an average of 34% solids.

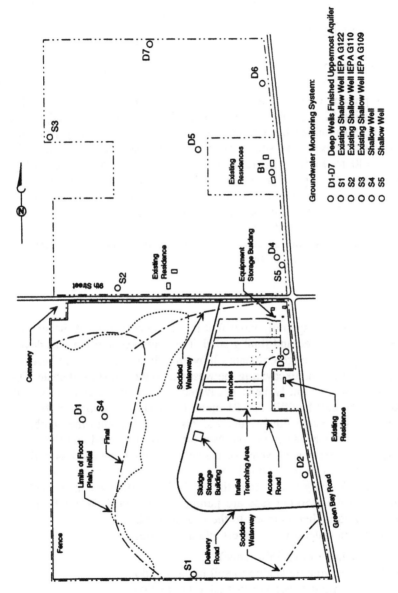

Figure 7.3 Landfill operating site plan—North Shore Sanitary District.

Groundwater Monitoring System:

- ○ D1-D7   Deep Wells Finished Uppermost Aquifer
- ○ S1   Existing Shallow Well IEPA G122
- ○ S2   Existing Shallow Well IEPA G110
- ○ S3   Existing Shallow Well IEPA G109
- ○ S4   Shallow Well
- ○ S5   Shallow Well

Dewatered sludge is conveyed to small storage ponds and then to dump trucks, which haul the sludge approximately 1206 meters (3/4 mile) to the southern part of the site. Sludge is dumped, mixed with soil, and placed in a sludge-only landfill. Landfilling began in 1976. Recently, a second landfill area, which is also on plant property, was developed adjacent to the present landfill.

Site Description

The present Central Plant site is slightly over 121.4 hectares (300 acres), with 52.6 hectares (130 acres) dedicated to two sludge-only landfills. Surrounded by a flood plain, the site is completely protected by earthen levees. The site is an alluvial deposit some 9.14 to 16.76 meters (30 to 55 feet) thick. This alluvial material ranges from clays and silts in the upper layers to sands and gravel at the base. Eagle Ford shale 50 to 100 meters (several hundred feet) thick lies below the alluvial deposits. The shale is unweathered, relatively unjointed and unfractured, and has extremely low permeability (1×10^{-8} cm/sec and lower), making it an effective barrier to vertical seepage. Figure 7.4 presents the plant site layout.

Design

The original sludge-only landfill at the southeast corner of the site has recently been capped and final graded. The southwest portion was developed in 1989 as a sludge-only landfill by providing a barrier levee between the two landfills and excavating the west site to the Eagle Ford shale. The west site had previously been excavated for soil material to mix with sludge in the east landfill. A 0.91-meter (3-foot) thick slurry cutoff wall, keyed into the Eagle Ford shale, was also placed around the combined landfill site. After placement, the west site was dewatered and the initial working face excavated. Much of the soil remains to be used as bulking material. A lined leachate pond handles a twenty-five-year storm condition and limits discharge back into the landfill. All drainage and leachate collected at the working face are pumped to the pond. Water from the leachate pond is then pumped to the treatment plant.

The design criteria for the west site are as follows:

- 34% solids (30 to 55% range)
- 1:1.5 wet volume sludge to soil bulking ratio
- 3 to 5% annual increase in sludge production
- yield of 10.56 kilograms of solids per cubic meter per second (1.02 pounds of solids per MGD) of plant flow (including chemicals) to landfill

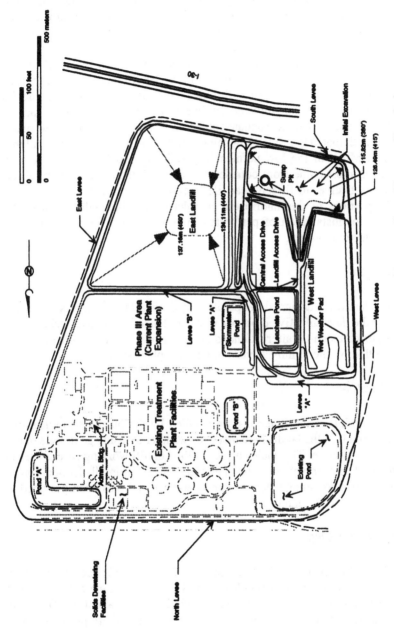

Figure 7.4 Trinity River Authority of Texas—Central Regional Wastewater System Treatment Plant site layout.

421

Because the landfill operations were deemed to be efficient, the present use of construction vehicles to dry, mix, and place the sludge and soil would be continued. A sludge to soil ratio of 1:1.5 is achieved by windrowing the soil and sludge for mixing. Ten groundwater monitoring wells surround the landfill site to monitor the effectiveness of the slurry cutoff wall. Because the water level outside the site is higher than the level inside, and because the dewatering wells within the landfill site continue to keep the water level low, any flow through the slurry cutoff wall should be into the fill site. Contaminated surface water near the working face is collected, pumped to the leachate collection pond, and returned to the plant.

It is anticipated that sufficient soil for mixing with the sludge is available on-site. Much of the low-permeability soil required for the final cover, however, must come from outside sources. With the design criteria listed earlier, a twelve- to fifteen-year life is expected from the west site.

As portions of the landfill are completed, trenches will be cut in the surface for installation of perforated pipe and gravel for a passive-type gas collection system.

Operation

The landfill is designed to operate only during daylight hours, normally during the day shift, five days per week. Extended operations using over-time permit catch-up after extended rainy periods. The complete operation is handled by the Trinity River Authority's plant staff.

Site Facilities

All administrative functions are handled by the plant's administrative group. Roads in the landfill are made all-weather by using geotextile fabric and crushed stone. The shale surface at the landfill working face is spread with sand and gravel for safe operations after a rainfall. In addition, a concrete pad for sludge storage during wet weather has been provided at the entrance to the landfill area.

Sludge/Soil Mixing

Sludge hauled by 12.23-cubic meter (16-cubic yard) end-dump trucks is placed in a long narrow windrow parallel to the working face and about 18.29 meters (20 yards) away. These windrows are about 18.29 meters (20 yards) wide by 91.44 meters (100 yards) long and are placed twenty-four hours per day, seven days per week. Dry soil is placed by scrapers about 6.1 meters (20 feet) away in a slightly larger windrow. Bulldozers spread the sludge into a 101.6-mm (4-inch) deep lift and place a 152.4-mm (6-inch) lift of soil over it. This spreading operation allows for some additional drying. A large industrial disc/plow pulled by a bulldozer thoroughly mixes the soil and sludge. This mixing operation, which takes a day or more, helps to dry the mixture. An auger-type tractor/mixer has been used, but maintenance costs were excessive.

Operational Procedures

The mixture is placed on the working face in 304.8-mm (12-inch) lifts and then compacted with at least four passes of the bulldozer or by routing loaded sludge trucks or loaded scrapers over the filled area. The individual lifts should be stacked in benches about 3.05 meters (10 feet) thick with a fill face at a 6:1 slope. At least two adjacent benches parallel to the working face should be placed before stacking a bench vertically. Active daily mixing and fill areas do not require a daily cover. All other areas require a 152.4-mm (6-inch) interim cover.

The landfill working face moves from south to north, so mixing soil and shale must be removed progressively northward to open up new fill areas. Approximately 3.05 meters (10 feet) of shale is excavated to meet the twelve-to fifteen-year expected life. The shale should be stored for use as final cover material. A ripper is needed on the bulldozer to excavate the shale.

The working face and all excavations must be sloped at least one percent toward the drainage sump. Drainage on the working face and excavations is pumped to the leachate pond by a single hydraulically-driven submersible trash pump. A backup pump is available.

As sections of the fill are completed, the final cover is placed. It consists of 609.6 mm (24 inches) of very low permeability soil, 203.2 mm (8 inches) of low-permeability material, and 101.6 mm (4 inches) of topsoil. The material is spread in lifts of 203.2 mm (8 inches) or less before compaction. Then, the low-permeability soils are broken up with a plow and compacted by tamping or pneumatic roller to 95% standard proctor maximum dry density.

When working face operations are stopped during wet weather, dewatered sludge is placed on an uncovered concrete pad. When the weather improves, this sludge is also placed in windrows. However, more soil, or more sludge mixing, may be required to mix and dry the sludge. Stockpiled soil is covered to keep it dry during wet weather.

Staffing

The Trinity River Authority operates the complete facility, including all trucking operations. A forty-hour-per-week operation requires a staff of nine full-time people, including the landfill supervisor. In addition, one trucker per shift hauls sludge to the fill. The trucking is included under solids operations, not landfill operations.

Equipment

The present equipment includes two bulldozers (with plow and ripper attached), two front-end loaders, two scrapers, two motor graders, one backhoe, one Brown Bear auger, and one dump truck for miscellaneous landfill operations. The two motor graders, one front-end loader, and the backhoe are used as required by a single operator. Operators are assigned full-time to the other pieces of equipment. Two 12.23-cubic meter (16-cubic yard) end-dump trucks are used to haul dewatered sludge to the landfill.

Problems

This landfill has been operated in the same manner for many years, so the staff is well trained. However, four basic problems affect operations.

(1) *Problem:* Surface water/drainage slows operations after a rainfall. *Solution:* This problem plagues any landfill operation. The only solution is to make sure there are good surface slopes and no places for water to pond.

(2) *Problem:* An odor problem occurs during and after wet weather. *Solution:* "Rewetting" sludge causes an odor. Intermediate cover operations must be kept up and the final cover placed as soon as possible so that as little sludge/soil mixture as possible is exposed to rain. Masking agents are also used. If the odors become very prevalent, liquid lime is sprayed on the surface.

(3) *Problem:* Because the sludge is raw and chemically conditioned, "sink-holes" sometimes occur on the working face or final cover. *Solution:* Better compaction improves this situation. Frequent compaction tests assure good compaction. Placing the final cover as soon as possible is the best solution, along with keeping heavy vehicles off the final cover.

(4) *Problem:* Equipment breakdowns hamper landfill operations. *Solution:* This problem occurs in any landfill because of the severe duty. A rigid protective maintenance program and good operator training are necessary to reduce problems. Backup equipment is helpful, but expensive.

Monitoring

Monitoring is required for operations control and to satisfy permit requirements. The monitoring wells and piezometers are sampled monthly, a schedule similar to the requirements set forth for Dallas Water Utilities. The only differences are that total dissolved solids (TDS) are not required and PCBs must be analyzed annually.

Moisture-density testing using a nuclear density gauge is conducted for every 1070.44 cubic meters (1400 cubic yards) of compacted fill—about five tests per week. The goal is to achieve greater than 90% of maximum dry density as determined by the standard proctor test. Laboratory moisture-density relationship is performed at least monthly or when mixing soils change significantly. The site is surveyed and cross-sectioned annually to determine progress and as a check for consumption of soil and landfill space.

Gas vents require monitoring in a manner similar to that set forth for Dallas Water Utilities. One aspect of monitoring is simply observation of operations. Any difference observed must be logged and reviewed by the staff for potential problems.

Costs

The slurry cutoff wall, barrier wall, excavation for the west landfill, leachate pond, concrete pad for sludge storage during wet weather, and roads were constructed at a cost of $3,500,000 in 1989. Operations and maintenance costs, which are closely monitored by the Trinity River Authority, run $16.53 per dry metric ton ($15 per dry ton), including all labor and equipment for landfill operations not including sludge hauling. All equipment is depreciated over a ten-year period.

CASE STUDY C—DALLAS WATER UTILITIES' SOUTHSIDE WASTEWATER TREATMENT PLANT—LANDFILLING AND DEDICATED LAND DISPOSAL

Background and History

The Southside Wastewater Treatment Plant is located in the southeastern part of the city of Dallas and Dallas County. This 3.94 cubic meter per second (90 MGD) WWTP currently treats about 2.85 cubic meters per second (65 MGD) average annual flow. The 6.57 cubic meter per second (150 MGD) Central WWTP is the other city of Dallas's WWTP. Sludges from both plants are dewatered and disposed of at the 930.8-hectare (2300-acre) Southside WWTP site. Present disposal methods include an on-site DLD site and a sludge-only landfill, with the landfill taking about 80% of the more than 113.4 dry metric tons (125 dry tons) of sludge per day. Composting and agricultural land application being investigated may be implemented in five to ten years. Dedicated land disposal and sludge-only landfilling provide sufficient sludge disposal capability for the next fifteen or more years at projected growth rates.

The Central WWTP includes screening, grit removal, primary sedimentation, rock trickling filters, secondary sedimentation, medium-bubble activated sludge nitrification, final sedimentation, deep mono-media coal filtration, and chlorination/dechlorination using SO_2. Secondary sludge is returned to the primary clarifiers for removal. The resulting primary sludge is partially digested in existing digesters and pumped to a holding tank, where it is mixed with the waste-activated sludge. After dilution to 0.8 to 1.0% solids, the sludge is pumped 20.9 kilometers (13 miles) to an aerated storage basin at the Southside WWTP.

A relatively new facility, the Southside WWTP includes screening, grit removal, primary sedimentation, mechanically aerated activated sludge, final sedimentation, dual-media filtration, and chlorination/dechlorination using SO_2. Primary sludge is mixed with solid-bowl centrifugally thickened WAS, and then the mixture is anaerobically digested. The digested sludge is pumped to a second aerated storage basin.

Central- and Southside-generated sludges are kept separate to satisfy permit requirements. The digested/stabilized Southside sludge can be applied to the DLD fields or disposed of in the sludge-only landfill, whereas the raw/partially digested Central sludge can only be landfilled. From each 169,920 cubic meter (6 million gallon) aerated storage basin, sludge is pumped to its respective aerated holding tank, which serves as a wet well for the belt filter press feed pumps. Ten two-meter belt filter presses, which operate continuously, have exceeded design expectations. The 18 to 20% dewatered cake solids are either pumped 396.2 meters (1300 feet) to the landfill mixing facility or to dewatered sludge spreading vehicles at an adjacent loading area.

The dewatering facilities are operated by the Dallas Water Utilities. Dedicated land disposal has been used for the past twelve years, the dewatering facilities since late 1989. Since startup in early 1990, the landfill facilities have been operated by a private contractor.

Site Description

The 73.3-hectare (181-acre) sludge-only landfill, which is wholly within the Southside WWTP site, is bounded on the north by the on-site dedicated land disposal facility, on the south and east by gravel quarries, and on the west by a flood protection levee on the Trinity River. The nearest residential area is an unincorporated area 457.2 meters (1500 feet) from the landfill boundary. A 6.10- to 12.2-meter (20- to 40-foot) layer of alluvial material, primarily silty and clayey sands, overlays the Taylor Marl formation. This material becomes more granular with depth so that at the interface with the Taylor Marl, it becomes coarse sand and gravel. The landfill site, an old gravel quarry, has been excavated below the groundwater table. Relatively flat across the site and 15.24 to 30.48 meters (50 to 100 feet) thick, the Taylor Marl is a dense dark clay shale with a permeability of less than 1 × 10^{-7} cm/sec. Underlying it is 182.88 meters (600 feet) of a light gray limestone imbedded with a clay formation Austin Chalk, below which lies a 91.44-meter (300-foot) layer of Eagle Ford shale. Although the groundwater is within 3.05 meters (10 feet) of the surface, this perched water is not used for drinking water. The state considers the Taylor Marl an effective horizontal barrier to vertical seepage.

The present site has three DLD fields with a total of 129.5 hectares (320 acres). Each field drains to a small pond. The ponds are connected and any excess water is pumped to the head of the treatment plant. The city acquired additional land immediately north of the site and is adding 66.78 hectares (165 acres) of fields. The new fields do not have hydrants, so only dewatered sludge application is possible. In the future, another field of 16.19 hectares (40 acres) will be leveled and prepared for spreading operations. The site plan is shown on Figure 7.5.

Design

The design development consisted of selecting the best location for the landfill. All potential sites were in the southern area of the site because of existing or planned expansions in other areas. The selected site was preferred for these reasons:

• farthest from the public
• closest to dewatering facilities
• ample cover material available at short distances
• levee construction material available at short distances
• location consistent with current Special Use Permits for the site
• least effect on existing DLD fields and stormwater detention basins

The design considerations were primarily surface- and groundwater-related. Another goal was to obtain the maximum area without limiting future growth. The design criteria were:

• 20% solids (18 to 25% range)
• 90.72 dry metric tons (100 dry tons) per day
• 1:1 up to 1:3 wet volume sludge/soil bulking ratio
• 949.2 kilograms per cubic meter (1600 pounds per cubic yard) compacted density

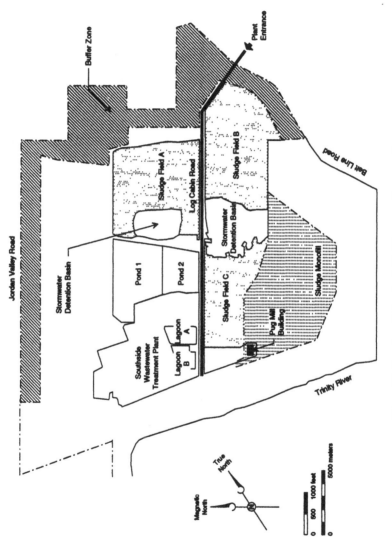

Figure 7.5 Site plan—Dallas Southside Wastewater Treatment Plant.

To achieve the highest sludge to soil ratio and thus the longest landfill life, an efficient pug mill mining facility was selected over construction vehicles.

Conventional construction vehicles for mixing sludge and soil can rapidly use up available landfill volume because it is easier to mix and place a material high in soil content. With a premixed material, however, construction vehicles only place the solids into the fill. The mixing system is shown schematically in Figure 7.6. A key element for success is the quality of the material to be mixed with the sludge. The soil available at the site is a fine "sugar" sand, with some silt and clay. Test work in the design showed that a 1:1 wet volume mixture of sludge and soil would be possible, but a 1:3 ratio was used as a worst case because of prior test work in Fort Worth. No full-scale compaction tests were done, but they are recommended because conventional soil compaction testing gives questionable results with a wet, sticky sludge.

The landfill area itself included new levees with impermeable clay cores and a 0.91-meter (3-foot) thick slurry cutoff wall keyed into the Taylor Marl. This design creates a bathtub that prevents any contamination from escaping the landfill site. In addition, because of the on-site dedicated land disposal, Dallas Water Utilities constructed another slurry cutoff wall around its 930.8-hectare (2300-acre) site. Although a comprehensive groundwater and soil monitoring program showed no groundwater contamination and minimal soil contamination due to DLD operations, a slurry cutoff wall would limit any potential contamination to the plant site. This second cutoff wall could be considered a "secondary containment" for the landfill. Groundwater monitoring wells are located between the cutoff walls and outside the plant perimeter cutoff wall.

The landfill site was graded to control surface water by directing it to a stormwater detention/evaporation pond. If the level in this pond becomes too high, the excess water is pumped to the treatment plant.

The soils within the confines of the levees/cutoff walls would be used as the bulking material. Calculations show that the existing material is not sufficient for the anticipated seventeen-to twenty-year landfill life, so borrow, which is available on the treatment plant site, will be needed in later years. One section of the landfill was excavated into the Taylor Marl for initial landfilling. The excess Taylor Marl removed will serve as intermediate and final cover, although a foam intermediate cover is desirable, but very costly, to extend the landfill life.

A complete passive-type methane gas control system was designed for the landfill site. Perforated pipe surrounded by gravel and geotextile fabric will be placed at mid-depth in the fill and close to the top of the fill, with the pipes at 15.24-meter (50-foot) centers. Risers at the site perimeter require monitoring and permit a conversion to an active system.

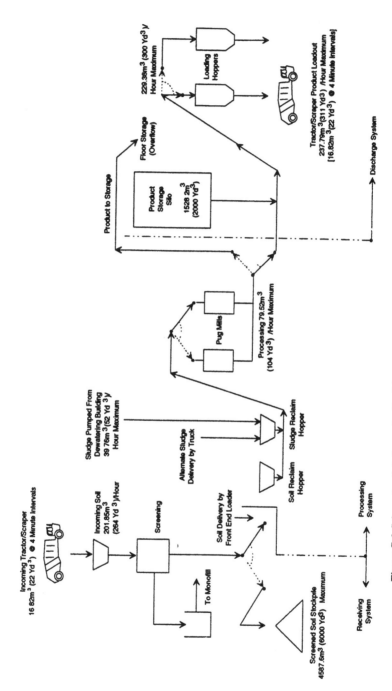

Figure 7.6 Schematic of sludge and soil mixing system—Dallas Southside Wastewater Treatment Plant.

Incoming Tractor/Scraper
16.82m³ (22 Yd³) @ 4 Minute Intervals

Incoming Soil
201.85m³
(264 Yd³)/Hour

Screening

To Monofill

Screened Soil Stockpile
4587.6m³ (6000 Yd³) Maximum

Receiving
System

Soil Delivery by
Front End Loader

Soil Reclaim
Hopper

Sludge Reclaim
Hopper

Alternate Sludge
Delivery by Truck

Sludge Pumped From
Dewatering Building
39.76m³ (52 Yd³) y
Hour Maximum

Pug Mills

Processing 79.52m³
(104 Yd³) /Hour Maximum

Processing
System

Product to Storage

Floor Storage
(Overflow)

Product
Storage
Silo
1528.2m³
(2000 Yd³)

229.38m³ (300 Yd³) y
Hour Maximum

Loading
Hoppers

Tractor/Scraper Product Loadout
237.79m³ (311 Yd³) /Hour Maximum
[16.82m³ (22 Yd³) @ 4 Minute Intervals]

Discharge System

430

Operation

The landfill is designed to operate during daylight hours only, normally eight hours per day, five days per week. This schedule permits longer work days and weekend work needed after rainy periods or periods of high sludge flow.

Site Facilities

Office facilities for landfill operations are in the pug mill mixing facility. Roads and ramps to the pug mill facility from the fill site are topped with Taylor Marl, which provides an excellent road bed, but it becomes very slippery in rainy periods. Because fill operations are suspended in wet weather, this has not been a problem. To prevent soil erosion on the levee slopes, natural grasses are maintained and watered, if needed, from the on-site detention pond. This water is also used for dust control.

Sludge/Soil Mixing

Sludge pumped to the pug mill facilities is mixed with screened, dry sand at a ratio of 1:3 sludge to soil. The mixture is loaded into 11.47-cubic meter (15-cubic yard) end-dump tractor/trailers and hauled to the fill site. Front-end loaders load field-screened sand into trucks that haul it to a large undercover storage area in the pug mill building. This covered area is required to keep the sand below 10% moisture, preferably below 5%. The design assumed the use of paddle wheel-type scrapers to haul sand to the facility and return the sludge/soil mixture to the fill site. The private contractor uses end-dump trucks.

The pug mill mixing facility requires very dry sand to work efficiently, leaving much unusable reject material, which the contractor mixes with sludge in the field at about a 1:3 sludge to soil ratio. Windrows are created near the fill face and a motor grader turns over the material, drying and mixing it. The mixture is pushed to the fill face or hauled, depending upon the proximity of windrows to the working face.

Operational Procedures

As a part of the landfill construction, an area was excavated for the initial mixture placement. The sequence of operation is as follows:

- Excavate area for future filling—includes clearing and grubbing, removal of bulking material, and removal of some Taylor Marl

for daily cover, road base, final cover and stockpile of materials.
- Provide for stormwater drainage.
- Place sludge/soil mixture using bulldozers. Use three 203.2- to 304.8-mm (8- to 12-inch) loose lifts approximately 15.24 by 24.38 meters (50 by 80 feet) per day.
- Compact the fill with "sheeps foot" rollers or bulldozers.
- Place daily cover.
- As fill nears top of levee, place 0.91-meter (3-foot) thick compacted layer of Taylor Marl and cover with 304.8 mm (one foot) of topsoil.

Figure 7.7 shows a typical cross section of the active face. Excess water is drained from the surface of the Taylor Marl using a small working face detention pond and a portable pump. The portable pump pumps into a force main, which is extended with the working face. The working face is extended eastward along the southern half of the site toward the detention pond. Upon completion of the southern half, work begins in the northern half from the detention pond westward toward the pug mill mixing facility. During wet weather, most working face operations stop because the material is very slippery. Also, as the moisture content of the mixture increases, the stability of the fill is reduced. At this time, sludge and soil can still be mixed and the mixture stored undercover. There is only enough room for one day's production of the mixture. A covered area adjacent to the dewatering facilities can store dewatered sludge for one to two days. Liquid sludge can be stored for up to five days. As a last resort, sludge can be stored within a small, confined area of the fill adjacent to the pug mill facility. If this area is used, the sludge must be field dried before use.

At the completion of filling operations and placement of the final cover and topsoil, the site will have a mound in the center of the fill providing clean, uncontaminated drainage across the surface. A large stockpile of Taylor Marl in the center will be used to repair surface failures. The permit states that the land use will be open space, but options for more dedicated land disposal fields are possible.

The DLD operations were originally injection with liquid hydrants, one for every 4.05 hectares (10 acres) in each of the three fields, and provided connections for the hoses to the injection vehicles. One to five percent solids were dredged from lagoons and pumped to a high-pressure [689.5 kilopascals (100 psi)] pump station, which fed the hydrants.

Since the operation of the dewatering facilities began in 1989, the two sludge lagoons are only used for emergency backup, and dewatered sludge spreading to the DLD fields is used almost exclusively. Dewatered sludge

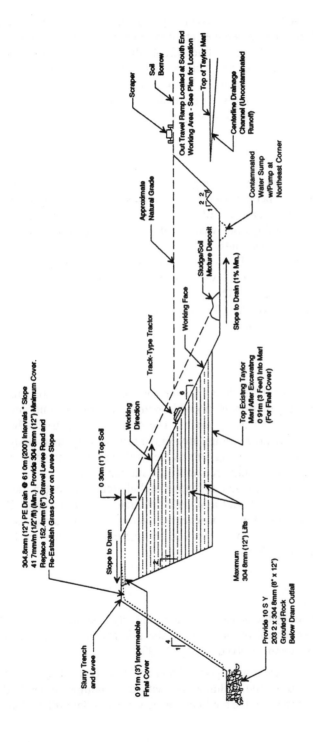

304.8mm (12") PE Drain @ 61.0m (200') Intervals * Slope
41.7mm/m (1/2"/ft) (Min.). Provide 304.8mm (12") Minimum Cover.
Replace 152.4mm (6") Gravel Levee Road and
Re-Establish Grass Cover on Levee Slope

Scraper

Soil
Borrow

Out Travel Ramp Located at South End
Working Area - See Plan for Location

Top of Taylor Marl

Centerline Drainage
Channel (Uncontaminated
Runoff)

Approximate
Natural Grade

Working
Direction

Track-Type Tractor

Working Face

Sludge/Soil
Mixture Deposit

Slope to Drain (1% Min.)

Contaminated
Water Sump
w/Pump at
Northeast Corner

0.30m (1") Top Soil

Top Existing Taylor
Marl After Excavating
0.91m (3 Feet) Into Marl
(For Final Cover)

Slope to Drain

Maximum
304.8mm (12") Lifts

* Provide Drainage Ditches Parallel to
Levee Road with Minimum 20.8mm/m (1/4"/Foot)
Slope to Drain Toward 304.8mm (12") Drain Line

Slurry Trench
and Levee

0.91m (3') Impermeable
Final Cover

Provide 10 S.Y.
203.2 x 304.8mm (8" x 12")
Grouted Rock
Below Drain Outfall

Figure 7.7 Working face containment area typical section—Dallas Southside Wastewater Treatment Plant.

433

is pumped to 9.18-cubic meter (12-cubic yard) spreading vehicles at the dewatering facilities. These vehicles are driven to the fields and sludge is applied about 12.7 mm (1/2-inch) thick [22.42 dry metric tons per hectare (10 dry tons per acre)]. After application to one of the three fields, a disc harrow is pulled by a dual rear-wheeled tractor over the fields to mix the sludge and soil. The field is plowed in one direction and then cross plowed. Normally this is repeated the following day.

Staffing

A private contractor operates the pug mill facility and the sludge-only landfill. Nine full-time people operate the complete facilities, plus several contract truckers. This number will shortly increase because the landfill is being operated five days per week, and a longer operating period is needed to meet current demands.

The city operates the DLD system. A total staff of fifteen is assigned to this operation, which is run seven days per week, normally two shifts per day during good weather. This crew is assigned to other duties in the plant during inclement weather. The DLD operation is the responsibility of the solids area supervisor.

Equipment

The private contractor has four front-end loaders, one backhoe, three bulldozers, a 0.15-meter (6-inch) portable pump, one tractor, one dump truck, and several contract dump trucks. The contract dump trucks haul sludge from the wet weather storage area to the field mixing area; sludge/soil mixture from the field mixing area to the working face; and soil from the site to the pug mill mixing area and field mixing area.

The vehicles used for sludge application include four 9.18-cubic meter (12-cubic yard) dewatered sludge application vehicles, one 736.6-mm (29-inch) yard spreader trailer, four dual-wheeled tractors, one disc harrow with 12.19-meter (40-foot) wingspread (pulled by tractors), and four tracked vehicles with injection plows.

Problems

Operations begun by the private contractor in early 1990 have improved rapidly. Initial mechanical problems were the primary difficulties, all of which should have been resolved by April 1991.

(1) *Problem:* Sludge cake pump poor reliability. The only way to get dewatered sludge out of the building was with three pumps. When both pumps were down, sludge could not be dewatered or landfilled. *Solu-*

tion: After months of rework, the pump manufacturer has significantly improved reliability. A conveyor for truck loading will also be added.

(2) *Problem:* High moisture content and clay/silt pockets in the soil plugged the pug mill facility. *Solution:* Soil is field screened to remove clay/silt lumps. Screened soil is field dried, if needed, and stored under-cover to reduce moisture content about 5%.

(3) *Problem:* Many sludge/soil mixtures clogged the pug mill equipment, such as conveyors and chutes. The equipment manufacturer went bank-rupt. *Solution:* Some redesign of chutes, screened soil, a 1:2 to 1:3 sludge to soil ratio, and better operations seem to have resulted in reli-able operation.

(4) *Problem:* Dewatered sludge from cake pumps is much stickier and more difficult to mix than anticipated. *Solution:* Drier soil and redesign of some of the pug mill facilities (to be implemented mid 1991) appear to solve the problem.

(5) *Problem:* In the DLD operations, injection maintenance becomes ex-cessive with the clayey, sandy soils. *Solution:* Changed from liquid to dewatered sludge injection.

(6) *Problem:* The DLD fields are poorly drained and liquid injection makes the fields even wetter. Usable time was down to 120 days per year. *Solution:* Use of dewatered sludge spreading reduced the water being applied to the fields. The slurry cutoff wall with dewatering should reduce the water table sufficiently to allow more than 200 days of opera-tion annually.

(7) *Problem:* The spreader trucks track mud all over plant roads when returning to the dewatering facilities. *Solution:* A wash area in each field will remove the mud.

Monitoring

Monitoring is conducted for two reasons: operations control and permit requirements. The permit requires groundwater and sludge analyses. Twelve groundwater monitoring wells surround the site. In addition, one well is inside the site, with four piezometers for measuring groundwater inside the landfill levees. The wells are analyzed quarterly and the analyses required are presented in Table 7.1.

Sludge must be monitored for the same constituents listed in Table 7.1 except for water level and TDS. PCBs must be analyzed annually. Monitoring for landfill operations control includes sludge/soil mixture preparation. To determine stability in the fill site and to assist in setting mixture proportions, the tests shown in Table 7.2 should be performed.

TABLE 7.1. Sludge-Only Landfill and DLD Site Monitoring Requirements.

Parameter	Sludge Units	Sludge Sampling Frequency	Groundwater[a] Units	Groundwater[a] Sampling Frequency	Soil[b] Units	Soil[b] Sampling Frequency
Total nitrogen	mg/kg	Monthly	mg/L	Quarterly	mg/kg	Quarterly
Nitrate nitrogen	mg/kg	Monthly	mg/L	Quarterly	mg/kg	Quarterly
Ammonia nitrogen	mg/kg	Monthly	mg/L	Quarterly	mg/kg	Quarterly
Phosphorus	mg/kg	Quarterly	mg/L	Quarterly	mg/kg	Two/month
Potassium	mg/kg	Quarterly	mg/L	Quarterly	mg/kg	Two/month
Cadmium	mg/kg	Quarterly	mg/L	Quarterly	mg/kg	Two/month
Lead	mg/kg	Quarterly	mg/L	Quarterly	mg/kg	Two/month
Zinc	mg/kg	Quarterly	mg/L	Quarterly	mg/kg	Two/month
Copper	mg/kg	Quarterly	mg/L	Quarterly	mg/kg	Two/month
Nickel	mg/kg	Quarterly	mg/L	Quarterly	mg/kg	Two/month
pH	Std	Monthly	Std	Quarterly	Std	Quarterly
Polychlorinated biphenyls	mg/kg	Annually	mg/L	Annually	mg/kg	Annually
Water level	NA	—	feet	Quarterly	NA	—
Cation exchange capacity	NA	—	NA	—	meq/100g	Quarterly

[a]One well per 50 acres of DLD fields
[b]One sample at 6-inch, 18-inch, and 30-inch for each 20 acres of DLD fields.
Values are from NPDES permit.

436

TABLE 7.2.   Landfill Operation Control Tests.

Test	Material	Where Tested
Atterberg limits	Soil	Lab
Standard proctor density	Mixture	Lab
Nuclear density	Mixture	Field
California bearing ratio	Mixture	Field
Moisture content	All	Field

Each gas collection riser must be monitored monthly after the final cover is placed. A portable combustible gas detector is used. With identification of final cover cracking or high gas levels, monitoring must be increased to weekly. An active gas collection system with fans may also be required.

Achieving the maximum operating life of the sludge-only landfill depends upon knowing the volumes of various soil materials used. Therefore, quarterly land surveys are required. Volume calculations are made using profiles and 30.48-meter (100-foot) cross sections. By comparing sludge volume and fill volume, a ratio can be back-calculated and remaining life determined. Quarterly aerial photographs also provide a history of operation.

For the DLD operation, a record of how much sludge is applied to each field is required. There are glandular monitoring wells, one for every 20.24 hectares (50 acres) of DLD area, and each requires the same level of sampling as the landfill monitoring wells. The soil in each field must also be sampled. One sample is required for each 8.09 hectares (20 acres) of DLD field at depths of 152.4, 459.2, and 762 mm (6, 18, and 30 inches). The required laboratory tests and frequencies are shown in Table 7.1.

Costs

As with any large facility, costs are difficult to determine. The construction costs were $10,500,000. Assuming a total sludge disposal capacity of 567,907.2 dry metric tons (626,000 dry tons), the capital cost is $18.49 per dry metric ton ($16.80 per dry ton). This cost increases if the fill volume is reduced because a greater soil to sludge ratio is required. Because the first two years of operation were contracted out, the bid price per metric ton (ton) of dry solids can be used as the "real" operating and maintenance cost. At this time, it is unknown whether the contractor is making or losing money. This bid cost of $61.51 plus $18.49 ($55.80 plus $16.80) yields a total cost of $80.00 per dry metric ton ($72.60 per dry ton) of sludge, not including land acquisition and dewatering. Currently,

DLD operations and maintenance are about $71.65 per dry metric ton ($65.00 per dry ton), not including capital, land, and dewatering cost.

CASE STUDY D—DUBLIN SAN RAMON SERVICES DISTRICT— SLUDGE STORAGE AND DEDICATED LAND DISPOSAL FACILITIES

Background and History

In 1980 the Dublin San Ramon Wastewater Treatment Plant was being expanded from 0.26 to 0.53 cubic meters per second (6 to 12 MGD). Nearby neighbors had been experiencing odors from the plant's anaerobic sludge-drying lagoons. These lagoons, located within the plant site, were filled with well-digested sludge but were not being emptied regularly. Therefore, whenever their surfaces went through dry to wet conditions, serious odor problems resulted.

As part of the plant's expansion, a study of sludge disposal alternatives resulted in the subsequent construction of a DLD site for the plant sludges in an abandoned oxidation pond wastewater treatment plant site. This site had served as a treatment plant for a large military installation during World War II and as a consequence was still available for public use. In addition, this 188-hectare (465-acre) site was adjacent to one corner of the treatment plant site so that liquid sludge transport was relatively simple and inexpensive.

In 1964 the California Department of Water Resources and the U.S. Geological Survey published a report evaluating soil, geologic, and hydrologic conditions beneath all wastewater treatment and discharge sites in the Alameda Creek watershed above Niles. Because the dedicated land disposal (liquid sludge only) site functioned at the time as the wastewater treatment and disposal ponds for the Camp Parks Military Reservation, a complete subsurface evaluation was made of the site.

This evaluation indicated that there was a 12.19-meter (40-foot) thick upper aquiclude of gray-to-black, essentially impermeable plastic clay underlining the whole site. The evaluation concluded that the coefficient of permeability of this aquiclude was as low as 11.52 mm (0.37 inch) per year (0.3×10^{-7} centimeters per second). This indicated the site would provide good groundwater protection.

The fact that Camp Parks Wastewater Treatment Plant used oxidation ponds for final treatment meant that the entire site was surrounded by dikes made of similar impermeable material. This meant that the surface water runoff could be fully retained as required and then returned to the treatment plant via local sewers for disposal.

The environmental impact report for the siting of the DLD site was accepted and the facility was constructed in 1983 and 1984. The first anaerobically stabilized sludge was discharged to the facultative sludge storage lagoons in 1985. The first facultative sludge storage lagoon harvesting operation took place in 1989, was continued in 1990, and will be finished in 1991.

Design

DLD sites which use liquid sludge require dry weather for maximum application rates. To achieve this condition in the Dublin San Ramon area of central California, it is necessary during normal years to limit the application period to the dry late spring, summer, and early fall months. This period of time usually falls between May 15th and October 15th of each year. To limit sludge disposal to these six months, it is necessary to include an environmentally acceptable liquid storage capability between the daily production of stabilized sludge from the digesters and the summer disposal operation.

The Dublin San Ramon plant was provided with four facultative sludge lagoons to provide this storage. Each lagoon has a surface area of approximately 1.62 hectares (4 acres), is 4.57 meters (15 feet) in depth, and is equipped with two surface mixers. These lagoons are designed to accept dry volatile solids loadings up to one metric ton per hectare (20 pounds per 1000 per square feet) per day and still maintain an aerobic surface layer capable of supporting an active growth of algae. The surface mixers help to dissipate the scum surface buildup and maintain the surface layers in an aerobic condition. Figure 7.8 provides a profile of the zones of facultative sludge lagoons.

The bottom anaerobic layer of these lagoons provides up to an additional 45% destruction of volatile solids during the first year of storage of the solids. Design anticipated that the 4.57-meter (15-foot) depth of the lagoons could store up to four years of solids production. Thus they would allow the liquid sludge disposal system to be operated to clean one lagoon each year or all lagoons every four years.

Harvesting of the stored sludge was designed to take place by dredge. The lagoons are designed to never be dewatered. The aerobic surface layer is to be maintained during the entire harvesting operation. Each lagoon is provided with a harvesting pipe hookup designed to allow the dredge to cover the entire lagoon without changing its discharge connection. Each lagoon is provided with cable anchors at each end designed to guide the dredge as it works its way back and forth across the surface of the lagoon. During harvesting the lagoon is taken out of storage service, although its liquid level is maintained by adding the supernatant from the other lagoons. Figure 7.9 shows a profile of a portable dredge removing sludge from the bottom of a lagoon.

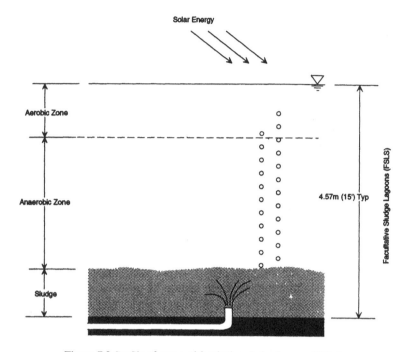

Figure 7.8 Profile of zones of facultative sludge lagoons (FSLs).

The net 22.26-hectare (55-acre) DLD site is divided by the harvesting pipe system into four areas. Ditches divide the areas and a central storm drainage retention basin bisects the entire site. Harvesting sludge hydrants are strategically located in each area to assure complete access of the umbilical hose connected tractor. The Central storm drainage retention basin is provided with a connection to a 1066.8-mm (42-inch) sewer that runs directly to the treatment plant. Figures 7.10 and 7.11 provide a flow diagram and a site layout of the entire Dublin San Ramon sludge disposal system. Table 7.3 provides the design data.

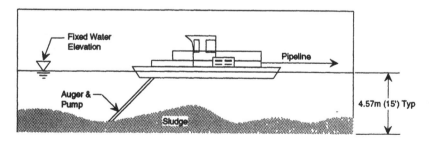

Figure 7.9 Profile of portable dredge removing sludge from an FSL.

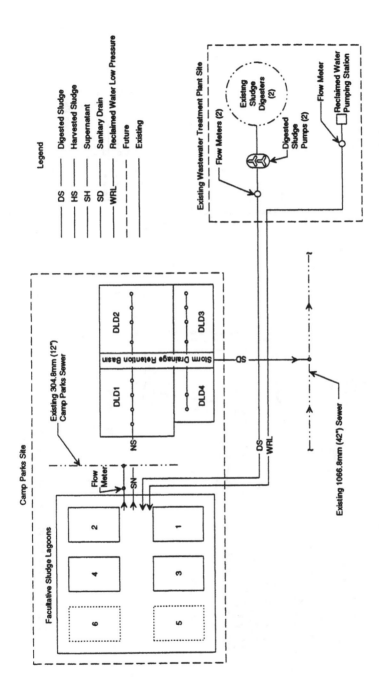

Figure 7.10 Solids treatment and disposal facilities flow diagram—Dublin San Ramon Services District.

441

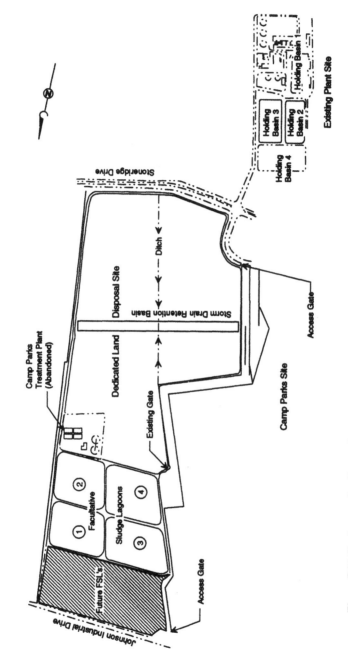

Figure 7.11 Site layout—San Ramon Services District—wastewater treatment plant and sludge treatment and disposal facilities.

TABLE 7.3. Design Data.

Description	Existing	Future
Plant flow, cubic meters per second (million gallons per day)	0.39 (9)	0.79 (18)
Sludge production:		
Average annual TS, kg/day (lb/day)	4854 (10,700)	9707 (21,400)
Average annual VS, kg/day (lb/day)	3429 (7560)	6858 (15,120)
Average annual flow:		
Cubic meters per day (gal/day)	324 (85,530)	647 (171,060)
Liters per second (gal/min), 24 hour pumping	3.72 (59)	7.51 (119)
Liters per second (gal/min), 10-1/2 hour pumping	8.58 (136)	17.16 (272)
Peak month TS, kg/day (lb/day)	7122 (15,700)	14,243 (31,400)
Peak month VS, kg/day (lb/day)	4749 (10,470)	9498 (20,940)
Peak month flow:		
Cubic meters per day (gal/day)	475 (125,500)	950 (251,000)
Liters per second (gal/min), 24 hour pumping	5.49 (87)	10.98 (174)
Liters per second (gal/min), 10-1/2 hour pumping	12.55 (199)	25.11 (398)
Average annual TS to DLD, kg/day (lb/day)	3384 (7460)	6768 (14,920)
Average annual VS to DLD, kg/day (lb/day)	1950 (4300)	3901 (8600)

(continued)

443

TABLE 7.3. (continued).

Equipment Parameters	Values
Digested sludge pumps:	
Number	2
Type	Rotary lobe
Power, watts (horsepower)	5593 (7-1/2)
Capacity, liter/sec (gpm)	12.6 (200)
Facultative sludge lagoons (FSLs):	
Number	4 to 6
Water surface area, hectares/lagoon (acres/lagoon)	1.62 (4.0)
Loading rate, metric tons/hectare (lb/1000 sq ft)	1 (20)
Water depth, meters (feet)	4.57 (15)
Levees:	
Top width, meters (feet)	7.62 (25)
Slopes, interior and exterior	3 to 1

TABLE 7.3. (continued).

Equipment Parameters	Values
Roadway width, meters (feet)	4.57 (15)
Roadway surfacing:	
Thickness, mm (inches)	152 to 304 (6 to 12)
Type	Crushed rock
Slope protection:	
Thickness, mm (inches)	152 (6)
Type	Rock rip rap
Freeboard, meters (feet)	0.19 (3)
Levee foundation key, location	Exterior levees only
Surface mixers:	
Number	8
Power, watts (horsepower)	3728.5 (5)
Dedicated land disposal (DLD) site:	
Net area for sludge injection, hectares (acres)	22 3 (55)
Maximum loading rate, dry metric tons/hectare (dry tons/acre)	244 (100)
Solids concentration, percent	4 to 6
Harvesting rate, liters/sec (gpm)	31.5 to 44.2 (500 to 700)

Operation

The Dublin San Ramon Services District Wastewater Treatment Plant is presently operating at about 0.35 cubic meter per second (8 MGD). The operation of the facultative storage lagoons (FSLs) and the dedicated land disposal site at Dublin San Ramon have been very successful. The lagoons have maintained an aerobic top layer with green algae ever since startup. On one occasion the lagoons did lose some of their healthy appearance; however, the rapid determination that the electrical conductivity (specific conductance) of the surface layer must be maintained at approximately 2000 microohms $\cdot$ cm^{-1} to assure a healthy green algae top layer quickly restored this important parameter.

At the present time the plant staff is flushing the surface layer of each lagoon with the equivalent of 378.5 cubic meters (100,000 gallons) of effluent each day to maintain this specific conductance. This is done on a four-day cycle so that the operators do not have to manipulate numerous valves each day. This means that each FSL receives 1514 cubic meters (400,000 gallons) of effluent every four days. This effluent is added approximately 0.91 to 1.22 meters (3 to 4 feet) below the surface of each FSL at the end opposite the lagoon's overflow. FSL overflow is returned to the treatment plant. Earlier work at Sacramento indicated that the sludge loading of these lagoons can also be done intermittently on up to a four-day cycle.

Costs

In 1989 the plant staff took bids on the harvesting of the sludge accumulation from the four facultative sludge lagoons. The bids ranged from a low bid of $230,000 to a high bid of $360,000. This low bid was based on a unit price of $3.45 per cubic meter ($2.64 per cubic yard). Sludge volume is determined by measuring depth of the sludge in the lagoons before and after sludge harvesting and multiplying by the average lagoon surface area. Sludge depth is measured at a minimum of twenty locations randomly located over the full area of each lagoon. Measuring tools used are: (1) a "sludge judge," and (2) a PVC pipe with gauze taped along the full length of the pipe. Depth used is the average of the two measuring methods.

The harvesting schedule has been as follows:

Original contract: 1989—FSLs 1 and 3

FSL 1	Start work	August 4
	Stop work	September 27
	Volume sludge	33,335 cubic meters
		(44,906 cu yd)
	Compensation	$118,552

FSL 3	Start work	September 27
	Stop work	October 24
	Volume sludge	10,420 cubic meters
		(13,628 cu yd)
	Compensation	$35,978

Contract CO 1: December 1989—Includes FSLs 2 and 4 and reduces price to $3.32 per cubic meter ($2.54/cu yd)

FSL 3	Start work	May 1
	Stop work	August 15
	Volume sludge	42,090 cubic meters
		(55,048 cu yd)
	Compensation	$139,822
FSL 4	Start work	August 15
	Stop work	October 9
	Volume sludge	47,456 cubic meters
		(62,067 cu yd)
	Compensation	$157,650

Contract CO 2: January 1991—Extend contract to September 30, 1991

FSL 2	Start work (approx)	June 1
	Stop work (approx)	September 30
	Volume sludge (est)	52,751 cubic meters
		(68,991 cu yd)
	Compensation (est)	$175,237

The estimated sludge removed and applied to the DLD sites over the past three years is 187,052 cubic meters (244,640 cubic yards). Indications are that the sludge concentration over this period of time ranged between 6 and 9% solids. Total costs for this harvesting are expected to be about $629,240. Total harvesting time is expected to be about eleven months. A great deal of difficulty was experienced in harvesting FSL 3. Apparently during its early loading it received bottom sludge from the old drying lagoons. This contained a large number of rocks, which made its removal from the bottom of FSL 3 very time-consuming.

Harvesting operations have consisted of removing between 63.09 and 75.71 liters per second (1000 and 1200 gallons per minute) of bottom sludge through the harvesting pipeline and umbilical hose to the direct injection equipment on a Caterpillar tractor. The disposal site was plowed by the plant staff to a depth of 457.2 mm (18 inches) prior to the start of the harvesting operation. This plowing of the approximately 24.28 hectares (60 acres) cost an additional $5000. In August of 1989 the harvesting operations resulted in about a six-day cycle between sludge loadings to the soil.

Monitoring

The DLD site is immediately bordered on the east by white collar offices and on the southeast by single family homes. Figure 7.12 provides an aerial view showing this close-by development. The facultative sludge lagoons lie within 152.4 meters (500 feet) of a large hotel. It is obvious that such neighbors make odors the chief concern of the storage and disposal system. The FSLs are monitored almost on a daily basis for pH, DO, and specific conductance. Special precautions are taken to assure that each basin is loaded with anaerobically digested sludge and flushed with plant effluent equally. To date both the storage and disposal operations have been carried out with no odor complaints from any of the neighbors.

At the present time the plant is conducting new soil analysis to assure the state of California that the dedicated land disposal site is completely sealed off from the local groundwater. Monitoring wells to date have indicated no contamination.

CASE STUDY E—CITY OF COLORADO SPRINGS' HANNAH RANCH SOLIDS HANDLING AND DISPOSAL FACILITIES—DEDICATED LAND DISPOSAL

Background and History

In 1978 the city of Colorado Springs implemented a long-range sludge management study. At the time their sludge processing and disposal system at the existing Colorado Springs Wastewater Treatment Plant within the city limits was plagued with problems associated with the stabilization and disposal of all sludges produced on-site. These operational problems also resulted in the failure of the liquid treatment process to consistently comply with the effluent quality requirements due to the return of high-strength waste from the sludge heat treatment process. This process was retired by the City in 1977 due to both its impact on the liquid treatment process and its huge odor impact on the surrounding neighbors.

The long range study was completed in 1980, making the following major recommendations:

(1) All sludge processing and disposal are to be moved from the waste-water treatment plant site to the Hannah Ranch some 32.2 kilometers (20 miles) south of the plant.

(2) Raw thickened sludge is to be transported by a dual pipeline system to the Hannah Ranch site.

(3) An anaerobic digestion facility is to be constructed at the Hannah Ranch site.

Figure 7.12 Aerial view of Dublin San Ramon Services sludge disposal system.

449

(4) Facultative sludge basins will be built near the anaerobic digesters for storing anaerobically digested solids.

(5) Land application facilities for disposal of the stabilized sludge by subsurface injection [dedicated land disposal (DLD) sites] are to be located around the facultative sludge basins.

(6) Supernatant disposal from the facultative storage basins will be retained in evaporation ponds.

The key element that made this such a practical solution to Colorado Springs' sludge treatment and disposal problems was the existence at the Hannah Ranch of a hazardous waste disposal site for fly ash from the City's power plant already accepted by all agencies. An added incentive was the lack of space at the wastewater treatment plant and the strong feeling by the plant's neighbors that the odor-producing solids treatment and disposal systems should be off-site.

The Hannah Ranch site includes some 2024 hectares (5000 acres) and was purchased in early 1970 by the City's Utilities Department. It is the home of the R. D. Nixon Power Plant. The ash disposal site is located in a section of the ranch that is underlined with a thick Pierre shale formation and is sealed off from surface water contamination by a large earthen dam. These conditions had been located and implemented during the environmental impact mitigation for the ash disposal site. The ultimate 121.4 hectares (300 acres) of sludge disposal area is located along the ridge top marking the northern edge for the ash disposal site and just south of this location in an area that was being used from 1978 through 1983 as the permanent disposal site for raw thickened and partially digested sludge from the wastewater treatment plant.

Design

As with other operations involving liquid sludge, a key part of the Colorado Springs design involves the use of storage basins. The facultative sludge basins (FSBs) at the Hannah Ranch Solids Handling and Disposal Facilities consist of six 2.02-hectare (5-acre) units 4.57 meters (15 feet) deep. Each FSB is equipped with a surface aerator, two digested sludge loading locations, and a harvested sludge valve pit, and is designed for a loading of 1 metric ton of volatile solids per hectare (20 pounds of volatile solids per 1000 square feet) per day. A dredge pumps the well-stabilized and thickened sludge from the bottom of the FSBs and discharges it through the harvested sludge valve pit connection to the wet wells of the harvested sludge pumping station. FSB supernatant is disposed of on-site in a 6.07-hectare (15-acre) evaporation lagoon.

The harvested sludge pumping station and FSBs 1 and 2 were built in

1981 and 1982, several years before the completion of the 32.2-kilometer (20-mile) dual raw sludge pipeline and the Hannah Ranch anaerobic digestion facilities. FSB 2 was placed in service in December 1981 storing partially digested sludge hauled by truck from the treatment plant. FSBs 1 and 2 were used for partially digested sludge storage and harvesting in 1982 and 1983. When the digestion facilities were placed in service in 1984, the rest of the six existing FSBs were also put in operation. Figure 7.13 provides a flow diagram for the solids handling and disposal facilities, and Table 7.4 provides the design data.

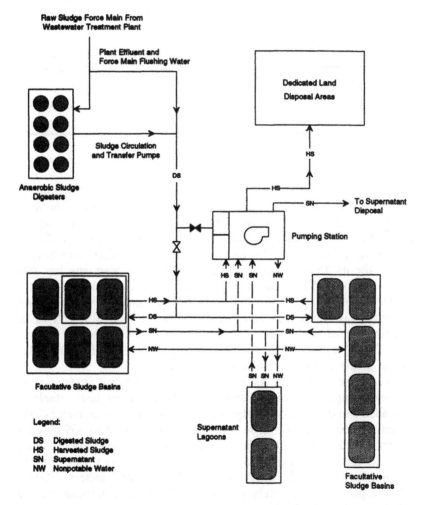

Figure 7.13 Flow diagram ultimate development—city of Colorado Springs—Hannah Ranch Solids Handling and Disposal Facilities.

TABLE 7.4. **Design Data.**

Description	Existing	Ultimate
Facultative sludge basins:		
Number	6	11
Surface area, hectares (acres)	12.1 (30)	22.3 (55)
Average depth, meters (feet)	4.57 (15)	4.57 (15)
Freeboard, meters (feet)		
Minimum	0.61 (2)	0.61 (2)
Average	0.91 (3)	0.91 (3)
Side slope, horizontal to vertical	3:1	3:1
Minimum dike width, meters (feet)	6 10 (20)	6 10 (20)
Digested sludge volume, liters/sec (gal/day)	10.19 (232,600)	18.10 (413,100)
Total solids, percent	2 06	2.06
Volatile solids, percent	64	64
Dedicated land disposal areas:		
Number	3	4
Total area, hectares (acres)	36.4 (90)	66.8 (165)
Assumed sludge data		
Total solids applied, percent	8	8
Application rate, dry metric tons/hectare/yr (dry tons/acre/yr)	193 (86)	193 (86)

The harvested sludge pumping station is designed to pump the harvested sludge to remote harvested sludge outlet stations to the north, south, east, and west dedicated land disposal sites. The north site has two outlets for its 20.2 hectares (50 acres), while the east site has three outlets for its 20.2 hectares (50 acres). The south side has two outlets for its 16.2 hectares (40 acres), while the west site has three outlets for its 16.2 hectares (40 acres). Each outlet is designed to service an injector tank truck. Flow from the pumping station through the outlets is continuous and circulates back to the wet well of the pumping station when the trucks are not being filled, thereby assuring rapid filling of the vehicle and the continuous mixing of the harvested sludge in the pumping station wet well.

Use of the four areas for injection of the harvested sludge assures maximum application rates, for it allows four injector tank vehicles to be loaded simultaneously. Design expected the 13.63-cubic meter (3600-gallon) injector vehicles to be capable of injecting up to 439.06 cubic meters (116,00 gallons) per day over 2.63 hectares (6.5 acres). Average injection rates were expected to be 1.89 cubic meters (500 gallons) per minute with each truck injecting 378.5 cubic meters (100,000 gallons) per day. Ultimately it is expected that five injector vehicles will be used at the DLD site.

Operation

In 1990 the Colorado Springs Wastewater Treatment Plant operated at an average flow of 1.42 cubic meters per second (32.5 MGD). Little harvesting from the six FSBs took place in the years between 1984 and 1986. In 1987 almost 63,345 cubic meters (17 million gallons) of 5.3% solids were harvested to the DLD sites. In 1988 this was reduced to just over 22,710 cubic meters (6 million gallons) of 5.7% solids. However, in 1989 this harvesting increased to 58,667 cubic meters (15.5 million gallons) of 5.6% solids, and in 1990 to 81,377 cubic meters (21.5 million gallons) of 5.2% solids.

Since 1978 the DLD sites have received over 1,109,000 cubic meters (293 million gallons) of raw, partially digested, and FSB harvested sludge at an average concentration of 5.2% solids.

The volume of raw sludge pumped to Hannah Ranch via the dual 32.2-kilometer (20-mile) pipeline in 1990 was 276,748 cubic meters (73,117,130 gallons). The average solids content of this sludge was 2.75% solids. Percent solids from the digesters to the FSBs ranged from 1.4 to 2.1%, with its volatile content ranging from 51.8 to 72.0%. The FSBs are loaded at about 0.5 to 0.75 metric ton of dry volatile solids per hectare (10 to 15 pounds of dry volatile solids per 1000 square feet) per day. In addition, 2239 cubic meters (591,600 gallons) of solids from publicly owned treatment works (POTW) waste sludge haulers was discharged into FSB 1.

In 1990, 81,378 cubic meters (21.5 million gallons) of harvested sludge was distributed to the four DLD areas. The north 20.23 hectares (50 acres) received 31,585 cubic meters (8,344,800 gallons), the east 20.23 hectares (50 acres) received 3870 cubic meters (1,022,400 gallons), the south 16.19 hectares (40 acres) received 18,450 cubic meters (4,874,400 gallons), and the west 16.19 hectares (40 acres) received 27,473 cubic meters (7,258,400 gallons). The yearly average percent solids was 5.2%, with a total of 4221 dry metric tons (4653 dry tons) injected. The average loading for 1990 was 57.83 dry metric tons per hectare (25.8 dry tons per acre). Only FSBs 1 and 6 were harvested, and harvesting took place between April and October.

Costs

Operating costs for the harvesting/injection disposal operation are currently running about $63.93 per dry metric ton ($58 per dry ton) of harvested sludge. These costs include all labor and normal O&M, salary fringe benefits, vehicle insurance, amortization costs, and other small miscellaneous items. The estimated capital cost to construct each FSB is approximately $500,000.

Monitoring

Due to the isolated nature of the entire Solids Handling and Disposal Facilities, little monitoring is required at the site. Laboratory analysis of the raw sludge, digested sludge, and FSB sludge indicated considerable concentration of nutrients and metals through the system. Only potassium shows any decrease in concentration in the FSBs. All the other nutrients show significant increases, with total phosphate almost tripling and total nitrogen almost doubling between the digested sludge and the harvested sludge content. Specific conductance in the Hannah Ranch FSBs is at the 6000 microohms · cm⁻¹ level.

All the metal levels either double, triple, or more between the raw sludge and FSB harvested sludge. Cadmium ranges from 5.1–6.0 to 21–26 mg/kg. Copper ranges from 330–395 to 840–980 mg/kg. Nickel ranges from 25–45 to 110–200 mg/kg. Lead ranges from 71–96 to 250–340 mg/kg. Zinc ranges from 630–880 to 2300–3100 mg/kg. With ample levels for manganese, ammonia, and phosphate, special care must be exercised to control the formation of Struvite.

CASE STUDY F—SACRAMENTO REGIONAL COUNTY SANITATION DISTRICT—SACRAMENTO REGIONAL WASTEWATER TREATMENT PLANT—DEDICATED LAND DISPOSAL

Background and History

The successful development of facultative sludge lagoon storage took place at the future home of this major wastewater treatment plant in the mid-1970s. This development is documented in an eleven-volume report by Sacramento Area Consultants published in 1979 under the title, "Sewage Sludge Management Program," and available through the NTIS [1] and in a paper entitled "Seven Years of Low-Cost Sludge Disposal at a 150-MGD Plant" [2]. The documents contained in these references provide a complete description of the background, history, and design of this sludge disposal system.

Operation

The 1990 operation of the Sacramento Regional Wastewater Treatment Plant sludge disposal system injected 13,970 dry metric tons (15,400 tons) of dry solids from their facultative sludge lagoons [solids storage basins (SSBs)] to their DLD sites. Their harvest operations for the year (April 16 through December 10) included 239 calendar days, 100 working days, and 1541 hours of harvested sludge pumping totalling 155,185 cubic meters

(41 million gallons). The average concentration of the harvested sludge was 8.7%, one-tenth of a percent above the concentration level achieved in 1989.

It is estimated the umbilical hoses were dragged 1969 kilometers (1224 miles) in 1990, with six new hoses being placed in service. Hoses currently cost approximately $6000 each and are replaced after being dragged about 402 kilometers (250 miles). A total of 2595 metric tons (2860 tons) of Dolomite 10 lime was applied to the DLD sites to maintain the soil pH at 7.0. This level of pH restricts the movement of metals through the soils.

The harvesting operation in 1990 involved four SSBs and all five DLDs. DLD 1 was loaded at the rate of 143 dry metric tons/hectare (64 dry tons/acre), DLD 2 at 132 dry metric tons/hectare (59 dry tons/acre), DLD 3 at 220 dry metric tons/hectare (98 dry tons/acre), DLD 4 at 157 dry metric tons/hectare (70 dry tons/acre), and DLD 5 at 202 dry metric tons/hectare (90 dry tons/acre). In their eight years of operation, all DLDs have received an accumulative loading of over 1121 dry metric tons/hectare (500 dry tons/acre). DLD 3's accumulative loading is 1363 dry metric tons/hectare (608 dry tons/acre).

It is interesting to note that during the past eight years of operation, the sludge into the digesters has averaged just about 80% volatile solids. The volatile solids destruction in the digesters has averaged 56% and the volatile solids destruction in the SSBs has averaged 46%. This means that the digesters and SSBs have together destroyed (or disposed of) over 60% of the total solids being produced by the treatment plant. This leaves only 40% of the total solids to be harvested into the DLDs.

The harvesting crew in 1990 consisted of twelve employees. Two were permanent personnel and the other ten, temporary maintenance helpers. For the first time since startup, a number of the floating hoses from the dredges to the harvested sludge pipeline will require replacement in 1991. The Sacramento harvesting sludge pipeline is designed to allow the complete system to be flushed with plant effluent when it is taken out of service each year. Problems with the dredge flow meters may have affected the measured harvested sludge output of the 1990 season.

Costs

Capital costs for the Sacramento Regional Wastewater Treatment Plant sludge disposal system (includes the SSBs and DLDs) was $42,520,000. When annualized at twenty-five years at 8%, this cost is $3,980,000, or $179 per dry metric ton ($162 per dry ton) based on the system's design capacity of 22,272 dry metric tons (24,550 dry tons) of digested sludge per year. This cost was considerably reduced for the local agency because about 80 to 85% was funded by state and federal grants. Estimated capital cost for the local agency is $30.86 per dry metric ton ($28 per dry ton) of digested sludge.

Operation and maintenance costs for the sludge disposal system for 1990 were $1,343,000. This is broken down as follows:

Labor	$427,000
Services and supplies	325,000
(including fuel)	
Power	241,000
Monitoring	300,000

This amounts to a cost of $97 per dry metric ton ($88 per dry ton) of harvested sludge.

Monitoring

Extensive monitoring is part of the normal routine for the Sacramento Regional Wastewater Treatment Plant sludge disposal system. This monitoring includes maintaining quantities on all parameters of the sludge through the system, keeping track of the SSB inventories and operating parameters, analyzing the DLD soil parameters and maintaining an inventory of surface water runoffs, supporting twenty-five groundwater monitoring wells and numerous surface water sampling stations in on- and off-site streams, continuous monitoring of on-site weather and atmospheric conditions, and the recording and response documentation of all odor complaints.

Monitoring data continue to show no problem with the area's surface water or groundwater. However, recent new geotechnical work required by new state of California solid waste disposal regulations indicate that there has been movement of nutrients below the soils beneath the DLDs. At this time there has been no definitive report indicating what the permanent effect will be of these new findings. It is not expected that the method of disposal would be abandoned even if the DLDs had to be lined. During the eight-year history of this sludge disposal system there have been no odor complaints attributed to its operation.

UTILIZATION OF MUNICIPAL SEWAGE SLUDGE AS DAILY AND FINAL COVER FOR MUNICIPAL SOLID WASTE LANDFILLS

INTRODUCTION

One of the management practices for municipal sewage sludge that has been less publicized is its use as daily and final cover at municipal solid waste landfills. Municipal solid waste landfills require large amounts of material to provide daily cover for the disposal of municipal solid waste at a landfill. Final cover material is needed after a landfill is closed in order to establish a vegetative cover on the landfill. Soil (agricultural or clay) is normally used for daily and final cover, but municipal sludge can be used for these purposes as well. Information will be presented on the beneficial use of municipal sewage sludge for daily and final cover at municipal solid waste landfills.

UTILIZATION OF MUNICIPAL SEWAGE SLUDGE AS DAILY COVER AT MUNICIPAL SOLID WASTE LANDFILLS

Benefits of Daily Cover

At municipal solid waste landfills in the United States, the municipal solid waste delivered to the site is routinely covered at the end of daily operations. This is a requirement of many states such as Illinois (Title 35, Subtitle G, Part 807, Illinois Pollution Control Board Regulations) and the Federal Government (40 CFR, Part 241, Solid Waste Regulations, United States Environmental Protection Agency). This cover material serves many useful purposes including:

(1) Reduction of vectors such as rodents and flies
(2) Reduction of odor emissions
(3) Control of blowing litter
(4) Enhancement of aesthetics
(5) Reduction of the chance and spread of fires
(6) Reduction of the potential for surface and groundwater pollution

In the sections that follow, it will be demonstrated that municipal sewage sludge is a perfectly acceptable alternative product to soil for use as a daily cover at municipal solid waste landfills.

Reduction of Vectors

Municipal sewage sludge with a solids content of 50% or greater looks and functions much like soil (clay or agricultural). Municipal sewage sludge acts as a physical barrier and prevents vector penetration into the landfill provided that the volatile solids content of the sludge has been previously reduced by biological stabilization or other means. Municipal sewage sludge with a solids content of 50% or greater has a high moisture adsorption capacity and reduces the moisture content of municipal solid waste, thereby helping to further control such vectors as files, mosquitoes, and rodents, which can thrive under wet conditions.

Reduction in Odor Emissions

Municipal sewage sludge has high odor-absorbing abilities, just like soil. Also, municipal sewage sludge with a solids content greater than 50%, acts as a physical barrier and reduces the emission rate of odorous gases simply by reducing the surface area of solid waste exposed to the atmosphere. Dr. M. Finstein [3] has found that municipal sewage sludge has high odor-absorbing abilities in experiments he has conducted.

Control of Blowing Litter, and Enhancement of Aesthetics

Municipal sewage sludge with a solids content of 50% or greater has the same properties as soil in the control of blowing litter. Such sludge acts as a physical barrier and reduces blowing litter. Municipal sewage sludge with a solids content of 50% or greater has a soil-like appearance. Such sludge, when applied as daily cover to municipal solid waste, improves the aesthetic appearance of municipal solid waste landfills.

Reduction in the Chance and Spread of Fires

Fire is a possible hazard associated with landfilling of municipal solid waste. A daily cover of municipal sewage sludge will reduce the fire hazard associated with landfilling municipal solid waste provided that the volatile content of the sludge has been reduced by biological stabilization or other means.

In laboratory tests, municipal sewage sludge with a volatile content of 50 to 55% normally has a flash point of about 250°C. This flash point allows municipal sludge to be used as a fire control agent at municipal solid waste landfills.

Reduction in the Potential for Surface and Groundwater Contamination

One of the concerns of utilizing municipal sewage sludge as a daily cover at municipal solid waste landfills is the impact on the quality of leachate from these landfills. However, there has been a study of the quality of leachate where both municipal sewage sludge and municipal solid waste were simultaneously landfilled.

Farrell et al. [4] of the U.S.EPA Office of Research and Development, Cincinnati, Ohio reported that the addition of municipal sewage sludge to landfills improved the quality of leachate. During a twenty-month study, test cells containing municipal sewage sludge and municipal solid waste produced a leachate exhibiting a chemical oxygen demand (COD) of 1500 mg/L in comparison to a leachate COD of 30,000 mg/L produced from test cells that did not have the municipal sludge. In addition, concentrations of metals such as Cd, Cr, Cu, Pb, Ni, Fe, and Zn were lower in the leachate from the cells containing municipal sludge than those that did not. Farrell et al. [4] concluded from their study the following:

> It is a common misconception that introducing sludge into landfills degrades leachate quality. This study shows the reverse to be true. Results of this investigation should be made widely available to EPA and state authorities concerned with landfill regulations to improve the scientific basis for their decisions.

The use of municipal sludge for daily cover reduces the potential for leachate contamination or surface and groundwater at municipal solid waste landfills, and can actually improve the quality of any leachate generated.

Use of Municipal Sewage Sludge for Daily Cover—State Regulations

Most state regulations do not require that soil be used as daily cover material at municipal solid waste landfills. For example, landfill regulations in the state of Illinois (Title 35, Subtitle G, Part 807, Illinois Pollution Control Board Regulations) define daily cover as a "compacted layer of at least six inches of suitable material." Soil is not even mentioned in these regulations.

In the state of Illinois, the Illinois Environmental Protection Agency (IEPA) allows municipal sewage sludge to be used as a daily cover on a site-specific basis as long as the particular sludge is considered suitable. In general, sewage sludge used for this purpose would have a solids content greater than 50% and have undergone a process designed to reduce its volatile solids content.

Three landfills in the state of Illinois currently utilize municipal sewage sludge for daily cover. These landfills range in size from 21 to 29 ha (52 to 72 acres). Two of these landfills are owned and operated by Land and Lakes Inc. (Northbrook, Illinois) and the other is owned and operated by Waste Management Inc. (Oak Brook, Illinois). These landfills are currently operating under permits issued by the IEPA.

The Metropolitan Water Reclamation District of Greater Chicago (MWRDGC) currently has an agreement with Waste Management of Illinois (Oak Brook, Illinois) to utilize 226 mt (250 tons) per day of air-dried municipal sewage sludge (60% solids) at the firm's CID landfill site. This sludge will be delivered to the landfill site five days per week and used for daily cover. The agreement allows the MWRDGC to send nearly 22,675 mt (25,000 tons) of sludge to the CID site in any given year. This amounts to about 13% of the annual sludge production from the MWRDGC.

UTILIZATION OF MUNICIPAL SEWAGE SLUDGE AS A FINAL COVER

General Benefits of Final Cover

When a municipal solid waste landfill has reached the end of its useful life, it is a requirement of both state and federal regulations that it receive a final cover, and that vegetation be grown on this final cover material. This placement of final cover and the growing of vegetation serve both aesthetic and environmental purposes.

Obviously, it is important that a completed landfill site be compatible with its surroundings. Therefore, the placement of a final cover and the growing of vegetation including grass and trees allow the landfill site to be aesthetically acceptable to nearby residences and businesses. In addition, the final cover with vegetation helps prevent rainfall from entering the landfill and thereby reduce leachate generation. This reduction in leachate generation will significantly reduce the potential for surface and ground-water contamination at the landfill site.

It should be noted that the term *final cover* as used here is the final layer of soil placed upon a landfill, which is used to grow a vegetative cover. Sometimes the term *final cover* is also used to describe the nonpermeable soil cover such as clay designed to prevent rainfall from entering the landfill. However, the term *final cover* used herein refers to the soil or sludge placed on top of the nonpermeable clay cover.

In the section that follows there is a review and discussion of the benefits of the use of municipal sewage sludge as a final cover for municipal solid waste landfills.

Benefits of Municipal Sewage Sludge as a Final Cover

The U.S.EPA in its Technology Transfer Document (EPA-625/4-78-012, 1978) entitled "Sludge Treatment and Disposal," discusses the use of municipal sewage sludge as a final cover for landfills. This document provides general guidelines for the use of sludge as a final cover for landfills and states: "This is not strictly a sludge landfilling method since the sludge is not buried. However, it is a viable option for the disposal of sludge at refuse landfills which has been performed and should be used in many cases."

The MWRDGC has successfully used its municipal sewage sludge for final cover at the CID municipal solid waste landfill operated by Waste Management of Illinois in Chicago, Illinois and at the city of Chicago refuse disposal site at 103rd Street and Doty Avenue in Chicago, Illinois. The MWRDGC's municipal sewage sludge applied at both these sites produced a dense vegetative cover that greatly improved the aesthetic appearance of these sites. This has resulted in reduced erosion from these sites and a reduction of rainfall intrusion, thus reducing leachate production and the potential for surface and groundwater contamination.

BENEFITS OF THE USE OF MUNICIPAL SEWAGE SLUDGE AS DAILY AND FINAL COVER AT MUNICIPAL SOLID WASTE LANDFILLS

In the preceding sections, the technical benefits of the use of municipal sewage sludge as daily and final cover at municipal solid waste landfills was discussed. In the following sections, other benefits of such use will be discussed.

Economics

Daily and final cover material represents one of the major costs of the operation of municipal solid waste landfills. Daily cover, usually clay soils, must be excavated from the landfill site or imported from other sites. Final cover almost always consists of good-quality topsoil imported to the site.

The Pigeon Point landfill in New Castle County, Delaware, U.S.A. is a 48.5 ha (120 acre) site that receives a daily volume 1269 mt (1400 tons) of refuse from the city of Wilmington, Delaware. In order to provide daily cover to the municipal refuse received, cover material must be imported at an annual cost of $300,000 [5].

Glebs and Juszczyk [6] estimated that importation and placement of a 15.2 cm (six-inch) topsoil layer as final cover to a 2.8 ha (seven acre) landfill site would cost $5415.00. For a 40.4 ha (100 acre) site, the cost would be $77,357.14.

Obviously the cost for daily and final cover at landfills represents a substantial economic burden for operators. The use of municipal sewage sludge for daily and final cover material would substantially decrease the cost for such material since most sludge generators could elect to provide sludge at no cost to the landfill operator.

Of course the municipal sewage sludge generator too would benefit economically if its sludge were used as daily and final cover at municipal solid waste landfills. The generator would not have to bear the costs associated with the ownership and operation of the landfill site.

The presence of municipal sewage sludge has been shown by Farrell et al. [4] to increase the gas production from municipal solid waste landfills. Since many operators collect, utilize, and sell the gas produced from their municipal solid waste landfills, sludge used as a daily cover would enhance the economic return from their ongoing gas collection and production facilities.

Operational Benefits

Since the municipal sewage sludge generator would deliver its sludge to the landfill site, the landfill operator would not have to tie up equipment in order to import cover material. This would reduce equipment and manpower requirements, ease daily operations, and substantially reduce operating costs. Application of municipal sludge rather than organic topsoil as a daily cover also represents more prudent use of increasingly scarce agricultural soil.

Cooperation between Public and Private Sector

The use of municipal sludge as a daily and final cover at municipal solid waste landfills would represent cooperation between the public and private sectors in a beneficial sludge use program. The municipal sewage sludge generator would benefit because such a use would lower its cost of operation and help to keep local taxes down. The private sector would not have to incur costs for importing cover material and would benefit from increased gas production from the landfill site.

Thus it appears that from this type of cooperation between the public and private sectors, the environment, local government, the private sector, and the paying community would all emerge as winners.

Institutional Benefits

The municipal sludge generator needs to have as many management options for its sludge as possible. The use of sludge as daily and final cover in municipal solid waste landfills would provide yet another beneficial use option for municipal sludge.

Daily and final cover use at landfills represents a beneficial use and therefore supports the concept of municipal sludge as a resource. Municipal sludge generators who utilize their sludge for daily and final cover at landfills show their willingness to support programs where sludge can be used as a community resource and not be viewed as a liability.

Topsoil and other soils used for daily and final cover at landfills are becoming increasingly scarce commodities in urban centers. Since less of these materials would be used at landfills if municipal sludge were used as an alternative, the community could use these materials for other purposes.

REGULATORY CONTROL OF THE USE OF MUNICIPAL SLUDGE AS A DAILY AND FINAL COVER AT MUNICIPAL SOLID WASTE LANDFILLS

Regulatory Control at the State Level

The states now control municipal solid waste landfills by issuing permits to landfill operators. For example, the Illinois Environmental Protection Agency controls and regulates the use of the MWRDGC's sludge at landfills through a permit issued to the municipal solid waste landfill operator. This insures that all operations are carried out in strict adherence to the state of Illinois regulations.

Superfund Regulatory Control

Under the U.S.EPA's superfund regulations, municipalities who place their sewage sludge in municipal solid waste landfills can potentially be liable for superfund cleanup costs if the landfill is later found to have received (knowingly or unknowingly) hazardous wastes. This potential liability exists even though the municipal sludge meets all the U.S.EPA's regulatory requirements associated with disposal of sewage sludge and solid waste. This liability of course could extend to sludge placed in the landfill for daily cover and sludge placed on the landfill as final cover.

However, the U.S.EPA has recently released a draft of its "Interim Municipal Settlement Policy" for the superfund program in the *Federal Register* (pages 51,071–51,076), on December 12, 1989. While this policy does not exempt municipalities from liability under the superfund program, it indicates that the U.S.EPA does not intend to pursue liability cases against municipalities that generate sewage sludge and legally place this sludge at landfills permitted to accept this sludge.

U.S.EPA Part 503 Sludge Regulations

The U.S.EPA proposed Part 503 Regulations for the management of sewage sludge do not apply to the use of sludge as a daily cover at landfills since this practice is defined as co-disposal and is currently regulated under 40 CFR Part 258. The preamble to the Part 503 proposed regulations, however, does discuss beneficial uses of sludge such as daily and final cover for landfills in conjunction with "U.S.EPA's policy of strongly supporting the beneficial reuse of sewage sludge."

DESIGN CONSIDERATIONS

Use of Sewage Sludge as Daily Cover

The important functions of daily cover, whether it is soil or sludge, are to control odors, vectors, litter, fire, and moisture at the working face of the municipal solid waste landfill. Generally, a 15.2-cm (six-inch) layer of sludge will perform these functions if it is applied at the end of the working day to the municipal solid waste being landfilled on that day.

Obviously, the sewage sludge being used as daily cover must be sufficiently dry to perform the required functions of daily cover. Since the physical properties of various sludges can differ widely despite the fact that they may have the same moisture content, it is difficult to state what the maximum moisture content should be for sludge used as daily cover. In general, however, it is recognized that sludge with a moisture content not higher than 50% can be used for daily cover.

Since odor and vector control are two prime functions of daily cover, the sludge being used as daily cover must have been subjected to a process designed to reduce its volatile solids content. Normally, anaerobic digestion is used for reducing the volatile content of sewage sludge and this process provides a sludge product suitable for daily cover. However, other processes provide perfectly satisfactory reduction in volatile solids. These include:

(1) Aerobic digestion

(2) Lime stabilization

(3) Chlorine stabilization

(4) Composting

(5) Lagoon storage

(6) Wet-air oxidation

Use of Sewage Sludge as Final Cover

In order to establish a vegetative cover on a closed landfill, normally about 0.3 to 0.9 m (one to three feet) of sewage sludge is applied. The sewage sludge being used for final cover should have a solids content greater than 20% and should have previously undergone a process to reduce its volatile solids content. To prevent the sludge from sliding off the top layer of soil on the side slopes, it should be combined with the surface soil so that the resulting mixture of soil and sludge is about 1:1.

The closed landfill site will require a vegetative cover since plants help to control erosion and infiltration and are visually appealing. Winter rye grass is an excellent vegetative cover material. It is quick-growing and provides early erosion control. If a rye grass/bermuda grass mixture is used, this will further help to stabilize slopes and reduce runoff. Other useful grasses are bunch and canary. Trees also can be used. Since local weather conditions strongly influence selection of vegetation, it is wise to consult with county extension agencies and local universities before a final selection of cover vegetation is made.

GROUNDWATER TRANSPORT MATHEMATICAL MODELS

INTRODUCTION

Groundwater is a major natural resource that more than half of the U.S. population uses for its drinking water supply. Furthermore, approximately 75% of the major cities in the U.S. depend on groundwater to supply their potable water [7]. Although most deep groundwater sources meet legal requirements for purity, there is a great deal of concern over contamination, particularly regarding shallow groundwater resources and the eventual contamination of deeper resources if contamination remains unabated. Lehr [8] has estimated that between 0.5 to 2% of the groundwater in this country is contaminated and urges that we eliminate the initiation of new sources of contamination and create a public ethic and morality that will put an end to practices that cause contamination of our groundwaters. The uncontrolled or unexpected migration of dissolved contaminants from waste disposal sites into potable groundwater is a major environmental concern facing both the public and private sectors, and the regulatory agencies.

BACKGROUND

The government response to these growing public concerns has been increased regulation of potential sources of groundwater contamination, including land disposal of municipal sewage sludges. The purpose of this section is to provide a background and concepts related to computer-modeling and groundwater transport of contaminants from sewage sludge. This section will be structured to provide an overview of the basics of mathematical modeling, the uses of models in the regulation of potential groundwater contamination from sewage sludge disposal, and a brief summary. The focus will be on monofills (sludge-only landfills) since the practices of disposal in area-fills, diked containment, narrow trenches, and land application are largely variations on the application of the models and the principles discussed. Monofills are emphasized because landfilling of sludge is reported to be the lowest-cost disposal option [9], but creates the highest potential loading rates to an area along with the minimum unsaturated zone transport distance to an aquifer. As such, monofills appear to represent the ''worst case'' scenario for groundwater contamination for land disposal of sludges.

The U.S. Environmental Protection Agency (U.S.EPA) is charged with the responsibility to develop and issue regulations regarding land disposal of sludge. The Agency has estimated that there are a little over 200 monofills in the U.S. and that this sludge disposal practice accounts for about 2% of the volume of sludge disposed. About half of the POTW's surveyed by the Agency reported that the depth to groundwater below their monofills was less than 2 meters [10]. Contaminants in sludge disposed of in monofills can leach into the groundwater under a site. These leached contaminants can then migrate to wells from which drinking water is obtained as illustrated in Figure 7.14. Thus contaminants leaching from sludge potentially pose risks to human health through the groundwater pathway.

In recent years, there has been an increase in the use of computer-assisted, solute transport models to estimate the geometry and chemical composition of leachate plumes based on waste leaching characteristics, soil-material properties, hydrogeological data, and aqueous chemistry of the waste-leachate-site system. The basic theorem that underlies this type of assessment has been given by Griffin and Roy [11] and Griffin et al. [12]. The concept is that a given land disposal site has a finite capacity to attenuate contaminants in solution to environmentally acceptable levels. If the attenuation capacity of the site is exceeded, then the site will fail to be protective of the environment.

There are four major factors that will ultimately determine the success of a sludge disposal site in protecting human health and the environment with respect to groundwater contamination by sludge contaminants:

- the toxicity of the contaminant
- the attenuation characteristics of the contaminant and the soil/ aquifer materials
- the mass loading rate, i.e., the amount of contaminant entering the subsurface per unit time
- the total amount of contaminant in the sludge available to leach into the groundwater

The physicochemical processes that attenuate contaminants include dilution by groundwater, adsorption by soil materials, and dispersion. Degradation of the contaminant, representing the combined effects of chemical and biological degradation is also an important attenuation mechanism. These processes are illustrated in Figure 7.15 and have been discussed in previous studies [11,12]. These studies concluded

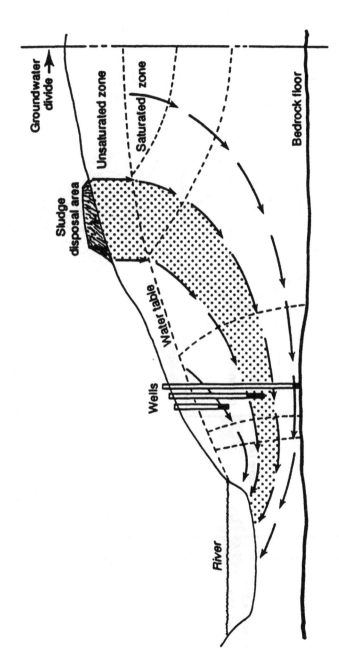

Figure 7.14 Flow of contaminants in a water table aquifer (humid region, adapted from Miller [47]).

467

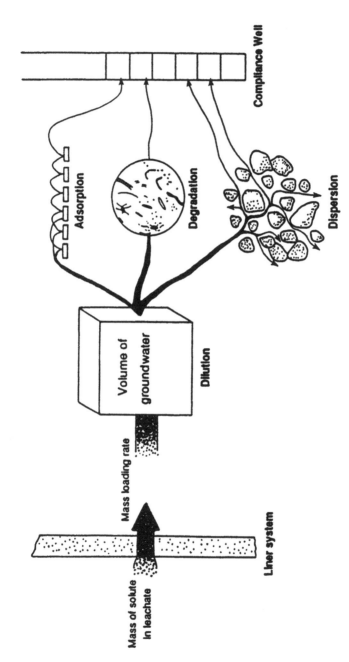

Figure 7.15 Basic components of attenuation at a waste disposal site [11].

that if the attenuation capacity of a site can be estimated, then these data can be used as criteria to make decisions as to what wastes can be landfilled, and what quantities of contaminants in a given waste can be safely accepted for disposal. For example, given a water quality standard at a receptor well, mathematical models can be used in a back-calculation mode to determine the allowable leachate concentration and hence the maximum allowable sludge concentration that can be monofilled at a site with certain known characteristics.

The Agency has developed detailed risk assessment methodologies for several sludge disposal options, including monofilling. These methodologies consist of a set of computer models, based in large part on the concepts presented above. For the monofill disposal practice, the groundwater pathway has been identified by U.S.EPA as one of the most critical. The groundwater pathway occurs in many other reuse/disposal practices and thus the discussion presented here is also relevant in those practices.

BASICS OF MATHEMATICAL MODELING

Contaminant Transport

The major mechanisms of contaminant transport in groundwater are advection, dispersion, adsorption, and degradation. Advection refers to the movement of groundwater in response to a hydraulic gradient. Dispersion (sometimes called hydrodynamic dispersion) is a physical process where a liquid moves through a porous medium at different velocities due to the variation in pore sizes available to conduct the liquid. This attenuation process has the effect of spreading contaminants in the porous material, thereby reducing their concentration at a given point over time.

Adsorption is a physical-chemical process whereby contaminants dissolved in solution are concentrated at solid-liquid interfaces and are hence removed from solution. In comparison, degradation represents a transformation of the compound and hence a reduction of its concentration in solution.

These qualitative generalizations of attenuation can be quantified and used in contaminant transport analyses. However, in order to make this transition, it is necessary to introduce some concepts and associated mathematical expressions. In saturated homogeneous materials that are subjected to steady-state flow conditions (the rate of water moving through a given area does not change over time), the change in solute concentration as a function of time may be generalized [13–16] as:

$$\frac{\partial C}{\partial t} = Dx \frac{\partial^2 C}{\partial x^2} - Vx \frac{\partial C}{\partial x} - \frac{\rho \partial S}{\theta \partial t} \qquad (7.1)$$

where

C = the concentration of the solute in solution (mass/volume)

Dx = the diffusion-dispersion coefficient along a flow path x (distance^2/time)

Vx = the mean flow velocity of the water along flow path x (distance/time)

ρ = the dry bulk density of the aquifer materials (weight/volume)

θ = the volumetric water content of the aquifer materials (volume/volume)

S = the amount of solute adsorbed per mass of adsorbent (mass/mass)

x, t = distance and time variables

Equation (1) may be rearranged as

$$R \frac{\partial C}{\partial t} = Dx \frac{\partial^2 C}{\partial x^2} - Vx \frac{\partial C}{\partial x} \qquad (7.2)$$

where

$$R = 1 + \frac{\rho}{\theta} \frac{\partial S}{\partial C}$$

The reduced term R is called the retardation factor. In order to solve Equation (7.2), a functional relationship for $\partial S/\partial C$ must be determined. The approaches used in solving this relation range from simple approximations to complex numerical computations. An adsorption isotherm equation is a mathematical expression that describes how the amount of solute adsorbed (S) varies with the concentration of the solute in equilibrium with the adsorbed phase. The adsorption behavior of many organic solutes conforms to the Freundlich equation, viz.,

$$S = K_f C^N \qquad (7.3)$$

where C is the equilibrium concentration of the solute, and K_f and N are constants.

At sufficiently low solute concentrations, the Freundlich exponent is often equal to 1, and the adsorption behavior is characterized as being a linear isotherm. In this case, the Freundlich adsorption constant (K_f) may be expressed as a partition coefficient (often written as K_d):

where

$$\text{where } S = K_d C \tag{7.4}$$

$$\text{and thus } \frac{\partial S}{\partial C} = K_d$$

$$\text{hence } R = 1 + \frac{\rho K_d}{\theta} \tag{7.5}$$

In other words, in the case of a linear isotherm, the Freundlich constant (K_f) reduces to a simple partition coefficient (K_d) which is a single-valued number that can be used to estimate solute-adsorbate partitioning. Because of its mathematical simplicity, this approach (the linear isotherm assumption) has been widely used. Partition coefficients may also be derived for nonlinear isotherms as described elsewhere [17,18].

A partition coefficient (K_d) for a specific solute-adsorbent system is usually determined experimentally using batch adsorption procedures. Specific laboratory procedures are given by Roy et al. [19] and are necessary to obtain K_d values for metals. However, in the absence of such data, partition coefficients for the adsorption of given hydrophobic organic compounds by soil materials can be determined from the amount of organic carbon present in the adsorbent soil material [20].

A partition coefficient for a specific solute-adsorbent system may be estimated by Equation (7.6), viz.,

$$K_d = K_{oc} \times f_{oc} \tag{7.6}$$

where f_{oc} is the organic carbon fraction, and K_{oc} is the soil-water adsorption constant for the particular organic compound. K_d values can also be estimated from octanol-water partition coefficients (K_{ow}) [20].

Preliminary K_d values for sludge contaminants are given by U.S.EPA [21]. If tabulated values are not available, and the organic carbon fraction of the soil or aquifer material is known, then the partition coefficient (K_d) can be estimated, which in turn can be used in application of Equation (7.2).

The mean flow velocity of water (V) may be calculated from Darcy's Law as:

$$V = \frac{K_{sat} i}{\eta_e} \tag{7.7}$$

where

K_{sat} = the unsaturated hydraulic conductivity
η_e = the effective (water conducting) porosity
i = the hydraulic gradient

This relationship [Equation (7.7)] can be coupled with Equation (7.1) and Equation (7.5) and rearranged (after Faust [22]) if dispersion is neglected to yield a relatively simple relation that can be used to estimate the distance of migration (x) of a contaminant after a given time (t):

$$x = \frac{tK_{sat}i}{\eta_e R} \tag{7.8}$$

where all variables have been defined previously.

The dispersion coefficient Dx along flow path x is:

$$Dx = \alpha V + D* \tag{7.9}$$

where α is the dispersivity (cm), and $D*$ is the diffusion coefficient in water (cm²/sec).

The dispersivity term is a proportionality constant relating the mean velocity of the water to the dispersion coefficient. The diffusion coefficient is a proportionality constant used to determine the spontaneous movement of solutes due to positive changes in the entropy of the system. Diffusion can also be thought of as being caused by the thermal motion of molecules (Brownian Motion), which, in the presence of a concentration gradient, causes a net migration of molecules toward the region of lower concentration.

The dispersivity has been found to be scale dependent, and can be estimated to be about 10% of the distance measurement of the analysis [23]. The diffusion coefficient for a contaminant would have a negligible effect on the dispersion coefficient when the hydraulic conductivity is much greater than the diffusion, and consequently:

$$\alpha V \gg D*$$

hence,

$$Dx \sim \alpha V$$

where α is approximately 0.1 × distance of migration analysis (meters).

Landon [24] generalized that diffusion is not a significant transport process in soil systems, although if the hydraulic conductivity is sufficiently low, the movement of solutes through a compacted clay liner may be a diffusion-controlled process [25]. While data are not available to include this factor in sludge assessments, it is a point that may warrant future research and possible inclusion in more complete assessments.

Solute degradation during transport can be an important attenuation mechanism. Degradation per se represents the combined effects of hydrolysis and biodegradation. Hydrolytic reactions are chemical reactions whereby solutes react with water to form weak acids or bases. Biodegradation is a biochemical process whereby microorganisms (e.g., bacteria, actinomycetes, fungi, and algae) transform compounds by oxidation, reduction, or immobilization.

The degradation of contaminants is commonly treated by an exponential decay model [26], viz.,

$$Ct = C_0 \exp(-0.693 t / t_{1/2}) \tag{7.10}$$

where
Ct = the solute concentration at time t
C_0 = the initial solute concentration at time = 0
$t_{1/2}$ = the degradation half-life
t = time

This exponential decay can be incorporated into the decay term of transport models and illustrates that the importance of decay as an attenuation mechanism depends mainly on the half-life of the contaminant solute in the system.

Types of Models

Contaminant transport models applicable to sludge disposal problems may be classified in a number of ways. A few of the common ones will be described here to provide a framework for further discussions. A comprehensive compilation of models, their documentation, and software is available from the International Ground Water Modeling Center (Holcomb Research Institute, Indianapolis, Indiana 41208). More detail on groundwater modeling and transport processes is given by Freeze and Cherry [16], Mercer and Faust [27], Wang and Anderson [28], and Bedient et al. [29].

Saturated vs. Unsaturated

As their names imply, one common way of describing models is with respect to their capability of handling unsaturated flow conditions as opposed to saturated flow. Saturated zone models use the simplifying assumption that the pores of the soil or aquifer media are completely saturated with fluid and consequently that the hydraulic conductivity is constant over the domain being computed. In comparison, the hydraulic conductivity varies as a function of the moisture content in the unsaturated zone. Consequently, unsaturated zone models are more complicated and require more input data for solution of a comparable application.

Model Dimensionality

Contaminant transport models are commonly described as one-dimensional (1-D), two-dimensional (2-D), or three-dimensional (3-D). As the names imply, they describe the ability of the model to predict contaminant migration in the x direction for the 1-D models, in the x and y directions for the 2-D models, and in the x, y, and z directions for the 3-D models.

The 1-D models are the simplest and typically require the least amount of input data. They however require a number of simplifying assumptions such as uniform, homogenous, and steady-state flow in one direction. The need for these types of assumptions necessarily limits their general applicability. However, these models perform well for cases where these assumptions are closely met, such as prediction of contaminant migration in soil columns. These models are also excellent tools for teaching the fundamentals of contaminant transport modeling.

Two-dimensional models are quite popular because they strike a practical balance between simplicity and realism. They are frequently the most useful tools for modeling contaminant transport in field situations. They are much more versatile than the 1-D models while not suffering from the long computational times and large data input requirements of 3-D models.

The 2-D models are capable of analyzing flow and contaminant transport in both the x and y directions. This capability to handle two-directional flow and transport makes 2-D models capable of predicting spatial distributions of contaminants, and are consequently applicable to a wider range of problems than 1-D models. A horizontal 2-D analysis is the most common application because monitoring wells and other observation points are typically distributed in a horizontal fashion. The major weakness of the 2-D models are that the vertical distribution of contaminants and soil material properties is assumed to be uniform. If detailed information is available on the vertical and horizontal variations in contaminant distribution and soil material properties, then 3-D modeling may be applicable and appropriate.

Three-dimensional models are capable of analyzing groundwater flow and contaminant transport in all three dimensions, x, y, and z. These models are capable of the most realistic simulations, but also require the most computational time and input data. The 3-D models have not been as popular as the 1-D and 2-D models for a number of reasons. As model complexity increases, the demands on the skill and judgement of the modeler also increase. In addition, the large 3-D codes can require expensive or specialized computing equipment not readily available to everyone. The quality of the results from any model are heavily dependent on the quality and nature of the input data. In many cases the quality and amount of data available simply do not warrant the effort and expense associated with a comprehensive 3-D modeling exercise. The choice of model should be based on the available data and the level of modeling required by the application.

The saturated-unsaturated zone and dimensionality model classifications are basically descriptive. To utilize any model, you must first solve the governing transport equation. The techniques for developing a solution include analytical and numerical methods. This forms the basis for perhaps the most fundamental classification of models, analytical and numerical.

Analytical Models

Analytical models have been developed for 1-D, 2-D, and 3-D applications as well as both saturated and unsaturated conditions. Their main distinction is the way they solve the governing transport equation. These models utilize simplifying assumptions, particularly regarding boundary and initial conditions. Example assumptions would be uniform, steady-state, and isotropic (hydraulic conductivity is uniform in all directions) flow conditions. Processes that may be included in these models are advection, dispersion, adsorption, and degradation.

Analytical models are characteristically simpler, faster, and easier to use than numerical models, and generally have more relaxed data input requirements. While these models generally require simple geometries and boundary conditions, they typically will yield useful insights into contaminant migration problems and are quite valuable for modeling hypothetical situations.

Numerical Models

Numerical models solve the governing transport equation using an iterative procedure. These solutions are generally more flexible than analytical solutions and allow for solution of more complex problems and site geometries. Consequently, they are preferred when modeling an actual field site. The general method is to break the problem into a network of cells. The solution to the governing equations are approximated by differences between the values of the various parameters over the network at a given time, then new values are predicted for the next time increment. The three most common techniques for approximating the solutions to the equations are known as finite-difference, finite-element, and method-of-characteristics.

FINITE-DIFFERENCE

The finite-difference technique was generally used in the early numerical groundwater models. The method divides space into rectangular cells along the coordinate axis. Values at a single node represent the parameter values within each cell. The set of equations resulting from approximating the partial differentials by differences can then be solved by iteration. The main disadvantage of the finite-difference method is the problem known as numerical dispersion. The truncation error in approximating the partial differential equations can result in errors of the same order of magnitude as does the physical process of dispersion [29,30].

FINITE-ELEMENT

The finite-element technique works in a manner similar to the finite-difference, except that the cells or elements can vary in size and shape. The hydraulic head and contaminant concentration are computed at each node of the element and values within the element are computed by interpolation. The truncation errors common in the finite-difference method are significantly reduced by this technique. The use of elements of variable size and shape allow these models the flexibility to analyze problems of great complexity. Their main disadvantage is the generally higher computational costs and the need for the modeler to have a higher level of expertise [28,29,31].

Method-of-Characteristics

The method of characteristics takes a somewhat different approach whereby idealized particles are tracked through the flow field. The first step is to assign a mass of contaminant to each particle which can then be used to compute concentrations in each cell. The particles move according to the flow velocity, and the effects of dispersion are added onto the particle movement. This procedure is computationally efficient and minimizes the problems of numerical dispersion. The most widely used 2-D groundwater transport model is the U.S. Geological Survey Method of Characteristics (MOC) model. This model uses a combination of finite-difference and method-of-characteristics techniques [29,32].

Validation of Mathematical Models

An important concept in the application of groundwater transport models used to make predictions is their validation. Validation can be defined as the three steps of (1) sensitivity analysis, (2) calibration, and (3) verification.

Sensitivity Analysis

Sensitivity analysis involves running the model while varying the input parameters. This is done to determine which input parameters have the greatest impact on the final results. This can be of great value in interpreting the results and in guiding data collection efforts.

The values of input parameters are normally varied over the range expected to occur or that are important to the particular application. Results of sensitivity analysis typically indicate that hydraulic conductivity is the most sensitive parameter, while parameters such as porosity are much less sensitive. This is usually because hydraulic conductivity can vary by several orders of magnitude over relatively short distances, especially with depth, while the natural range in porosity values is relatively small. With this knowledge, conclusions can be tempered based on the reliability and amounts of data available for sensitive parameters. Data collection efforts can be allocated so that proportionately more financial and manpower resources are used to collect data for sensitive parameters than insensitive parameters.

Calibration

After the model has been set up, run, and sensitivity analyses performed for the particular application, it must be tested against actual field data. This process is known as calibration. Calibration is achieved by adjusting the various input parameters of the model until the discrepancies between the observed and computed values are minimized. At least two sets of data are required for an adequate calibration, and preferably representing the full range of expected conditions to be simulated.

Verification

When the model has been calibrated, it is then ready to make predictions. The process of verification is the comparison of predictions with actual field data. This data must be different than the data set used in the calibration process. When a model has satisfactorily predicted a field event, the model is said to be verified and the validation process is completed.

Limitations of Models

The validity of any solute transport model depends on the quality of the input parameters and on a number of assumptions. Advection-dispersion models of the types discussed previously have been questioned on several bases [29,30,33]. One major difficulty is that dispersivity has little physical significance and varies with the scale of the problem. Tracer tests can be used to determine dispersivity, but laboratory-determined values have little meaning to field predictions, and large-scale field tracer tests are expensive and often impractical. The presence of spatial heterogeneities and other mechanisms can make estimation of dispersivities in different planes difficult and uncertain. At the present time, properly performed calibration studies are among the most useful ways to obtain reliable dispersivities. Where coefficients of dispersion have been derived from proper calibration studies, prediction can be quite accurate [33].

One of the basic assumptions commonly employed is that of steady-state conditions. Gradients, contaminant leaching concentrations, hydraulic conductivities, and biodegradation rates are examples of parameters commonly assumed constant over time ,but which may actually vary.

The assumption of equilibrium conditions is subject to some error and may contribute to inaccurate predictions in some cases. The use of equilibrium adsorption constants assumes that the reaction between the solute and the soil materials is rapid relative to groundwater flow rates. While groundwater flow rates are normally slow, and adsorption reactions are rapid, the extent that this assumption holds can affect the accuracy of predicted results.

Biodegradation rates are particularly difficult to predict in field settings. Laboratory-measured degradation rates tend to greatly overpredict those observed in the field, where conditions are not as uniform and ideal as those in the laboratory. Conditions such as acclimation of the microbial population, nutrient status, oxidation-reduction potential, pH, and competing reactions are examples of parameters that must be well understood to make accurate predictions.

Hydraulic conductivity has been suggested as probably the single most important parameter for accurately predicting contaminant transport, while also being one of the most difficult to measure [29]. A large variation in hydraulic conductivity values determined at a field site is quite common. Because of the large variations in hydraulic conductivity that commonly occur in most aquifers, even extensive testing may not accurately represent the spatial heterogeneities.

In unsaturated flow conditions, moisture release curves are nonlinear and hysteresis effects may produce a different curve for the wetting cycle and the drying cycle. If these are not properly represented or if simplifying assumptions are made, such as ignoring hysteresis or that of a linear moisture release curve, then the accuracy of calculations of hydraulic conductivity as a function of moisture content may suffer. As a consequence, predictions of flow velocities and subsequent contaminant migration under either saturated or unsaturated conditions may be subject to some error at difficult problem sites.

Despite all of the above-mentioned problems, plus others not discussed, contaminant transport models are still very valuable tools. They offer the best means presently available to organize data, predict contaminant migration, and provide insights into complex interactions. They also give guidance for data collection needs, monitoring, site selection, and design of disposal facilities.

EPA SLUDGE CONTAMINANT TRANSPORT MODELS

The simulation of contaminant transport in groundwater can be accomplished by models of varying degrees of sophistication, as has been described above. Comprehensive three-dimensional models can provide the best simulations available of contaminant transport beneath a sludge disposal facility. However, these models are complex and usually require more input data than is available at most sludge management facilities. These models also require considerable modeling expertise for proper execution and interpretation. One-dimensional models, on the other hand, are simpler and faster to use, require fewer input data, are less expensive, and can be applied to a wide variety of sites.

Unsaturated Flow

The method proposed by the U.S.EPA to simulate contaminant movement through the unsaturated zone relies on a one-dimensional, finite-element model known as VADOFT [34,35]. Three candidate models for predicting unsaturated zone flow and transport were considered. They were the CHAIN model [36] as used in U.S.EPA [37], the LEACHM model [38,39] and the VADOFT module from the RUSTIC model [34,35]. CHAIN was rejected as a result of a critical review by the Peer Review Committee [40]. LEACHM and VADOFT were recommended for consideration by the PRC [40] and after review, the U.S.EPA selected VADOFT [34,35,41]. VADOFT was selected in part because technical support is available from the U.S.EPA office of Research and Development, Environmental Research Lab, Center for Exposure Assessment Modeling, Athens, Georgia [34,35]. VADOFT has been compared in benchmark tests to the more sophisticated two-dimensional finite-element codes UNSAT2 [42] and SATURN [43] with good results [34].

The VADOFT model is a one-dimensional, finite-element unsaturated zone flow and transport model. Details of the model, the governing equations, and the mathematical solution techniques are discussed in U.S.EPA [34,35]; utilization in sludge disposal applications is discussed in O'Neal et al. [41]. The model considers advection, dispersion, linear equilibrium sorption, and both chemically and biologically induced first-order degradation. The model also allows for consideration of changes in water table level and the changes in physical and chemical processes with depth in a vertical soil column. The code, however, ignores the effects of hysteresis. The mass flux of contaminant through the soil computed with VADOFT is evaluated at the water table based on the derived concentration distribution and Darcy flow velocity. The mass flux resulting from the VADOFT simulation is used as input to the saturated zone model.

Saturated Zone

Flow and transport through the saturated zone is computed with the analytical model AT123D [44]. The saturated zone module of RUSTIC, known as SAFTMOD, also computes the transport of contaminants through the saturated zone and could have been used, but was rejected in favor of AT123D for consistency with other sludge methodologies. Prior to selecting the AT123D model, its performance was compared with that of a fully 3-D numerical model and with that of another analytical model. The tests were run against SEFTRAN, a finite-element code available from Holcomb Research Institute that has been benchmarked by Huyakorn et al. [45]. The alternative analytical code is currently used by the U.S.EPA Office of Solid Waste and uses a "patch source" to represent the source term [43,46]. Over the range of conditions tested, the AT123D model compared quite favorably with the 3-D model and generally out-performed the patch source analytical model [41].

The AT123D model, "Analytical Transient One-, Two-, and Three-Dimensional Simulation of Waste Transport," is described in detail by Yeh [44], U.S.EPA [37], and O'Neal et al. [41]. The general description of the model is that it uses an analytical solution to the basic advection-dispersion transport equation and can be run in either 1-D, 2-D, or 3-D mode, hence its name. The model is known for its flexibility with up to 450 options, giving a wide range of source configurations and boundary conditions. It considers the transport mechanisms of advection, dispersion, adsorption, decay/degeneration, and waste losses to the atmosphere.

Calculation of Criteria

The calculation of criteria for maximum sludge contaminant levels for the groundwater pathway is illustrated in Figure 7.16. The figure diagrams the overall methodology and shows how contaminant losses are divided among the environmental pathways. Figure 7.16 also diagrams the relation between the various model modules as used to determine criteria for sludge concentration and annual or cumulative loadings. The purpose of the methodology is to determine the maximum concentrations of contaminants that can be present in sludge and still not exceed allowable groundwater quality standards at a point-of-compliance well. A mass balance calculation is used to divide contaminants among the various media, and a square-wave approximation (which conserves a mass balance) is used to compute the pulse of contamination that is used as input to VADOFT. The results from VADOFT, the unsaturated zone flow and transport module, are then passed as input to AT123D, the saturated zone module.

For a given leachate concentration and duration of leaching, the groundwater model computes the concentration of contaminant at a specific well location. For determination of criteria for sludge disposal, a reverse calculation or back-calculation is performed in which the maximum allowable concentration in the well is used to derive the maximum allowable leachate concentration. Contaminant partitioning equations that describe the relation between leachate concentrations and sludge levels are then used to determine the maximum allowable levels of contaminant that can be contained in sludge to be disposed under the particular site conditions.

The U.S.EPA-proposed methodology links two well-accepted models

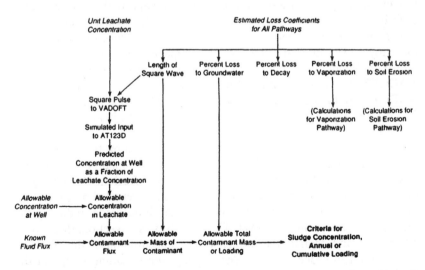

Figure 7.16 Calculation of criteria for sludge disposal via the groundwater pathway (personal communication, E. Hausman, Abt Associates, Inc., Cambridge, Mass.).

for unsaturated (VADOFT) and saturated (AT123D) zone flow and transport. The combination of these two codes creates a model that is considered to be relatively simple to use and appropriate for deriving sludge disposal criteria [41].

SUMMARY

Groundwater is a vital national resource that must be protected from excessive levels of contamination. Groundwater models can be used to compute the transport of contaminants and to evaluate the attenuation capacity of a site to prevent excessive loadings of sludge contaminants and thus prevent the pollution of the groundwater beneath a disposal site.

The major mechanisms of contaminant transport in groundwater are advection, dispersion, adsorption, and degradation. The models used to describe these processes can be lumped into various categories: unsaturated and saturated zone models; analytical and numerical models; and one-, two-, and three-dimensional models. Models used to make predictions should be validated. The process of validation consists of three steps: (1) sensitivity analysis, (2) calibration, and (3) verification. The validity of any model's predictions also depend on the quality of the input data and the validity of the assumptions used in the development of the model.

Sludge disposal criteria development is based on a back-calculation scheme utilizing groundwater transport models. Some of the basic concepts used in the analysis are briefly reviewed here. There exists a given mass of contaminant, expressed as its concentration per volume of leachate that discharges over time at some defined seepage rate. The mass loading rate is the amount of contaminant entering the unsaturated zone model VADOFT as a square wave pulse per time increment. The results from VADOFT are then passed as input to the saturated zone model AT123D. The mass per volume is reduced during advective flow in the groundwater beneath the site.

The extent of contaminant movement will depend on the thickness of the aquifer and the flow rate. During movement, the contaminant is attenuated due to spreading caused by dispersion and diffusion. Depending on the type of contaminant, its movement may also be retarded by adsorption onto the soil materials. The contaminants may biodegrade, reducing the mass per volume while in transit. The processes of advection, dispersion, adsorption, and degradation influence the concentration of the solute at any position in space or time. The toxicity, mass loading rate, and the total mass of compound disposed are major factors that should be carefully considered in sludge land disposal practices.

If the attenuation capacity of the disposal site is carefully evaluated, site-specific contaminant loading rates and total masses disposed can be determined such that the attenuation capacity of the site is not exceeded and thus avoids disposal scenarios leading to serious groundwater contamination.

REFERENCES

1 Sacramento Area Consultants. 1979. *Sewage Sludge Management Program.* Eleven Volumes. NTIS PB 80 166721.

2 Kido, W. and W. R. Uhte. 1990. "Seven Years of Low-Cost Disposal at a 150-MGD Plant," presented at *IAWPRC 1990 Sludge Management Conference,* Loyola Marymount University, Los Angeles, California.

3 Finstein, M. 1990. Personal communication, Rutgers University, January 1990.

4 Farrel, J. B., G. K. Dotson, J. W. Stamm and J. J. Walsh. 1988. "The Effects of Sewage Sludge on Leachates and Gas from Sludge Refuse Landfills," presented at the *Residuals Conference of the Water Pollution Control Federation,* Atlanta, Georgia.

5 Madora, A. W. and J. F. Duffield. 1978. "Alternative Sources for Landfill Cover," *Public Works,* May.

6 Glebs, R. T. and T. Juszczyk. 1990. "Closure and Post-Closure Costs," *Waste Age,* March.

7 Griffin, R. D. 1988. *Principles of Hazardous Materials Management.* Chelsea, Michigan: Lewis Publishers, Inc., p. 55.

8 Lehr, J. H. 1985. "Calming the Restless Native: How Ground Water Quality Will Ultimately Answer the Questions of Ground Water Pollution," in *Ground Water Quality,* C. H. Ward, W. Giger, and P. L. McCarty, eds., New York: J. Wiley & Sons, pp. 11–24.

9 Zenz, D. R. and C. Lue-Hing. 1983. "Municipal Sludge Management," in *Environment and Solid Wastes,* C. W. Francis and S. I. Auerbach, eds., Woburn, MA: Butterworth Publishers, pp. 13–24.

10 U.S.EPA. 1990. "National sewage sludge survey," *Federal Register* 40 CFR Part 503, Part III, Vol. 55, No. 218, Fri. Nov. 9, 1990, pp. 47210–47283.

11 Griffin, R. A. and W. R. Roy. 1986. "Feasibility of Land Disposal of Organic Solvents: Preliminary Assessment," Environmental Institute for Waste Management Studies, Report No. 10, The University of Alabama, Tuscaloosa, Alabama, 57 pp.

12 Griffin, R. A., W. R. Roy, J. K. Mitchell and R. A. Mitchell. 1987. "Feasibility and Limitations of Land Disposal of Organic Solvents," Environmental Institute for Waste Management Studies, Report No. 19, The University of Alabama, Tuscaloosa, Alabama, 47 pp.

13 Ogata, A. 1970. "Theory of Dispersion in a Granular Medium," U.S. Geological Survey Professional Paper 411-I. U.S. Department of Interior, Washington, D.C.

14 Bear, J. 1972. *Dynamics of Fluids in Porous Media.* New York: American Elsevier.

15 Boast, C. W. 1973. "Modeling the Movement of Chemicals in Soil by Water," *Soil Science,* 115:224–230.

16 Freeze, R. A. and J. A. Cherry. 1979. *Groundwater*. Englewood Cliffs, New Jersey: Prentice-Hall, 604 pp.

17 Rao, P. S. C. 1974. "Pore Geometry Effects on Solute Dispersion in Aggregated Soils and Evaluation of a Predictive Model," Ph.D. Dissertation, University of Hawaii, Dissertation Abstracts International 36:527-B.

18 van Genuchten, M. Th., P. J. Wierenga and G. A. O'Connor. 1977. "Mass Transfer Studies in Sorbing Porous Media III. Experimental Evaluation with 2,4,5-T," *Soil Sci. Soc. Am. J.* 41:278–285.

19 Roy, W. R., I. G. Krapac, S. F. J. Chou and R. A. Griffin. 1990. "Batch-type Procedures for Estimating Soil Adsorption of Chemicals," Technical Resource Document U.S.EPA/530-SW-87-006-F. Cincinnati, Ohio: U.S. Environmental Protection Agency, 96 pp.

20 Griffin, R. A. and W. R. Roy. 1985. "Interaction of Organic Solvents with Saturated Soil-Water Systems," Environmental Institute for Waste Management Studies, Tuscaloosa, Alabama. Report No. 3, The University of Alabama, 86 pp.

21 U.S.EPA. 1989. "Landfilling of Sewage Sludge," Technical Support Document, Office of Water Regulations and Standards. Washington, D.C.: U.S. Environmental Protection Agency, pp. 3–27.

22 Faust, C. R. 1982. "Uncertainty in Contaminant Migration Predictions," Unpublished final report submitted to the U.S. Environmental Protection Agency, Contract No. 68-01-6464.

23 Gelhar, L. W. and G. J. Axness. 1981. "Stochastic Analysis of Macrodispersion in 3-Dimensionally Heterogeneous Aquifers," Report No. 8, Hydrologic Research Program. New Mexico Institute of Mining and Technology, Soccoro, New Mexico.

24 Landon, R. A. 1978. "Pollution Prediction Techniques for Waste Disposal Siting— A State of the Art Assessment," in *Proceedings of the First Annual Madison Waste Conference*, University of Wisconsin, Madison, Wisconsin, pp. 169–198.

25 Gray, D. H. and W. J. Weber. 1984. "Diffusional Transport of Hazardous Waste Leachate Across Clay Barriers," in *Seventh Annual Madison Waste Conference*, University of Wisconsin, Madison, Wisconsin, pp. 73–389.

26 Bumb, A. C., C. R. McKee, J. M. Reverand, J. C. Halepaska, J. I. Drever and S. C. Way. 1984. "Contaminants in Groundwater: Assessment of Containment and Restoration Options," *Conference on Management of Uncontrolled Hazardous Waste Sites*. Hazardous Materials Control Research Institute, Washington, D.C.

27 Mercer, J. W. and C. R. Faust. 1981. *Ground-Water Modeling*. Worthington, Ohio: National Water Well Assoc.

28 Wang, H. F. and M. P. Anderson. 1982. *Introduction to Groundwater Modeling*. San Francisco, California: Freeman Publishers.

29 Bedient, P. B., R. C. Borden and D. I. Leib. 1985. "Basic Concepts for Groundwater Transport Modeling," in *Ground Water Quality*, C. H. Ward, W. Giger and P. L. McCarty, eds., New York: John Wiley & Sons, New York, pp. 512–531.

30 Anderson, M. P. 1979. "Using Models to Simulate the Movement of Contaminants through Groundwater Flow Systems," *Critical Reviews Environmental Control*, 9:97–156.

31 Pinder, G. F. and W. G. Gray. 1977. *Finite Element Simulation in Surface and Subsurface Hydrology*. New York: Academic Press.

32 Konikow, L. F. and J. D. Bredehoeft. 1978. "Computer Model of Two-Dimensional Solute Transport and Dispersion in Groundwater," in *Book 7: Automated Data*

Processing and Computations, Techniques of Water Resources Investigations of the U.S. Geological Survey. Washington, D.C.

33 Sternberg, Y. 1985. "Mathematical Models of Contaminant Transport in Groundwater," Environmental Institute for Waste Management Studies, Open File Report No. 4. Tuscaloosa, Alabama: University of Alabama.

34 U.S.EPA. 1989. "Risk of Unsaturated/Saturated Transport and Transformation Interactions for Chemical Concentrations (RUSTIC), Volume 1: Theory and Code Verification," prepared by Woodward Clyde Consultants, HydroGeologic, and AQUA TERRA Consultants for the Office of Research and Development, Environmental Research Laboratory, Athens, GA. Contract No. 68-03-6304.

35 U.S.EPA. 1989. "RUSTIC Documentation, Volume II: User's Guide," prepared by Woodward-Clyde Consultants, Hydro-Geologic, and AQUA TERRA Consultants for the Office of Research and Development, Environmental Research Laboratory, Athens, GA. Contract No. 68-03-6304.

36 van Genuchten, M. T., 1985. "Connective-Dispersive Transport of Solutes Involved in Sequential First-Order Decay Reactions," *J. Computers Geosci.,* 11:129–147.

37 U.S.EPA. 1986. *Development of Risk Assessment Methodology for Municipal Sludge Landfilling,* Office of Health and Environmental Assessment, U.S. Environmental Protection Agency, Cincinnati, Ohio.

38 Wagenet, R. J. and J. L. Hutson. 1987. "LEACHM: Leaching Estimation And CHemistry Model. A Process-Based Model of Water and Solute Movement, Transformation, Plant Uptake and Chemical Reactions in The Unsaturated Zone," *Continuum, Vol. 2 (Version 1.0).* Ithaca, New York: Water Resources Institute, Cornell University.

39 Wagenet, R. J. and J. L. Hutson. 1989. "LEACHM: Leaching Estimation And CHemistry Model. A Process-Based Model of Water and Solute Movement, Transformation, Plant Uptake and Chemical Reactions in the Unsaturated Zone," *Continuum. Vol. 2 (Version 2.0).* Ithaca, New York: Water Resources Institute, Cornell University.

40 PRC. 1989. "Peer Review—Standards for the Disposal of Sewage Sludge," Peer Review Committee organized by Cooperative States Research Service Technical Committee W-170.

41 O'Neal, K., J. Weisman, V. A. Hutson, S. E. Keane, S. G. Buchberger and B. H. Lester. 1990. *Development of Risk Assessment Methodology for Surface Disposal of Municipal Sludge,* Cincinnati, Ohio: Environmental Criteria and Assessment Office, U.S. Environmental Protection Agency.

42 Davis, L. A. and S. P. Neuman. 1983. *Documentation and User's Guide: UNSAT2— Variably Saturated Flow Model.* U.S. Nuclear Regulatory Commission Report, NUREG/CR-3390, Washington, D.C.

43 Huyakorn, P. S., A. G. Kretschek, R. W. Broome, J. W. Mercer and B. H. Lester. 1984. "Testing and Validation of Models for Simulating Solute Transport in Ground-Water," International Ground Water Modeling Center, Holcomb Research Inst., HRI No. 35, Indianapolis, Indiana.

44 Yeh, G. T. 1981. "AT123D: Analytical Transient One-, Two-, and Three-Dimensional Simulation of Waste Transport in the Aquifer System," ORNL-5602. Environmental Sciences Division, Pub. No. 1439. Oak Ridge National Laboratory, Oak Ridge, TN.

45 Huyakorn, P. S., M. J. Ungs, E. A. Sudicky, L. A. Mulkey and T. D. Wadsworth. 1985. *RCRA Hazardous Waste Identification and Land Disposal Restrictions: Groundwater*

Screening Procedures. Technical Report prepared for the Office of Solid Waste, U.S.EPA, under contract No. 68-01-7075.

46 Lester, B. H., P. S. Huyakorn, H. O. White Jr., T. D. Wadsworth and J. E. Buckley. 1986. *Analytical Models for Evaluating Leachate Migration in Groundwater Systems,* prepared for the Office of Solid Waste, U.S.EPA, Washington, D.C.

47 Miller, D. W. 1985. "Chemical Contamination of Ground Water," in *Ground Water Quality,* C. H. Ward, W. Giger and P. L. McCarty, eds., New York: J. Wiley and Sons, pp. 39–52.

CHAPTER 8

Incineration of Municipal Sewage Sludge

INCINERATION of sewage sludge has been determined to be an environmentally sound and cost effective means for the disposal of sewage sludge. Approximately sixteen percent (i.e., 865,000 dry metric tons) of the sewage sludge removed annually from wastewater in the United States is disposed-of through incineration.

On February 19, 1993, the United States Environmental Protection Agency (U.S.EPA) promulgated Standards for the Use and Disposal of Sewage Sludge (i.e., 40 CFR Part 503). Under Subpart E of the Part 503 Regulation, U.S.EPA established new requirements and limits governing the incineration of sewage sludge (biosolids).

With the promulgation of the Part 503 Regulation, many wastewater treatment agencies were forced to reevaluate emission controls on their incinerator installations and the manner in which their incinerators were operated, in order to achieve compliance with the new emissions limits and regulatory requirements. This chapter discusses several options for retrofitting incinerators to meet new regulatory requirements.

In addition, incinerator installations must also dispose of incinerator ash, which is often done by landfilling of the ash. This chapter also discusses some innovative beneficial uses for incinerator ash as an alternative to landfilling.

Allen Baturay, Carlson Associates, Manassas, VA; Cecil Lue-Hing, Bernard Sawyer, and David Zenz, Metropolitan Water Reclamation District of Greater Chicago, Chicago, IL; Robert Dominak, Northeast Ohio Regional Sewer District, Cleveland, OH.

FEASIBILITY AND RELATIVE COST OF RETROFITTING INCINERATORS FOR REGULATORY COMPLIANCE

INTRODUCTION

The Part 503 Regulation limits the emission of arsenic, beryllium, chromium, lead, mercury, nickel and total hydrocarbons (THC) from the sewage sludge incinerators.

Metal Emissions

The Part 503 Regulation contains equations to calculate site-specific numeric emission limits for arsenic, cadmium, chromium, lead, and nickel. These limits were determined by U.S.EPA not to cause an incremental carcinogenic risk level higher than one in 10,000 for the Highly Exposed Individual (HEI) [1]. The Part 503 emission limits for beryllium and mercury are identical to U.S.EPA's National Emission Standards for Hazardous Air Pollutants (NESHAPs).

Additional information concerning the techniques used by U.S.EPA to determine risk associated with the incineration of sewage sludge and other sludge disposal methods may be found in a U.S.EPA document titled ''A Guide to the Biosolids Risk Assessments for the EPA Part 503 Rule (EPA832-B-93-005)'' dated September 1995.

Total Hydrocarbon (THC) Emissions

At the time the Part 503 Regulation was promulgated, U.S.EPA concluded that it was infeasible to establish a limit that would protect public health for each organic pollutant contained in the incinerator exhaust stack gases. As a result, U.S.EPA determined that THC emissions should be used as a surrogate for the emission of organic pollutants and established an THC operational standard of 100 ppm, as propane, corrected to 7% oxygen and 0% moisture, on a monthly average basis.

In addition, U.S.EPA also noted that in the judgment of its Administrator, the operational standard for THC of 100 ppm, protects the public health from reasonably anticipated adverse effects of the organic pollutants in the incinerator stack exit gases [1].

In order to demonstrate compliance with the Part 503 THC limit, POTWs that practice incineration were required to install Flame Ionization Detectors (FIDs) to continuously monitor and record the THC concentration in the exit stack gases. (It should be noted that many POTWs are experiencing major problems with the operation of the Total Hydrocarbon

Continuous Emissions Monitoring Systems currently available on the market.)

However, U.S.EPA recently indicated that it is considering revising the Part 503 Regulation to allow the monitoring of carbon monoxide (CO) or incinerator exhaust gas temperature in lieu of THC in order to demonstrate compliance with the THC limit. It is currently anticipated that the U.S.EPA will formally proposed these changes, for public comment, during the fourth quarter of 1996. U.S.EPA also estimated that a number of multiple hearth type furnaces would be required to practice afterburning (either internal or external) in order to meet the THC limit.

It is important to note that in addition to the Part 503 Regulation, some states have already imposed more restrictive regulations that require the installation of air pollution control equipment meeting the Best Available Control Technology (BACT), or high-temperature afterburning or have established an exit stack gas CO limit of 100 ppm, as assurance for the total destruction of toxic organic compounds [2,3].

ORGANIC EMISSIONS

Organic emissions from sewage sludge incinerators include: (1) products of incomplete combustion; (2) compounds contained in sludge that are not completely burned during combustion; (3) products synthesized from products of incomplete combustion in low temperature zones of the incinerator.

In perfect combustion, carbon and hydrogen contained in organic material are oxidized to form carbon dioxide and water. However, when there is imperfect mixing of organic material with air or when there is not enough time to complete the oxidation reactions, or when the flame temperature is not adequate, perfect combustion can not be achieved.

If oxidation reactions are slowed down by the unavailability of oxygen or are prematurely terminated by flame quenching, certain intermediate compounds can form during the process and appear in the final effluent. These products of incomplete combustion usually are paraffins, olefins, aromatics, and acetylenes. The reactions of these hydrocarbons with other species may produce compounds like aldehydes, ketones, alcohols, and acids. In the presence of chlorides, sulfur, and nitrogen, more complex compounds may be formed. The compounds formed during combustion may include numerous species in parts-per-billion to parts-per-trillion concentrations which are impossible to identify.

However, researchers agree that regardless of the type of material being incinerated, certain compounds are most frequently found as the products of incomplete combustion in the flue gases. These compounds usually constitute about 10 to 15% of the total amount of organics emitted from all types of incinerators, and are represented in Table 8.1. The test data from sewage sludge incinerators indicate the presence of similar compounds [4,5]. It should be noted that the FID basically measures the concentration of carbon atoms (THC), excluding CO, in a gas sample without identifying the species.

THC EMISSIONS FROM SEWAGE SLUDGE INCINERATORS

The two most common types of sewage sludge incinerators in the U.S. are the multiple hearth furnace and the fluid bed combustor. The test data indicates that the THC emission from properly designed and operated fluid bed combustors are substantially less than the regulatory limit of 100 ppmv.

On the other hand, the average THC emissions from the multiple hearth furnaces, during testing conducted in late 1980s was 150 ppmv as propane at 7% oxygen [6]. However, by employing a series of THC emission reduction strategies, as discussed in this chapter, many if not all of the multiple hearth furnaces located within the United States are now in compliance with the Part 503 THC limit.

The THC measurements from thirty-three multiple hearth installations, in 1989, are presented in Figure 8.1 [6,7]. Although the data indicated that the THC emission level were highly dependent on the top hearth or afterburner temperature, the variations in THC emissions at a given temperature were widespread (one to two orders of magnitude). This wide range of variability and the shape of the envelope around the data points suggested that THC emission levels from multiple hearth furnaces were also dependent on other parameters, especially if the top hearth or afterburner temperature was below 1400 to 1500°F.

Below is an examination of the fundamentals of the processes involved in the drying and burning of sewage sludge in multiple hearth furnaces, which suggest some possible explanations for these conflicting results.

TABLE 8.1. Most Frequently Found Products of Incomplete Combustion.

Volatiles	Semivolatiles
Benzene	Naphthalene
Toluene	Phenol
Carbon tetrachloride	Bis(2-ethylhexyl)-phthalate
Chloroform	Diethylphthalate
Methylene chloride	Butylbenzylphthalate
Trichloroethylene	Dibutylphthalate
Tetrachloroethylene	
1,1,1-Trichloroethane	
Chlorobenzene	

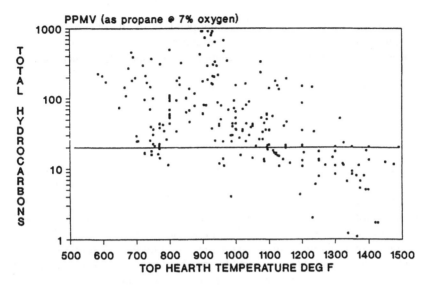

Figure 8.1 THC emissions from multiple hearth furnaces.

PROCESS DESCRIPTION

The combustible fraction in sludge solids is characterized as volatile or char (fixed carbon). This subdivision is not sui generis, but process dependent. The amount of char produced depends on the composition and volatile combustible burning characteristics of the sludge. While efficient combustion at high temperatures with plenty of air minimizes char production, poor combustion with substoichiometric amounts of air yields higher amounts of fixed carbon.

The multiple hearth furnace is a very versatile piece of equipment that can be operated in different modes. This inherent flexibility is one of the reasons for the apparent differences in performance between units. The combustible portion of sludge solids in a multiple hearth furnace is burned in two or more phases, each having many steps. The first phase is the burning of volatile combustibles in a so-called "burning hearth(s)" or "hottest hearth," which usually happens somewhere in the middle of the furnace after the feed sludge has substantially dried.

The second phase is the burning of char, which takes place in the hearths immediately following the "burning hearth." An optional third phase is an afterburner stage, which may be at the top hearth or in a separate external afterburner.

In addition to these phases, depending on the sludge characteristics and operating parameters, distillation of certain types of organics may take place at the drying hearths. If the temperature of the hearth above the burning zone is too low for the destruction of the distilled organics, a portion of the volatilized organic matter may escape the combustion zone unburned. These organics usually are the cause for odor problems, especially with polymer-conditioned sludges.

Pyrolysis Zone

In the "burning hearth(s)," substantially dried sludge is heated by the radiant energy provided by the orange luminous flame above the bed. There is very little or no oxygen present at the surface of the sludge cake and the thermal conductivity is very low, which restricts heating the interior of the sludge bed. However, in a multiple hearth furnace sludge is turned over and plowed by rabble teeth approximately every 30 seconds. Heat absorbed by radiation is dissipated by evaporating the remaining moisture and pyrolizing volatile combustibles on the surface of the bed until the sludge bed gets hot enough to volatilize the remaining volatile combustibles.

Sludge combustibles burn with a so-called "diffusion flame," which is graphically presented in Figure 8.2. The radiant energy transferred from the flame in the primary combustion zone provides the driving force for the pyrolytic precombustion reactions. The pyrolytic reactions, in the absence of oxidants, promote the nucleation and particulate growth that cause radiation in the form of an orange flame while passing through the combustion zone.

Because no oxidizing radicals or agents are present in the precombustion zone, pyrolytic reactions lead to the formation of relatively large and complicated molecules that may be difficult to incinerate in the later stages of the multiple hearth.

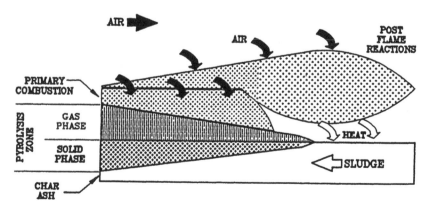

Figure 8.2 Burning sludge in a multiple hearth furnace.

At lower temperatures, pyrolytic reactions are likely to produce higher molecular weight compounds while higher temperatures encourage dehydrogenation and cracking of hydrocarbons to form lower molecular weight species. This step is suspected of producing most of the products of incomplete combustion (PICs) found in flue gases from multiple hearth furnaces. Most researchers agree that ''pyrolysis reactions are responsible for most incinerator emissions, even though the incinerator nominally operates with excess air'' [8].

Primary Combustion Zone

Volatilized combustibles are further heated and vaporized in the ''gas phase reaction zone'' and burned in the ''primary combustion zone'' with an orange luminous flame in the presence of air. The combustion with diffusion flame is controlled by mixing, turbulence, and the geometry of the system, not by the kinetics. The temperature and the length of the flame influence the effectiveness of combustion.

Higher flame temperatures with longer flames improve the effectiveness of combustion to oxidize volatiles and products of incomplete combustion coming from the primary combustion zone. However, it is important to note that the better combustion characteristics also cause evaporation of volatile metals and organometallic compounds that condense to form submicron particles in the postflame zone or in the scrubber.

The availability of combustion space (volume of burning hearth) can also be a factor influencing the combustion effectiveness. If the space above the burning hearth(s) is insufficient to accommodate the amount of heat released during the burning of volatiles, the combustion will be incomplete. The burning rate of the volatilized combustibles can be as high as 15 lb/hr/sq ft of hearth area. With the assumption of a heating value of 10,000 BTU/lb of volatiles and 3-ft hearth spacing, volumetric thermal loading can be as high as 50,000 BTU/cu ft/hr, which is rather high for burning dewatered sludge cake. Therefore, some multiple hearth incinerators are designed to have higher hearth spacing where the burning of volatilized combustibles takes place.

The amount of water vapor, type of sludge, excess air rate and air temperature, amount and type of auxiliary fuel used, location of active burners, rabbling pattern and efficiency, and the amount of sludge inventory in the furnace are additional factors that affect the flame temperature and combustion effectiveness.

Postflame Reactions

In the postflame reaction zone, hydrocarbons that leave the primary combustion zone continue to oxidize and emit luminous radiation. In this zone, both pyrolytic and oxidative reactions may occur. The type of hydrocarbon compounds found in this zone may originate in the flame if the combustion is rich. Some of the hydrocarbons may be synthesized by mixing the unburned hydrocarbons from the cold region with the hot hydrocarbons coming from the flame zone. Researchers suggest that "thermal degradation in postflame regions of the incinerator, as opposed to the flame zones, control the incinerability of hazardous materials" [9].

In the postflame reaction zone, most of the products of incomplete combustion, especially carbon monoxide, are oxidized to CO_2 with the radical ⁻OH. However, because this reaction is temperature-sensitive, and if gases in the post-combustion zone are quenched quickly by being exposed to wet sludge or mixed with large amounts of cold air (too much excess air), a high concentration of CO will be frozen into the flue gas. This may be a possible explanation for the high CO concentrations observed in flue gases from many multiple hearth furnaces.

Char Burnout Phase

The burning of char takes place in the hearths immediately following the volatile burning hearth(s) at a very slow rate compared to the volatilized combustibles. Burning char is different than oxidizing volatiles associated with gas to gas reactions. First oxygen diffuses into char to form carbon monoxide, producing heat in a manner similar to a charcoal broiler. Heat liberated by oxidizing carbon radiates from the glowing char to the refractory lining in the furnace, which is then cooled by the large amounts of air flowing upward.

Reducing conditions are prevalent near the surface of the char, and the bed is probably hotter at this point than at any other in its long spiral travel throughout the furnace. It is suspected that carbon monoxide produced during the char burning may not be totally oxidized in the upper hearths and shows up in the flue gases. The data collected at some installations shows that vaporization of volatile metals takes place mostly in this phase.

Afterburner Phase

The afterburner requirements for a multiple hearth furnace are dependent on the nature and the quantity of the unburned combustible species coming from the furnace. While some complex products of incomplete combustion are easily incinerated with the presence of a flame at moderate temperatures, others such as methane and carbon monoxide require high temperature and good mixing.

Researchers agree that "no organic can survive under oxidizing conditions at temperatures higher than about 1500°F, providing fuel and air have been perfectly mixed. In the real world combustion system, the minimum temperature requirement is around 1600°F, since perfect mixing is difficult to achieve" [8]. However, the risk assessment conducted by U.S.EPA has concluded that complete destruction of the organic compounds is not required and that the Part 503 THC limit of 100 ppm is protective of public health and the environment.

THC EMISSION REDUCTION TECHNIQUES

Operational Improvements

One technique for controlling THC emissions is to limit their formation in the furnace. The sludge burning temperature profile for the multiple hearth furnace is presented in Figure 8.3. Increasing the temperatures in the pyrolysis, primary combustion and postflame reaction zones should result in the reduction of THC emissions. Reducing the amount of excess air and preheating the combustion air at lower hearths will increase the "burning hearth" temperature and will produce longer and hotter primary combustion and postflame reaction zones.

In some cases, firing auxiliary burners at or above the combustion hearth should also be effective in keeping the postflame reaction zone hotter

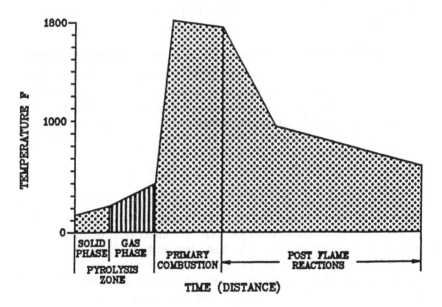

Figure 8.3 Sludge burning temperature profile.

and longer, thus destroying the distilled volatile organics and products of incomplete combustion formed in the burning hearth. However, it should be noted that hotter burning hearth temperatures usually cause slagging problems and increase the volatile metal emissions and oxidation of chromium from trivalent to the more toxic hexavalent form, as shown later in Figures 8.6 and 8.7 [10].

The amount and the type of auxiliary fuel used in the furnace, together with the moisture content of sludge, could be a key to controlled burning at high temperatures. The exceptional performance of a multiple hearth furnace, which routinely achieves an average THC emission level of 15 ppmv as propane with only an average top hearth temperature of about 750°F, is an example of this type of operation [11]. At this installation, sludge is lime conditioned and dewatered to about 16% solids with vacuum filters and burned in a multiple hearth furnace with large quantities of natural gas. The thermal load from the auxiliary fuel compared to sludge combustibles is so large that any product of incomplete combustion (PIC) from burning sludge is completely destroyed with the natural gas flame.

Because the thermal input from the natural gas behaves like a thermal flywheel and the presence of lime raises the ash fusion temperature, the process is very stable and not sensitive to changes in sludge characteristics or feed rate. These changes are easily and quickly compensated by burner adjustments. The much disputed "thermal jump" occurs in this type of operation. The stack is usually clean and void of the yellow haze associated with the operation of some multiple hearth furnaces. It appears that this multiple hearth furnace does not need an afterburner, but requires large quantities of auxiliary fuel.

On the other hand, the 1989 data [6,7] indicated that some multiple hearth furnaces were being operated under conditions that produce THC levels in the hundreds, which requires well-mixed and high-temperature (above 1000°F) afterburners (either internal or external) to meet the THC level of 100 ppmv. In addition, the analysis of the data indicated that there is a wide range of variability of THC emissions from the same installation. It appeared that the THC emissions can vary from day to day and probably are dependent on sludge characteristics, operational procedures, feed rate fluctuations, or unstable conditions.

Because the process of incinerating sewage sludge in a multiple hearth furnace is not as stable as that found for a fluidized bed combustor, control of the THC emissions to meet the Part 503 limits with normal sludges required operational changes. The best place to burn sludge, considered by most operators, is the middle of an "out-hearth" where the flame is less restricted and the temperatures are uniform and the controls (air and heat) are most effective (Figure 8.4). However, it is not always possible to keep the fire at this location all the time. With changing conditions and

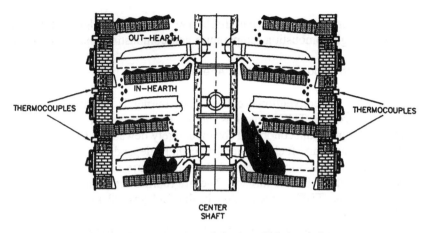

Figure 8.4 Best place to burn sludge in multiple hearth furnace.

operating parameters, fire often uncontrollably moves to other places where burning is less efficient (Figure 8.5).

When the fire moves to the periphery of an "in-hearth," the flame temperature is usually lower because of the wall effect and the necessity of preventing melting of the ash at the drop holes. Moving the fire to a more favorable location may take hours and cause prolonged unstable conditions.

The analysis of the data from case studies indicated that the THC emissions from multiple hearth furnaces can be reduced by operational improvements and that compliance can be readily achieved with the instal-

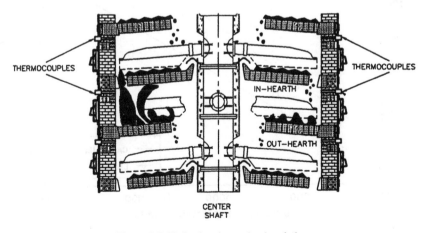

Figure 8.5 Sludge burning under drop holes.

lation of an internal or external afterburner. (See Figures 8.6 and 8.7 for effects of high combustion temperature and combustion characteristics.)

"Zero Hearth" and Conventional Afterburner

Although the conventional external afterburner is the most effective way of controlling THC emissions, the prohibitive cost of auxiliary fuel makes this option uneconomical for moderately dewatered sludges (20 to 25%). The auxiliary fuel requirement usually depends on the operating temperature of the afterburner and the temperature of gases exhausting from the multiple hearth furnace. The afterburner temperature in return depends on the THC concentration in the fuel gas exhausting from the multiple hearth.

The 1989 data [6,7] showed that the THC concentration in flue gas exhausting from the multiple hearth furnaces could be from 10 to 1000 ppmv as propane. As shown before, a multiple hearth furnace producing flue gas with low THC concentration may get by with 1000°F afterburning. A test conducted at a multiple hearth furnace equipped with a conventional afterburner indicated that THC reduction at 1200°F afterburning is about 50% [12]. The result of this test is presented in Figures 8.8 and 8.9.

CADMIUM and LEAD EMISSIONS
MULTIPLE HEARTH FURNACES

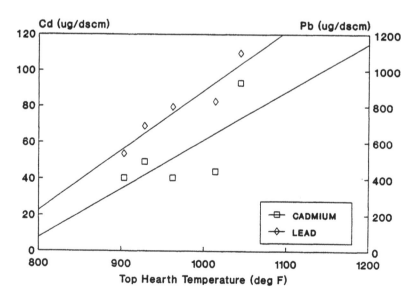

Figure 8.6 Effect of high combustion temperature on metal emissions.

Cr CONVERSION in MULTI-HEARTH FURNACE
Cr+6 Conversion vs THC Concentration

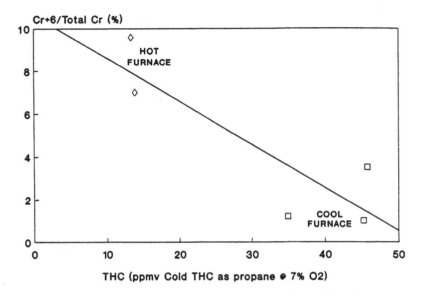

Figure 8.7 Effect of combustion characteristics on oxidation of chromium.

Although it is possible to convert the top hearth of a furnace into a so-called "zero hearth" afterburner, such conversion may not be completely successful in certain installations in limiting THC emissions due to poor mixing and limited retention time. As a result, a number of incinerator operators have converted their second or third hearths into afterburning zones.

Converting an existing hearth into an afterburning zone is feasible as long as the number of hearths can be reduced without a reduction in incineration capacity. If there is excess incineration capacity, then a hearth can be converted into an afterburning zone without adversely affecting the operation of the system.

If an external afterburner is constructed, a larger Induced Draft (ID) fan will be required to handle the increased gas flow due to the additional fuel burning in the afterburner. With conventional afterburning, waste heat recovery becomes a necessity. The waste heat can be recovered with a "direct recovery system," where the sludge is preheated or dried by flue gases, or "indirect recovery," where steam is generated with a waste heat boiler. At one incineration facility, the use of the waste heat boilers resulted in a $1 million per year savings in fuel costs.

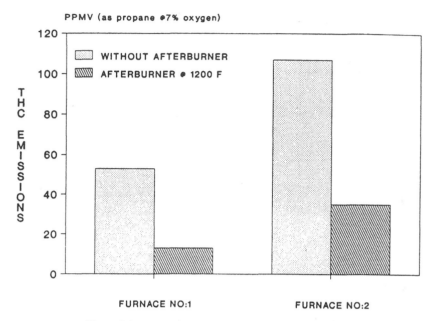

Figure 8.8 Conventional afterburner performance at 1200°F.

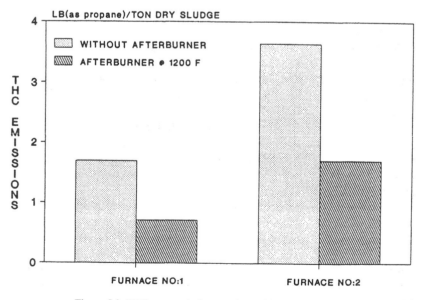

Figure 8.9 THC mass emission rate from 1200°F afterburning.

500

Depending on the size of the plant and the incineration system, steam can be used for indirect sludge drying, power generation, space heat, or hot water generation. In case of sludge drying, recently referred to as "thermal dewatering," the fuel demand for afterburning can be reduced and the furnace capacity may be increased. A typical conventional external afterburner arrangement with waste heat boiler is presented in Figure 8.10.

Anderson Recycle Configuration

The tall multiple hearth furnaces with extra hearths can be modified into one of the Anderson recycle configurations for burning dryer sludges (above 35% solids) [14].

If the sludge does not require extensive drying, the combustion products exhausted from the burning hearth can be split into two flows. While a portion of the gases are directed to the upper hearths for sludge drying, the remaining combustion products are directly exhausted from the burning hearth at relatively high temperatures to an afterburner and preferably followed by a waste heat recovery boiler. This type of configuration is more suitable for coupling with indirect sludge drying equipment. A variation of the Anderson recycle is presented in Figures 8.11 and 8.12.

Post-Scrubber Afterburner with Waste Heat Recovery

The regenerative and recuperative heat exchangers offer retrofit options for economic afterburning to reduce THC and CO emissions from multiple hearth furnaces [14]. Where required, these heat exchangers, coupled with

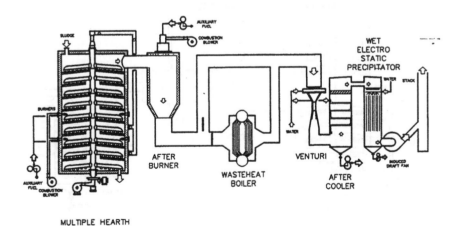

Figure 8.10 Conventional afterburner arrangement with heat recovery.

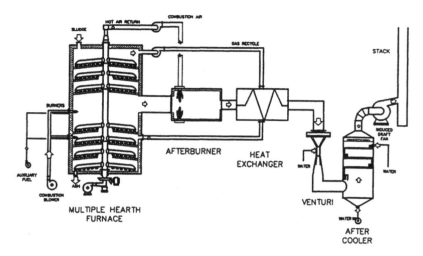

Figure 8.11 Anderson recycle arrangement with afterburner.

afterburners located at the downstream of the existing wet scrubbers, can operate at high temperatures (1800°F) without requiring prohibitive amounts of auxiliary fuel.

It should be noted, that as the afterburner temperature is increased, and as THC and CO emissions decrease, the emissions of Nitrogen Oxide (NOx) increase. This increase in NOx emissions is not acceptable at POTWs that are located either in a NOx non-attainment area or an Ozone

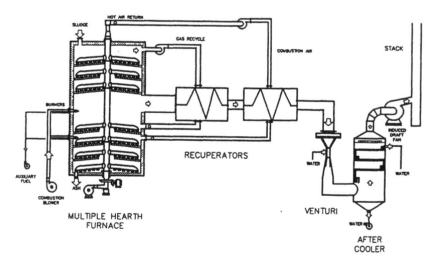

Figure 8.12 Anderson recycle with recuperative heat exchangers.

non-attainment area. Therefore, the afterburner temperature should be set at or near the minimum value required to ensure compliance with the THC limit.

The effectiveness of these systems is defined as:

$$\text{Effectiveness} = (T_{\text{exch out}} - T_{\text{gas in}})/(T_{\text{afterburner}} - T_{\text{gas in}})$$

where

$T_{\text{exch out}}$ = temperature of gas exiting heat exchanger
$T_{\text{gas in}}$ = temperature of gas from scrubber
$T_{\text{afterburner}}$ = afterburner temperature

The effectiveness of regenerative heat exchangers can be as high as 95% while recuperative exchangers are limited to about 80% due to the limitations of size and material of construction. The post-scrubber afterburning system configurations with recuperative and regenerative afterburners are presented in Figures 8.13 and 8.14. The temperature profile comparing conventional and post-scrubber afterburning with regenerative afterburner is presented in Figure 8.15.

An energy diagram for a multiple hearth furnace equipped with a conventional afterburner is presented in Figure 8.16. In this analysis it is assumed that the furnace is operating with 100% excess air on sludge solids and the gas outlet temperature is 700°F. The sludge is dewatered to 25% solid consistency having 72% combustibles with a heating value

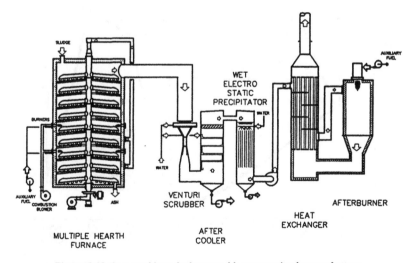

Figure 8.13 Post-scrubber afterburner with recuperative heat exchanger.

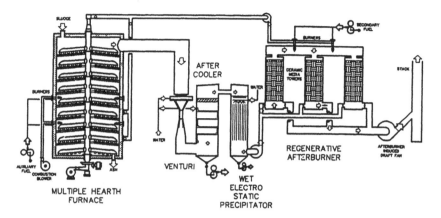

Figure 8.14 Post-scrubber regenerative afterburner arrangement.

of 9000 BTU/lb. The supplemental energy demand for this condition is 14 million BTU/dry ton of sludge incinerated.

The same sludge can be incinerated in a multiple hearth furnace equipped with an efficient post-scrubber regenerative afterburner operating at 1700°F with a supplemental energy requirement of only 2 million BTU/dry ton of sludge. An energy diagram for this condition is presented in Figure 8.17 [15]. As stated above, the afterburner temperature should be limited in order to limit the emissions of NO$_x$.

Operating conventional afterburners at higher temperatures as required

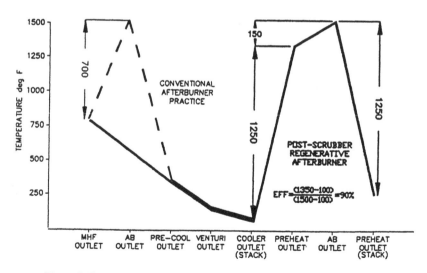

Figure 8.15 Temperature profile in conventional vs. post-scrubber afterburners.

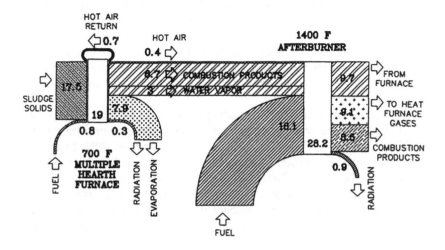

NUMBERS INDICATE MILLION BTU/HOUR

Figure 8.16 Energy diagram for conventional afterburner operating at 1400°F.

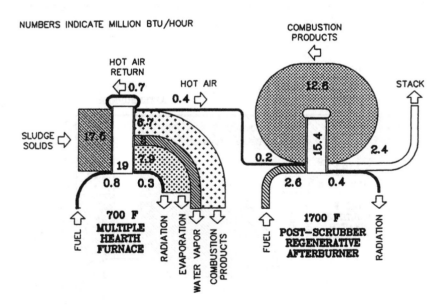

Figure 8.17 Energy diagram for post-scrubber regenerative afterburner operating at 1700°F.

505

by some states will further increase the auxiliary fuel demand as presented in Figure 8.18. The analysis indicates that for sludges with moderate (20 to 28%) solid concentrations, a multiple hearth furnace with a post-scrubber regenerative afterburner is more fuel efficient than a fluid bed combustor.

An energy diagram for a hot-windbox fluid bed combustor using 1200°F preheated air incinerating sludge with the same characteristics is presented in Figure 8.19. The fuel demands for the four options are comparatively presented in Figure 8.20.

The analysis indicates that depending on the sludge characteristics, the effectiveness of the heat exchanger and the afterburning temperature, the auxiliary fuel demand can be reduced by as much as 90% compared to conventional afterburners without heat recovery, and multiple hearth furnaces can be as fuel efficient as fluid bed combustors [16].

In addition to fuel saving, post-scrubber afterburning offers certain other advantages. Because the amount of additional NOx production in afterburners is in direct proportion to the amount of fuel consumption, post-scrubber afterburning with heat recovery offers less NOx emission than conventional afterburners.

Because most of the chlorides are expected to be removed in wet scrubbers located upstream of the afterburner, the possibility of forming chlorinated compounds in the afterburner is minimized. Two-stage combustion with a wet scrubber located between the furnace and the afterburner also offers a solution to the conflict between the necessity to achieve

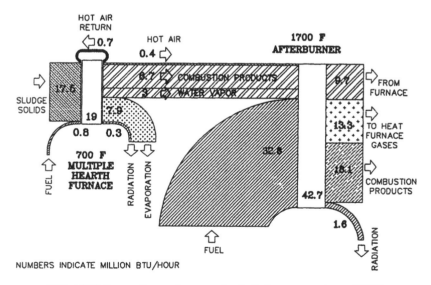

Figure 8.18 Energy diagram for conventional afterburner operating at 1700°F.

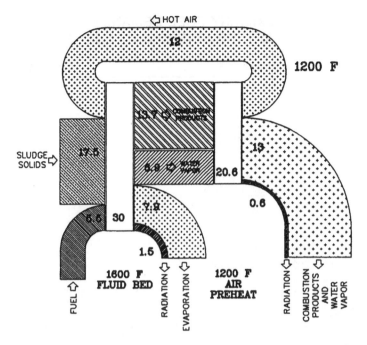

Figure 8.19 Energy diagram for hot-windbox fluid bed combustor with air preheat.

substantial destruction of organic compounds at high temperatures while minimizing the volatilization of metals.

The fact that the afterburner is downstream of the scrubber also eliminates the need to replace the existing scrubbers and off-gas handling equipment. Because these afterburners are installed after the existing wet scrubbers, the installation should be relatively simple. However, gases entering the heat exchangers should be clean to prevent fouling.

While both recuperative and regenerative heat exchangers can be used for all sizes of multiple hearth furnaces, the regenerative heat exchangers are more economical in larger sizes. Flue gases from several multiple hearth furnaces can be combined and afterburned efficiently in one regenerative afterburner system.

Catalytic Afterburner

The catalytic afterburner also offers a solution for the reduction of THC emissions. However, because of the heavy particulate loading in the flue gas from the incinerator, the catalytic burner has to be installed downstream of the air pollution control equipment to prevent problems with the catalyst.

AUXILIARY FUEL DEMANDS
MHF W/AFTERBURNERS and FLUID BED

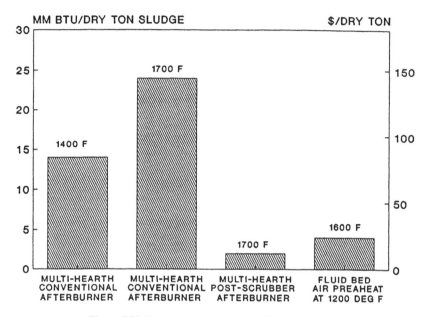

Figure 8.20 Energy demand comparison between systems.

Cleaned gases leaving the wet scrubber require preheating prior to entrance into the catalytic afterburner.

Wet Electrostatic Precipitator (ESP)

In addition to being a fine-particulate scrubber, the wet ESP is also a THC collector. Wet ESPs have been successfully used in the wood drying industry for the elimination of the blue haze plume associated with high-temperature wood dryers. In this application, gases leaving the dryers are first cooled with water to condense hydrocarbon vapors in order to form droplets that are collected in the wet electrostatic precipitator.

A test conducted with a pilot unit at a multiple hearth furnace installation indicated that the yellow haze plume, associated with the operation of some multiple hearth furnaces, can be eliminated with a wet ESP at some installations [17]. The test conducted at another installation indicated a reduction of 50% of THC's in stack gases (7 to 3.5 ppmv as propane) while the multiple hearth furnace was operating in a steady-state mode with good combustion characteristics [18]. (It should be noted that the yellow haze can also be reduce with the use of either internal or external afterburners.)

Corrosion has been a major problem with the use of ESPs. Therefore, proper corrosion resistant alloys must be used in the construction of these units.

Non-Thermal Systems

There are non-thermal systems that may be capable of selectively reducing the THC emissions from multiple hearth furnaces. These are carbon adsorption, chemical scrubbing, and ozone oxidation. However, no reliable data are available to determine the capability of such systems to reduce the type of organics and the products of incomplete combustion (PIC) emitted from sludge incinerators. It is currently anticipated that the Water Environment Research Foundation (WERF) will conduct testing of non-thermal systems during 1997 to 1998.

METAL EMISSIONS

The U.S.EPA's Part 503 numerical limits for the emission of arsenic, cadmium, chromium, lead, and nickel are based on the maximum annual ground level concentrations and exposure of the highly exposed individual (HEI) to these levels over 70 continuous years. These limits are site specific since they are dependent on the efficiency of the air pollution control equipment, the annual sludge feed rate, the metal concentrations in the sludge feed, the height of the exhaust stack, meteorlogical conditions, and the stack dispersion factor.

On the other hand, the following Part 503 emission limits for beryllium and mercury are identical to the NESHAPs limits, and therefore protective of human health and the environment:

- beryllium = 10 grams/24 hours
- mercury = 3200 grams/24 hours

Through formulae contained within the Part 503 Regulation, each POTW that practices incineration is required to establish maximum concentrations in the sewage sludge feed to its incinerators.

The Part 503 Regulatory limits for arsenic, cadmium, chromium, and nickel are determined by the following formula:

$$C = \frac{\text{RCS} \times 86{,}400}{\text{DF} \times (1 - \text{CE}) \cdot \text{SF}}$$

where

(1) C = Daily concentration of arsenic, cadmium, chromium, and nickel in the sewage sludge being fed to the incinerator in milligrams per kilogram of total solids (dry weight basis)

(2) CE = Sewage sludge control efficiency in hundredths. This value is determined by the following formula:

$$CE = \frac{\text{Mass in} - \text{Mass out}}{\text{Mass in}}$$

where

Mass in = Mass of metal in the sewage sludge being fed to the incinerator (in grams per hour)

Mass out = Mass of metals contained within the incinerator exhaust stack gases, after the air pollution control device, (in grams per hour)

(3) DF = Dispersion factor in micrograms per cubic meter per gram per second. This is a value that must be calculated using air dispersion modeling.

(4) RSC = Risk specific concentration in micrograms per cubic meter. The risk specific concentrations for arsenic, cadmium and nickel are as follow: (1) arsenic = 0.023; (2) cadmium = 0.057; and (3) nickel = 2.0. The risk specific concentration for chromium can be calculated by using either the values in the following table or equation:

- RSC table

Incinerator Type	RSC
Fluidized bed w/wet scrubber	0.65
Fluidized bed w/wet scrubber and wet electrostatic precipitator	0.23
Other types w/wet scrubber	0.064
Other types w/wet scrubber and wet electrostatic precipitator	0.016

- Equation

$$RSC = \frac{0.0085}{r}$$

where r = decimal fraction of the hexavalent chromium concentration in the total chromium concentration measured in the exit gas from the sewage sludge incinerator stack in hundredths.

(5) SF = Sewage sludge feed rate in metric tons per day (dry weight basis). This can either be the daily annual average feed rate or the design capacity of the incinerator.

(6) 86,600 = Conversion factor

The formula to calculate the Part 503 limit for lead is:

$$C = \frac{0.1 \times \text{NAAQS} \times 86,400}{\text{DF} \times (1 - \text{CE}) \cdot \text{SF}}$$

where:

NAAQS = National Ambient Air Quality Standard for lead. This value equals 1.5 micrograms per cubic meter.

METAL EMISSION REDUCTION TECHNIQUES

Pretreatment Program

One of the simplest manners to comply with the calculated metal limits is to reduce the quantity of the Part 503 regulated metals entering the plant, thereby reducing the quantity of metal in the sludge feed to the incinerator. This is done by determining who is discharging the regulated metals into the sewer system and ensuring that the quantities being discharged by each business or industry is below prescribed local limits.

A recent inspection of one wastewater treatment system revealed that a single plating firm was illegally discharging 20 pounds of cadmium into the sewer system each day. After the illegal discharge was eliminated, the concentration of the cadmium in the Plant's sludge dropped from 100 micrograms/dry kilogram (sludge) to 20 micrograms/dry kilogram.

Operational Improvements

Although the reduction of combustion temperatures in "burning hearth(s)" may reduce the emission of volatile and certain other metals, this approach may also cause an increase in the formation and emission of THC's [5,10]. Changes in sludge characteristics and conditioning methods may also reduce emissions of certain metals. Stack tests indicate that emissions of certain metals are affected by the presence of chlorides in sludge. Cadmium and lead emissions increase with the increase of chloride concentration in the sludge being incinerated.

Afterburning

Testing has determined that the installation of either an internal or external afterburner will decrease particulate emissions, thereby reducing metal emissions.

Extending Stacks

Most sewage sludge incinerators are equipped with stacks shorter than good engineering practice (GEP) heights. The Part 503 Regulation allows the extending of the existing stacks to the GEP heights to improve dispersion factors. The formula used to determine GED stack height is as follows:

$$H_g = H + 1.5L$$

where:

H_g = GEP stack height, measured from the ground level elevation at the base of the stack, in meters

H = Height of nearby structure(s), measured from the ground level elevation at the base of the stack, in meters

L = Lesser dimension, height or projected width, of nearby structure(s), in meters

In addition, the dispersion factor will also be improved if the temperature or velocity of the exit stack gas are increased. It should be noted that the dispersion factor is inversely proportional to the dispersion of the exit stack gases, i.e., as dispersion increases, the dispersion factor used to calculate the Part 503 metal limits decreases.

Improve Performance of Existing Scrubbers

Almost all of the sewage sludge incinerators in the U.S. are equipped with wet scrubbers (Venturi and/or impingement plate, cyclonic or packed tower). The performance of these scrubbers can be optimized by improving gas preconditioning and/or increasing pressure drops and/or replacing existing impingement plates with small-hole trays.

Increasing the Pressure Drop Across the Venturi

By increasing the pressure drop across the Venturi, particulate emissions and therefore the associated metal emissions are reduced. As a result, POTWs that are slightly above their metal limits should attempt to increase the pressure drop across the Venturi in order to achieve compliance.

Retrofit with Fine-Particulate Scrubbers

The particle size analysis data generated at one installation shows that, as expected, most of the volatile metal species penetrating the Venturi scrubbers installed at the sewage sludge incinerators are in submicron size, as presented in Figure 8.21.

Adding fine particulate scrubbers to the existing air pollution control equipment to collect these submicron size particulates is a practical and economical way to improve performance. Wet electrostatic precipitators (WESP), ionizing wet scrubbers (IWS), and electro dynamic Venturi (EDV) scrubbers offer this promise, but they are extremely expensive to install, operate and maintain. It should be noted that the use of Electrostatic Precipitators could potentially increase dioxin/furans in the exhaust stack gases.

IWS's have been used on a limited number of hazardous waste incinerators for fine particulate and metal removal. An IWS has been installed at a sewage sludge incinerator to replace a Venturi scrubber that did not meet the current particulate emission limit of 1.3 lb/dry ton sludge incinerated [20]. Performance tests indicated that the IWS performed better than the Venturi scrubber and achieved an emission rate of 0.6 lb/dry ton sludge.

Performance tests conducted at a fluid bed combustor equipped with a

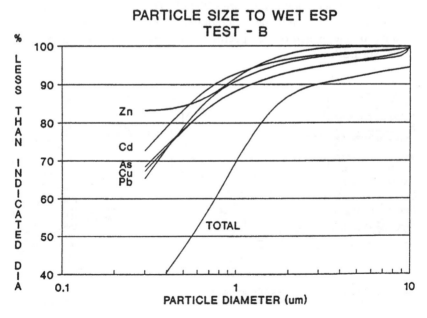

Figure 8.21 Typical metal particle size analysis exiting Venturi scrubber.

combination of a dry electrostatic precipitator and EDV scrubber indicates that a controlled particulate emission rate of 0.0006 gr/scf dry at 7% oxygen is achievable with such a combination, versus the current limit of approximately 0.03 gr/scf [21].

If gases leaving a wet scrubber are reheated to a temperature above their dew point, a fabric filter can be used as a second stage fine particulate scrubber following an existing Venturi or impingement type of scrubber.

In the United States, tests conducted with pilot and full-size units indicate that the wet electrostatic precipitator is capable of removing substantial amounts of submicron particulates and especially volatile metals found in stack gases from sewage sludge incinerators. The test data show that a combination of a wet ESP and a Venturi scrubber installed in series as shown in Figure 8.22 can achieve a controlled particulate emission rate of about 0.003 gr/scf dry at 7% oxygen [18].

A summary of these test results are presented in Figures 8.23 through 8.28. The tests identified as SITE-A, B, and C are pilot tests conducted at three different multiple hearth furnaces, and the test identified as EPA-9 represents a full-size wet ESP installation [16,22].

The data developed from the pilot tests and a full-size wet ESP unit provide the necessary information for sizing the precipitators for the required performance. The sizing criteria for a specific wet ESP manufacturer for an installation is shown in Figure 8.29 [22]. The SCA refers to "specific collection area," which is the area required at the collection electrode for the unit volume, which is site-specific and depends on the design features of the wet ESP.

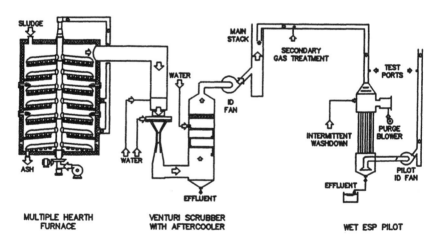

Figure 8.22 Wet ESP pilot test arrangement.

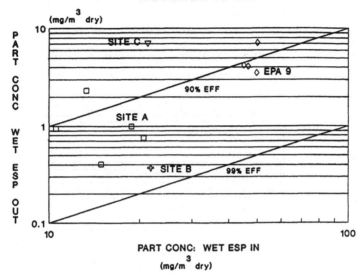

Figure 8.23 Wet ESP performance for total particulate removal.

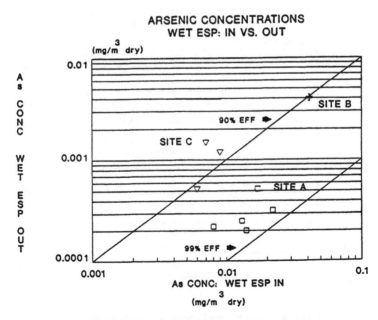

Figure 8.24 Wet ESP performance for arsenic removal.

515

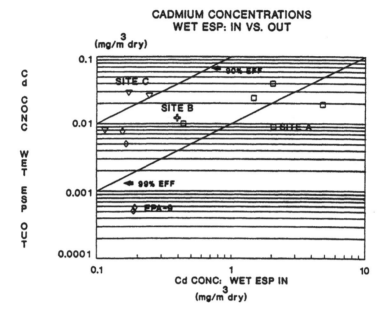

Figure 8.25 Wet ESP performance for cadmium removal.

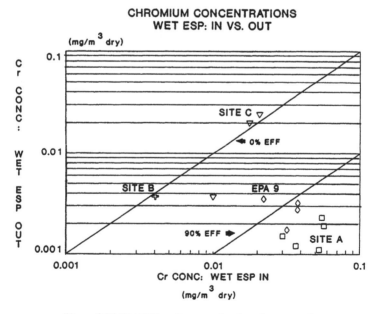

Figure 8.26 Wet ESP performance for chromium removal.

516

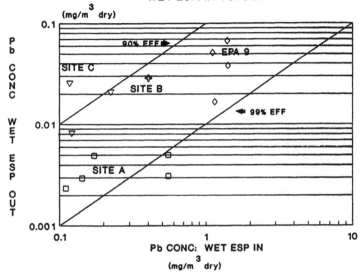

Figure 8.27 Wet ESP performance for lead removal.

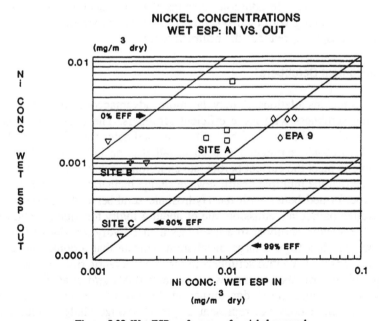

Figure 8.28 Wet ESP performance for nickel removal.

517

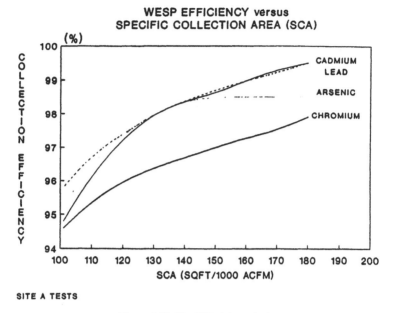

Figure 8.29 Wet ESP sizing criteria.

As previously reported, the wet ESP must be constructed out of corrosion resistant alloy. In addition, their installation, operation and maintenance costs, along with their potential to increase dioxin/furan emissions to the atmosphere, reduces their attractiveness.

Replacement Scrubbers

At some installations it may be more advantageous to replace the existing scrubbers with more efficient scrubbers. The Calvert scrubber, Hydrosonic scrubber, and fabric filter are candidates for this approach.

A Hydrosonic scrubber has been installed at an installation that was not able to meet the current particulate emission regulations. The test data indicate that the multiple hearth sewage sludge incineration system at this plant performed better with the Hydrosonic scrubber and improved the controlled particulate emission rate of about 4 lb/dry ton sludge achieved with the Venturi scrubber to a level less than 1 lb/dry ton sludge [23].

The test data from a large installation indicate that a combination of two low-pressure Venturi scrubbers in series and a fabric filter can achieve a controlled particulate emissions rate of about 0.001 gr/dscf at 7% oxygen. However, because the sludge burned at this plant is about 98% solids, this installation does not represent a typical sewage sludge incinerator installation [24].

RETROFIT COSTS

The estimating of the retrofit costs generally applicable to all existing sewage sludge incinerators is difficult. Because each installation requires a different degree of improvement in emission controls to satisfy the Part 503 Regulation, each retrofit situation is different. While some installations may require improvements in only a specific metal or organic emissions, others may require reduction in the emissions of both types of pollutants.

Some installations may reduce the maximum annual ground level concentration of metals by extending stack heights to improve dispersion coefficients, while other might not see an improvement in dispersion due to downwash or a complex topography.

It should also be noted that the second round of the Part 503 Regulation, which is currently scheduled to be published for public comment in 1999, may include requirements that the deposition pathway other than inhalation be considered for possible regulatory limits on dioxin/furan emissions.

The Association of Metropolitan Sewerage Agencies (AMSA) conducted a survey in 1995 to determine the dioxin/furan emissions from sewage sludge incinerators. The findings revealed that sewage sludge incinerators account for less that eight-one hundredths of a percent (i.e., 7 grams) of the total amount of dioxin/furans emitted to the atmosphere in the United States each year (i.e., 9200 grams).

The average emission factor for dioxin/furans from sewage sludge incinerators was found to be 7.61×10^{-2} µg TEQ/kg of dry sewage sludge incinerated. The 95% upper limit and lower confidence limits on the emission factor were 1.45×10^{-2} µg TEQ/kg and 2.20×10^{-2} µg [25].

Some researchers have indicated that the deposition pathway may be more demanding than the inhalation pathway, which is the basis for the current Part 503 Regulation. However, U.S.EPA stated in the Preamble to the Part 503 Regulation that the regulated metals were only found to be carcinogenic through the inhalation pathway.

The retrofit costs for meeting the regulatory requirements have to be reasonable to keep incineration a viable sludge management option. Over-conservative and unnecessary improvements to the existing incineration systems may cause incineration to be economically uncompetitive against other sludge management alternatives.

Value of Existing Equipment

The condition and status of the existing incinerator and ancillary equipment are very influential factors in the determination of the most suitable and economical retrofit option. If the major components of an incineration system are at the end of their useful life, it may be more advantageous to replace the old incinerator with more efficient equipment.

Replacement may be appropriate for some aging multiple hearth furnaces where fluid bed combustors offer a viable alternative. On the other hand, even if the incinerator requires extensive rehabilitation, replacement with new units may be more costly and risky because of the uncertainties involved in obtaining new permits to install and operate, and the length of time required to design, procure and install the new equipment.

At some installations the cost of replacement or alteration of supporting equipment, such as sludge conveying and feed systems, to accommodate different types of incinerators, may offset the advantages of new incinerators. Also, there is an intangible cost associated with achieving successful operation of a new incineration system.

After the installation of new incinerators, it usually takes a substantial amount of time to achieve a routine operation, to debug the system, to train operators and maintenance personnel, to conduct compliance tests and obtain operating permits. It can take as long as four years to design a system, prepare plans and specifications, select a contractor, install, start up and debug the equipment.

Improvements in Sludge Characteristics

Because incineration operation is influenced by sludge conditioning and dewatering methods, at some installations retrofit may include changes in these practices. Depending on the nature and the degree of improvement required to meet the regulatory emission limits, in some cases certain changes in sludge characteristics may provide enough improvement to satisfy these limits.

It has been suggested that lime in sludge influences the conversion of non-carcinogenic and non-water-soluble trivalent chromium to a more toxic carcinogenic and water-soluble hexavalent form. Therefore, those installations that do not comply with the health-risk criteria because of the hexavalent chromium emissions should consider changing their sludge conditioning practices if sludge is conditioned with lime.

However, it should be noted that elimination of lime from sludge conditioning causes an apparent decrease in ash softening and fusing temperatures. This may necessitate lowering of the "hottest hearth" temperature, which may result in increasing THC emissions. The elimination of lime will also cause distillation of certain volatile organic and odorous compounds in the drying zone of multiple hearth furnaces. This increase in THC emissions may require higher afterburner temperatures to control organic and odor emissions. Because lime conditioning of sludge makes ash hygroscopic, the elimination of lime will cause deterioration of the wet scrubber performance.

It has been demonstrated that burning grease along with the sewage sludge causes the formation of submicron size organic and organometallic particulate matter that is difficult to capture in the scrubbers [27]. Therefore, the reduction of the grease content in sludge should improve scrubber performance and lower controlled emissions.

The cost of improvements that can be achieved by changing the sludge characteristics is site-specific and process-dependent.

Retrofit Cost to Reduce Risk from Metal Emissions

Because the risk associated with the Part 503 metal emissions is based on the exposure of the highly exposed individual (HEI) to the metals of concern, the risk can be reduced by improving dispersion characteristics, reducing emissions, or a combination of both.

Extending Stacks

The Part 503 Regulation allows for the extension of stack heights to GEP heights in order to improve dispersion characteristics. Extending stack heights alone or in combination with other improvements may be less costly than a simple retrofit option for some installations.

Because most of the sewage sludge incinerators are equipped with stacks shorter than GEP heights, increasing stack heights is a feasible option to improve dispersion characteristics. A sensitivity study conducted by U.S.EPA shows that in the vast majority of cases dispersion factors can be improved considerably by extending stacks [1].

The cost of extensions or replacement stacks is dependent on the choice of material and configuration. Because of the usual presence of chlorides and sulfur in sludge, the stacks for sewage sludge incineration systems are usually made from type 316L stainless steel, more corrosion resistant material (i.e., Inconel or Hastelloy), or fiberglass material. However, taller stacks may require both steel and acid-resistant liners, aircraft warning lights, access ladders, and platforms. The cost of installing new stacks or extending the height of the existing stacks is site-specific.

Retrofit with Fine-Particulate Scrubbers

The cost of retrofitting the wet scrubbers depends on the degree of improvements required and the material of construction for the fine-particulate scrubber. In the selection of the material of construction, the acidic nature of the gases leaving the wet scrubbers requires consideration.

A cost estimate of retrofitting existing scrubbers with wet ESPs is shown in Figure 8.30. The cost of installing ionizing wet scrubbers (IWS)

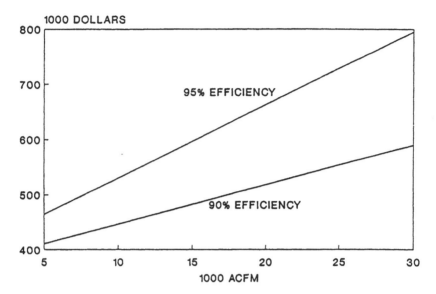

Figure 8.30 Capital cost for wet electrostatic precipitator.

and electro dynamic Venturi (EDV) scrubbers is expected to be comparable.

Internal Afterburing

The installation of an internal afterburner will decrease particulate emissions, thereby decreasing metal emissions.

Replacement Scrubbers

The cost of replacing the existing scrubbers with a Hydrosonic scrubber is shown in Figure 8.31. While the cost for a Calvert scrubber should be comparable to a Hydrosonic scrubber, the fabric filter is expected to be the most expensive option to install and operate.

Retrofit Cost to Reduce THC Emissions

It has been determined that fluid bed combustors do not have any problems meeting the THC emission limit of 100 ppmv as propane corrected to 7% oxygen and 0% moisture.

On the other hand, testing in 1989 [6,7] revealed that multiple hearth furnaces most likely would have to retrofitted in order to reduce THC emissions. It is generally agreed that either internal or external afterburning at temperatures of 1000°F would be necessary to consistently achieve this THC emission limit. The greatest cost to achieve the required THC limit is associated with the cost of fuel for the afterburners.

Conventional Afterburners

The capital cost for a conventional afterburner is also site-specific and depends on many factors, including space availability and building configuration. In most cases, addition of a conventional afterburner to a multiple hearth furnace necessitates the replacement of the existing scrubber and the induced draft fan to handle the increased combustion products due to burning additional auxiliary fuel in the afterburner. In most cases the cost of new scrubber and the ID fan, and the relocation of equipment would cost considerably more than the afterburner.

Post-Scrubber Afterburners

Although the equipment purchase cost of the regenerative afterburner is higher than that of the conventional or zero-hearth afterburner, the ease of installation and savings from auxiliary fuel consumption make this approach the most economic for most of the multiple hearth furnace installations burning sludge with 20 to 30% solids.

Because this afterburner is located downstream of existing equipment,

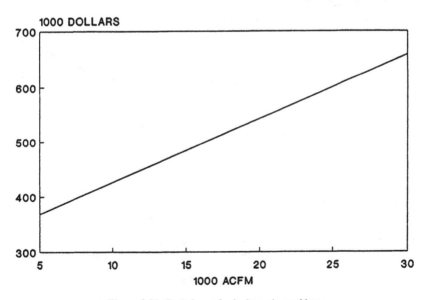

Figure 8.31 Capital cost for hydrosonic scrubber.

the installation of post-scrubber afterburner can be conveniently accomplished as an "add-on" without interrupting ongoing operations.

A cost estimate for installing post-scrubber afterburners with regenerative heat exchangers is presented in Figure 8.32. Although the cost of recuperative heat exchangers are about 70% of the cost of the regenerative heat exchangers, the effectiveness of these systems is limited to about 80%.

BENEFICIAL USE OF ASH FROM MUNICIPAL SEWAGE SLUDGE INCINERATION

INTRODUCTION

As previously discussed, the incineration of municipal sewage sludge has been determined to be environmentally sound and cost effective. However, incineration does not result in the ultimate disposition of municipal sewage sludge since an ash residue will always remain.

Generally, the mass of dry ash produced by municipal sewage sludge incineration ranges from 15 to 25% of the wet weight of sludge solids fed to the incinerator.

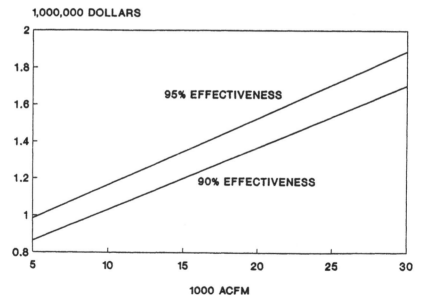

Figure 8.32 Capital cost for post-scrubber regenerative afterburner.

MUNICIPAL AGENCIES UTILIZING INCINERATOR ASH

There are a number of wastewater treatment agencies that currently utilize incinerator ash for beneficial purposes.

Palo Alto, California

The city of Palo Alto owns and operates a sewage treatment plant with an average dry weather flow of 113,550 m³/day (30 MGD). All sludge from the treatment facility is incinerated in six multiple hearth incinerators. These incinerators produce about 18.7 mt (20.7 tons) of incinerator ash per week.

The incinerator ash from the city of Palo Alto has relatively high levels of gold and silver. Gold and silver is found in concentrations of 32 and 680 mg/kg, respectively. These levels are found because of industrial inputs from nearby computer circuit board manufacturers. The city of Palo Alto is in the region of California commonly known as "silicon valley."

The incinerator ash from the city of Palo Alto is currently being shipped in 815 kg (1800 pound) bags to the Cyprus Miami Mining Company (Claypool, Arizona). This company has a smelter where the silver and gold are extracted from the ash. The company pays the city of Palo Alto about $60,000 per year for its incinerator ash. Transportation of the ash to the smelter is paid for by the Cyprus Miami Mining Company.

Nashville, Tennessee

The city of Nashville is currently utilizing its incinerator ash for daily cover of municipal solid waste at the Bordeaux Landfill in Nashville, Tennessee. This landfill is owned and operated by the city of Nashville. Clay soil is scarce in Nashville and the ash represents a good cover material for municipal refuse. This operation is considered to be very successful and has been in place for several years.

Columbus, Ohio

The city of Columbus, Ohio markets its incinerator ash as "flume sand," which is sold at $1.00 per cubic yard unscreened and $4.00 per cubic yard screened. The flume sand is placed on baseball diamonds and horse arenas and is used as bedding materials for patios.

Minneapolis/St. Paul, Minnesota

Ninety-five percent (95%) of the ash removed from the Metropolitan Council's (Minneapolis/St. Paul, Minnesota) wastewater treatment plants is used in a cement manufacturing process. The balance is mixed with lime sludge from a water treatment plant and sold as a soil amendment.

Other Options

Several firms are presently investigating the feasibility and cost effectiveness of making bricks from ash or using the ash in asphalt mixes.

USE OF INCINERATOR ASH FOR AGRICULTURE

Fertilizer Value of Incinerator Ash

Sewage sludge ash has been described to be a good fertilizer because of its phosphorus and lime content [28]. In fact, incinerator ash has been sold to commercial fertilizer firms in Japan by the cities of Tokyo and Nagoya. This ash had an N, P_2O_5, and K_2O content of 0.2%, 6.0%, and 1.0%, respectively.

The composition of sewage sludge ash from multiple hearth incinerators in Washington County, Oregon has been reported by Mellbye et al. [29]. They found this ash to contain no nitrogen but a phosphorus content of 4.1 to 5.8% and a lime content (CaO equivalence) of 13 to 14%. Mellbye concluded that the ash could serve as a source of phosphorus and lime for application to crops.

Furr and Parkinson [30] conducted a study of the composition of sewage sludge ashes from ten cities in the United States. They reported the following fertilizer values for incinerator ash from these ten cities:

Nutrient	Range (% Dry Weight)
N	0.06 to 0.1
P	1.0 to 5.2
K	0.23 to 1.05

Plant Uptake of Incinerator Ash Components

The enhancement of plant growth and plant uptake in soil amended with sewage sludge incinerator ash has not been studied extensively. Most studies have focused on the effect of sewage sludge on plant growth and plant uptake. However, there have been a few studies involving incinerator ash.

Mellbye [29] studied the plant growth and plant uptake of sweet corn grown with and without sewage sludge ash application. With ash application up to 63 mt/ha (28.1 T/A) they observed no toxic or nutrient deficiency symptoms in the sweet corn. The Zn concentration in corn leaves increased with ash application but Cu, Cd, Cr, Mo, Mg, Ca, and Pb in the leaves and corn kernel remained unaffected.

Jakobsen and Willett [31] compared the fertilizing properties of sewage sludge with that of its incinerated ash. The ash was found to be a good source of lime (CaO) and was more effective than sludge in raising soil pH. The ash was not a good source of P despite its high phosphorus content (2.14 to 2.73%). The authors concluded that the phosphorus in the ash was not available for plant growth while the phosphorus in the sewage sludge was.

REFERENCES

1 EPA. February 19, 1993. "40 CFR Parts 257 and 503, Standards for the Disposal of Sewage Sludge: Final Rule," *Federal Register.* United States Environmental Protection Agency.

2 Communications with Department of Environmental Protection, State of New Jersey, 1990.

3 Communications with Department of Environmental Resources, Commonwealth of Pennsylvania, 1990.

4 Bostian, E. H. and E. P. Crumpler et al. 1988. "Emissions of Metals and Organics from Four Municipal Wastewater Sludge Incinerators—Preliminary Data," *Proceedings of the National Conference on Municipal Sewage Treatment Plant Sludge Management.*

5 Ohio Air Quality Development Authority. 1989. "Toxic Air Emissions from Sewage Sludge Incinerators in Ohio."

6 Baturay, A. 1991. "Total Hydrocarbon Emissions from Multiple Hearth Furnaces," *Proceedings of the 84th Annual Meeting and Exhibition of Air and Waste Management Association.*

7 Waltz, E. W. 1990 (unpublished). "Technical Discussion Paper of Proposed EPA Hydrocarbon Regulation for Sludge Incinerators."

8 Lee, K. C. 1988. "Research Areas for Improved Incineration System Performance," *JAPCA,* 38(12):1542–1550.

9 Edwards, J. B. 1974. *Combustion—Formation and Emission of Trace Species.* Lancaster, PA: Technomic Publishing Co., Inc.

10 DeWees, W. G. and S. A. Davis, et al. 1990. "Sampling and Analysis of Municipal Wastewater Sludge Incinerator Emissions for Metals, Metal Species, and Organics," *Proceedings of the 83rd Annual Meeting and Exhibition of Air and Waste Management Association.*

11 Knisley, D. R., L. M. Lamb and A. M. Smith. 1987. "Site 1 Draft Emission Test Report—Sewage Test Program," EPA 68-02-6999, a report prepared for U.S. Environmental Protection Agency.

12 Baturay, A. 1991. "THC Emissions at Plant D," unpublished test data developed by Carlson Associates, Manassas, Virginia.

13 Anderson, J. May 1976. U.S. Patent 3,958,920.

14 Albertson, O. E. and A. Baturay. February 1990. U.S. Patent 4,901,654.

15 Coker, C. S. and A. Baturay, et al. 1989. "Multiple-Hearth RHOX Process—A Response to the New Regulations," *Proceedings of the National Conference on Municipal Sewage Treatment Plant Management.*

16 1989. "RHOX Process," a technical bulletin prepared by RHOX International, Inc., Salt Lake City, Utah.

17 Baturay, A. and J. M. Bruno. 1990. "Reduction of Metal Emissions from Sewage Sludge Incinerators with Wet Electrostatic Precipitators," *Proceedings of the 83rd Annual Meeting and Exhibition of Air and Waste Management Association.*

18 Hentz, L. H., A. Baturay, A. and F. B. Johnson. 1990. "Air Emission Studies of Sewage Sludge Incinerators at the Western Branch Wastewater Treatment Plant," *Proceedings of the Water Pollution Control Federation Specialty Conference: The Status of Municipal Sludge Management for the 1990s.*

19 Baturay, A. 1990. "A Common Sense Risk Assessment for Sewage Sludge Incinerators—A Case Study," presented at the *Pre-Conference Workshop, 63rd Annual Conference and Exposition,* Water Pollution Control Federation, Washington, D.C.

20 Communications with Ceilcoat/Air Pollution Control, Berea, Ohio, November 1990.

21 Personal communications with Mr. Gerwyn Jones, Belco Technologies Corporation, Parsippany, New Jersey, August 1990.

22 Personal communications with Mr. J. M. Bruno of Sonic Environmental Systems, Parsippany, New Jersey, May 1991.

23 Personal communications with Mr. Arthur L. White, T. Z. Osborne Wastewater Treatment Facility, Greensboro, North Carolina, October 1990.

24 Haug, R. T. and F. M. Lewis, et al. 1989. "Air Emissions from State-of-the-Art Sludge Combustion," *Proceedings of the 19th National Conference on Municipal Sewage Treatment Plant Sludge Management.*

25 Cambridge Environmental Inc., January 12, 1995; Updated May 11, 1995. "Comments on Estimating Exposure to Dioxin-Like Compounds."

26 Hattemer-Frey, H. A. and C. C. Travis. 1991. "Assessing the Extent of Human Exposure through the Food Chain to Pollutants Emitted from Municipal Solid Waste Incinerators," *Health Effects of Municipal Waste Incineration.* Boca Raton, Florida: CRC Press, Inc.

27 Guillory, J. L. 1977. "Particulate Emissions Resulting from Combustion of Municipal Sewage Skimmings," *Proceedings of the 83rd Annual National Meeting of the American Institute of Chemical Engineers,* Houston, Texas.

28 Sebastion, F. P. 1972. "Waste Management in China: Ancient Traditions and High Technology," *Ambio,* 1:209–216.

29 Mellbye, M. E., D. O. Hemphill and V. V. Volk. 1982. "Sweet Corn Growth on Incinerated Sewage Sludge-Amended Soil," *Journal of Environmental Quality,* 11 (2).

30 Furr, A. K. and T. F. Parkinson. 1987. "Multi-Element Analysis of Municipal Sewage Sludge Ashes. Adsorption of Elements Grown in Sludge Ash-Soil Mixture," *Environmental Science and Technology,* 13 (12).

31 Jakobsen, P. and I. R. Willett. 1986. "Comparisons of the Fertilizing and Liming Properties of Lime-Treated Sewage Sludge with Its Incinerator Ash," *Fertilizer Research,* (9):187–197.

CHAPTER 9

Sludge Application to Dedicated Beneficial Use Sites

THE intent of this chapter is to provide information needed to select and design a suitable land area that can be used as a dedicated site for municipal sewage sludge. The chapter describes: (1) the characteristics of a dedicated site; (2) the factors needed in choosing, designing, and preparing dedicated sites; and (3) the case history of an existing site, which has been operating since the early 1970s.

INTRODUCTION

Application of municipal sewage sludge to dedicated land is commonly practiced by publicly owned treatment works (POTW). Differences in application practices and management of the dedicated land have given rise to two distinct classifications of dedicated land: Dedicated disposal sites and dedicated beneficial use sites. Dedicated disposal sites are parcels of land set aside specifically for sewage sludge disposal, with the land being utilized solely as a treatment system. The sludge solids are retained at the top of the soil profile where soil microorganisms and chemical processes degrade the sludge organic matter. Organic pollutants are degraded or, along with metals and phosphorus, are bound to soil colloids thereby becoming largely immobile. These sites are normally designed and operated to maximize the amount of sludge applied to the land each year, and sludge loading rates of 220 to 900 Mg/ha/yr (98.1 to 401 tons/acre/yr) are typically achieved [1]. There is generally no attempt to produce

Thomas C. Granato and Richard I. Pietz, Metropolitan Water Reclamation District of Greater Chicago, Chicago, Il.

a vegetative cover or to utilize the nutrient content or soil conditioning properties of the sludge. This is because the repeated applications that are necessary to produce the high sludge loading rates normally attained at dedicated disposal sites do not permit establishment of vegetation on the application fields.

Generally, these sites are either owned or are leased for long terms by the treatment works that is using the parcel for sludge disposal. This allows the treatment works to limit public access to the site, thereby minimizing the risk of adverse human health effects that are potentially elicited by exposure to sludge-borne pollutants or pathogens. This is advantageous for treatment works that receive heavy industrial inputs and produce sludges with relatively high levels of organics and metals. These sludges are not well suited to other management options that incur greater public exposure to the sludge-borne pollutants. The fact that the treatment works owns the dedicated site is also advantageous because it insures availability of land for a long-term period. Treatment works that select other land application management options, such as application to agricultural fields, generally must constantly seek new parcels of land for application of sludge.

Dedicated beneficial use (BU) sites differ from dedicated disposal sites in that the objective of achieving maximal sludge loading rates is balanced in the former with the objective of deriving beneficial use of the land-applied sludge. This beneficial use normally involves growing vegetation. Typically, crops such as corn are produced and sold for animal feed or alternative fuel production (ethanol), or the land is used to produce sod or nursery stock. Usually crops that are ingested directly by humans are not produced on dedicated BU sites because cumulative soil pollutant loadings become high after many years of operation. The sludge is said to be used beneficially at these sites because it is applied to the land to improve poor soils or to reclaim disturbed soils, and it supplies nitrogen and phosphorus as well as other macro and micronutrients to the vegetation being grown.

Like dedicated disposal sites, dedicated BU sites are either owned or are leased for long terms by the treatment works that is using the parcel for sludge application. Dedicated BU sites are advantageous because, due to limited public exposure to the sludge-borne pollutants applied to the land, afforded by restricted site access and production of vegetation not directly consumed by humans, higher sludge loading rates are normally permissible than for sludge application to privately owned agricultural fields. While loading rates will not be as high at dedicated BU sites as they are at dedicated disposal sites, dedicated BU sites are advantageous because land use at these sites is usually favored, by the general public and by regulatory agencies, over land use at dedicated disposal sites, and revenues can be generated by marketing the crops produced. Dedicated BU sites are the management option of choice for treatment works that adopt a beneficial use of sludge policy but either produce great quantities of sludge or produce sludge with too high a level of pollutants to be managed by application to privately owned agricultural fields.

CHOOSING, DESIGNING, AND PREPARING DEDICATED SITES

SITE SELECTION

General Considerations

In selecting a parcel of land for development into a dedicated site, several factors must be considered. The land should be located as close to the POTW as is possible. This will allow the POTW to hold costs of transporting sludge from the treatment works to the application fields to a minimum. The sludge is normally transported by truck, rail, or barge, and transportation costs will be further minimized if direct routes exist for at least one of these transportation modes between the POTW and the dedicated site.

The POTW must also consider the politics of public perception when acquiring land for dedicated sites. In general, dedicated sites that are located within the jurisdictional boundary of the wastewater treatment agency are more readily accepted by surrounding communities than dedicated sites located across municipal, county, or state boundaries. This is because the general public is more accepting of waste perceived to be self-generated than of waste perceived to have been imported from other jurisdictions. This poses a dilemma for many of the larger POTWs that service major urban municipalities. Often, there is not sufficient land available in these urban environments to construct dedicated sites, and even if sufficiently large parcels are available, they are usually in close proximity to residential developments, making them unsuitable for dedicated site development due to high real estate value and resistance of residents to siting. This situation usually forces the wastewater treatment agency to look beyond its own boundaries in acquiring land for dedicated sites. Thus, agencies servicing large urban centers must often acquire land from nearby rural counties to locate and build dedicated sites. The rural public is often not very receptive to accepting wastes imported from large municipalities, and public relations becomes an important issue in choosing and designing a site.

For this reason, dedicated sites are often located on lands that are largely composed of disturbed or naturally unproductive soils. This lessens the public perception that productive lands will be spoiled or polluted. Obviously, dedicated BU sites go farther in combatting negative public opinion of land application of sludge than dedicated disposal sites. This is because sludge is used at dedicated BU sites to increase productivity of the disturbed or unproductive soils of the site. The reclamation activities that occur at dedicated BU sites improve environmental quality and soil productivity, and these tangible benefits help counter the initial negative public perceptions of land application of sludge.

Soil Considerations

Soil Surveys

In selecting a dedicated site, soil surveys are an important source of information in making preliminary judgement on the suitability of potential sites for sludge application [2]. Soil surveys, prepared by the U.S. Soil Conservation Service and its local cooperators, provide detailed soil maps on photographic background, a general soil map, description of the soil by series and mapping unit, data on engineering and agronomic properties of soils, and interpretive tables. All of these components provide information on the suitability of a potential site.

An important component of the soil survey is the land capability or suitability classification. Land is classified according to the most suitable sustained use that can be made of it while providing adequate protection from erosion or other means of deterioration. Under the system set up by the U.S. Soil Conservation Service, eight land capability classes are recognized. These are described in soil survey reports and need to be consulted in evaluating a potential site.

Another component of soil surveys is the description of soil physical and chemical properties, soil engineering properties, and soil and water features. The physical and chemical properties listed include permeability, available water holding capacity, pH, and shrink-swell potential. The engineering properties include soil texture, and the soil and water features include hydrologic group, flooding characteristics, and water table information. All of these survey components provide useful information needed to evaluate a site.

Soil surveys are concerned primarily with the top 2 m (6.6 ft) of the regolith, which is the unconsolidated mantle of weathered rock and soil material on the earth's surface. Consequently, it is important to have additional information on the geology and hydrology of the site. Some of this information may be available from geologic maps and geologic literature of the area. However, on-site studies are usually needed to provide additional information to characterize the site. This information will help determine if the location is suitable for sludge application and will be used in the design of the sludge application system. Consequently, on-site investigations by a soil scientist, hydrologist, and geologist are needed to determine the actual suitability of the potential site for receiving wastes [3].

Landscape Topography and Soil Hydrology

The selection of a dedicated site needs to include an evaluation of landscape topography along with the soil and underlying geologic layers. The landscape can be looked upon as a surface transport system for the applied sewage sludge, while the soil can be looked upon as an internal transport system for sludge constituents. The interaction of both of these components determines the effectiveness of a dedicated site for sludge utilization [4].

The importance of landscape topography is noted by Larson and Schuman [4], who describe six combinations of landscape and soil conditions for consideration at sites receiving high rates of wastes. The two landscape types that need evaluation are open and closed. An open landscape is typically one where runoff is away from the site watershed being impacted by sludge application. A closed landscape is typically one where runoff moves toward the low point in the site watershed being impacted by sludge application.

Table 9.1 lists the six types of landscape and soil conditions for consideration in selecting a dedicated site. Larson and Schuman [4] give a graphical depiction of each of these conditions. The landscape and soil conditions can be engineered to create more suitable conditions for the high rates of sludge application at dedicated sites. Open landscapes can be converted to closed landscapes by use of water detention structures and field berms. Closed landscapes with slowly permeable surface and subsurface horizons, with grading and the addition of drainage systems, can be made into very acceptable dedicated sites. Larson and Schuman [4] give examples of innovative engineering designs that may be applicable to dedicated sites.

An important component of topography for site selection is the slope. The steepness, length, and shape of slopes influence the rate of runoff from a site. Rapid surface runoff can readily erode sludge-soil mixtures and transport them to surface waters. Specific guidance on maximum slopes allowable for sludge application sites under various conditions, such as sludge physical characteristics, application techniques, and application rates, should be obtained from the designated regulatory agency. For general guidance, the U.S.EPA [5] suggests the limitations shown in Table 9.2.

Soil Physical Properties

Soil physical properties often determine the suitability of a site for the assimilation of applied sewage sludge. These physical properties are interrelated, and the application of sludge will affect these soil properties in both a beneficial and negative manner [6].

Texture, defined as the relative proportions of various sized particles (sand, silt, clay) in a soil, can have a significant impact on the suitability of a soil for sludge application. Texture influences the tillage, soil water retention, permeability, infiltration, and drainage of soils. The soil profile

TABLE 9.1. Hydrologic Conditions for Consideration in Selecting a Dedicated Site.

Hydrologic Condition	Remarks
1 Open land type— deep permeable soil	Precautions may be necessary to limit surface water runoff and movement of contaminated water into deep layers and groundwater. Deep permeable soils favor rapid infiltration and provide a large soil surface area to interact with contaminants.
2 Closed land type— deep permeable soil	Reduces concern about the spread of contaminants outside the watershed by runoff. The soil permeability will favor deep percolation of water and contaminants.
3. Open land type— permeable surface—slowly permeable subsurface	The same remarks as "condition 1" with the added hazard that runoff is likely to be increased and the movement through the profile will be decreased.
4. Closed land type— slowly permeable soil	Runoff is collected at the low point in the watershed and lateral flow above a constricting layer may also go toward the low point. This may be an advantage because water can be intercepted and collected for further renovation if needed.
5. Open land type— slowly permeable surface and subsurface	Runoff will be increased during heavy storms and contaminants will be spread outside the watershed Infiltration of applied constituents and water will be reduced.
6. Closed land type— slowly permeable surface and subsurface	Runoff is collected at the low point in the watershed. Infiltration of applied constituents and water will be reduced. Ponding of water will be a problem in low areas unless it is intercepted and collected for further renovation if needed.

TABLE 9.2.   Recommended Slope Limitations for Land Application
of Sludge [5].

Slope	Comment
0–3%	Ideal; no concern for runoff or erosion of liquid sludge or dewatered sludge.
3–6%	Acceptable, slight risk of erosion, surface application of liquid sludge or dewatered sludge okay.
6–12%	Injection of liquid sludge required for general cases, except in closed drainage basin and/or extensive runoff control Surface application of dewatered sludge is usually acceptable.
12–15%	No liquid sludge application without effective runoff control, surface application of dewatered sludge acceptable, but immediate incorporation recommended.
Over 15%	Slopes greater than 15% are only suitable for sites with good permeability where the slope length is short and is a minor part of the total application area

horizons at a site may range in texture from sands to clays. Coarse-textured soils are easier to till and manage than are fine-textured soils. These soils usually have higher saturated hydraulic conductivities and high water infiltration rates unless a seal develops at the soil surface. Fine-textured soils have low saturated hydraulic conductivities and lower infiltration rates. These soils have a low volume of air-filled pores that can readily lead to anaerobic conditions if the soils have high loadings of liquid sewage sludge. The low content of oxygen and high content of gas products from sludge decomposition, such as methane, ethylene, and carbon dioxide in these soils can reduce root growth, nutrient uptake, and plant growth [7].

Soil texture influences the soil water retention curve. Clay soil holds much more water at a given soil water potential than does loam or sand. Brady [8] shows typical soil moisture tension curves for three representative mineral soils. Consequently, as the soil water content decreases, water is held more tenaciously in clay as compared to loam and sandy soils.

The addition of sewage sludge will shift the water retention curves in soils of various textures. Increases in soil water holding capacity at both field capacity (−0.033 MPa or −0.33 bars) and wilting point (−1.5 MPa or −15 bars) occur in both fine-textured and coarse-textured soils. Research data suggest that increases in water retention for coarse-textured soils are larger than those for fine-textured soils [9,10]. Consequently, soil texture at a dedicated site is important in determining water relations.

Water transmission properties of soils are important in selecting a dedicated site for sludge application. These properties (hydraulic conductivity, infiltration, and permeability) affect the amount of water in runoff and the amount leached through the soil profile. Hydraulic conductivity is a transmission coefficient that describes the ease with which water moves through the soil, and it is dependent upon the size of soil pores. The saturated hydraulic conductivity is dependent primarily on the largest pore size because the velocity is proportional to the pore radius raised to the fourth power. For land application of sludge, a moderate to moderately rapid, but not excessive, soil hydraulic conductivity is usually desirable [4]. Soils with very low or excessive hydraulic conductivity should be avoided because of the impact on permeability, drainage, and runoff.

Infiltration, the downward entry of water into soil, is the principal means by which dissolved salts and organics are transported into and through soils. Infiltration of water into soils depends on the initial water content, soil water potential, texture, structure, and the uniformity or homogeneity of the profile [7]. Generally, coarse-textured soils have higher infiltration rates than fine-textured soils.

Soil permeability, the ease with which water and air are transmitted through a layer or mass of soil, is an important component to be evaluated when selecting a site. Permeability is determined by soil pore space and size, shape, and distribution. Fine-textured soils generally possess slow or very slow permeability, while coarse-textured soils range from moderately rapid to very rapid. A medium-textured soil, such as a loam or silt loam, tends to have moderate to slow permeability. The U.S. Soil Conservation Service [11] has defined permeability classes for use in describing and evaluating soils (Table 9.3).

In evaluating soil physical properties for accepting sewage sludge, Witty and Flach [3] proposed several soil limitations. Table 9.4 describes these factors and the degree of limitation.

The effect of sewage sludge on soil water transmission properties has been studied by several researchers. Metzger and Yaron [12] indicate that with few exceptions, such as temporary plugging of the uppermost soil layer from liquid sludge application, the organic components of sludge do not directly affect the water transmission properties of soils. The saturated hydraulic conductivity can increase in soils after sludge application because

TABLE 9.3. Permeability Classes for Saturated Soil [10].

Soil Permeability (cm/hr)	Permeability Class
<0.15	Very slow
0.15 to 0.5	Slow
0.5 to 1.5	Moderately slow
1.5 to 5.1	Moderate
5.1 to 15 2	Moderately rapid
15.2 to 51	Rapid
>51	Very rapid

TABLE 9.4. Soil Limitations for Accepting Biodegradable Sludges and Solids [3].

Item Affecting Use	Degree of Soil Limitations for Accepting Sludge		
	Slight	Moderate	Severe
Permeability of the most restricting layer above 150 cm	Moderately rapid and moderate, 1.5–15 cm/hour	Rapid and moderately slow, [a] 15–50 and 0.5–1.5 cm/hour	Very rapid, slow, and very slow, >50 and <0.5 cm/hour
Soil drainage class[b]	Well drained and moderately well drained	Somewhat excessively drained and somewhat poorly drained	Excessively drained, poorly drained, and very poorly drained
Runoff[c]	None, very slow, and slow	Medium	Rapid and very rapid
Flooding	None	None	Soils flooded
Available water capacity from 0 to 150 cm or to a limiting layer	> 15 cm	8–15 cm	<8 cm

[a]Moderate and severe limitations do not apply for moderately slow, slow, and very slow permeability unless the waste is plowed or injected into the layers having these permeabilities or if evapotranspiration is less than water added by rainfall or irrigation
[b]For class definition see Soil Survey Manual, pp. 169–172 [11].
[c]For class definition see Soil Survey Manual, pp. 166–167 (amended to use "None" for "Ponded") [11]

537

of a decrease in bulk density and an increase in total porosity [13]. Infiltration through the soil surface, which depends upon the hydraulic conductivity of the underlying soil, may be improved by sludge applications [13]. These soil physical changes are more likely to occur at sites that receive sludge continuously and at high rates [12].

Soil structure, aggregation and bulk density are factors that are not usually studied extensively in selecting a site. Soil structure is the arrangement of soil particles, and aggregation is defined mainly by the size-distribution of aggregates and their stability under wetting. Soil structure and aggregation, through their effect on pore space, influence the movement of air and water through soils, biological processes, mechanical impedance, and root distribution. An evaluation of these soil physical properties will provide additional information in selecting a site.

Bulk density is the mass of a unit volume of dry soil, including the air-filled pores. A measurement of this property helps to characterize the relationship between pore space and solids in soils. Bulk density is a useful physical characteristic of soils because it can be used to determine the amount of pore space through which percolation may occur. The bulk density of soils can range from 1.00 g/cm^3 in soils with a large amount of pore space, to 1.60 g/cm^3 in soils with a low amount of pore space. In very compact subsoils, bulk densities can be 2.0 g/cm^3 or greater.

Sewage sludge applications affect the aggregation and bulk densities of soils. The aggregation status is improved by sludge application, so this improves the seedling emergence of planted crops and reduces soil particle detachment during rains [12]. The effect of sewage sludge applications is to reduce soil bulk densities, regardless of soil texture [9,14,15]. The decrease in bulk density is due to mixing of the sludge-applied organic matter with the more dense mineral fraction of the soil, and to increased aggregation creating more pore space.

Soil Chemical Properties

Soil chemical properties need to be evaluated in assessing the suitability of a site for the application of sewage sludge. These properties help to determine the rate at which sewage sludge can be applied because of their effect on the chemical reactions that may occur with sludge-applied components in the soil.

The soil pH is an important soil chemical property that needs to be evaluated. The soil pH at a selected site is important because it affects the chemical and microbial reactions in sludge-amended soils and the uptake of ions by plants. Sludge-applied inorganic constituents, with time, become soluble and a part of the soil solution. Soil pH affects the solubility of the inorganic constituents and their availability for exchange reactions, sorption and precipitation, plant uptake, leaching, reactions with soil organic matter, and utilization by soil microorganisms.

Soil pH is the parameter most consistently identified as controlling the solubility of sludge-applied metals. Almost all metals, except molybdenum and selenium, are more soluble at a low pH and their solubility decreases as the soil pH increases. Liming is a standard agronomic practice used to control soil pH, and it is used to control pH at sludge application sites. Most researchers agree that a soil pH between 6.0 and 6.5 is effective in reducing plant metal uptake on sludge-amended soils [16,17,18]. A pH of 6.5 is currently required by the U.S.EPA [19] for growing food chain crops.

Cation exchange capacity is another soil chemical property that needs to be determined in soils that receive sewage sludge. This property is simply the sum total of the exchangeable cations that a soil can absorb. Fine textured soils tend to have higher cation exchange capacities, while coarse textured soils tend to have lower cation exchange capacities. Within a textural group, the organic matter content and the amount and kind of clay affect the cation exchange capacity.

Soils with a higher exchange capacity have a higher buffering capacity against changes in soil pH. This concept is recognized and used in agricultural practices because heavier textured soils need more lime to change the soil pH. Consequently, at sludge application sites it will take longer for the pH to decline in soils with a higher cation exchange capacity.

Many researchers [16–18,20,21] recognize cation exchange capacity as one of the soil properties that is related to soil retention of metals. Because of this, cation exchange capacity is used by the U.S.EPA and many state agencies to determine cumulative loading limits for sludge-applied metals. Soils were arbitrarily divided into three groups by the U.S.EPA based on cation exchange capacity. These groups were soils with cation exchange capacities of <5, 5 to 15, and >15 cmol/kg.

Many researchers currently feel that the use of cation exchange capacity to predict the plant uptake of sludge-applied metals is overstated and not supported by long-term field experiments. The CAST Report [18] concluded that cation exchange capacity is best viewed as a general but imperfect indicator of the soil components that limit the solubility of sludge-applied metals. These soil components include organic matter, clays, and hydrous oxides of iron, aluminum, and manganese.

In selecting a dedicated site, knowledge of the soil organic matter content and soil mineralogy is desirable. These two components affect the cation exchange capacity of the soil and the ability of the soil to retain sludge-applied metals.

Soil organic matter can bind metals through cation exchange and the formation of organometallic complexes. Sludge-applied organic matter can also affect soil metal retention through similar mechanisms. In soils receiving sewage sludge continuously over a period of time, the solubility of sludge-applied metals will be determined by the composition of the applied sludge, including its organic matter [22].

The mineralogy in soils is important because along with organic matter, this soil fraction helps to determine the solubility and availability of sludge-applied metals. Clay minerals along with hydrous oxides of iron, aluminum, and manganese affect the retention of metals in soils. Adsorption of metals by these soil components is considered to be a major mechanism of removal [18]. Additional reactions by the soil mineral fraction include precipitation, occlusion in other precipitates, and diffusion into soil minerals. Consequently, a knowledge of the soil mineralogy (e.g., clay types, hydrous oxides) at a site can provide useful information on the ability of the soil to retain sludge-applied metals. At sites with continuous applications of sewage sludge, the inorganic component of the applied sludge will be a major factor in determining metal availability in amended soils [22].

Surface and Groundwater Considerations

Proximity to Surface Water

The number, size, and nature of surface water bodies on or near a potential sludge application site are significant factors that need to be evaluated in site selection. These surface water bodies have the potential to be contaminated by site runoff or flooding. In general, areas subject to frequent flooding have severe limitations for utilization of sewage sludge.

Typical surface waters that need to be considered in site selection are ponds and lakes, springs and streams, rivers and creeks. Table 9.5 shows suggested setback distances for sludge applied by either injection or surface application. In addition, many state regulatory agencies have their own specified setback distances for surface waters. For example, the state of Illinois [23] requires a minimum of 61 m (200 ft) from surface waters, waterways, or flood plains for sludge applied by low-pressure sprayers of <0.345 MPa (<50 lb/in^2) and a minimum of 304 m (1000 ft) for high-pressure sprayers >0.345 MPa (>50 lb/in^2).

The uses of surface waters on or near a sludge application site also need to be considered in site evaluation. The U.S.EPA [5] suggests, as an ideal condition, setback distances greater than 305 m (1000 ft) from any surface water body, pond, or lake used for recreational or livestock purposes, or any surface water body classified under state law. A distance of greater than 61 m (200 ft) is suggested for intermittent streams.

If runoff is to occur from a sludge application site, the water quality needs to be within acceptable limits for immediate discharge to a receiving

TABLE 9.5. Suggested Setback Distances for Sewage Sludge Application Areas [24].

| | Distance from Feature to Sludge Application Site | | | | |
| Feature | 15 to 90 m | | 90 to 460 m | | >460 m |
	Injection[1]	Surface	Injection[a]	Surface	Injection and Surface
Residential development	No	No	Yes	No	Yes
Inhabited dwelling	Yes	No	Yes	Yes	Yes
Ponds and lakes	Yes	No	Yes	Yes	Yes
Springs	No	No	Yes	Yes	Yes
10-year high water marks of streams, rivers, and creeks	Yes	No	Yes	Yes	Yes
Water supply wells	No	No	Yes	Yes	Yes
Public road right-of-way	Yes	No	Yes	Yes	Yes

[a]Injection of liquid sludge or surface application of dewatered sludge
Metric conversion: 1 m = 3 28 ft.

stream or watercourse. For nonacceptable runoff, provisions must be made to detain the water in a holding structure so the runoff can be reapplied to land, held until it is acceptable for discharge, or treated so that it can be discharged off-site. In many situations, this may require designing surface runoff improvements such as ditches, terraces, and berms, depending upon the slope and soil physical characteristics at the site.

It is beyond the intent of this discussion to describe all the hydrological calculations needed to make an accurate assessment of the maximum runoff that can be expected at a specific site. These calculations need to be made by an experienced hydrologist for the site area. Based upon the curves developed for a maximum precipitation year, an estimate can be made of the runoff storage volume and surface area needed at a site.

Depth to Groundwater

Information on groundwater is necessary to consider a potential site for sewage sludge application. Critical information required includes: (1) depth to groundwater, which includes historical highs and lows; (2) groundwater quality and use classification as designated by regulatory authorities; and (3) an estimate of groundwater flow patterns. When a specific site or sites have been selected for sludge application, a detailed field investigation may be necessary to determine the above information. During preliminary screening, published general resources may be obtainable from local USGS or state water resource agencies. This information can be valuable in selecting a potential site or sites for sludge application.

Generally, the greater the depth to the water table, the more desirable the site is for sludge application. Sludge should not be placed where there is potential for direct contact with the groundwater table. The actual thickness of the unconsolidated material above a permanent water table constitutes the effective soil depth [5]. The desired depth may vary according to sludge characteristics, soil texture, soil pH, method of sludge application, and sludge application rate. Table 9.6 summarizes recommended criteria for the various sludge application options.

The nature and condition of consolidated material above the water table is an important factor at high-rate sludge application sites. Fractured rock may allow leachate to move rapidly with little opportunity for contaminant removal. Unfractured bedrock at shallow depths will restrict water movement, with the potential for groundwater mounding, subsurface lateral flow, or poor drainage. Limestone bedrock is of concern where sinkholes may exist. A sinkhole, like fractured rock, can accelerate the movement of leachate to groundwater. Potential sites with potable groundwater in areas underlain by fractured bedrock at shallow depths, or sites containing sinkholes, should be avoided. Groundwater recharge zones that recharge

TABLE 9.6. Recommended Limits for Depth to Groundwater [5].

Type of Site	Drinking Water Aquifer	Impacted Aquifer[a]
Agricultural	1 m	0.5 m
Forest	2 m[b]	0 7 m
Drastically disturbed land	1 m[c]	0.5 m
Dedicated land disposal	At least 3 m	0 5 m

[a]Clearances are to allow for vehicle traffic on the surface and not for groundwater protection
[b]Seasonal (springtime) high water and/or perched water less than 1 m is not usually a concern Design chapter discusses these limits [5]
[c]Assumes no groundwater contact with leachate from operation
Metric conversion 1 m = 3 28 ft

major aquifers that are currently used or have potential use for drinking water should be eliminated from consideration.

By definition, dedicated sites are designed to contain contaminants within the site or manage their movement off-site in a controlled, environmentally acceptable manner. In this regard, an ideal site is one where leachate is of no concern because of favorable site conditions. However, if the site is located where it could contaminate potable groundwater aquifers, the site needs to be designed so the percolating leachate is intercepted. This may require the installation of subsurface drainage systems when natural drainage is restricted by relatively impermeable layers in the soil profile near the surface or by high groundwater. A description of groundwater leachate collection and control is given by the U.S.EPA [5]. If a subsurface drainage collection system is installed, the leachate collected needs to be stored, treated, and disposed of in an effective manner.

SITE DESIGN AND PREPARATION

The ultimate goal of the dedicated site user is to manage sludge disposal (for dedicated disposal sites) and sludge utilization (for dedicated BU sites) in an environmentally sound manner. Since management of the site must provide adequate capacity to accommodate the municipality's annual production of sewage sludge, it is important to set utilization goals prior to choosing, designing and preparing the dedicated site so that the utilization goals can be met by the capacity of the site without exceeding regulatory loading limits or jeopardizing the quality of the surrounding environment. It is relatively simple for the municipality to compute its annual sludge production, and if land application at the dedicated site is to be the sole sludge management practice, the utilization goal is immediately apparent. If other sludge management options are available such that only a portion of the municipality's sludge will be land applied at the dedicated site, then the municipality may wish to set immediate utilization goals as well as projecting long-term utilization goals.

If the municipality projects increased dependence on land application of sludge at the dedicated site in the long term, it may be advantageous to acquire a site large enough to accommodate the anticipated future increases and to develop the site as needed, rather than attempt to expand it by purchasing additional parcels of land at a later date. This is because adjacent lands may not be available for purchase when needed and the economy of centralization and consolidation inherent to large sites is lost if the municipality is forced to utilize many small, scattered sites to meet its needs. The space required for housing equipment and personnel, storing the transported sludge prior to land application, and for providing proper design of application fields should also be determined prior to selecting and acquiring a site.

For dedicated disposal sites, the primary utilization strategy is to design the site so that high sludge loading rates can be achieved without impacting the environment. Usually, runoff of pollutants to local waters and leaching of pollutants to groundwater are the principal environmental concerns. Typically, sites will be selected and engineered to minimize these impacts. Sites are normally divided into fields of various sizes, all of which must be large enough to allow for efficient operation of application equipment. The fields can be graded to reduce any naturally occurring slopes, thereby reducing runoff. To further prevent runoff, application fields are often bermed. The berms are soil or clay hills usually 0.5 to 1.0 m high that are built around the perimeter of each field. They are usually grassed for stability, which prevents their erosion.

The fields are gradually sloped toward drains where the runoff may be collected after significant rainfalls. If the dedicated site is located near a POTW within the municipality's jurisdiction, the runoff can be treated prior to discharge. If the dedicated site is at a remote location, retention basins may be built to hold and treat the collected runoff. Since the pollutants in runoff water are predominantly associated with the eroded solids, the treatment of runoff water usually consists of retention in the basin for long enough periods to facilitate settling of the solids. After settling of the sediment load, the water may be discharged to surface waters. If nitrate pollution is a big concern, the water from the retention basins may be reapplied to the soil during dry periods.

During the course of operating a dedicated disposal site, it is necessary to store sludge that has been transported from the POTW to the site. This is normally done in holding basins, which are similar to sludge lagoons. They are large excavations that are several meters deep. Care should be taken to design the basins so that they are not below the upper limit of the water table. If the water table rises above the bottom of the basin, groundwater contamination will result unless the basins are built with impermeable liners on all sides. These basins should be designed to hold enough sludge to accommodate at least one full year of land application. This will insure that operations will proceed without interruption and will make it easier to coordinate sludge treatment and transportation with the land application operations at the dedicated site.

Dedicated BU sites differ from dedicated disposal sites in that sludge applications are made to the land to improve soil conditions and supply nutrients to crops that are grown and harvested. The harvested crops may then be used for animal feed or may be marketed as raw materials for processes such as alternative fuel production (methanol, ethanol, etc.). However, the site design for dedicated BU sites is essentially identical to that for dedicated disposal sites, although the sludge utilization and application strategy differs.

SITE MANAGEMENT AND MONITORING

Selecting Crops

In managing a dedicated BU site, crops must be chosen that are suitable for growth on the soil types that predominate at the site and which thrive under the expected climatic conditions. Furthermore, the crop must yield some material that is harvested and removed from the fields each year. The process of crop selection should also include consideration of desired quantities of sludge to be applied to the land. Likewise, the quantity of sludge applied and the timing of the application must be suited to the cropping system chosen. If the dedicated BU site is on marginal land or mine spoil, the soil may require some initially high sludge loading rates in excess of the crop's nutrient requirements to improve the tilth of the soil. These initially high application rates may be necessary to increase soil organic carbon, soil cation exchange capacity, soil pH, or plant available water holding capacity.

Once the soil tilth has been improved sufficiently, the loading rates of sludge should be calculated to supply the crop's nitrogen requirement. Most, if not all, environmental regulatory bodies in the United States require that annual sludge application rates not supply more nitrogen than is removed in the harvested portion of the crop. This is done so that nitrate-N does not build up in soil and leach to groundwater. Where high sludge loading rates are desirable, it becomes necessary to select crops that have high nitrogen requirements. These crops may be identified as producing a large quantity of harvestable biomass that is rich in nitrogen. The crop must be well-suited for growth at the dedicated BU site or the quantity of harvestable biomass will be sub-optimal and the nitrogen requirement may not be as high as desired.

After several potential candidate crops are selected, the dedicated BU site operator should begin running management trials. These trials consist of determining best suited crop varieties and will provide preliminary yield targets that can be used as goals for the large-scale operation that will follow. These trials may be run during the initial start-up and soil conditioning phase. In the United States, local county agricultural extension agents can be a valuable source of information in proposing crops and varieties that may be most appropriate for trials. Good site management should entail use of more than one crop so that crop rotations can be utilized to help avert infestation of fields with undesirable pests.

Upon selecting crops and target crop yields, the dedicated BU site manager can calculate the nitrogen supply required to replenish the nitrogen that will be removed with each harvest. The sludge loading rate can then be calculated as an amount of sludge that has enough plant available nitrogen (PAN) to meet the crop nitrogen requirement. The only difficulty in this calculation is the estimation of the PAN content of sludge. All environmental regulatory agencies that require that sludge loading rates be restricted so that the PAN supplied does not exceed the nitrogen requirement of the crops grown, can provide mathematical formulas for calculating the PAN supplied by various sludges. The PAN content of sludge is normally assumed to be comprised of ammonium-N and the fraction of the sludge organic-N that can be mineralized to ammonium after land application. The rate at which sludge organic-N is mineralized to ammonium in soil is highly dependent on sludge, soil, and climatic variables, and discussion of its determination is beyond the scope of this text.

However, most regulatory agencies have developed guidelines for computation of sludge PAN, which are average values across many sludges and soil types. For example, the Illinois Environmental Protection Agency (IEPA) recommends that sludge PAN be calculated by assuming that 20% of the sludge organic-N is mineralized in the first year of application and becomes plant available. Subsequently, in the second through the fifth year after application 10%, 5%, 2.5%, and 1.25%, respectively, of the remaining organic-N is mineralized and becomes plant available. No mineralization of organic-N is assumed to occur after the fifth year.

The IEPA has also made recommendations for calculating the amount of sludge ammonium-N that is plant available. This quantity is dependent on soil texture and sludge application methodology as shown in Table 9.7. The IEPA further assumes that no ammonium-N is available after the first year. Thus, for fields receiving continuous annual sludge applications, the PAN can be computed during any year from the organic-N and ammo-

TABLE 9.7. Illinois Environmental Protection Agency Guidelines for Calculation of Plant Available Sludge Ammonium-N [23].

Soil Type	Application Method	Plant Available Sludge Ammonium (%)
Sandy	Surface	50
	Surface w/incorporation	50
	Subsurface	50
Non-sandy	Surface	50
	Surface w/incorporation	80
	Subsurface	100
Clay	Surface	25

nium content of the sludge applied in that year and the mineralization of organic-N from sludge applications in the previous five years. This amount of PAN is then equated with the crop nitrogen requirement and the annual sludge loading rate is computed.

Planning a Monitoring Program

Groundwater

A groundwater monitoring program will usually be required by state regulatory agencies for sludge applications at a dedicated site. The intent of the program is to insure that the project is not contaminating useful groundwater aquifers in the sludge application site or sludge storage area [5]. The goal of a groundwater monitoring program is to establish a monitoring network that provides representative groundwater samples. The U.S.EPA [5] recommends that a hydrogeologist be consulted during the initiation and implementation of the program. A detailed groundwater monitoring procedures manual is available from the U.S.EPA [25].

In most instances, detailed site investigations will have been conducted prior to initiating the permitting process for a dedicated site. These investigations should give information on the following factors when developing a groundwater monitoring program [26]:

(1) Soil and rock formations existing on the site

(2) Direction of groundwater flow and anticipated rate of movement

(3) Depth of seasonal high water table, and an indication of seasonal variations in groundwater depth and direction of movement; this should not be a problem with dewatered sludge or liquid sludge at agronomic rates

(4) Nature, extent, and consequences of groundwater mounding, which may occur above the naturally occurring water table

(5) Depth of impervious layers

The number of monitoring wells and their proper placement depends on the location of the water table and the direction of groundwater flow. If several aquifers could be affected, the U.S.EPA [5] recommends a set of monitoring wells for each aquifer. The depth of the monitoring wells is dependent on the depth of the aquifer being sampled, and the predicted pathway of potential migrating contaminants. The U.S.EPA [5] recommends that a qualified hydrologist be consulted in making these decisions.

Barcelona and Morrison [27] state that monitoring network designs should include the integration of both unsaturated and saturated zone sampling. However, unsaturated zone monitoring is rarely required or used, except occasionally for dedicated disposal sites [5]. Barcelona and Morrison [27] indicate that the common elements of sampling protocols for groundwater contamination are:

(1) Sampling locations within the likely pathways of water and contaminant movement

(2) Sampling points designed, constructed, and operated with minimal disturbance of ambient condition

(3) Efficient sample retrieval so as to preserve the *in situ* condition, minimal sample handling, and the need to begin analytical procedures as soon as possible

The establishment of a groundwater monitoring network generally requires wells to be installed in three major areas. The first area is a background zone with one or more wells located upstream or at a location not affected by sludge application. The second area is the placement of one or more wells off-site downgradient from the site. These wells are used to detect leachate migration, and the number required depends on site size and hydrogeological factors. The third area is within the application region, where one or more wells are placed in the zone of maximum leachate concentration.

At a dedicated site, it is certainly desirable to establish baseline groundwater quality prior to the application of sewage sludge. The monitoring should be initiated six months to a year before startup to establish background water quality and any seasonal fluctuations. This may be done by installing simple observation wells prior to installation of the permanent monitoring wells, using wells installed during site selection and design investigations, or using the installed permanent monitoring wells.

Groundwater sampling methods are discussed by Barcelona and Morrison [27], Fenn [25], Gillham et al. [28], Patrick et al. [29], and U.S.EPA

[5]. The sample collection frequency is dependent upon the goals of the groundwater monitoring program, whether short- or long-term. The estimated rate of pollutant travel in a given hydrogeologic setting will indicate intervals of time required to show a change in water quality. The U.S.EPA [5] indicates that arbitrary selection of sampling frequency may not reveal an accurate picture of groundwater quality. Barcelona and Morrison [27] suggest a quarterly sampling frequency for major ionic constituents because more frequent sampling can entail a high degree of redundancy with little additional information.

The constituents included in the analyses of groundwater samples are dependent on factors such as monitoring goals, budgetary restrictions, waste composition, uses of groundwater, and regulatory requirements. There is no single list of parameters applicable to all cases. A listing of parameters that are often analyzed in groundwater samples at sludge application sites is shown in Table 9.8.

Surface Waters

Surface water monitoring is usually minimal at a properly designed site

TABLE 9.8. Chemical and Physical Parameters Typically Determined in Monitoring Samples from Sewage Sludge Application Sites.

Source	Chemical and Physical Parameters
Groundwater	pH, Electrical Conductivity, Total Hardness, Total Dissolved Solids, Chlorides, Sulfates, Total Organic Carbon, Nitrate Nitrogen, Total Phosphorus, Methylene Blue Active Substances (surfactants), Selected Metals or Trace Organics where Applicable, Indicator Organisms
Surface Waters	Fecal Coliforms, Total Phosphorus, Total Nitrogen (Kjeldahl), Dissolved Oxygen, BOD or TOC, Temperature, pH, Suspended Solids
Soils	Exchangeable Ammonium Nitrogen and Nitrate-Nitrite Nitrogen, Available Phosphorus, pH, Electrical Conductivity, Organic Carbon, Exchangeable Cations (Calcium, Magnesium, Potassium, Sodium), Extractable Metals—DTPA or 0.1 N HCl (Cadmium, Copper, Nickel, Zinc), Cation Exchange Capacity, Particle Size Distribution (texture), Other Constituents[1] (Persistent Organics)
Crops	Heavy Metals (Cadmium, Copper, Nickel, Zinc), Macro-Nutrients—Optional (Nitrogen, Phosphorus, Potassium), Other Constituents[1] (Antimony, Arsenic, Chromium, Iron, Mercury, Molybdenum, Selenium, PCBs)

[1]The other constituents are analyzed only if there are significant quantities of those contained in the sludge applied Analysis for trace organics, such as PCBs, DDT, dieldrin, etc is not normally done except in unusual situations where analytical costs are warranted

that is located, constructed, and operated to minimize the chance for surface water runoff of sludge-applied constituents. The U.S.EPA [5] indicates that surface water monitoring is done only in one or more of the following situations:

(1) Surface water runoff from the site is collected, stored, and discharged to surface waters outside the application area under an NPDES permit.

(2) The sludge application site is in close proximity to surface waters that are sensitive (e.g., drinking water supplies, swimming areas, etc.), and monitoring is required by a regulatory agency to insure that migration of sludge constituents to these waters is not occurring.

(3) It is desirable for public acceptance purposes to moderate community concern about surface water impacts.

The implementation of surface water monitoring procedures should follow a systematic plan for selection of surface water sampling stations. The sampling frequency is usually determined by the state regulatory agency and it can vary from monthly to quarterly. Surface water sampling locations should be established in areas that have the greatest potential for contamination. These sampling locations can be determined after examining the pathways available for runoff to enter a surface water body. Flow patterns and seasonal variations should be noted when applicable in establishing a surface water monitoring network.

The sampling procedures for rivers, streams, and reservoirs is discussed by the U.S.EPA [5]. Briefly, the sampling locations for rivers and streams should be established at locations where water composition is relatively uniform. Such locations are possible on small-and medium-sized streams, but they may be impossible to find on large rivers.

In sampling lakes and reservoirs, the U.S.EPA [5] recommends sampling along a three-dimensional grid pattern with samples being collected at different depths at each grid intersection. A more economical approach is to sample a different depth along selected cross section and sampling points. If only one sample is to be collected, it should be collected near the center of the water mass. The collection of one sample to characterize the composition of a lake is considered completely inadequate by the U.S.EPA [5] because it provides only an approximation of average water quality in the best case. For reservoirs that discharge water to potential users downstream, the sampling site should be located at or near the point of discharge.

Surface water parameters of concern at sludge application sites are either those that may affect public health, or those that may contribute to eutrophication such as nitrogen and phosphorus. An example of the parameters typically determined in surface water samples from a sludge application site is shown in Table 9.8.

Soils

Soils can be sampled and analyzed at potential sludge application sites as a part of the site selection process. After the site has been selected, more extensive soil sampling is desirable in order to establish baseline data. Once the sludge application program is underway, it is usually necessary at dedicated sites to monitor the changes occurring in the soil characteristics. This is especially true for those sites that are receiving sewage sludge on a continuing basis. The U.S.EPA [5] notes that periodic soil monitoring of a sludge application site is done when one or more of the following situations exist:

(1) The sludge contains significant quantities of one or more heavy metals or priority persistent organics.

(2) Heavy sludge application rates are used, as with a dedicated disposal site, and there is concern about the impact on vegetation grown on the site.

(3) The regulatory agency requires certain periodic soil monitoring.

(4) Demonstration projects, test plots, are being implemented to increase the knowledge of the interaction between sludge-applied constituents and soil systems.

Soil expertise is needed to conduct and interpret an adequate soil sampling program because of the potential variables involved. Some of these variables include the spatial and vertical variations in soil type, size of the application site or sites, and the objectives of the soil sampling program. Advice should be obtained from the University Cooperative Extension Service, County Agricultural Agents, and others with expertise in sampling and analysis of soils at sludge application sites.

The number and location of samples necessary to adequately characterize soils prior to sludge application is primarily a function of the spatial variability of the soils at the site. An in-depth discussion of these items is beyond the intent of this review. The U.S.EPA [5] discusses some of the factors involved in making these determinations. Also, in many instances, the state regulatory agency stipulates the minimum number of soil borings that must be analyzed.

The depth to which the soil profile is sampled and the extent to which each horizon is vertically subdivided depend largely on the parameters to be analyzed, the vertical variations in soil profile, and the objectives of the soil sampling program. These determinations need to be made by those with soils expertise. The U.S.EPA [5] discusses some of these items. Usually, as a minimum at sludge application sites, soil samples are collected on an annual basis from the upper soil layer, e.g., 0 to 15 cm (0 to 6 in). Soil samples taken from specific fields or site locations with a homogeneous soil are normally composited for several borings. Soil samples from deeper depths may be collected to obtain additional information on the movement of sludge-applied constituents. The state regulatory agency may require soil profile sampling and may specify the actual sampling depths required.

The specific soil chemical parameters to be determined are variable because they relate to the composition of the sludge being applied and the loading rates. It is important to consider sludge chemical properties when planning soil physical and chemical analyses so that the same analyses are done on soil samples collected before and after sludge applications are made. Typically, the determinations made on soil samples from sludge application sites are shown in Table 9.8. The U.S.EPA [5] and Ellis et al. [30] provide guides for soil sampling and analysis.

Crops

Crop monitoring at sewage sludge application sites is not normally done unless heavy sludge application rates are used, such as at a dedicated site. Crop monitoring at dedicated BU sites is conducted to evaluate the impact of sludge application on crops. The main concern is about heavy metals, particularly cadmium, in food chain vegetation.

Plant tissue may be sampled during several growth stages, although mature leaves or stalks on main branches or stems are generally preferred. The appropriate plant part and stage of development to be sampled must be established with a specific purpose in mind, such as determining heavy metal accumulation in the food chain or predicting possible phytotoxic levels [30].

The extreme variability in the systems being sampled makes obtaining a representative sample very difficult. The results obtained from one grab sample may be of little value. Consequently, a composite sample representing several individual samples collected at different places is preferred. Walsh and Beaton [31] and Ellis et al. [30] provide detailed procedures for plant sampling and analyses. These should be used as a guide in implementing an annual plant monitoring network. The quantity and type of samples to be collected depends upon the goals, economics, and regulatory requirements.

Table 9.8 presents a list of potential monitoring parameters for agricultural crops. The actual parameters monitored may vary from the list presented, depending upon the sludge constituents of concern.

A CASE STUDY: THE METROPOLITAN WATER RECLAMATION DISTRICT OF GREATER CHICAGO'S DEDICATED LAND RECLAMATION SITE AT FULTON COUNTY, ILLINOIS

BACKGROUND

Management Needs

The Metropolitan Water Reclamation District of Greater Chicago (MWRDGC) serves the city of Chicago and 124 adjacent suburban communities covering an area of approximately 2330 km² (900 mi²). The MWRDGC's sewered population is 5.5 million and the commercial and industrial population equivalent is 4.5 million, bringing the total population equivalent served by the MWRDGC to 10.0 million people. From this residential and industrial complex, the MWRDGC is collecting and treating an average of 1.4 billion gallons of sewage every day. The solids produced from the treatment of the MWRDGC's sewage for ultimate disposal amount to approximately 473 dry Mg/day (520 dry tons/day) [32].

Before 1971, processed sewage solids were either stored in lagoons in liquid form, heat-dried and sold in bulk as a dry fertilizer, or were air-dried (Imhoff digested sludges only) and stored on MWRDGC property. Additional storage lagoon space in or around Chicago was not available. Possible options included incineration, dewatering, drying, and improving sludge-disposal operations. In 1967, the MWRDGC's governing board formally adopted the environmental policy of putting sludge solids on land with the concept of beneficially recycling stabilized solids.

Several years of surveys, feasibility studies, laboratory experimentation, and field demonstrations were conducted to ascertain the most effective and environmentally safe method of using sewage sludge on land. The most attractive and reasonable alternative was land application of sewage sludge at a dedicated site where agricultural crops could be grown. With this policy, the MWRDGC started looking for a suitable site.

Site Selection

After approximately two years of looking for sites with suitable acreage, the MWRDGC was approached by officials from Fulton County, Illinois, who wanted to reclaim some of the county mine spoils. The strip-mined areas represented 7.24% of the total county land area [33]. The MWRDGC agreed to help and purchased 2856 ha (7052 acres) in 1970 from private landowners. Additional purchases were made and by 1976, the District owned 6289 ha (15,528 acres). The site is approximately 306 km (190 miles) southwest of Chicago.

In 1971, after the purchase of the 2856 ha (7052 acres), an extensive soil and geologic survey was conducted [34]. The survey consisted of thirty-five borings being taken throughout the purchased acreage. One boring sample was taken approximately every 81 ha (200 acres). The boring samples were taken at 3.05 m (10 ft) intervals, and the depth varied from 6.1 to 15.2 m (20 to 50 ft). Analyses of the 179 boring samples showed that the spoils were calcareous with a mean profile pH of 7.7, fine textured (silt clay loam to clay loam), low in nitrogen, and low in organic matter. Infiltration of water was low with infiltration rates of 2.26 cm/hr (0.89 in/hr) being observed.

The application of liquid sewage sludge to the mine spoils was considered desirable, the reasons being that sludge would provide essential plant nutrients, improve the soil tilth, increase water infiltration, increase the organic matter content, and thereby increase crop yields. The calcareous spoil would reduce the availability of sludge-applied metals.

Site Design and Preparation

The prime goal of site design and development was to prevent contamination of surface and groundwaters. Fields to receive sewage sludge were leveled to a maximum 5% grade to prevent rapid runoff. A berm was constructed around each field to contain the applied sludge and any runoff. Retention basins designed to capture the runoff from an equivalent 100-year storm for the region were constructed to collect the runoff from each field. A desiltation area with a slotted standpipe was constructed adjacent to each runoff retention basin. This was done so storm runoff was retained in the field to allow time for silt to settle before the runoff washed into the retention basin.

From 1972 to March 1983, anaerobically digested waste-activated sewage sludge was barged from the Stickney Water Reclamation Plant down the Illinois River 290 km (180 miles) to Fulton County. The sludge was pumped from barge unloading facilities 17 km (10.6 miles) to holding basins at the site. During the application season, sludge was dredged from the holding basins and pumped to the application fields comprising 1033 ha (2551 acres).

Sewage sludge barged to Fulton County was stored in four holding basins with a total storage capacity of 6 million m^3 (8 million yd^3). The basins were constructed to permit year-round shipment of sludge to the Fulton County site. In addition, the basins would allow the sludge to age and reduce the odor potential.

Sludge from the holding basins was applied to the fields using traveling sprinklers from 1972 until 1975, when the sprinkler method was phased out. By 1977, the major method of application was by incorporation with a heavy duty off-set disk. The application procedure is described by Peterson et al. [35]. Since 1986, dewatered, centrifuged, and lagooned sludge has been hauled from barges on the Illinois River directly to a specific field. The stockpiled sludge is then loaded into a side-shoot manure spreader for surface application, and incorporated into the soil by disking. Because of land reclamation, the annual allowable loading rate for sludge started at 168 Mg/ha (75 tons/acre) the first year. The application rate tapered down to 56 Mg/ha (25 tons/acre) by the fifth year and has continued at that rate.

Utilization of sludge-applied nutrients, especially nitrogen and phosphorus, at a dedicated site is essential to reduce the potential for contamination of surface and groundwaters. Crops can also provide an economic return for the site. The two major crops grown on an annual basis in fields receiving sludge are corn and winter wheat. Since 1979, when the U.S.EPA [19] published regulations for sludge application projects, the MWRDGC has been using grain harvested from crops produced for either animal feed or ethanol production.

Monitoring

Prior to applying any sludge on the property, the MWRDGC met with state and local regulatory officials to design a monitoring program to be instituted at the Fulton County site. A program was instituted to monitor groundwaters, surface waters, soils, and crops. An important requirement of the monitoring program was the establishment of baseline, preexisting conditions for groundwaters, surface waters, soils (surface and subsurface), and crops.

Groundwaters

Part of the environmental protection system established in 1971 was the installation of twenty-four groundwater monitoring wells. The wells were sealed, steel cased units with submersible electric pumps installed at depths of 10.7 to 21 m (35 to 70 ft). Water samples were collected monthly from 1971 to 1986 and then quarterly after 1987. The collected samples were analyzed for twenty-five chemical and physical characteristics and static water elevations were determined. Table 9.9 shows the 1981 yearly summary for two monitoring wells located in mine soil adjacent to a field receiving sewage sludge. The field adjacent to well 14 had received a cumulative total of 291 Mg/ha (130 tons/acre) of sludge, and the field adjacent to well 15 received 299 Mg/ha (133 tons/acre) of sludge prior to 1981.

TABLE 9.9. 1981 Yearly Summary for Wells 14 and 15 at the Fulton County Site.

Well No.:	14			15		
Constituent	Mean[a] (mg/L)	Max. (mg/L)	Min. (mg/L)	Mean[a] (mg/L)	Max. (mg/L)	Min. (mg/L)
pH	6.63	6.85	6.30	6.75	7.01	6.40
Total P	0.06	0.10	<0.01	0.07	0.18	0.02
Cl⁻	32.33	42.00	25 00	57 50	342.00	24.00
SO₄⁼	1279	1720	581.00	679.08	933 00	13.00
N-Kjeldahl	0.56	1.90	<0.10	1.55	2.20	1.00
NH₃	0.42	1.50	0.10	1.43	2.20	1.10
N—NO₂ + NO₃	0.50	1.75	0.02	0.03	0.13	0.01
Alk as CaCO₃	779.67	887.00	29 00	489.00	863.00	367.00
EC (µmhos/cm)	3428	3890	3000	2343	2550	2150
K	3.50	15.00	<1.00	4.17	12.00	<1.00
Na	74.17	130.00	4.00	72.33	210.00	7.00
Ca	356.00	442.00	310.00	180 00	280.00	140.00
Mg	239.00	295.00	197 00	120.33	200.00	98.00
Zn	0 69	1.20	0.30	1.35	2.90	0.70
Cd	<0.02	0.03	<0 02	<0.02	0.02	<0.02

TABLE 9.9. (continued).

Well No.:	14			15		
Constituent	Mean[a] (mg/L)	Max. (mg/L)	Min. (mg/L)	Mean[a] (mg/L)	Max. (mg/L)	Min. (mg/L)
Cu	<0.02	<0.02	<0.02	<0.02	0.03	<0.02
Cr	<0.02	0.05	<0.02	<0.02	0.05	<0.02
Ni	<0.10	0.10	<0.10	<0.10	0.10	<0.10
Mn	2.81	5.58	2.07	0.91	2.74	0.39
Pb	<0.02	0.04	<0.02	<0.02	0.02	<0.02
Fe	47.16	250.00	0.40	51.71	126.00	0.90
Al	<1.00	<1.00	<1.00	<1.00	<1.00	<1.00
Hg (µg/L)	0.16	0.40	<0.10	0.14	0.50	<1.00
Se	<0.10	0.30	<0.10	<0.10	0.30	<0.10
F.coli/100 mL	<1.00	<1.00	<1.00	<1.00	5.00	<1.00
St H2O el (ft)[b]	636.05	637.70	633.60	657.80	660.80	665.20

[a]In computing the mean, values below the minimum detection limit were considered as zero. The geometric mean was used for fecal coliforms, and values below the MDL were considered as 0.1.
[b]Static water elevations are determined by subtracting the water depth from the surface elevation of the monitoring well

557

TABLE 9.10. 1981 Yearly Summary for Reservoirs R3 and R4 at the Fulton County Site.

Reservoir No.:	R3			R4		
Constituent	Mean[a] (mg/L)	Max. (mg/L)	Min. (mg/L)	Mean[a] (mg/L)	Max. (mg/L)	Min. (mg/L)
pH	7.95	8.43	7.20	7.93	8.23	7.61
Total P	0.10	0.22	0.04	0.12	0.22	0.04
Cl-	22.17	27.00	14.00	17.33	27.00	12.00
$SO_4^=$	560.17	885.00	445.00	362.48	456.00	266.00
N-Kjeldahl	1.20	1.70	0.70	0.52	0.70	0.40
NH_3	<0.10	0.20	<0.10	<0.10	0.10	<0.10
$N—NO_2 + NO_3$	0.95	2.55	0.01	0.12	0.23	0.03
Alk as $CaCo_3$	285.75	371.00	157.00	178.75	316.00	100.00
EC (µmhos/cm)	1632	2125	750.00	953.50	1040	810.00
K	3.67	15.00	<1.00	1.50	4.00	<1.00
Na	45.25	170.00	4.00	7.25	25.00	2.00
Ca	136.83	210.00	85.00	124.58	150.00	90.00
Mg	100.17	150.00	69.00	56.67	67.00	40.00
Zn	<0.10	0.20	<0.10	<0.10	<0.10	<0.10
Cd	<0.02	0.03	<0.02	<0.02	0.02	<0.02

TABLE 9.10. (continued).

Reservoir No.:	R3			R4		
Constituent	Mean[a] (mg/L)	Max. (mg/L)	Min. (mg/L)	Mean[a] (mg/L)	Max. (mg/L)	Min. (mg/L)
Cu	<0.02	0.02	<0.02	<0.02	0 02	<0.02
Cr	<0.02	0.02	<0.02	<0 02	0.02	<0.02
Ni	<0.10	0.20	<0.10	<0.10	0 20	<0 10
Mn	0 10	0.19	<0.02	0 21	0.49	<0.02
Pb	<0.02	0.10	<0 02	<0.02	0.10	<0.02
Fe	0.12	0.40	<0.10	<0 13	0.40	<0.10
Al	<1.00	<1.00	<1 00	<1.00	<1.00	<1.00
Hg (µg/L)	0.21	1.30	<0.10	0.18	0.50	<0.10
Se	<0.10	0 20	<0.10	<0.10	0 20	<0.10
F coli /100 mL	3 16	32.00	<1.00	<36 320	600.00	<1 00
Dissolved oxygen	10 04	14.50	<7.60			
Temp (°C)	14 99	28.00	1.00			
Total suspended solids	10.67	22 00	3.00	17.50	104 00	2.00
Total dissolved solids	1623	1821	1282	827 33	898.00	615.00

[a]In computing the mean, values below the minimum detection limit were considered as zero. The geometric mean was used for fecal coliforms, and values below the MDL were considered as 0 1

559

TABLE 9.11. Discharge from Two Runoff Retention Basins
at the Fulton County Site.

| Constituent | Runoff Retention Basins[a] (mg/L) | |
	11-1	25-1
pH	8 20	8 00
Total P	1 21	0.85
Cl⁻	68 0	80.0
SO₄⁻	244	390
N-Kjeldahl	28.0	13 5
N—NH₃	18.0	6.6
N—NO₂ + NO₂	18 5	26 9
Alk as CaCO₃	138	95.0
EC (μmhos/cm)	960	1220
K	13.0	7 0
Na	11.0	6 0
Ca	132	184
Mg	46.0	79.0
Zn	0.00	0 00
Cd	0.00	0.02
Cu	0 03	0.02
Cr	0.00	0.00
Ni	0 00	0 00
Mn	0.12	0.48
Pb	0 00	0 00
Fe	0 20	0 10
Hg (μg/L)	0 00	0.50
Se	0.00	0.00
F coli./100 mL	280	310
Biological oxygen demand	15.0	15.0
Total suspended solids	36 0	26 0
Total volatile suspended solids	34.0	23.0

[a]Grab samples taken during discharge on July 22, 1981.

Sampling of the unsaturated (vadose) zone is also done at the site.
Fifteen porous cup samplers were installed at depths of one, four, and
fifteen feet at selected locations throughout the site. Soil water samples
are collected at the same frequency as are those for groundwaters.

Surface Waters

Surface water sampling is an extensive part of the monitoring network
at the Fulton County site. Part of the surface water network is twelve

TABLE 9.12. pH and Metal Contents of 1986 Corn Field Soils at the Fulton County Site.

Field No.[a]	Cumulative Sludge Applied[b] Dry Solids (Mg/ha)	Cumulative Sludge Applied[b] Dry Solids (tons/acre)	pH	0.1 M HCl Extractable (mg/kg) Zn	Cd	Cu	Cr	Fe	Ni	Pb	Mn
3	545	(243)	6.86	780	56.6	324	232	5123	62.6	160	320
10	615	(275)	6.72	1,094	73.8	457	360	7137	86.5	210	548
16E	639	(285)	6.86	864	62.8	317	188	3842	73.8	152	366
16W	639	(285)	6.80	890	67.1	336	185	4136	64.4	155	493
18	0	(0)	7.00	16.9	1.2	6.5	5.9	409	2.8	4.9	188
19	644	(287)	6.93	795	57.7	296	194	4106	55.6	142	319
20	531	(237)	6.72	752	53.8	288	154	3336	59.6	132	286
21	618	(276)	6.71	697	49.9	256	158	3634	48.6	128	403
22	455	(203)	6.61	743	58.4	296	196	4074	56.5	136	363
23	473	(211)	6.71	828	61.4	319	166	4085	64.6	152	358
25	569	(254)	6.50	880	64.6	321	194	4192	63.2	120	287
29	0	(0)	7.16	24	1.8	4.8	3.8	714	9.7	7.5	312
		LSD[c]	N.S.	196[d]	14.8[d]	98[d]	53[d]	805[d]	4.3[d]	33[d]	118[d]

[a] Fields 3, 16E, 16W, 18, 25, and 29 are on strip-mined land.
[b] Amount applied through 1985
[c] Least significant difference (LSD) procedure was used to compare two treatment means when analysis of variance (ANOVA) was significant at the 0.05 and 0.01 levels, respectively. Two means differ significantly when the difference between them exceeds the LSD value.
[d] Significant at the 0.01 level
N.S. = Not statistically significant.

TABLE 9.13. Metal Analysis of 1986 Corn Leaves from Fulton County Site Fields.

Field No.	Cumulative Sludge Applied[a] Dry Solids (Mg/ha)	(tons/acre)	[mg/kg (oven dried, 65°C)] Zn	Cd	Cu	Cr	Fe	Ni	Pb	Mn
3	545	(243)	163	16.0	13.8	0.30	66	0.53	0.75	37
10	615	(275)	208	15.2	15.0	0.29	72	0.40	0.72	41
16E	639	(285)	116	10.0	13.1	0.26	79	0.94	0.60	38
16W	639	(285)	201	12.1	17.7	0.28	74	1.85	0.70	47
18	0	(0)	29	0.7	10.2	0.28	70	0.32	0.72	39
19	644	(287)	134	7.8	13.1	0.48	74	3.96	0.75	39
20	531	(237)	92	7.1	12.1	0.86	93	6.64	0.72	31
21	618	(276)	137	8.5	12.0	0.29	81	0.46	0.66	28
22	455	(203)	153	8.4	15.3	0.28	84	0.52	0.69	41
23	473	(211)	98	9.0	11.7	0.28	82	0.56	0.69	46
25	569	(254)	274	15.9	18.4	0.36	82	2.75	0.70	44
29	0	(0)	25	0.7	7.9	0.25	68	0.89	1.39	47
		LSD[b]	57[c]	3.9[c]	2.5[c]	0.40[c]	14[c]	2.70[c]	0.31[c]	N S

[a] Amount applied through 1985
[b] Least significant difference (LSD) procedure was used to compare two treatment means when analysis of variance (ANOVA) was significant at the 0 05 and 0.01 levels, respectively. Two means differ significantly when the difference between them exceeds the LSD value.
[c] Significant at the 0.01 level
N S = Not statistically significant.

streams and ten reservoirs that were sampled monthly from 1971 to 1981 and then three times per year since 1982. In addition, there are fifteen surface water sites that have been sampled three times per year since 1976. Sixty runoff retention basins were constructed to collect runoff from the sludge application fields. Runoff water is released only after it meets the applicable state of Illinois effluent standards of ≤99 mg/L total suspended solids, ≤33 mg/L biological oxygen demand, ≤494 fecal coliform counts/ 100 mL, and pH within the limits of 6.0 to 10.0.

Table 9.10 shows the 1981 yearly summary for two reservoirs that received drainage from major watersheds at the site. Watershed reservoirs, R3 and R4, are discussed by Peterson et al. [36]. Drainage into these reservoirs in 1981 was from fields where sludge was incorporated.

Table 9.11 shows a typical discharge from two runoff retention basins during 1981. Basin 11-1 drained into the R3 watershed and basin 25-1 drained into the R4 watershed.

Soils and Crops

Soil sampling to determine the effects of applied sewage sludge on the chemical composition of application fields is part of the monitoring program. Soil sampling of site fields was initiated in 1972. Soil characteristics of the site prior to sludge application is described by Peterson et al. [36].

Table 9.12 shows the pH and metal contents of 1986 corn fields at the Fulton County site. Additional analyses conducted on each sample include electrical conductivity, organic carbon, available phosphorus, cation exchange capacity, exchangeable NH_4–N and NO_3 and NO_2–N, and exchangeable Ca, Mg, K, and Na. Agricultural lime is applied, as required, to raise the soil pH to 6.5.

Crop sampling of sludge-amended fields is done on an annual basis as a part of the monitoring program. The intent is to determine the accumulation of sludge-applied metals in crops grown at the site. Plant tissue sampling at the site was initiated in 1972. Typically, grain samples are collected from all harvested crops, and for corn, leaf samples are also collected. Table 9.13 shows the metal content in corn leaves collected from sludge-amended fields in 1986.

REFERENCES

1 U.S.EPA. 1989. "40 CFR Parts 257 and 503 Standards for the Disposal of Sewage Sludge: Proposed Rule. Monday, February 6, 1989," *Federal Register* 54(23):5746–5902.

2 Flach, K. W. and F. J. Carlisle. 1974. "Soils and Site Selection," in *Factors Involved in Land Application of Agricultural and Municipal Wastes*, ARS, USDA, National Program Staff, Soil, Water, and Air Sciences, Beltsville, MD, pp. 1–17.

3 Witty, J. E. and K. W. Flach. 1977. "Site Selection as Related to Utilization and Disposal of Organic Wastes," in *Soils for Management of Organic Wastes and Waste Waters*, L. F. Elliot and F. J. Stevenson, eds., Madison, WI: Soil Science Society of America, American Society of Agronomy, and Crop Science Society of America, pp. 327–345.

4 Larson, W. E. and G. E. Schuman. 1977. "Problems and Need for High Utilization Rates of Organic Wastes," in *Soils for Management of Organic Wastes and Waste Waters*, L. F. Elliot and F. J. Stevenson, eds., Madison, WI: Soil Science Society of America, American Society of Agronomy, and Crop Science Society of America, pp. 589–618.

5 U.S.EPA. 1983. *Process Design Manual for Land Application of Municipal Sludge*, U.S.EPA Rep. EPA-625/1-83-016. Cincinnati, OH: U.S. Environmental Protection Agency.

6 Stewart, B. A. and L. R. Webber. 1976. "Consideration of Soils for Accepting Wastes," in *Land Application of Waste Materials*. Ankeny, IA: Soil Conservation Society of America, pp. 8–21.

7 Epstein, E. 1974. "The Physical Processes in the Soil as Related to Sewage Sludge Application," in *Recycling Municipal Sludges and Effluents on Land*, July 9–13, 1973, Champaign, IL, National Association of State Universities and Land Grant Colleges, Washington, D.C., pp. 67–73.

8 Brady, N. C. 1984. *The Nature and Properties of Soils*, Ninth Edition. New York, NY: MacMillan Publishing CO.

9 Gupta, S. C., R. H. Dowdy and W. E. Larson. 1977. "Hydraulic and Thermal Properties of a Sandy Soil as Influenced by Incorporation of Sewage Sludge," *Soil Sci. Soc. Am. J.* 41:601–605.

10 Unger, P. A. and B. A. Stewart. 1974. "Feedlot Waste Effects on Soil Conditions and Water Evaporation," *Soil Sci. Soc. Am. Proc.*, 8:954–957.

11 U.S. Soil Conservation Service. 1951. *Soil Survey Manual, USDA Handbook No. 18*. Washington, D.C.: U.S. Government Printing Office.

12 Metzger, L. and B. Yaron. 1987. "Influence of Sludge Organic Matter on Soil Physical Properties," in *Advances in Soil Science*, B. A. Steward, ed. New York: Springer-Verlag, 7:141–163.

13 Khaleel, R., K. R. Reddy and M. R. Overcash. 1981. "Changes in Soil Physical Properties Due to Organic Waste Applications: A Review," *J. Environ. Qual.*, 10:133–141.

14 Kladivko, E. J. and D. W. Nelson. 1979. "Changes in Soil Physical Properties from Application of Anaerobic Sludge," *J. Water Pollut. Control Fed.*, 51:325–332.

15 Chang, A. C., A. L. Page, and J. E. Varneke. 1983. "Soil Conditioning Effects of Municipal Sludge Compost," *J. Environ. Eng. (ASCE)*, 109:574–583.

16 Chaney, R. L. 1974. "Crop and Food Chain Effects of Toxic Elements in Sludges and Effluents," in *Recycling Municipal Sludges and Effluents on Lands*, July 9–13, 1973, Champaign, IL, National Association of State Universities and Land Grant Colleges, Washington, D.C., pp. 129–141.

17 CAST. 1976. "Application of Sewage Sludge to Cropland: Appraisal of the Potential Hazards of Heavy of Heavy Metals to Plants and Animals," Council for Agricultural Science and Technology, Report No. 64, Ames, IA.

18 CAST. 1980. "Effects of Sewage Sludge on the Cadmium and Zinc Content of Crops," Council for Agricultural Science and Technology, Report No. 83, Ames, IA.

19 U.S.EPA. 1979. "Criteria for Classification of Solid Waste Disposal Facilities and Practices," Thursday, September 13, 1979. *Federal Register* 44(179):53438–53464.

20 Haghiri, F. 1974. "Plant Uptake of Cadmium as Influenced by Cation Exchange Capacity, Organic Matter, Zinc, and Soil Temperature," *J. Environ. Qual.* 3:180–183.

21 Latterell, J. J., R. H. Dowdy and G. E. Ham. 1976. "Sludge-Borne Metal Uptake by Soybeans as a Function of Soil Cation Exchange Capacity," *Commun. Soil Sci. Plant Anal.* 7:465–476.

22 Corey, R. B., L. D. King, C. Lue-Hing, D. S. Fanning, J. J. Street and J. M. Walker. 1987. "Effects of Sludge Properties on Accumulation of Trace Elements by Crops," in *Land Application of Sludge: Food Chain Implications*, A. L. Page, T. J. Logan and J. A. Ryan, eds., Chelsea, MI: Lewis Publishers, Inc., pp. 25–51.

23 State of Illinois. 1984. "Rules and Regulations, Title 35: Environmental Protection, Subtitle C: Water Pollution, Chapter II: Environmental Protection Agency, Part 391, Design Criteria for Sludge Application to Land," Chicago, IL.

24 Sommers, L. E., O. W. Nelson and C. D. Spies. 1980. "Use of Sewage Sludge in Crop Production," Report No. AY20. Purdue University, Cooperative Extension Service, West Lafayette, IN.

25 Fenn, D. G. 1977. *Procedures Manual for Ground Water Monitoring at Solid Waste Disposal Facilities*, U.S.EPA Rep. EPA/30/SW-611. Wehran Engineering, Mahwah, NJ.

26 Blakeslee, P. A. 1976. "Site Monitoring Considerations," in *Application of Sludges and Wastewaters on Agricultural Land: A Planning and Educational Guide*, B. D. Knezek and R. H. Miller, eds., Wooster, OH: Ohio Agricultural Research and Development Center, pp. 11.1–11.5.

27 Barcelona, M. J. and R. D. Morrison. 1988. "Sample Collection, Handling and Storage: Water, Soils and Aquifer Solids," in *Methods for Ground Water Quality Studies, Proceedings of a National Workshop;* Arlington, Virginia, November 1–3, 1988, D. W. Nelson and R. H. Dowdy, eds., Agricultural Research Division, University of Nebraska, Lincoln, NE, pp. 49–62.

28 Gillham, R. W., M. J. L. Robin, J. F. Barker and J. A. Cherry. 1983. "Ground Water Monitoring and Sample Bias," American Petroleum Institute Report No. 4367, Environmental Affairs Department, American Petroleum Institute, Washington, D.C., 206 pp.

29 Patrick, R., E. Ford and J. Quarles. 1987. *Groundwater Contamination in the United States*, 2nd Edition. Philadelphia, PA: University of Pennsylvania Press.

30 Ellis, R. J., Jr., J. J. Hanway, G. Holmgren, D. R. Keeney and O. W. Bidwell. 1975. "Sampling and Analysis of Soils, Plants, Waste Waters, and Sludge. Suggested Standardization and Methodology," Research Publication 170, Agricultural Experiment Station, Kansas State University, Manhattan, KS.

31 Walsh, L. M., and J. D. Beaton (eds.). 1973. *Soil Testing and Plant Analysis*, Revised Edition. Madison, WI: Soil Science Society of America.

32 Lue-Hing, C., B. T. Lynam, J. R. Peterson and J. G. Gschwind. 1976. "Chicago Prairie Plan—A Report on Eight Years of Municipal Sewage Sludge Utilization," Report No. 76-16. Department of Research and Development, The Metropolitan Sanitary District of Greater Chicago, Chicago, IL.

33 Haynes, R. J. and W. D. Klimstra. 1975. "Illinois Lands Surface Mined for Coal," Cooperative Wildlife Research Laboratory, Southern Illinois University at Carbondale, IL.

34 Peterson, J. R. and L. Papp-Vary. 1972. "Soil and Geological Study of the Fulton

County Site Prior to the Application of Liquid Fertilizer," Research Report, Department of Research and Control, The Metropolitan Sanitary District of Greater Chicago, Chicago, IL.

35 Peterson, J. R., C. Lue-Hing, J. Gschwind, R. I. Pietz and D. R. Zenz. 1982. "Metropolitan Chicago's Fulton County Sludge Utilization Program," in *Land Reclamation and Biomass Production with Municipal Wastewater and Sludge*, W. E. Sopper, E. M. Seaker and R. K. Bastian, eds., University Park, PA: The Pennsylvania State University Press, pp. 322–338.

36 Peterson, J. R., R. I. Pietz and C. Lue-Hing. 1979. "Water, Soil, and Crop Quality of Illinois Coal Mine Spoils Amended with Sewage Sludge," in *Utilization of Sewage Effluent and Sludge on Forest and Disturbed Land*, W. E. Sopper and S. N. Kerr, eds., University Park, PA: The Pennsylvania State University Press, pp. 359–368.

CHAPTER 10

Application of Municipal Sewage Sludge to Soil Reclamation Sites

T HE intent of this chapter is to provide information needed in using municipal sewage sludge for land reclamation sites. The chapter describes: (1) benefits of land reclamation, (2) properties of municipal sewage sludge, (3) detailed site investigation, (4) site preparation, (5) monitoring requirements, (6) sludge handling and application methods, (7) application quantities, and (8) reclamation site responses at existing sites.

INTRODUCTION

The surface mining of coal, exploration for minerals, generation of mine spoils from underground mines, and tailings from mining operations have created over 1.5 million ha (3.7 million acres) of disturbed lands in the United States [1]. The properties of these disturbed and marginal lands vary considerably from site to site. These lands are usually a harsh environment for seed germination and plant vegetation. Typically, this is the result of a lack of nutrients and soil organic matter, low pH, low water-holding capacity, low rates of water infiltration and permeability, poor physical properties, reduced biological properties, and the presence of toxic levels of trace metals.

The reclamation of these lands has historically been accomplished by grading the slopes to minimize erosion and facilitate revegetation. Soil amendments such as lime and fertilizer are added, and grasses, legumes, and/or trees are planted. The U.S.EPA [1] notes that although these methods

John Gschwind and Richard I. Pietz, Metropolitan Water Reclamation District of Greater Chicago, Chicago, IL.

are sometimes successful, numerous failures have occurred because of the very poor physical, chemical, and biological properties of these disturbed lands.

The use of sewage sludge for rebuilding disturbed lands offers a material that cannot be duplicated by other amendments. Sludges have several characteristics that make them suitable for reclaiming and improving disturbed and marginal soils. The organic matter in the applied sludge improves the soil physical properties by improving granulation, reducing plasticity and cohesion, and increasing water-holding capacity. In addition, the sludge increases the soil cation exchange capacity, supplies plant nutrients, and buffers soil pH [1].

There have been a number of successful land reclamation projects using sewage sludge or sludge compost. Operational experience is available for handling systems, application systems, amounts required per hectare, and the response of various types of vegetation. The single most important cost factor for most utilization sites is the cost of transportation for moving the sludge from the source of production to the site of utilization [2].

BENEFITS OF LAND RECLAMATION

IMPROVED SITE UTILIZATION

There are numerous reasons for land reclamation. A major one is to revegetate the site so it will reduce water and wind erosion. A revegetated site offers the potential for other uses such as agricultural production, animal grazing, and reforestation for lumber and pulp production. Additional reasons for land reclamation include: (1) improving surface topography to permit good drainage, (2) modifying the chemical conditions of the soil to prevent acidic leaching of toxic salt conditions, and (3) removing debris to provide for agricultural usage [2]. These benefits all relate to the fact that land reclamation with sewage sludge offers the potential to increase agricultural and forest utilization of disturbed lands and to reduce environmental contamination from these lands.

ENVIRONMENTAL PROBLEM CORRECTION

Contamination from a site can be caused by a number of factors. Wind or water erosion can remove particulate matter from a site. Altering the site conditions to provide a more favorable environment for tilth and vegetative growth by using sludge and grading to improve the topography can correct the problem.

The chemical composition of water leaving the site, either as overland flow or as groundwater, could be such that it causes undesirable ecological responses or environmental contamination. Excessive plant nutrients such as nitrogen, phosphorus, and potassium can cause undesirable plant growth in streams and lakes and degrade drinking water quality. Acidic water leaving coal refuse piles, pyritic mine wastes, and other types of acidic sites, because of acidity and soluble metals, can degrade surrounding streams and lakes. Sewage sludge, because of its buffering capacity, can be an effective amendment in reclamation activities at these sites.

PROPERTIES OF MUNICIPAL SEWAGE SLUDGE

CHEMICAL CHARACTERISTICS

Sewage sludge is derived from the organic and inorganic matter removed from wastewater. The influent wastewater varies tremendously depending on the nature of the inputs to the sewage treatment plant. Generally, the more industrialized a community, the greater will be the content of heavy metals and organics.

For purposes such as land reclamation, it is the fertilizer value of sludge that is most important. Data from the National Sewage Sludge Survey indicates that the national average concentration of nitrogen and phosphorus in United States sludges is 5.27% and 0.5%, respectively, on a dry weight basis. Sludge potassium levels are generally too low to be considered as an adequate source for this element. Thus, it can be said that sludge is an excellent source of nitrogen and phosphorus.

BIOLOGICAL PROPERTIES

Sludge stabilized by digestion can be safely left on the soil surface without concern of odor or vermin infestation. Hinesly et al. [3] reported a 99% decrease of fecal coliform after thirty days of desiccation on the soil surface. Burd [4] reported a 99.8% reduction after thirty days of mesophilic anaerobic digestion, and also reported that pathogenic organisms die within seven to ten days of digestion.

Research by Meyer et al. [5], with the porcine enterovirus (ECPO-1) using germfree ten-day piglets to test virus viability, indicated that no porcine enterovirus survived in the sewage sludge digester after five days. Reed et al. [6] reported on an incubation study using digested sludge supernatant in which all inoculated echovirus were inactivated in two days, Coxsackie B-4 in three days, and poliovirus type 1 in five days. In studies by Bertucci et al. [7], 250 mL laboratory anaerobic digesters containing 200 mL of digesting sludge were inoculated with poliovirus type 1, Coxsackie virus type A-9, Coxsackie virus B-4, and echovirus type 11. After twentyfour hours there was an inactivation of 93.8%, 97.5%, 89.5%, and 58%, respectively. At the end of forty-eight hours of digestion, the respective inactivations were 98.5%, 99.7%, 98.6%, and 92.5% for the previously listed viruses.

In most programs that use sewage sludge as a nutrient source and soil amendment, there will be many times when the sludge must be held until the weather or other farming operations allow for its application to the fields. During this storage, further reduction in virus and bacterial counts can be expected. Berg [8] determined the time in days required for 99.9% reduction in the number of virus and bacteria by storage of untreated sludge at different temperatures. His data clearly showed that time and increased temperature decreased virus survival.

PHYSICAL PROPERTIES

The physical properties of sewage sludge depend on the untreated wastewater, the type and extent of the wastewater treatment, and the method of sludge stabilization. A dense, granular sludge may be produced from primary sludge, while waste activated sludge from secondary treatment results in a sludge containing mostly bacterial cells that are viscous and difficult to dewater. The distribution of water in sewage solids was estimated by Bjorkman [9] to be 70% between cells, 22% adhesion and capillary water, and 8% absorption and intracellular fluids. Peterson et al. [10] reported that the particle size of waste activated sludge was as follows: 99% <9 mm and 60% <3 mm. The density of the sewage sludge was determined by McCalla et al. [11] to be 0.58 g/cm^3 for heat-dried waste activated sludge, 1.01 g/cm^3 for liquid digested sludge, 1.08 g/cm^3 for lagoon digested sludge, and 1.2 gm/cm^3 for aged Imhoff sludge with a solids content of 25 to 60%.

Most sewage sludges behave as a thixotropic pseudoplastic (non-Newtonian) fluid when pumped. That is, they become less viscous when mixed. This makes accurate calculations of frictional losses quite difficult. Rimkus and Heil [12] found that the plastic viscosity and yield stress varied as an exponential function of sludge solids content. For a lagoon sludge having a 13% solids content, with a density of 1.08 g/cm^3, Rimkus and Heil reported a plastic viscosity of 0.79 poise and yield stress of 232 dyn/cm^2.

DETAILED SITE INVESTIGATION

Disturbed or marginal land areas differ in their physical, hydrological, and soil chemical characteristics. These differences are the result of variations in mining operations, ore extraction processes, length of time since the area was disturbed, climate, soil and geological variations, and other factors [1]. On land that has been disturbed, extensive site investigations are often necessary to characterize the site so that an effective reclamation plan can be developed.

There may be areas at disturbed sites where the physical, hydrological, or chemical properties are unsuitable for sludge application. These areas, along with those suitable for sludge application, need to be surveyed and boundaries marked [1].

Federal and many state mining regulations in effect require that areas disturbed by mining operations be restored to the approximate original contour and productivity [13]. However, many prelaw sites mined before reclamation laws came into effect have never been reclaimed. The U.S.EPA [1] recommends that an accurate topographic contour map of the site area be developed to provide a basis for: (1) delineating the areas with slopes that are too steep for sludge application; (2) regrading the areas if the expense is cost-effective; and (3) designing surface runoff water improvements, e.g., ditches, terraces, berms, etc. Slope limitations need to be considered and appropriate regulatory agencies should be consulted to determine applicable slope criteria for the site.

GROUNDWATERS

Groundwaters at a reclamation site need to be evaluated during a site investigation. The site investigation should determine the following:

(1) Depth to groundwater, including seasonal variations

(2) Quality of existing groundwater

(3) Present and potential future use of groundwater

(4) Existence of perched water

(5) Direction of groundwater flow

The general regulatory philosophy is that sludge application to a site should not degrade useful groundwater resources beyond the boundary of the sludge application site [1]. Occasionally, groundwaters adjacent to the disturbed site are already degraded by previous mining operations and the aquifer can be "exempted" from nondegradation regulations. The suggested minimum depth to groundwater is 1 m (3.3 ft) for reclamation sites.

SOIL SAMPLING

The chemical and physical characteristics of the soil at reclamation sites need to be characterized. U.S.EPA [1] indicates that soil sampling and analysis are necessary to: (1) establish sludge application rates, both periodic and accumulative; (2) determine amounts of supplemental fertilizer, lime, or other soil amendments required to obtain desired vegetative growth; (3) determine the infiltration and permeability characteristics of the soil; (4) determine background levels of soil characteristics prior to sludge application.

In many cases, the soil profile at reclamation sites is a mixture of soil and geologic materials. A field inspection will be needed to determine the number and location of samples necessary to characterize the materials. The specific analyses may vary from location to location based on the state and local regulations covering both the reclamation and sludge utilization aspects [1].

DRAINAGE WATER

Water pollution problems, such as acid mine drainage, are frequently associated with mining activity. Consequently, it is necessary to document the quality of both surface and groundwater prior to the use of sludge in reclamation activities at a site. The water quality on or adjacent to the reclamation site in many instances has already been adversely affected. Procedures for implementing a surface and groundwater monitoring network are described by U.S.EPA [1].

SITE PREPARATION

TOPOGRAPHIC MODIFICATION

The amount of topographic modification required at a reclamation site depends primarily upon the initial site conditions and the type of utilization at the site. In addition, the sludge application system at the site can influence the topography changes.

Agricultural utilization of the site, following reclamation, generally requires very restrictive slope conditions. Halderson and Zenz [2] indicate that short slopes of 5 to 10% can be readily cropped on most types of soil, but slopes of 5% or less are more desirable. The U.S.EPA [1] lists the recommended slope limitations for land application of sludge. The land should not be graded flat in most instances, except for soils with a high infiltration rate, because surface ponding of water can be a problem. Erosion control is important and the site needs to be designed to minimize erosion. Information on proper slopes, lengths of slopes, and suitable types of vegetation can be obtained from the USDA Soil Conservation Service (SCS) for any basic type of soil at the reclamation site.

DEBRIS REMOVAL

Site preparation for sludge application may require the removal of debris left behind by previous operations. This may include rocks, trees, steel cable, and other foreign material. Halderson and Zenz [2] note that the degree to which these items need to be removed depends upon how the site is to be utilized after reclamation. If the site is to be used for agricultural activities, then the top 60 cm (24 in) of soil needs to be completely free of foreign material of any significant size. Halderson and Zenz [2] discuss the equipment that can be used in debris removal.

Revegetation of a site for erosion control will also require debris removal. The depth of debris removal will be determined by the methods of sludge application [2]. Dry sludge application will require debris removal from the top 30 cm (12 in) of soil. Liquid sludge application will require debris removal from the same depth, and if the liquid sludge is to be applied by equipment that trails an irrigation hose, then extensive rock removal will be required to prevent extensive wear of the hose.

HYDROLOGY CONTROL

One of the major environmental tasks at the application site is to prevent contamination of surface and subsurface waters by sludge applied constituents. Nitrogen and phosphorus in surface waters and nitrate in groundwaters are normally the constituents of most concern.

Halderson and Zenz [2] indicate that two basic approaches can be taken to control the problem. Sludge can be applied by incorporation or injection to minimize the amount of sludge that can come into contact with rain. In most instances, these application methods help minimize the potential problem. Other important variables that affect potential contamination at the site include sludge application rate, climatic conditions, soil type, and topographic conditions.

The control of surface runoff at a reclamation site is important in protecting the environmental integrity of the area. Surface waters in the area need to be considered. In addition, if runoff is to occur from a sludge application site, the water quality needs to be within acceptable limits for immediate discharge to a receiving stream or watercourse. If the runoff has nonacceptable water quality, the water needs to be detained in a holding structure so the runoff can be reapplied to land, held until it is acceptable for discharge, or treated so it can be discharged off-site. In many instances, like a dedicated site, this may require designing surface water improvements such as ditches, terraces, and berms, depending upon the slope and soil physical characteristics at the site.

The prevention of groundwater contamination at a reclamation site is important in protecting the groundwater for either public or other designated uses. Like dedicated sites, the potential reclamation site needs to be surveyed to determine soil type and depth and groundwater characteristics, as described previously. Depending upon the application rate and whether sludge is applied one time or continuously, the restriction of natural drainage by relatively impermeable layers near the surface or a high groundwater table may require subsurface water control to prevent groundwater contamination. The U.S.EPA [1] gives a detailed discussion on evaluating these items.

MONITORING REQUIREMENTS

At a reclamation site, in order to comply with state, local, and federal requirements for land application of sludge, both the sludge to be utilized and the site characteristics must be evaluated [1]. A detailed discussion of these items is given by the U.S.EPA [1].

Background sampling prior to sludge application is essential at a reclamation site. Soil samples need to be collected from the site for the determination of pH, liming requirements, cation exchange capacity, available nutrients, and trace metals prior to sludge application. Water samples from surface waters such as lakes, streams, and reservoirs need to be collected and analyzed for nutrients, trace metals, and fecal coliform prior to the application of sludge. Groundwater samples from installed monitoring wells, lysimeters, and private wells in the area need to be analyzed for nutrients, trace metals, and fecal coliform. The U.S.EPA [1] provides guidelines on sampling and analyses of soils, surface waters, and groundwaters prior to sludge application. State and local regulators may also indicate site-specific monitoring requirements.

Monitoring of the sludge application site after sludge has been applied can vary from none to extensive. The requirements will vary depending on state and local regulations and site-specific conditions. Table 10.1 gives a minimum list of parameters that should be included in the routine analyses of water, soils, and vegetation at reclamation sites. Some sites may require more extensive and longer-term sampling than described in Table 10.1, especially if sludge is applied annually for reclamation. For these sites, the monitoring requirements can be similar to those required at a dedicated site.

SLUDGE HANDLING AND APPLICATION METHODS

Considerable knowledge is available on sludge handling and application

TABLE 10.1. Minimum Sampling Procedure at a Reclamation Site [1].

Sample Collection	Procedures
Water	1. A minimum of three samples are collected from each groundwater well and lysimeter station prior to sludge application on the site. 2. After sludge application, water samples are collected monthly for a period of one year. 3. Samples collected prior to sludge application and for the first three months following sludge application are analyzed for pH, Cl, NO_3—N, NH_4—N, Org—N, Fe, Al, Mn, Cu, Cr, Co, Pb, Cd, Ni, Zn, and fecal coliforms 4 Water samples collected from the fourth month to the eleventh month following sludge application are analyzed only for pH, nitrogen forms (NH_4—N, NO_3—N), trace metals (Zn, Cu, Pb, Co, Ni, Cd, Cr), and fecal coliforms 5 Water samples collected during the twelfth month following sludge application are analyzed for constituents listed in No. 3, above. 6 Water sampling is terminated after one year unless results of the third quarterly report indicate a need to continue sampling. If further sampling is required, samples are collected quarterly until sufficient data are collected to formulate a conclusion on the problem. 7 The monitoring well is maintained past the initial year of sampling to allow for the collection of samples at a later date, if deemed necessary
Soil	1. Soil samples are collected on the site prior to sludge application. Surface soil samples of the topsoil material are collected throughout the site and analyzed for buffer pH to determine lime requirements to raise the soil pH to 6.5 and to determine the cation exchange capacity. Samples from the complete soil profile are collected from the pits excavated to install the lysimeters. Soil samples are collected from 0 to 15 (0 to 6 in), 15 to 30 (6 to 12 in), 30 to 60 (12 to 24 in), and 60 to 90 cm (24 to 36 in) soil depth. 2. Soil samples are again collected one year following sludge application. Samples are collected at the 0 to 15 (0 to 6 in), 15 to 30 (6 to 12 in), and 30 to 60 cm (12 to 24 in) depth 3 All soil samples are analyzed for pH, Bray P, Ca, Mg, K, Na, Fe, Al, Mn, Cu, Zn, Cr, Co, Pb, Cd, Ni, and Kjeldahl nitrogen. 4. At the end of the second year after sludge application, surface soil samples are collected and analyzed for pH to determine if it is still at pH 6.5.

(continued)

TABLE 10.1. (continued).

Sample Collection	Procedures
Vegetation	1. Vegetation samples are collected for foliar analyses at the end of the first growing season following sludge application. Separate samples are collected for each of the seeded species All samples are analyzed for N, P, K, Ca, Mg, Fe, Al, Mn, Cu, Zn, Cr, Co, Pb, Cd, and Ni. 2. For sites seeded in the fall, vegetations samples are collected at the end of the following growing season.

methods. Available sources of information include Halderson and Zenz [2], Miner and Hazen [14], and the U.S.EPA [1]. Many variables need to be considered when selecting sludge handling and application methods.

SLUDGE CLASSIFICATION

Sludge handling and application methods are determined by the moisture content. Sludge with a solids content of 30% or more can be handled with conventional end loading equipment and applied with agricultural manure spreaders. Liquid sludges, typically those with a solids content of <6% are managed and handled by normal hydraulic equipment. Slurries are sludges with intermediate moisture content and they typically range from 6 to 15% solids. These may be handled hydraulically. However, specialized equipment is required.

CLIMATIC AND SCHEDULING CONSIDERATIONS

The climatic conditions of the area influence the choice of sludge transport and application methods. Miner and Hazen [14] note that climatic conditions may dictate that sludge be applied to land only during a restricted period of the year. It is not generally advisable to apply sludge to frozen or snow-covered ground because it cannot be immediately incorporated and seeded. Sludge applied to frozen ground has an increased chance of surface runoff as the snow melts or if a heavy rainstorm occurs.

Miner and Hazen [14] divided the United States into climatic zones for management selection. They indicate that handling facilities must differ according to location. Climatic zones can dictate whether sludge is applied on a seasonal, sporadic, or continuous basis. Sludge should not be applied during periods of prolonged extreme heat or dry conditions because considerable nitrogen will be lost before the vegetation has a chance to establish itself.

Applications of sewage sludge should be scheduled to accommodate the growing season of the selected plant species. Heavy vehicle traffic on wet soils during sludge application may damage the soil structure, increase the bulk density, and decrease infiltration. These changes in the soil physical characteristics may increase the possibility of soil erosion and surface runoff.

TRANSPORTATION

Transportation modes for sewage sludge include truck, pipeline, railroad, barge, or various combinations of these modes. The U.S.EPA [1] notes that the method of sludge transportation chosen and its costs are dependent on numerous factors, which include: (1) characteristics and quantity of sludge to be transported; (2) the distance from the POTW plant to the application site; (3) the availability and proximity of the transportation modes to both origin and destination; (4) the degree of flexibility required in the transportation method chosen; (5) the estimated useful life of the sludge application site and site characteristics (topography, vegetative cover, soil type, area available); and (6) environmental and public acceptance factors.

Liquid sludges should be transported in closed tank systems to minimize the danger of spills. Stabilized, dewatered sludges can be transported in open vessels, such as dump trucks and railroad gondolas if equipped with watertight seals and anti-splash guards [15].

Sewage sludge can be transported by several modes. Trucks are widely used for transporting both liquid and dewatered sludges, and are generally the most flexible means of transportation. The U.S.EPA [1] has an in-depth discussion of vehicle types available for hauling liquid and dewatered sludge. Pipeline transport is generally available to liquid sludge with a total solids content of 8% or less. However, sludges with higher solids concentrations can be pumped with special equipment. An extensive review of pipeline transport is given by the U.S.EPA [1]. Other modes of sludge transport include using rail cars and barges. These methods are normally only considered by large cities for long-distance transport to land application sites.

STORAGE

Sludge storage is needed to accommodate fluctuations in sludge production rate, breakdowns in equipment, agriculture cropping patterns, and adverse weather conditions that prevent immediate application of sludge to the land. Storage can potentially be provided at either the treatment plant, the land application site, or both. The U.S.EPA [1], in its process design manual for sludge treatment and disposal, presents methods for estimating sludge storage capacity and the types of storage facilities available. The U.S.EPA [1] discusses storage facility design considerations for dedicated land disposal sites, which may be useful in designing a storage facility at a land reclamation site.

APPLICATION METHODS

The technique used to apply sludge to land can be influenced by the method used to transport the sludge to the application site. The U.S.EPA [1] describes the following commonly used methods:

(1) The same transport vehicles haul the sludge from the POTW to the application site and apply sludge to land.

(2) One type of vehicle, usually with a large volume capacity, hauls sludge from the POTW to the application site. At the site, the sludge haul vehicle transfers the sludge either to an application vehicle or into a storage facility, or both.

(3) Sludge is pumped and transported by pipeline from the POTW to a storage facility at the application site. Sludge is subsequently transferred from the storage facility to sludge application vehicles.

Generally, sludge application methods involve either surface or subsurface application. The advantages and disadvantages of these methods are discussed by the U.S.EPA [1]. Sludge is applied either in liquid or dewatered form. The methods and equipment used are different for land application of these two sludge forms. The U.S.EPA [1] discusses the advantages and disadvantages of using each of these sludge forms.

The application of liquid sludge to land is attractive because of its simplicity. Dewatering processes are not required, and the liquid can be readily pumped. Liquid sludge application systems include:

(1) Vehicular surface application

- tank truck spreading
- tank wagon spreading

(2) Subsurface application

- plow furrow or disking methods
- subsurface injection

(3) Irrigation application

- spray application
- gravity flooding

The spreading of dewatered sludge is similar to that used for surface application of solid or semisolid fertilizers, lime, or animal manure. Dewatered sludge cannot be pumped or sprayed. Sludge spreading is done by box spreaders, bulldozers, loaders, or graders and then incorporated into the soil by plowing or disking. The U.S.EPA [1] covers in depth the methods for sludge application to land.

APPLICATION QUANTITIES

The amount of sludge to be applied at a reclamation site is dictated by the purpose for utilizing sludge at the site. Typical reasons include applying sludge for organic matter, nutrients, and buffering capacity. An evaluation also needs to be made as to whether the sludge is going to be applied as a single, one-time application or in continuous applications and how the reclaimed site will be used after sludge applications are terminated. A major factor in determining whether sludge is applied in a single application versus annual application is whether topsoil is present or not. On sites with topsoil, an agricultural utilization rate may be used, with small amounts of sludge being applied annually. However, on sites without topsoil replacement, much larger amounts of sludge may be needed in order to establish vegetation and improve soil physical properties. These factors will determine how much sludge needs to be applied in order to make the reclamation effective and to provide nutrients for a vegetative cover.

Utilizing large amounts of sludge in a one-time application may result in short-term leaching of excess nitrates to the groundwater, and it may also result in contamination of surface water runoff while the vegetative cover becomes established. State and local regulatory agencies may require monitoring of surface and groundwaters to quantify the short-term impact.

ORGANIC MATTER

Organic matter in the soil is necessary to maintain good soil structure, increase the cation exchange capacity to reduce elemental leaching, improve water relations, and provide for a continuous supply, though limited, of plant nutrients through mineralization. A high level of soil organic matter should not be the end objective, however; a combination of sound management practices, fertilizers, lime, and cultural practices is needed to accomplish the end objective, be it agricultural production for a profit or erosion control to stabilize the soil.

When using sewage sludges to increase the organic matter content, it is necessary that some measurements be made of the portion of the total dry sludge solids that are organic. Depending upon the type and stability of sludge, this figure can vary considerably. Unstabilized sludges can have upwards of 75–80% of the dry matter as organic, whereas anaerobically digested sludges might typically have 40–50% of the dry solid as organic matter. Most soils range from 1 to 6% in organic matter depending upon soil texture, soil temperature, type of vegetation, cultural practices, and extent of soil drainage.

NUTRIENT APPLICATION RATES

A factor in applying sewage sludge for reclamation is to apply enough plant nutrients in sludge to meet the nutrient demands of the vegetation being grown. In this context, calculations of nutrient loading for nitrogen is usually done. The applicant needs to determine the amount and type of nitrogen in the sludge, the plant available nitrogen content of the existing soils, and the fertilizer nitrogen requirements of the vegetation planned for the site. This information is utilized to determine a sludge application rate that supplies sufficient nitrogen to meet the needs for vegetation and minimizes the contamination of groundwater by nitrate leaching.

The use of land after reclamation also needs to be considered when determining the amount of nitrogen needed to supply the vegetative needs. If the vegetation grown is to be harvested and removed from the site, supplemental applications of sludge or nitrogen may be needed to maintain adequate productivity. Sludge is a slow-release fertilizer that supplies some nitrogen for three to five years.

At reclamation sites where long-term vegetative growth is desired, the amount of sludge utilized in one-time application needs to be higher than agronomic rates calculated for nitrogen. If the vegetation is to be harvested and enter the human food chain, then it is advisable to determine the maximum amount of sludge that can be applied based on metal loadings. The limiting metal is used to calculate the maximum amount of sludge that can be applied to a site. State regulatory agencies will usually determine during the permitting process what sludge application rates are acceptable for reclamation.

If the site is to be reforested or planted in vegetation not entering the human food chain, the metal accumulation is limited by the potential phytotoxicity to the trees and vegetation. Copper, zinc, and nickel are the elements of most concern in plant phytotoxicity. Maintaining the soil pH above 6.0 to 6.5 will minimize the amount of these elements taken up by the vegetation. In this context, having site-specific information to give state or local regulatory agencies on the desired application rates and the rationale for the rates is usually very beneficial. The U.S.EPA [1] and state regulatory agencies provide examples on how to calculate sludge application rates based on nutrient and metal loadings.

BUFFERING CAPACITY

Sewage sludge has been used as an amendment to reclaim acid mine spoils [17–19]. In addition, sludge has been used to reclaim acidic coal refuse materials [20–22]. Sewage sludge has a pH buffering capacity resulting from an exchange capacity and an alkalinity that is beneficial in the reclamation of acidic sites. Pietz et al. [22] and Peterson and Gschwind [23] also observed that sludge is effective in reducing metal percolation in acidic drainage waters.

Calculations on the amount of sludge to be used in reclamation of acidic materials on a long-term basis (>5 years) may be difficult. Pietz et al. [22] concluded that the actual amounts of lime and sludge required for reclamation of acidic coal refuse material were about twice that calculated. The actual amounts of lime and sludge needed for long-term reclamation (>5 years) at a coal refuse pile were 189 Mg/ha (84 tons/acre) lime and 1050 Mg/ha (468 tons/acre) sludge as compared to the applied amounts of 542 Mg/ha (242 tons/acre) sludge and 89.9 Mg/ha (40 tons/acre) lime. Pietz et al. [22] concluded that because of the difference between the theoretical and the actual amounts of lime and sludge needed, a built in safety factor of one and a half to two would be prudent for long-term reclamation of coal refuse material.

RECLAMATION SITE RESPONSES

The following discussion will concentrate on responses that have been obtained from the use of sewage sludge at reclamation sites. Data from some of the sites has been published in scientific literature, while in other instances, the available data has yet to be published.

ORGANIC MATTER ADDITION

A previous section of this chapter on "Application Quantities" discusses the detailed aspects involved in predetermining the amount of sludge required to achieve a certain increase in soil organic matter content. Table 10.2 shows two fields that have received sludge over a twelve- to thirteeny-ear period at the Fulton County, Illinois land reclamation site of the Metropolitan Water Reclamation District of Greater Chicago. Field 3 is a strip-mined field that was in pasture prior to sludge application, while field 20 is place land that has been in row crops continuously for many years. Through 1984, fields 3 and 20 had received accumulative totals of 496.2 and 531.0 dry Mg/ha (221.3 and 236.8 dry tons/acre) of sludge. The organic carbon content in strip-mined field 3 increased 3.79% above the base line levels in 1972, while the organic carbon content in place land field 20 increased 3.81% above the base line level in 1972. Assuming a sludge organic carbon content of 26%, calculations of the total amounts

TABLE 10.2. Response of Soil to Sludge Application at Fulton County, Illinois Land Reclamation Site.

Sludge Applied Mg/ha	Year	Organic Carbon (%)	pH
	Field 3—Strip-Mined Soil		
12.9	1972	1.24	7.7
2.7	1973	0.86	7.8
54.7	1974	0.96	8.0
22.8	1975	1.23	6.8
83.3	1976	1.28	7.2
0.0	1977	2.62	6.6
83.3	1978	3.08	6.5
13.9	1979	3.80	5.6
90.5	1980	4.37	6.3
54.9	1981	5.38	6.0
0.0	1982	4.80	6.4
43.6	1983	5.14	6.6
33.6	1984	5.30	6.8
	1985	5.03	6.6
	Field 20—Undisturbed Soil		
0 0	1972	1.89	6.2
0.8	1973		
52.4	1974	2.25	6 4
20.4	1975	2.46	5.7
39.2	1976	2.03	5.3
31 8	1977	2.84	5.3
58.0	1978	3.37	6.1
69.0	1979	4.42	5.2
53 8	1980	2 89	6.7
56.7	1981	4.62	6.3
82.9	1982	4.82	6.2
0.0	1983	3.20	6.6
66.0	1984	5.46	6.8
	1985	5.70	6.5

applied suggest that the decomposition rate of sludge applied organic matter was about 34% in field 3 and 38% in field 20 over the sampling period ranging from 1972 to 1985. Decomposition rates such as these need to be taken into account when applying sludge to build up the organic matter content of soils.

SOIL pH

One of the concerns at reclamation sites is the soil pH. The pH needs to be controlled if a vegetative cover is to be maintained. Table 10.2 shows the effect of sludge decomposition on soil pH at the Fulton County, Illinois reclamation site. On both calcareous (field 3) and slightly acid (field 20) soils, the pH dropped with continuous applications of sludge. Periodic applications of lime were carried out after 1979 to maintain the soil pH above 6.5 [16].

An acid soil pH can be a problem at many sites requiring reclamation with sludge. Several researchers [17–19] used sludge as an amendment to reclaim acid mine spoils. Table 10.3 shows the effect of sludge and lime applied at rates of 542 and 89.6 Mg/ha on the reclamation of acidic coal refuse material from 1976 to 1981.

VEGETATIVE RESPONSES

Many reclamation sites use sewage sludge as an amendment to enhance the establishment of vegetation. Topper and Sabey [24] reported the plant growth response on coal mine spoils in Colorado due to various sludge amendments (Table 10.4). The sludge was applied in 1982 and vegetative samples were collected in August 1983 and 1984. Pietz et al. [21] used sewage sludge and lime to reclaim acidic coal refuse material. The vegetative response to the treatments applied in 1976 and 1977 are shown in Table 10.5.

Crop yields at reclamation sites where sludge is applied for agricultural production can be variable. Limiting factors can be climatic conditions and shallow rooting depths confined to the zone of tillage. Peterson et al. [25] reported corn yields in selected sludge-amended fields at a strip-mine reclamation site (Table 10.6). They reported that adequate moisture and essential elements for crop needs on strip-mined soil were found to be critical for yields of crops grown immediately after land leveling. Pietz et al. [26] concluded that spoil factors such as shallow rooting depth, soluble salts, moisture stress, and element interactions in plant tissue, sludge, and soil appeared to be important parameters, which either alone or in combination with existing climatic conditions affected yields and element composition.

Many species and varieties of plants, including agricultural crops, are available for use at reclamation sites. The selection of the proper species for any planting should be based on the climate and soils, and on the use and management planned [27]. Each site should be considered to be unique, and plant species or seed mixtures to be used need to be carefully selected. Local authorities should be consulted for recommendations on crops or appropriate species and varieties of vegetative materials. Revegetation suggestions for various regions of the United States are presented by the U.S.EPA [1]. Plant materials for humid and dry regions of the United States are discussed by Vogel and Curtis [28], Thornburg and Fuchs [27], and Packer and Aldon [29].

TABLE 10.3. The pH, Water-Soluble Al and Fe, and Total Acidity of Coal Refuse Samples from Four Treatments in 1976 and 1981 [20].

Treatment	Depth cm	pH			Water-Soluble Al (mg/kg^{-1})			Water-Soluble Fe (mg/kg^{-1})			Total Acidity[b] (cmol/kg^{-1})		
		1976	1981	t[a]	1976	1981	t	1976	1981	t	1976	1981	t
Control	0–15	2.8	2.6	c	279	105	c	283	31	d	9.6	7.3	NS
	15–30	2.8	3.1	NS	222	139	NS	231	44	c	7.7	7.1	NS
	30–45	4.2	2.6	NS	108	189	NS	117	62	NS	3.5	8.6	c
	45–60	4.8	2.4	NS	10	221	d	1	104	c	0.5	10.0	c
	85–100	6.4	5.9	NS	1	<1	NS	1	1	NS	<0.1	<0.1	NS
Lime	0–15	3.4	2.9	NS	219	103	NS	64	61	NS	6.7	7.6	NS
	15–30	3.8	2.7	NS	81	135	NS	22	143	NS	3.2	8.2	NS
	30–45	3.4	3.3	NS	108	168	NS	119	366	NS	5.3	9.3	NS
	45–60	3.8	3.2	NS	75	200	NS	40	209	NS	4.0	8.7	NS
	85–100	6.4	6.0	c	2	<1	NS	1	2	NS	<0.1	1.0	NS

TABLE 10.3. (continued).

Treatment	Depth cm	pH			Water-Soluble Al (mg/kg⁻¹)			Water-Soluble Fe (mg/kg⁻¹)			Total Acidity[b] (cmol/kg⁻¹)		
		1976	1981	t[a]	1976	1981	t	1976	1981	t	1976	1981	t
Sludge	0–15	5.0	3.3	NS	7	134	NS	4	13	NS	1.1	6.3	d
	15–30	5.2	3.0	NS	17	196	d	4	33	d	1.0	8.2	c
	30–45	5.5	3.4	NS	6	118	NS	4	13	NS	0.4	5.4	d
	45–60	4.0	3.9	d	49	108	NS	49	15	NS	2.6	5.1	NS
	85–100	6.2	3.6		4	175	c	3	51	d	0.1	4.5	c
Sludge and lime	0–15	5.0	4.9	NS	9	2	NS	7	<1	NS	0.7	1.1	NS
	15–30	4.2	3.6	NS	13	20	NS	10	2	d	1.0	2.6	d
	30–45	4.6	4.1	NS	19	29	NS	10	7	NS	1.3	2.8	c
	45–60	4.0	4.2	NS	29	52	NS	25	13	NS	1.5	2.7	NS
	85–100	6.0	6.5	NS	7	<1	NS	3	2	NS	0.2	<0.1	NS

[a]Paired t test used to compare two sample means at the same depth for years 1976 and 1981.
[b]Combined acidity of sequentially extracted water-soluble and 2 M KCl extracts.
[c]Significant at 0.01 probability.
[d]Significant at 0.05 probability.
NS = not significant.

TABLE 10.4. Plant Form Composition of Sewage Sludge Amended Colorado Coal Mine Spoil from 1983 to 1984 [24].

Plant Form	Level (Mg/ha)	Plant Form Composition[a] (%) 1983	1984
Seeded grasses	0	3 1	84.7
	14	3.3	94.0
	28	3.4	76 5
	55	4.3	66.7
	83	2.3	61.3
Invading grasses	0	0	0.2
	14	T[b]	1 2
	28	T	1.1
	55	0.2	0.1
	83	T	0.0
Invading annual forbs	0	96.9	15 1
	14	96.7	4.8
	28	96 6	22 5
	55	95.5	33.2
	83	97 6	38 7

[a]Aboveground biomass values were used to calculate life-form composition
[b]Trace amount < 0 1% of composition.

TABLE 10.5. Plant Cover and Yields on Amended Acidic Coal Refuse Material [21].

Treatment[a]	Plant Cover (%) 1978	1979	1980	Dry Matter Yields (Mg/ha^{-1}) 1978	1979	1980
Control	0	0	0	0.0	0 0	0.0
Lime	8	14	35	0.53	0.96	2.38
Sludge	14	31	33	0.93	2.11	2.25
Sludge and lime	14	42	89	0.94	2.81	6.00
Gympsum	0	0	0	0.0	0.0	0.0
Gypsum and lime	3	4	6	0.19	0.25	0.38
Gypsum and sludge	13	21	32	0.89	1.41	2.13
Gypsum and sludge and lime	14	39	82	0.96	2.65	5.56
LSD	NS	22[b]	46[c]	NS	2.03[c]	3.17[c]

[a]Lime, sludge, and gypsum were applied at rates of 89 6, 542, and 112 Mg/ha dry wt., respectively.
[b]Significant at 0 05 probability.
[c]Significant at 0 01 probability.
NS = not significant

TABLE 10.6. Corn Yields in Selected Sewage Sludge Amended Fields[a] at Fulton County, Illinois Land Reclamation Site [25].

Field Number[b]	Corn (mg/ha)							
	1972	1973	1974	1975	1976	1977	1978	1979
Mine-Spoil								
1						0.85		3.64
2	3.24	2.86	0.55	2.95			1.37	
3				3.99		2.01		5 60
5						1.16		
7					1.62			2.97
25					2.39			4.10
26				2.80				4.22
28						0.70		
30				1.61			1.98	
34		4.12			5.32			
W$_x$	3.24	3.49	0.55	2.84	3.11	1.18	1.68	4.12

(continued)

587

TABLE 10.6. (continued).

Field Number[b]	Corn (mg/ha)							
	1972	1973	1974	1975	1976	1977	1978	1979
Placeland								
10	4.40	0.76	1.60	3.75				
19	5.72	1.44	1.25				6.26	
20	4.19	0.54	0.58					
21	5.12	3.72			6.46			
22	4.66	2.16			3.85			
23	7.45	4.47			3.30			
31		5.35		3.37				
35		5.50						
37								
40								
W_x	5 26	2 99	1.14	3.56	4.54		6.26	

[a]Commercial fertilizer was applied to fields 10 and 23 in 1972 and to fields 31, 34, 35, and 37 in 1973
[b]Weighted mean, W_x, is based on yield data reported for each year in the table.

588

In selecting plant species, whether they be crops, grasses, legumes, or trees, it is important to choose those that are compatible and will also grow well when sludge is used as a fertilizer. In many instances, plant species to be used should be selected because of their ability to grow under drought conditions and they should have a tolerance for either acid or alkaline soil material. A salt tolerance is also desirable.

REFERENCES

1 U.S.EPA. 1983. *Process Design Manual for Land Application of Municipal Sludge,* U.S.EPA Rep. EPA-625/1-83-016, U.S. Environmental Protection Agency, Cincinnati, OH.

2 Halderson, J. L. and D. R. Zenz. 1978. "Use of Municipal Sewage Sludge in Reclamation of Soils," in *Reclamation of Drastically Disturbed Lands,* F. W. Schaller and P. Sutton, eds., Madison, WI: Aamerican Society of Agronomy, Crop Science Society of America, and Soil Science Society of America, pp. 355–357.

3 Hinesly, T. D., O. D. Braids, R. I. Dick, R. L. Jones and J. E. Molina. 1974. *Agricultural Benefits and Environmental Changes Resulting from the Use of Digested Sewage on Field Crops,* U.S.EPA Rep. SW-30d, University of Illinois, Urbana.

4 Burd, R. S. 1968. "A Study of Sludge Handling and Disposal," *Water Pollut. Control Fed. Res. Ser.* Pub. No. WP20-4.

5 Meyer, R. C., F. C. Hines, H. R. Isaacson and T. D. Hinesly. 1971. "Porcine Enterovirus Survival and Anaerobic Sludge Digestion," in *Proc. Int. Symp. on Livestock Wastes,* Ohio State University, Columbus, OH, St. Joseph, MI: Am. Soc. of Agric. Eng., p. 183.

6 Reed, J. M., J. D. Fenters and C. Lue-Hing. 1975. "The Effect of Ammonia on Poliovirus Type 1," *Abstr. of the Annual Meeting,* Am. Soc. of Microbiol., Section Q, p. 206.

7 Bertucci, J., C. Lue-Hing, D. R. Zenz and S. J. Sedita. 1977. "Inactivation of Viruses During Anaerobic Digestion," *J. Water Pollut. Control Fed.,* 49:1642–1651.

8 Berg, G. 1966. "I. Virus Transmission by the Water Vehicle. II. Virus Removal by Sewage Treatment Procedures," *Health Library Sci.,* 3(2):90.

9 Bjorkman, A. 1969. "Heat Processing of Sewage Sludge," in *4th Congr. of the Int. Res. Group on Refuse Disposal,* June 2–5, 1969, Basel, Switzerland, pp. 670–686.

10 Peterson, J. R., C. Lue-Hing and D. R. Zenz. 1973. "Chemical and Biological Quality of Municipal Sludge," in *Recycling Treated Municipal Wastewater and Sludge through Forest and Cropland,* W. E. Sopper and L. T. Kardos, eds., University Park: Pennsylvania State Univ. Press, pp. 26–37.

11 McCalla, T. M., J. R. Peterson and C. Lue-Hing. 1977. "Properties of Agricultural and Municipal Wastes," in *Soils for Management of Organic Wastes and Waste Water,* L. F. Elliott and F. J. Stevenson, eds., Madison, WI: Soil Sci. Soc. Am., pp. 2–43.

12 Rimkus, R. R. and R. W. Heil. 1975. "The Rheology of Plastic Sewage Sludge," in *Proc. 2nd Natl. Conf. on Complete Water Reuse,* May 4–8, 1975, Chicago, IL: Am. Inst. of Chem. Eng. and U.S.EPA, pp. 722–740.

13 USDI. 1979. *Permanent Regulatory Program Implementing Section 501(b) of the Surface Mining Control and Reclamation Act of 1977: Final Environmental Statement,* OSM-EIS1, U.S. Department of Interior, Washington, D.C.

14 Miner, J. R. and T. E. Hazen. 1977. "Transportation and Application of Organic Wastes to Land," in *Soils for Management of Organic Wastes and Waste Waters*, L. F. Elliott and F. J. Stevenson, eds., Madison, WI: Soil Science Society of America, American Society of Agronomy, and Crop Science Society of America, pp. 379–425.

15 Metcalf and Eddy. 1979. *Wastewater Engineering: Treatment Disposal, Reuse*. New York: McGraw-Hill.

16 U.S.EPA. 1979. *Process Design Manual for Sludge Treatment and Disposal*, U.S.EPA Rep. EPA-625/1-79-011, MERL, ORD, U.S. Environmental Protection Agency, Washington, D.C.

17 Hinkle, K. 1982. *Reclamation of Toxic Mine Waste Utilizing Sewage Sludge—Contrary Creek Demonstration*, U.S.EPA Rep. EPA-600/2-82-061, U.S. Environmental Protection Agency, Cincinnati, OH.

18 Sopper, W. E. and E. M. Seaker. 1984. *Strip Mine Reclamation with Municipal Sludge*, U.S.EPA Rep. EPA-600/2-84-035, U.S. Environmental Protection Agency, Cincinnati, OH.

19 Stucky, D. J., J. H. Bauer and T. C. Lindsey. 1980. "Restoration of Acidic Mine Spoils with Sewage Sludge: I. Revegetation," *Reclam. Rev.* 3:129–139.

20 Pietz, R. I., C. R. Carlson, Jr., J. R. Peterson, D. R. Zenz and C. Lue-Hing. 1989. "Application of Sewage Sludge and Other Amendments to Coal Refuse Material: I. Effects on Chemical Composition," *J. Environ. Qual.*, 18:164–169.

21 Pietz, R. I., C. R. Carlson, Jr., J. R. Peterson, D. R. Zenz and C. Lue-Hing. 1989. "Application of Sewage Sludge and Other Amendments to Coal Refuse Material: II. Effects on Revegetation," *J. Environ. Qual.*, 18:169–173.

22 Pietz, R. I., C. R. Carlson, Jr., J. R. Peterson, D. R. Zenz and C. Lue-Hing. 1989. "Application of Sewage Sludge and Other Amendments to Coal Refuse Material: III. Effects on Percolate Water Composition," *J. Environ. Qual.*, 18:174–179.

23 Peterson, J. R. and J. Gschwind. 1972. "Leachate Quality from Acidic Mine Spoil Fertilized with Liquid Digested Sewage Sludge," *J. Environ. Qual.*, 1:410–412.

24 Topper, K. F. and B. R. Sabey. 1986. "Sewage Sludge as a Coal Mine Spoil Amendment for Revegetation in Colorado," *J. Environ. Qual.*, 15:44–49.

25 Peterson, J. R., C. Lue-Hing, J. Gschwind, R. I. Pietz and D. R. Zenz. 1982. "Metropolitan Chicago's Fulton County Sludge Utilization Program," in *Land Reclamation and Biomass Production with Municipal Wastewater and Sludge*, W. E. Sopper, E. M. Seaker and R. K. Bastian, eds., University Park, PA: The Pennsylvania State University Press, pp. 322–338.

26 Pietz, R. I., J. R. Peterson, T. D. Hinesly, E. L. Ziegler, K. E. Redborg and C. Lue-Hing. 1982. "Sewage Sludge Application to Calcareous Strip-Mine Spoil: I. Effect on Corn Yields and N, P, K, Ca, and Mg Compositions," *J. Environ. Qual.*, 18:685–689.

27 Thornburg, A. A. and S. H. Fuchs. 1978. "Plant Materials and Requirements for Growth in Dry Regions," in *Reclamation of Drastically Disturbed Lands*, F. W. Schaller and P. Sutton, eds., Madison, WI: American Society of Agronomy, Crop Science Society of America, and Soil Science Society of America, pp. 411–423.

28 Vogel, W. G. and W. R. Curtis. 1978. "Reclamation Research on Coal Surface-Mined Lands in the East," in *Reclamation of Drastically Disturbed Lands*, F. W. Schaller and P. Sutton, eds., Madison, WI: American Society of Agronomy, Crop Science Society of America, and Soil Science Society of America, pp. 379–397.

29 Packer, P. E. and E. F. Aldon. 1978. "Revegetation Techniques for Dry Regions," in *Reclamation of Drastically Disturbed Lands*, F. W. Schaller and P. Sutton, eds., Madison, WI: American Society of Agronomy, Crop Science Society of America, and Soil Science Society of America, pp. 425–450.

CHAPTER 11

Production and Distribution of Municipal Sewage Sludge Products

MANY municipalities have found that there is a demand for municipal sewage sludge as a fertilizer and soil conditioner due to its content of nitrogen, phosphorus, and organic matter. Sludge is also considered to be a slow-release fertilizer, thus avoiding the "burning" problems sometimes encountered with commercial inorganic fertilizers. This chapter illustrates successful brand name sludge products, and describes processes used to produce a saleable sludge product using lime conditioning and pelletization, and in-vessel composting.

BRAND-NAME MUNICIPAL SEWAGE SLUDGE PRODUCTS

INTRODUCTION

In the United States, municipal sewage sludge is often marketed and sold to the general public in a manner similar to commercial fertilizers. Because of the public's need to buy identifiable products, municipal sewage sludge is often marketed and sold under a brand name. Commercial fertilizers are marketed with names like "green power" or "turf builder" while municipal sewage sludge products are marketed with names like "Milorganite" (heat-dried sludge from the Milwaukee Metropolitan Sewerage District Commission).

The impact of brand-name sewage sludge products on the commercial

Jeffrey C. Burnham, Medical College of Ohio, Toledo, OH; John F. Donovan, Camp Dresser & Mckee, Inc., Cambridge, MA; Jane Forste, Bio Gro Systems, Annapolis, MD; John Gschwind and David Zenz, Metropolitan Water Reclamation District of Greater Chicago, Chicago, IL; Terry J. Logan, Ohio State University, Columbus, OH.

fertilizer market is not significant. Table 11.1 compares the amounts of N, P, and K that are currently used and those that are potentially available from sewage sludge with those consumed in commercial fertilizer [1]. As can be seen, the amounts of N, P, and K that are potentially available from sewage sludge products represents but a small fraction of that consumed in commercial fertilizers.

The amount of revenue received from brand-name sewage sludge products is obviously but a small portion of that from commercial fertilizers. However, the monetary return from selling brand-name sewage sludge products can reduce operating costs or even allow elimination of more expensive sludge management methods. Also, sludge users can often obtain brand-name sewage sludge products at a fraction of the cost of commercial fertilizers.

COMPOST PRODUCTS

Many brand-name sewage sludge products are the result of composting operations. Municipal sewage sludge compost is a natural organic product with a high humus content similar to peat moss. It has a musty odor and is usually moist and dark in color. It can be easily bagged, especially if it has been carefully screened to remove aesthetically objectionable materials.

Compost made from municipal sewage sludge increases the water holding capacity of sandy soils, and increases the porosity of clay soils. The organic matter in the compost makes the soil more "workable," allowing the roots of plants to penetrate easily. Since sewage sludge compost contains only small amounts of nitrogen, phosphorus, and potassium, its chief usefulness is as a soil amendment, not as a fertilizer.

Table 11.2 contains data on some brand-name municipal sewage sludge compost products marketed in the United States. This list is not inclusive of all the sewage sludge compost products in the United States, but does serve to show the variety of types produced and their nutrient content.

LIQUID SLUDGES

Liquid municipal sewage sludge can be a desirable product for sludge utilization because of its simplicity of application. Dewatering is not required and inexpensive sludge transfer systems can be used.

The nutrient content of liquid sludges is usually the highest of that found in municipal sewage sludges. This is because the sludge has not been subjected to processes such as heat-drying, dewatering, etc., which

TABLE 11.1. Comparison of Current and Potential Sludge Utilization with Commercial Fertilizer Consumption in the United States.

Nutrient	(A) Nutrients in Currently Used Sludges (1000 tons/year)	(B) Nutrients in Potentially Usable Sludges (1000 tons/year)	(C) Nutrients in Commercial Fertilizers (1000 tons/year)	(A) as % of (C)	(B) as % of (C)
Nitrogen (N)	21.6	65.3	10,642	0.2	0.6
Phosphorus (P)	21 5	80.7	2,453	0.9	3.2
Potassium (K)	4.3	18.9	4,841	0.1	0.4

TABLE 11.2. Examples of Municipal Sewage Sludge Brand-Name Products—
Based upon 1996 Survey of the Association of Metropolitan Sewerage
Agencies (AMSA).

Product Brand-Name	mg/kg			
	Total Kjeldahl-N	NH$_3$–N	Total P	K
Alkaline Stabilized Sludge				
AG Lime	26,300	5,400	8,080	460
Meadow Life	13,200	664	1,260	3,330
Dewatered Sludge				
Tagro	7,500	2,100	5,800	800
Alcosil	33,500	2,700	9,100	2,216
Gro Mulch	43,000	5,300	20,000	1,100
Top Gro	45,579	10,097	24,726	1,105
Metro Cake	64,800	10,100	23,300	1,940
Compost				
Nitro System	70,000	N/A*	30,000	1,100
Nutri Green	41,750	15,000	14,375	1,687
Garden Care	13,650	4,050	6,550	1,200
Metro Class A Fine Compost	28,000	48,000	21,300	2,620
Metro Fine Screened Compost	26,100	4,400	19,100	2,610
Philadelphia Mine Mix	24,115	6,037	647	14,445
Comtil	37,547	11,815	16,245	5,147
Earth Mate Compost	28,227	8,526	18,790	2,324
Heat Dried Sludge				
Hou-Actinite	63,000	N/A	41,400	3,690
Milorganite	64,700	N/A	N/A	3,900
Liquid Sludge				
Metro Gro	79,000	25,000	28,000	6,000
Bio Rich	51,684	24,000	20,579	1,950
Ocean Grow	40,904	16,158	8,396	805
Tagro	61,000	30,000	36,000	33,000
Air Dried Sludge				
Hi-K	--------------- Custom Blend[b]---------------			
666	--------------- Custom Blend ---------------			

[a]Not available
[b]Nutrient content is adjusted for individual end user.

can significantly reduce the nitrogen content of the sludge. In addition, as is often true of heat-dried and composted sewage sludge, additives are not normally used which can cause dilution of the nutrient content of the sludge.

Table 11.2 contains nutrient data for some liquid municipal sewage sludge products marketed in the United States. Again this list is not inclusive but represents an example of the liquid sludge products available.

HEAT-DRIED SLUDGES

Perhaps the most well-known brand-name municipal sewage sludge products are the heat-dried sludges. Almost everyone in the field of sludge management knows that the Milwaukee Metropolitan Sewerage District Commission markets the heat-dried municipal sewage sludge known as "Milorganite." This bagged product is sold throughout the United States and has even found its way into such chain stores as K-Mart.

Heat-dried municipal sewage sludges have been dried to a solids content of over 90%. Sludge at this solids content can be handled and applied easily. The most common use for heat-dried sludge is turf fertilization, where the sludge is usually applied using broadcast or drop spreaders.

Table 11.2 shows the uses and nutrient content typical of heat-dried sludge marketed in the United States. These products are used principally for turf fertilization.

AIR-DRIED SLUDGES

Municipal sewage sludges are often dried on beds in the open air. Air-dried sludge normally has a solids content greater than 50%. At this solids content, the sludge can be applied using broadcast and similar spreaders.

Table 11.2 contains examples of air-dried sludges marketed in the United States. These products are used principally for turf fertilization.

N-VIRO SOIL: ADVANCED ALKALINE SLUDGE STABILIZATION

INTRODUCTION

Technologies for advanced sewage sludge treatment remained largely unchanged until the last decade. Since then, large-scale composting has emerged as a viable technology, as has the N-Viro Soil Process for Advanced Alkaline Stabilization. Developed by Dr. Jeffrey Burnham and N-Viro Energy Systems, Inc., this technology uses a combination of microbiological stresses to achieve virtually complete pathogen destruction, and has been certified by the U.S.EPA as a process to further reduce pathogens (PFRP). The basis for the technology is the use of cement kiln dust (CKD), an alkaline by-product of the cement industry, to achieve alkaline pHs (>12), rapid drying, and temperature rise. The final product is a solid, granular, odor-free material with many of the desirable properties of soil. This, together with its PFRP designation, give N-Viro Soil great flexibility in terms of its beneficial end use.

In this paper, we describe the N-Viro process for sludge treatment, we identify major physical, chemical, and biological properties of N-Viro Soil, and we discuss the use of N-Viro Soil for various beneficial end uses.

THE N-VIRO PROCESS

The N-Viro process is summarized in Figure 11.1. The basis of the process is to destroy pathogens through a combination of the following stresses:

- alkaline pH
- accelerated drying
- high temperature
- high ammonia
- salts
- indigenous microflora

These stresses are produced in the sludge/CKD mixture through the unique properties of the CKD. Typical properties of CKDs are summarized in Tables 11.3 and 11.4. Of particular interest are its alkali content (stresses 1, 3, and 4), fine particle size (stress 2), and low moisture content (stress 2). Not apparent from the analysis is the high content of inert material, which keeps the concentration of soluble salts from being excessive, but contributes to pathogen kill. As with compost, mesophilic temperatures (52–62°C) and a soil-like environment contribute to the growth of indigenous microorganisms and suppress the regrowth of pathogens or putrefying organisms.

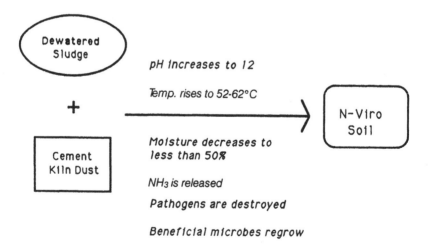

Figure 11.1 Basic reactions in the N-Viro process.

TABLE 11.3. Characteristics of Cement Kiln Dust and the N-Viro Soil from Which It Was Made [2].

Parameter	CKD	N-Viro Soil[a]
pH (1.1 water)	13.0	11.1
Elements (% by wt.)		
N	<0.01	1.40
P	0.04	0 39
K	1.61	1 00
Ca	28 5	19 8
Mg	1 22	0.97
Na	0 60	0.20

[a]CKD was added to sludge at a 35% by weight of dry sludge solids

Raw primary, activated sludge, or digested sludge with solids content of 18–40% is used. The use of raw primary sludge is preferred to conserve nitrogen in the final product and to reduce NH_3 emissions from the alkali addition. Anaerobically digested sludges can be treated, but NH_3 loss is greatest. Sludge, CKD, and other additives are mixed with a pug mill or screw blender (Figures 11.2–11.4). If the CKD contains enough free lime [CaO, $Ca(OH)_2$ or other strong alkali] to give a pH rise to >12 and an exothermic reaction necessary to achieve desired temperatures (52–62°C), no other additive is needed. CaO is added to supplement the free lime content of the CKD if it is not "hot" enough. The ratio of CKD to sludge

TABLE 11.4. General Characteristics of N-Viro Soil [7].

Characteristic	Units	Value
CKD Content	% by weight	35–75
Solids Content	% by weight	50–75
Material > 2 mm	% by weight	32
Mean Granule Size	mm	0 66
Bulk Density	g/cm^3	0.7–1.0
Volatile Solids	% by weight	9.3
pH (1.1 water)		11–12
$CaCO_3$ Equivalent	% by weight	50–80
Organic-C	% by weight	12 2
Total Kjeldahl N (TKN)	% by weight	1–1.5
NH_3–N	mg/kg	200
NO_3–N	mg/kg	50
P	% by weight	0 39
K	% by weight	1.0
Ca	% by weight	20
Mg	% by weight	1 0
Na	% by weight	0.2

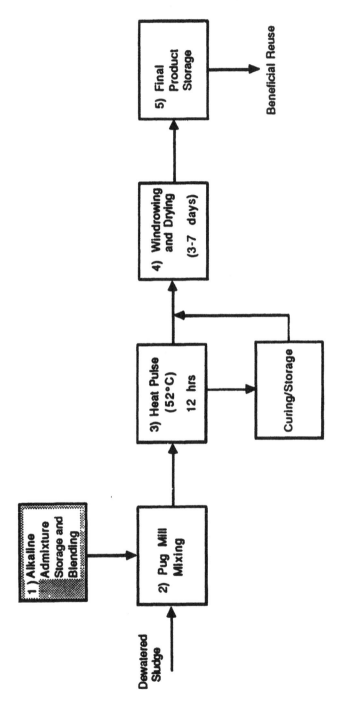

Figure 11.2 Advanced alkaline stabilization with subsequent accelerated drying (AASSAD) process steps to produce PFRP granular product.

Unit Process Operation	Description
1) Alkaline Admixture Storage and Blending	The N-Viro process requires AA dose rates ranging from 25%-50 % of the sludge cake wet weight. Approved admixtures are stored in silos and can be blended to produce the end product. Bulk storage silos, suitable for receiving and storing a highly alkaline material, typically receive 24 ton pneumatically discharged loads. Alkaline materials are normally conveyed from the silo to the mixing unit by screw conveyor. At the mixing unit, a metering screw and an optional weighing screw convey a measured amount of each alkaline material into the mixer. The alkaline admixture conveyors are totally enclosed to prevent release of dust.
2) Mixing	Alkaline admixture and dewatered sludge cake are blended in a mixer designed to provide complete and intimate contact between the materials. Design of the mixer is based on the requirement to properly blend the two materials without breaking the structure of the sludge cake and producing a paste-like material. The discharge from the blender should resemble small granular pellets or sand-like grainy particles. The mixing unit is enclosed to prevent release of dust, and to allow exhaust air containing ammonia fumes to be treated.
3) Heat Pulse	The blended material must be "cured" for at least 12 hours while the temperature is monitored and maintained above 52°C and below 62°C. All the blended material must be placed in Heat Pulse Containers (HPC), storage bins or enclosed piles Additional storage can also be provided for more than 12 hours of curing
4) Accelerated Drying and Windrowing	The blended material is transported and discharged into long piles forming windrows. The material is aerated and windrowed for 3-7 days, and is complete when the solids content of the material is above 50% TS, while the pH remains above 11. Further windrowing may be desirable to reduce volumes of material to be handled, bringing the solids content to 60-65% TS. Accelerated drying is performed on an asphalt or concrete pad, outdoors or inside a building, using an auger attachment, rotating drum, or lift and turn type of accelerated drying equipment.
5) Final Product Storage	The final product storage pad is used to hold material for distribution as required. The product is easily handled and can be stockpiled with stacking conveyors or front end loaders Because the material meets PFRP requirements, and odors are controlled, it can be stored on-site in accordance with the product distribution and beneficial reuse plan.

Figure 11.3 Process steps in the AASSAD process.

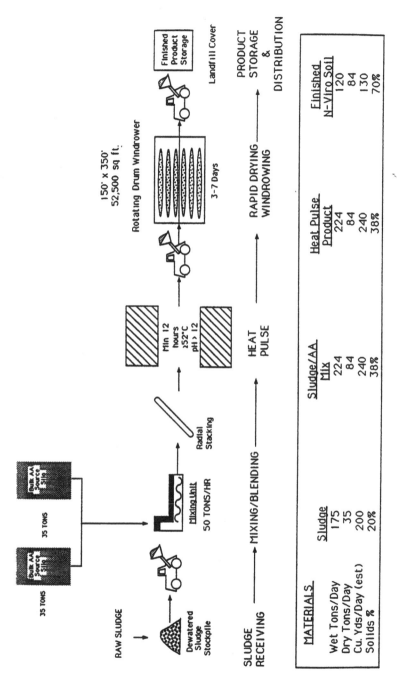

Figure 11.4 Flow diagram of N-Viro process for East Bay Municipal Utility District (Oakland, CA) located at the Redwood Landfill (Novato, CA).

solids varies primarily with the solids content of the sludge, with a higher ratio being used for sludges with lower solids content (Figure 11.5). There is a tradeoff here between the cost of sludge dewatering and the residual lime content of the final product. Beneficial use options for the N-Viro product should be considered together with sludge dewatering costs in designing a given system. With proper mixing speeds, the resultant product is a granular, easy-to-handle, soil-like material that is further processed by one of two methods:

- alternative 1: The sludge/CKD mixture is air dried while the pH remains above 12.0 for a least seven days. The N-Viro Soil must be held for a least thirty days and until solids content is at least 65% by weight. Ambient air temperatures during the first seven days of processing must be above 5°C.
- alternative 2: The sludge/CKD mixture is heated while the pH exceeds 12.0 using exothermic reactions from the alkali in the CKD and CaO, if required. Temperatures must be 52°C throughout the mixture. The material must be stored in such a way (e.g., in a bin) so as to maintain uniform minimum temperatures for a least twelve hours. Following this heat pulse, the N-Viro Soil is air dried (while pH remains above 12.0 for at least three days) by windrowing until the solids content is

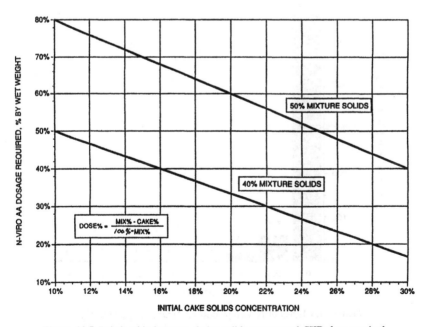

Figure 11.5 Relationship between sludge solids content and CKD dose required.

>50% by weight. This alternative has no ambient air
temperature requirements.

PATHOGEN KILL

The N-Viro process utilizes *pasteurization* rather than *sterilization* to
kill pathogens and stabilize the sludge. The distinction is important because
in sterilization there is an opportunity for recolonization by pathogenic
or putrefying organisms with subsequent health and odor problems. In
pasteurization, the mesophilic temperatures and other stresses are great
enough to kill the pathogens without destroying the more ubiquitous heterotrophic microorganisms found in soil. These continue to degrade sludge
compounds with the result that N-Viro Soil becomes more stable with
aging.

The N-Viro Soil process is an EPA-approved PFRP process. Not
only does it rapidly destroy pathogenic bacteria and viruses, but effectively kills *Ascaris* ova, the most resistant of sludge pathogens. *Ascaris*
eggs are reduced to <1 per 5 g sludge within six hours (Figure 11.6).
It should be noted that the N-Viro test for *Ascaris* kill involves seeding
the sludge with viable ova because most sludges are usually low in
this parasite.

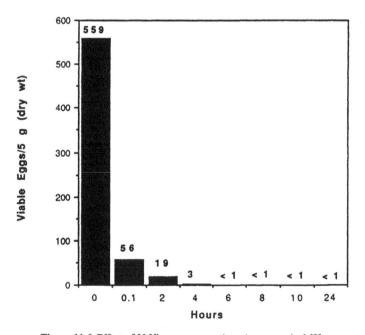

Figure 11.6 Effect of N-Viro process on *Ascaris* egg survival [3].

ODOR CONTROL

An important characteristic of N-Viro Soil is its ability to reduce sludge odors rapidly and to maintain a stable, odor-free product with prolonged storage. Initial odor control is achieved by the large surface area provided by the fine grained CKD, the rapid drying that occurs with windrowing, and pathogen kill. Long-term odor control is maintained by the continued degradation and stabilization of organic sludge solids by the remaining heterotrophic soil microorganisms. An odor test using a logarithmic scale was used to show that the N-Viro process rapidly reduced sludge odor and maintained odor control for extended periods (Figure 11.7).

Odor control and sludge stabilization are due, in large part, to the pasteurization process, which maintains a high population of indigenous organisms. Figure 11.8 shows that bacterial numbers in sludge cake were rapidly reduced in the N-Viro Soil, but numbers stabilized at about 10^4 per 5 g sludge. Total bacterial numbers in aged N-Viro Soil are similar to those in an agricultural soil (Figure 11.9).

PRODUCT UTILIZATION OF N-VIRO SOIL

Properties of N-Viro Soil

N-Viro Soil is a unique sludge product. Its physical characteristics (Table 11.4) can be summarized as follows: (1) Dry (50–75% solids); (2) soil-like density (0.7–1.0 g/cm^3); (3) granular (32% by weight >2 mm, mean granule size 0.66 mm). Its chemical characteristics are a combination of those of the CKD and of the sludge (Table 11.3). N-Viro Soil is an alkaline (pH 11–12) material in which the alkalinity is a combination of strong alkali [primarily $Ca(OH)_2$] and limestone ($CaCO_3$) (see discussion below on soil liming reaction). Most of the base cations (Ca, Mg, K, Na) are contributed by the CKD. These originally exist as oxides, but are hydrated, carbonated, precipitated by sludge constituents such as phosphate, or adsorbed/complexed by sludge organic matter. It is important to note that CKD contains significant amounts of K (1–1.5%), an essential macronutrient, that is usually low in sludge. This gives the N-Viro Soil a more balanced nutrient content than sludge alone. CKD contains very little P, so most of the P in N-Viro Soil is contributed by the sludge. Availability of P in N-Viro Soil is probably initially reduced by the high pH of the product, which converts part of the sludge P to Ca phosphates. As pH decreases on addition to soil, P availability will increase. Field studies have shown that crops were adequately supplied with P from N-Viro Soil at rates of 4–40 mt/ha [2].

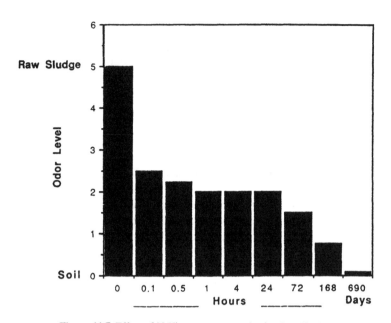

Figure 11.7 Effect of N-Viro process on reduction in soil odor.

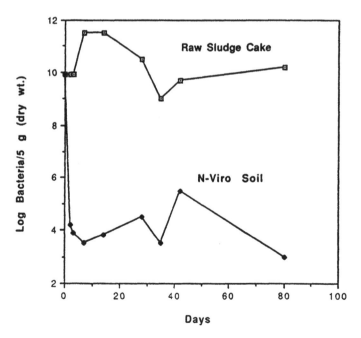

Figure 11.8 Effect of the N-Viro process on bacteria numbers compared to raw sludge [3].

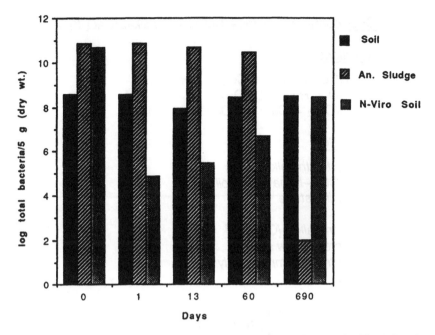

Figure 11.9 Bacteria numbers in field soil, anaerobically digested sludge, and N-Viro Soil with time [3].

The dry, biologically and chemically stable nature of N-Viro Soil provides one of the most important attributes: ease of storage. N-Viro Soil can be stored in the field if necessary without producing odors as long as water is not allowed to pond around the base of the pile. As previously discussed, N-Viro Soil improves with age, as pH falls with carbonation, as organic matter is further stabilized by decomposition, and as the solids content increases with drying.

Another important attribute of the material is its granular nature, which allows it to be spread with existing equipment such as manure spreaders, sludge cake spreaders, and lime and dry fertilizer applicators.

N-Viro Liming Reaction in Soil

As previously discussed, N-Viro Soil contains a combination of strong alkali and $CaCO_3$ (Figure 11.10), most of it as $CaCO_3$. Strong alkali compounds such as $Ca(OH)_2$ have an equilibrium pH in water of 12 or higher, but because they are strong bases they are rapidly neutralized. Limestone ($CaCO_3$), on the other hand, is relatively insoluble in water, and this property gives it its ability to buffer soil pH when added as agricultural limestone. Limestone has an equilibrium pH in water of 8–

CKD: contains limestone (CaCO₃) and alkali (CaO, K₂O)

Reactions

$$CaCO_3 = Ca^{2+} + CO_3^{2-} \quad pH = 8.5$$
$$CaO + H_2O = Ca(OH)_2$$
$$Ca(OH)_2 = Ca^{2+} + 2OH^- \quad pH = 12/5$$

N-Viro Soil

Most of the alkalinity is
CaCO₃—behaves in soil
like limestone

Strong alkali [Ca(OH)₂]
content is low and is
neutralized immediately in
soil

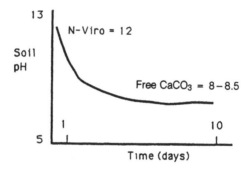

Figure 11.10 Liming reactions of CKD and sludge.

8.5. In soil with several percent or more of CaCO₃, the pH may be in the range of 7.5–8.2. We have hypothesized that when N-Viro Soil is added to soil, the strong base will be rapidly neutralized by acid in the soil (even a calcareous soil has sufficient acid to neutralize strong alkali) (Figure 11.10). Subsequently, soil pH will be determined by the liming reaction of the CaCO₃ in N-Viro Soil. In this respect, the liming reaction should be the same as that of agricultural limestone, and N-Viro Soil can be used as an agricultural lime substitute with application rates based on the lime requirement of the soil, which is determined by soil test, and the CaCO₃ equivalent content of the N-Viro Soil (determined by an acid neutralization test). Figure 11.11 shows that when N-Viro Soil was added to a fine-textured, calcareous soil at rates of 50 and 500 mt/ha in the laboratory,

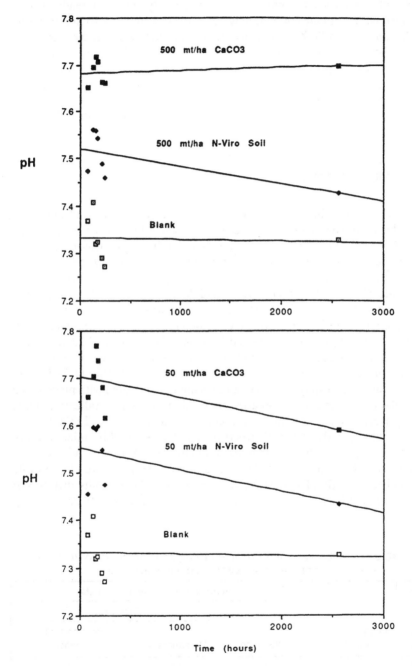

Figure 11.11 Change in pH of a calcareous soil with addition of N-Viro Soil or agricultural limestone.

607

the pH decreased to about 7.5 instantaneously (the first readings were taken within minutes of adding water to the dry soil/N-Viro Soil mixture). At times up to 3000 hours, pHs remained relatively constant, and were consistent in comparison with pure $CaCO_3$, which has a higher acid-neutralizing capacity on a mass basis. In field studies on the same calcareous soil, pHs were in the range of 7.5–8 within weeks of N-Viro Soil application (the soonest that soil samples were taken). Goodale et al. [4] found a similar pH response in his greenhouse study with N-Viro Soil (see below).

Nitrogen Availability

All of the N in N-Viro Soil is contributed by the sludge (Table 11.3), and normally it is about 1–1.5% organic N (Table 11.4) with minor amounts of NH_3 and NO_3. Most of the free NH_3 in the sludge is released as gaseous NH_3 in the initial stages of mixing with the alkaline CKD. The highest N content of N-Viro Soil is achieved with raw primary sludge because more of the N is in the organic form and less appears as NH_3. A primary consideration in the utilization of any organic waste as an N nutrient source is the rate and extent to which the organic N is converted (mineralized) to NH_3 and NO_3, the forms utilized by plants. Using a standard incubation procedure, Logan et al. [2] found that 12% of the total Kjeldahl N (TKN) was mineralized in thirty days and 16% in sixty days (Table 11.5). Mineral N produced in a sixty-day incubation is highly correlated with plant uptake in the field in the first crop after application. An N mineralization model [5] overpredicted mineral N production by about 100% (Table 11.5). It was subsequently shown that part of the NH_3 released by mineralization was strongly adsorbed or otherwise fixed by the N-Viro Soil. After correcting for this fixation, the model results more closely followed those of the incubation study (Table 11.5). It appears from these results that N-Viro Soil mineralized N at a rate of about 15–20% of TKN, a value which is

TABLE 11.5.   Observed and Predicted Nitrogen Mineralization from N-Viro Soil Based on Laboratory Incubation [2].

Incubation Time (days)	Nitrogen Mineralized (% of TKN)		
	Observed	Predicted	
		Uncorrected	Corrected for NH_3 Fixation
30	12	32	17
60	16	36	20

similar to that of composted sewage sludge and about 50% of that found for digested sewage sludge.

Trace Metal Immobilization

Cement kiln dust is usually low in trace metals, and N-Viro Energy Systems has a quality control program that screens potential CKDs to be used to make N-Viro Soil. Most of the trace metal in N-Viro Soil comes from the sludge. Therefore, if the N-Viro Soil product is to be used as a fertilizer amendment, it will have to meet metal limits set by the appropriate federal, state and local statutes. CKD will in most cases dilute the trace metal content of the sludge. In addition, reactions between the CKD and sludge result in decreased trace metal solubility in the N-Viro Soil. Some of these reactions, such as adsorption to mineral surfaces, precipitation as carbonates and hydroxides, and complexation by soil organic matter, favor cationic trace metal immobilization as pH increases. It appears, however, that trace metals are immobilized in N-Viro Soil even at acid pH. Bennett [6] studied the immobilization of trace metals from N-Viro Soil by use of a modified EPA Toxicity test (the EPTOX is a standard acid leaching procedure used to assess hazardous wastes). The samples were pretreated to neutralize the alkalinity in the N-Viro Soil and to give an initial pH of 5. Trace metals extracted from N-Viro Soil were greatly reduced compared to sludge. Using the same test procedure, Bennett extracted trace metals from the field study sites used by Logan et al. [2], and showed (Figure 11.12) that less metal was extracted from plots receiving 20 mt/ha N-Viro Soil than from untreated soil. This is in spite of the fact that significant amounts of the six metals were added with the N-Viro Soil.

Crop Response to N-Viro Soil

A number of greenhouse and field studies have been conducted to evaluate the use of N-Viro Soil as a soil amendment for crop production, floriculture, and reclamation. Results from several are summarized here. In a field study, Logan et al. [2] studied the response of corn and soybean to N-Viro Soil applied at rates of 4 to 40 mt/ha (dry weight) on a calcareous soil (Table 11.6). There was no response of corn yield to fertilizer or N-Viro Soil applications because of record drought that year, but there was a trend to higher yields at the 40 mt/ha rate of N-Viro Soil. Field observations suggested that the response might have been physical because the corn on the high rate plots appeared to be less drought stressed. Soybean yields were significantly increased with the higher rates of N-Viro Soil.

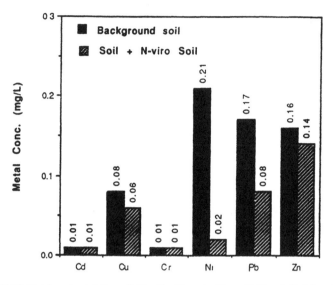

Figure 11.12 Metals extracted from field soil with and without N-Viro Soil (20 mt/ha) by a modified EP toxicity test [6].

In a greenhouse pot study of N-Viro Soil effects on vegetable and flower growth, Goodale et al. [4] found that N-Viro Soil rates as high as 25% of the media had little effect on pH, and increased soluble salts but below saline levels (Table 11.7). N-Viro Soil at a rate of 5% of the media increased growth of all species studied (Table 11.7), while higher rates were either ineffectual or decreased the growth of all species except oats.

In a greenhouse study of N-Viro Soil use for reclamation of acid mine spoil, Logan [8] found that N-Viro Soil at rates of 50, 100, and 200 mt/ha

TABLE 11.6. Yields of Corn and Soybeans with N-Viro Soil on a Calcareous Soil from NW Ohio [2].

Treatment	Crop Yield (bu/ha)	
	Corn	Soybeans
Check	96 2 ab[a]	30.7 b
Fertilizer	98.5 ab	33.5 ab
N-Viro (4 mt/ha)	96.1 ab	32.9 ab
N-Viro (4 mt/ha)	91 7 b	31 9 ab
N-Viro (9 mt/ha)	89 5 b	32.7 ab
N-Viro (16 mt/ha)	97.4 ab	34 2 a
N-Viro (20 mt/ha)	93 9 ab	34 1 ab
N-Viro (40 mt/ha)	105.6 a	35.1 a

[a]Means in the same column followed by the same letter are not statistically significant (p = 0 05)

TABLE 11.7. Effects of N-Viro Soil on Growth of Vegetables and Flowers in Pot Media [4].

N-Viro (% in media)	Media pH	Salts (mmhos/cm)	Plant Growth (grams dry weight)				
			Radish Root	Bean Tops	Rye Tops	Oat Tops	Impatiens Tops
0	7.5	0.35	0.6	1.0	0.7	0.8	0.5
5	7.3	0.58	2.2	1.7	1.8	1.8	1.3
10	7.4	0.99	2.3	1.9	2.6	5.0	0.4
15	7.6	1.03	2.2	1.2	3.3	4.6	0.8
20	7.9	1.25	2.0	1.1	4.1	6.9	0.4
25	7.4	1.40	1.1	0.5	4.0	6.7	0.4

611

gave aboveground biomass yields that were the same as equivalent rates of lime and NPK fertilizer (Figure 11.13). N-Viro Soil at 200 mt/ha produced more root biomass than the equivalent rate of lime and fertilizer. Field studies in Kentucky have shown that this material is an excellent amendment for reclamation of acid mine spoil.

CONCLUSIONS

The N-Viro process for alkaline stabilization of wastewater sludge has been shown to be highly effective in destroying sludge pathogens, reducing sludge odors, and in producing a dry, stable product, N-Viro Soil. Characteristics of N-Viro Soil make it an ideal, safe product for use in agriculture as a lime substitute, low analysis fertilizer, and soil amendment. Its physical, chemical, and biological properties make it equally suitable for revegetation, landfill cover, and as an ingredient of topsoil.

BIO GRO SYSTEMS' SLUDGE TREATMENT TECHNOLOGIES

Increasing interest in municipal sludge management practices for beneficial uses and nutrient recycling has resulted in a number of diversified

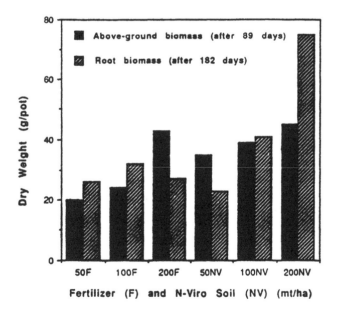

Figure 11.13 Grass biomass on acid mine spoil (initial pH 3) with N-Viro Soil and equivalent amounts of lime and fertilizer.

technologies. Bio Gro Systems, Inc., a comprehensive sludge management company, has developed two technologies to complement their existing land application/reclamation services. Both of these technologies produce a pathogen-free material that can be marketed for agricultural or other beneficial uses.

The BIO*FIX$_{SM}$[1] process provides a high-lime material with application as an agricultural liming agent or as a lime-containing organic amendment for the production of topsoil materials.

Bio Gro's drying and pelletization (Pelletech™[2]) unit provides an indirect thermal process that produces dried sludge pellets that can be sold as a fertilizer material (alone or as a component of a blended product). Both BIO*FIX and Pelletech processing are available through Bio Gro Systems as completely privatized sludge management systems, including product marketing.

BACKGROUND

Nationwide, more than 40% of municipal sludge volumes are applied to agricultural and nonagricultural lands. Production of sludge-based fertilizers and soil conditioners for distribution and marketing accounts for another 6% of the total.

To address the public health and environmental concerns associated with the above uses of wastewater sludge, Congress has given the U.S. Environmental Protection Agency the authority to regulate sludge under the Resource Conservation and Recovery Act and the Clean Water Act. Current (1979) U.S.EPA sludge regulations control chemical contaminants by placing a limit on the concentration of synthetic organics (e.g., PCBs) that can be present in sludges for land application. Similarly, U.S.EPA and individual State regulations limit the amounts of trace metals in sludges which are applied to soils. U.S.EPA technical regulations (to be promulgated in 1992) will address amounts of both trace metals and an expanded list of organics in detailed, risk-based limits for the application of sludges to land, distribution and marketing, and other sludge management practices [9,10].

The U.S.EPA has taken a different approach to the control of the concerns associated with the presence of pathogens in sludge. Current U.S.EPA regulations address pathogens by a combination of treatment (disinfection) requirements and specified management practices. Two levels of disinfection are described—processes to significantly reduce pathogens (PSRP) and processes to further reduce pathogens (PFRP). Management practices address issues such as public access, waiting periods before

[1] BIO*FIX is a service mark of Bio Gro Systems, Inc.
[2] Pelletech is a trademark of Bio Gro Systems, Inc.

cattle grazing, and use of sludges on crops for direct human consumption. Management practices specified for PSRP processes are more constraining than management practices specified for PFRP processes.

PATHOGEN REDUCTION PROCESSES

Currently, lime addition to achieve pH 12 after two hours is listed as a PSRP process, and the achievement of 70°C for thirty minutes qualifies as an add-on method to achieve PFRP (pasteurization), when used in conjunction with a PSRP. Bio Gro Systems offers this technology through completely privatized facilities provided for municipal and industrial clients.

The regulatory picture described above is changing. U.S.EPA's 1989 proposed 503 Rule for sludge disinfection describes three levels of sludge disinfection: Class A, B, and C [9]. Class A sludges have the greatest degree of pathogen inactivation and need the least restrictive management (similar to current PFRP designation). The additional flexibility and greater public acceptance offered by more stringent disinfection standards has resulted in the development of new treatment methods to meet these standards. The BIO*FIX process offered by Bio Gro Systems, Inc. uses the exothermic reaction of quicklime and water to achieve both current PFRP and proposed Class A (503) standards.

PART I: BIO*FIX SLUDGE PROCESSING

The proprietary BIO*FIX alkaline treatment processes include stabilization, chemical addition, PSRP and/or PFRP formulations, and enhancement of the end product. BIO*FIX processes have been developed as four separate stages (I through IV) to meet increasingly stringent operational and regulatory requirements.

Stages I and II BIO*FIX processes depend on the addition of lime or lime equivalents to reach the desired pH level (pH of 11 or greater for odor suppression; pH of 12 after two hours for a PSRP). For these process stage reactions, lime sources can be either $Ca(OH)_2$ (hydrated) or CaO (quicklime) material.

BIO*FIX Stage III processing meets the criteria for a Process to Further Reduce Pathogens (PFRP) by combining the alkaline and exothermic reactions of quicklime with water: $CaO + H_2O \rightarrow Ca(OH)_2$ + heat. Each pound of 100% quicklime theoretically produces 490 BTUs of heat and extracts nearly 1/3 of its weight in free water from the sludge. This reaction achieves temperatures in excess of 70°C for more than thirty minutes, as required by federal regulations for add-on pasteurization to meet PFRP criteria [9].

The above BIO*FIX Stage III pasteurization (PFRP) reaction must be carried out under carefully controlled and monitored conditions to ensure uniform sludge treatment and pathogen inactivation by the heat generated during the reaction. For all stages of BIO*FIX processing, odor control reagents and/or nutrient enhancements may also be incorporated as required.

The BIO*FIX PFRP process uses sufficient quantities of quicklime to insure that the pH will remain at 12 or greater, even when the material is stored for extended periods of time.

Stored BIO*FIX material can be monitored for pH and reprocessed through the BIO*FIX facility if necessary to maintain pH 12, thus preventing odors, regrowth, and other nuisance conditions.

BIO*FIX Stage IV processing includes specific additives to result in a soil-like material which is nonviscous and therefore not subject to liquefaction under mechanical stress. Varying the processing additives and mix ratios produces a range of waste-derived materials suitable for daily, intermediate, and final landfill cover or land reclamation.

Regulatory Approval

The BIO*FIX Stage II process is recognized by the U.S.EPA as a Process to Significantly Reduce Pathogens (PSRP) and BIO*FIX Stages III and IV as Processes to Further Reduce Pathogens (PFRP). The standard BIO*FIX formulations do not need specific approval from U.S.EPA's Pathogen Equivalency Committee (PEC) to meet state regulations. The BIO*FIX processes have already been approved for ongoing Bio Gro projects throughout the United States In addition, BIO*FIX processing can meet all proposed 503 technical regulations concerning sludge treatment for pathogen reduction.

In general, BIO*FIX Stage I and II materials will be applied to sites for which state permitting approval is required (e.g., land application, land reclamation). The more rigorous pathogen inactivation treatment of BIO*FIX Stages III and IV results in a material that can be distributed and/or marketed without the need for a site specific permit (e.g., as a liming material or topsoil amendment/substitute). Approval of the process as a PFRP will generally be obtained from the same state agency that regulates land application, composting, and other sludge management practices. In addition, the State Department of Agriculture or equivalent agency may also issue approval for sale of BIO*FIX as a liming material. BIO*FIX Stage III and IV products may be used in agriculture as substitutes for commercial lime. Nitrogen is not a limiting factor in application rates because the amount of nitrogen applied by BIO*FIX at typical agricultural liming rates is low. This is due to the large amounts of quicklime solids used in processing and nitrogen (ammonia) losses resulting from BIO*FIX processing.

Application rates for BIO*FIX as a liming material are developed based on recommendations for liming additions to specific soils developed by state agricultural extension specialists. Table 11.8 shows some typical application rates based on liming recommendations for both mineral and organic soils at different soil pH values.

Additional Properties of BIO*FIX Materials

Since it has been treated by a Process to Significantly Reduce Pathogens (PSRP), BIO*FIX material does not exhibit vector-attraction properties. This characteristic results both from residual alkalinity and from reduced levels of volatile solids. Odor production is suppressed by the high pH of BIO*FIX, which prevents biodegradation of organic matter and release of volatile odor-producing compounds. BIO*FIX material as landfill cover conforms readily to the contours of landfilled refuse, which further ensures odor control and prevents blowing litter or other nuisance conditions.

The 30 to 60% (or greater) solids content of BIO*FIX material provides cover material with stability. The compressive strength of a BIO*FIX

TABLE 11.8. Application Rates (Tons/Acre) for BIO*FIX Based upon Calcium Carbonate Equivalency (CCE).[a]

Buffer pH	Mineral Soils		Organic Soils	
	Plow Depth 6-2/3 inches	Plow Depth 9 inches	Plow Depth 6-2/3 inches	Plow Depth 9 inches
7.0	0	0	0	0
6.9	0	0	0	0
6.8	1	1.5	0	0
6.7	1.5	2	0	0
6.6	2.0	3	0	0
6.5	2 5	4	0	0
6.4	3.0	4.5	1	1.5
6.3	3.5	5	2	3
6.2	4.0	6	2.5	3.5
6.1	4.5	7	3	4 5
6.0	5.5	8	4	6
5.9	6.0	9	4.5	6.5
5.8	6.5	10	5	7.5
5 7	7.0	11	5.5	8
5.6	8.0	12	6	9
5.5	9 0	13	6.5	10

[a]CCE may be calculated by multiplying the Ca(OH)$_2$ content of specific BIO*FIX material by a factor of 1 35.

material can be adjusted by varying the BIO*FIX additive(s) (e.g., pozzolanic material) ratio. The ratio is determined by the physical characteristics of both the additive and the original sludge (e.g., primary, biological digested). Increasing the proportion of additive(s) can provide the compressive strength and workability needed for a specific landfill application.

Trafficability of BIO*FIX cover material will increase with time after treatment. The precise time required for the degree of solidification desired will depend on mix ratio, physical characteristics of additive(s), and original sludge properties. All mix ratios are designed to produce a BIO*FIX cover material that falls within designated USDA soil textural classifications.

Freezing of the water in BIO*FIX cover material temporarily affects its physical properties, just as freezing affects natural soil. The material returns to its normal state upon thawing. Repetitive freeze/thaw cycles will tend to break up larger clumps into smaller ones, again as it occurs with natural soils. No permanent changes in the BIO*FIX matrix result from freezing and thawing. Under very wet conditions, the BIO*FIX/additive ratio may need to be decreased to prevent stickiness and lowered granularity, which would cause difficulties for cover material use.

The BIO*FIX process results in a nutrient-rich material that will support vegetative growth when used in a final cover application. The natural weathering process that lowers the surface pH will provide a suitable medium for germination and rapid growth of plant species. The organic matter content (from the original sludge) of BIO*FIX cover material maintains a stable, nonerodible surface layer.

Equipment and Process Description

The BIO*FIX Stage III process provides simultaneous temperature and pH adjustment of the sludge to meet PFRP standards. Figure 11.14 shows the basic equipment used in the process. The processing or physical reactions that transform the sludge start in the blender. Quicklime (CaO) added to the sludge in the blender elevates pH to over 12. Quicklime also reacts with the water in the sludge to produce heat, generating and sustaining temperatures of 70°C or more for over thirty minutes. Typical contact time in the blender is two or three minutes, with a minimum of thirty seconds required for the reaction to occur. Cake sludge and quicklime are fed from input hoppers to the sludge blender by variable rate conveyors. The conveying rate is set to achieve the correct lime/sludge ratio for PFRP temperature (>70°C). As the percent solids of the input sludge varies, the lime feed is adjusted to maintain temperature. Figure 11.15 shows the theoretical pounds of lime needed to achieve PFRP. The lower line shows maximum pH requirements (for PSRP), and the upper line shows temperature requirements (for PFRP).

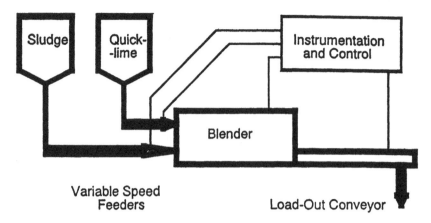

Figure 11.14 BIO*FIX process equipment flowchart.

The lime (CaO) requirement for PSRP treatment typically ranges from 15% to 25% of the dry weight of sludge. Figure 11.15 is based on a requirement of 25%. Note that, theoretically at least, while the lime requirement for BIO*FIX Stage III decreases with increased solids, the lime requirement for BIO*FIX Stage II increases with decreased solids. This

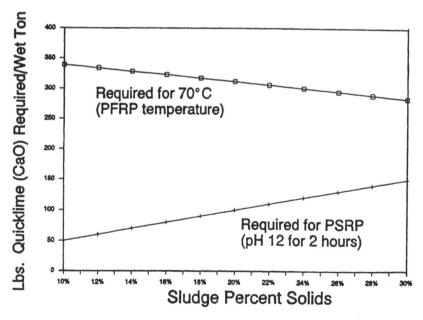

Figure 11.15 Theoretical lime requirements for PSRP and PFRP at varying sludge solids concentrations.

is due to the predominance of the heating requirements of the water for BIO*FIX Stage III versus the pH effects of the solids for BIO*FIX Stage II. The following assumptions were used for the PFRP temperature requirements in Figure 11.15:

- Sludge input temperature equals 20°C (58°F).
- 100% of the quicklime reacts with water in the sludge to produce heat (490 BTU per pound of quicklime).
- Quicklime is 100% calcium oxide (CaO).
- Specific heat of the sludge solids is 0.25.
- No heat loss from the sludge to the air or the equipment occurs.

In practice, these conditions rarely exist; therefore, the amount of quicklime used to achieve PFRP can be up to 50% greater than the PFRP temperature line (top) in Figure 11.15. A drier, more easily crumbled endproduct can be achieved by increasing the quicklime addition up to as much as twice the theoretical value shown.

The blender contains a weir-gate adjustment to regulate retention time during blending. The better the blending achieved, the quicker the limesludge reaction will produce heat. The critical blending time for this reaction in the BIO*FIX process is thirty seconds. An oversized blender can be used to insure that at full capacity the minimum retention time is one minute.

A load-out conveyor completes the process train.

Monitoring and Process Control

Each variable rate feeder in the process equipment train (shown in Figure 11.14) is connected to an instrument panel. From this panel the equipment operator can monitor all material flows.

The operator also monitors three physical characteristics of the process:

(1) Blender air temperature—an almost instantaneous indication of the state of the lime/sludge reaction used to adjust lime feed to the blender. It is measured by a temperature probe inserted at the process air connection from the blender.

(2) Finished product temperatures—measured on hourly grab samples by inserting a temperature probe that automatically records the information to a computer.

(3) Finished product consistency—visual inspection of the product as it comes off the load-out conveyor. This factor is unrelated to PSRP or PFRP processes, but may have importance to the user of the end product.

PFRP temperature is verified from hourly grab samples. The preferred method for recording these temperatures is connecting the temperature probe to a computer and automatically logging the results over time. Figure 11.16 shows the beginning portion of one daily log file on computer. The data are stored in a format easily imported into spreadsheet or database programs; Figure 11.17 shows the same data imported into Lotus 1-2-3™.

When the computer is not available, the backup procedure is used to record temperatures manually from the digital readout on the probe. Figure 11.18 shows the form used for backup temperature data.

The amount of quicklime required for pH adjustment to 12.0 for two hours (PSRP) is 25% (maximum) of the dry weight of the sludge, or approximately one-eighth of the amount of quicklime required to reach PFRP temperatures (see Figure 11.15). Since quicklime addition to meet temperature is the critical operating requirement for BIO*FIX PFRP, pH monitoring is less frequent than temperature monitoring.

```
"Accumet 915 Data Sampling Program"
"Project:
"Comment:
"Date:
"Time: 10:23:19"
" "
"Sample #","Time","°C","pH"
1,"10:24:25",98,7.00          <--- first temp sample
2,"10:25:25",98,7.00
3,"10:26:25",98,7.00
4,"10:27:25",98,7.00
5,"10:28:25",98,7.00
6,"10:29:25",97,7.00
7,"10:30:25",97,7.00
8,"10:31:25",97,7.00
9,"10:32:25",97,7.00
10,"10:33:25",96,7.00
11,"10:34:25",96,7.00
12,"10:35:25",96,7.00
13,"10:36:25",95,7.00
14,"10:37:25",95,7.00
15,"10:38:25",78,7.00          <--- temp probe removed
16,"10:39:25",,7.00
17,"10:40:26",27,7.00
18,"10:41:27",26,3.88
19,"10:42:28",25,5.09          <--- standard buffer
20,"10:43:28",24,6.78
21,"10:47:13",78,11.8
22,"10:48:15",76,10.7          <--- standard buffer
23,"10:49:15",42,6.52
24,"10:50:16",42,12.4          <--- first pH sample
25,"10:51:18",41,12.4
   . . .
```

Figure 11.16 Typical BIO*FIX computer log.

```
Accumet 915 Data Sampling Program
Project:
Comment:
Date:
Time: 10:23:19

Sample # Time      °C        pH
     1 10:24:25    98        7    <--- first temp sample
     2 10:25:25    98        7
     3 10:26:25    98        7
     4 10:27:25    98        7
     5 10:28:25    98        7
     6 10:29:25    97        7
     7 10:30:25    97        7
     8 10:31:25    97        7
     9 10:32:25    97        7
    10 10:33:25    96        7
    11 10:34:25    96        7
    12 10:35:25    96        7
    13 10:36:25    95        7
    14 10:37:25    95        7
    15 10:38:25    78        7    <--- temp probe remv'd
    16 10:39:25              7
    17 10:40:26    27        7
    18 10:41:27    26     3.88
    19 10:42:28    25     5.09    <--- standard buffer
    20 10:43:28    24     6.78
    21 10:47:13    78     11.8
    22 10:48:15    76     10.7    <--- standard buffer
    23 10:49:15    42     6.52
    24 10:50:16    42     12.4    <--- first pH sample
    25 10:51:18    41     12.4
            ...
```

Figure 11.17 BIO*FIX Lotus spreadsheet with data.

Grab samples are taken twice during the operating day. These samples, along with samples of any stockpiled end products, are measured for pH immediately and after two hours in the treatment plant laboratory [11]. With a pH probe attached to the computer sampling equipment, pH readings can also be obtained at the BIO*FIX processing site.

Like other chemical stabilization processes, BIO*FIX treatment results in increased mass of the final product. However, its competitive capital and operating costs, and the potential for rapid startup make BIO*FIX an economically attractive PFRP method when compared to thermal and composting options.

PART II: SIMULTANEOUS SLUDGE DRYING AND PELLETIZING

Removing water from liquid sludge (by gravity, mechanical, or thermal methods) results in a significant volume reduction in the material to be managed, and is therefore a common component of most sludge management systems. The thermal drying method also provides a Process to

```
                          BIO*FIXSM
                       Sludge Processing

                   Temperature Monitoring Log

Date: _____            Operator's Name:_____

Directions:

     - Collect approximately 2 to 3 gallons in a pail
     - Insert thermometer into the center of the sample
     - Use wristwatch with a second hand, or other timer, to
       measure time:
              - 0 = time sample was taken
              - Record temp. every 10 min. for 40 minutes
              - Repeat the process as necessary.


          Time: _____              Time: _____

    Minutes  Temp. (°C)        Minutes  Temp. (°C)

         0   _____             0   _____
         2   _____             2   _____
         4   _____             4   _____
         6   _____             6   _____
         8   _____             8   _____
        10   _____            10   _____
        12   _____            12   _____
        14   _____            14   _____
        16   _____            16   _____
        18   _____            18   _____
        20   _____            20   _____
        22   _____            22   _____
        24   _____            24   _____
        26   _____            26   _____
        28   _____            28   _____
        30   _____            30   _____
        32   _____            32   _____
        34   _____            34   _____
        36   _____            36   _____
        38   _____            38   _____
        40   _____            40   _____
```

Figure 11.18 BIO*FIX manual temperature monitoring log.

Further Reduce Pathogens (PFRP) as discussed in Part I along with the greatest volume reduction achieved by any process other than incineration. Unlike incineration, the sludge nutrients (nitrogen, phosphorus, potassium, organic matter, trace elements) are retained in the dried sludge product and can be used as a fertilizer. As an alternative, dried sludge can be integrated with sludge incineration to provide autogenous combustion when mixed with dewatered sludge.

Two general types of sludge heat-drying technologies are currently available: direct and indirect. Direct drying brings hot gas into direct contact with the sludge while indirect processes rely on conductive drying—the sludge contacts a metal surface heated by a heating medium such as steam or thermal oil. A major difference between the two technologies is the flue gas volume produced. Direct dryers generate up to twenty times greater flue gas volume than do indirect systems, with a corresponding increase in adverse environmental impact and significantly more expensive and less reliable air emission and odor control equipment.

Pelletech Sludge Drying

The central component of the thermal drying technology provided by Bio Gro Systems is a multi-stage indirect sludge dryer developed by Seghers Engineering Company (Belgium) from about 1975 through 1985; the technology is marketed by Bio Gro Systems in the United States under an exclusive license agreement. The Pelletech dryer simultaneously dries and pelletizes sludge in a one-step operation—no separate mechanical pelletization process is required, unlike other indirect sludge dryers currently available. The pellets can be marketed as a fertilizer source or combined with dewatered sludge for combustion (as is done at the Bruges, Belgium Wastewater Treatment Plant, shown in Figure 11.19).

The dryer portion of the thermal processing (without the fluidized bed incineration, as shown in Figure 11.20) illustrates the production of dried pellets.

The Pelletech dryer is a vertically oriented multi-stage unit that uses steam or thermal transfer fluid in a closed loop to achieve 90% or greater dry solids content in the product. This indirect dryer has a high overall thermal efficiency (up to 80%). Processing is accomplished in a gas and dust tight system—avoiding the potential for odors frequently associated with sludge treatment processes. The dryer's exhaust consists of water vapors, air, and some particulate and gaseous pollutants. After water vapors are condensed, only a small amount of noncondensable gas, mainly moist air, remains to be treated. This noncondensable gas is vented from the dryer to an odor control unit, such as the boiler, for thermal destruction of odor-causing compounds.

The sludge is exposed to a temperature of approximately 100°C for twenty to thirty minutes, inactivating or destroying all pathogens and other living organisms. Organic matter and nutrients remain in the dried sludge product.

Sludge is fed into the Pelletech dryer unit via the top inlet and moved by rotating arms from one heated tray to another in a spiralling zig-zag motion until it exits at the bottom as a dried, granular (pelletized) product at approximately 95% solids. The dryer trays are hollow and are internally heated by steam or recirculating thermal oil.

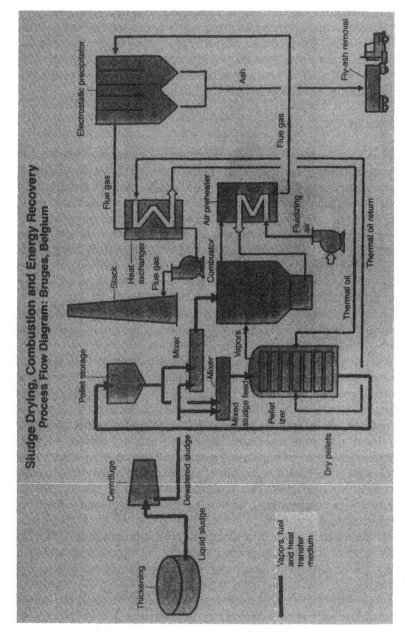

Figure 11.19 The process at the Bruges wastewater treatment plant includes sludge drying, combustion, and energy recovery.

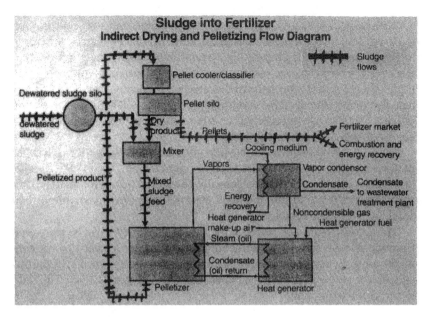

Figure 11.20 The indirect heat drying and pelletization process generates marketable products (natural fertilizer or fuel).

The rotating arms are equipped with adjustable scrapers that move and tumble the sludge in thin layers and small windrows over the heated trays, enhancing heat transfer and the escape of water vapor. The drying and pelletizing process begins with fine particles that gradually, layer by layer, build up, drying from the center of each particle to the outside. This process of recycling selectively sized products, coating them with wet sludge then simultaneously drying and compacting the wet sludge coating on the hard dry pellet core is termed "pearling." This process minimizes formation of dust and oversized chunks. The pearling method of pellet production also allows more freedom in adjusting the particular size of the product granules to meet the exact requirements of the end user. It also compacts the drying sludge to increase the product bulk density and durability.

The dried sludge is also recycled to maintain a moisture content between 60% and 70% total solids into the dryer, thus avoiding the glue-like phase inside the dryer and facilitating granulation. This glue-like or sticky phase can occur in municipal sludge at between 35% and 55% total solids.

The temperature of each tray or group of trays may be controlled separately to maximize the efficiency of the process. This also helps to produce sludge pellets that are homogeneous in moisture content and size distribution.

With respect to energy recovery, the water vapors and sweep air leaving the dryer contain significant amounts of heat. This energy can be recovered to further improve the already high overall energy efficiency of the unit. Recovered energy can be used to heat sludge digesters or to preheat the liquid sludge prior to mechanical dewatering. Preheating the liquid sludge can also reduce the need for conditioning chemicals. The Pelletech drying process can also use digester gas or other sources of inexpensive or waste energy to reduce operational costs.

When combined with a combustion unit, the dried sludge is mixed with dewatered sludge to achieve autogenous combustion. The 90–95% solids dry sludge, combined with mechanically dewatered (to approximately 20% total solids) sludge, results in a feed material of approximately 40% total solids entering the combustion unit.

The energy to dry and pelletize the sludge in the dryer can be recovered from the flue gas through a thermal oil heat exchanger located downstream of the air pollution control equipment. As shown in Figure 11.19, in this situation, water vapors from the Pelletech dryer may also be ducted directly to the combustion unit.

Properties of Sludge Pellets

Drying technology results in pathogen-free pellets with a nominal moisture content of approximately 5%. Nutrient content is primarily in the form of organic nitrogen and phosphorus, with smaller amounts of potassium and micronutrients. Many sludges also contain appreciable quantities of organic iron, which can be agriculturally beneficial. Marketability of the material will depend on both physical characteristics and the quality of the sludge being processed.

Thermally dried pellets generally have the following characteristics:

- dry solids content: greater than 90% total solids (nominally 90% total solids)
- organic material content: normally greater than 50%, depending on levels in the source sludge
- nitrogen content, total: 4% to 6%, depending on levels in the source sludge
- trace metals content: depends on levels in the source sludge
- durability: exceeds fertilizer industry standards
- nutrient release properties: slow release due to organic origin, temperature, and moisture dependence
- pellet size: variable between 0.4 mm and 6 mm, with a narrow or wide range of particle size distribution, depending on end use, nearly dust-free

Homogeneous pellets with a selectable particle size distribution in this range are optimum for diverse pellet marketing and are achieved through control of the drying process and an appropriate pellet screening and classification system. Moisture content should be maintained at a minimum in order to prevent growth of mold or other microorganisms on the surface and the heating of stored pellets, which can result from microbial growth.

The prices commanded by pelletized sludge fertilizer products, in 1989, ranged from $80 to $145 per ton of pellets. The price was largely determined by the overall nitrogen content of the sludge.

Pelletech Operating Parameters

The technical data sheet (Figure 11.21) for the Pelletech dryer provides information on some of the operating requirements for the unit at various levels of processing capacity.

The physical requirements for a 20–25 dry ton/day dewatering and drying facility include:

- site requirements: 20,000 square feet
- building size: 6000 square feet (approximately 60 feet by 90 feet in plan, and approximately 70 feet high)
- energy requirements: 110–115 therms/dry ton at 20% solids
- return streams: condensate 12–15 gpm with less than 180 mg/L BOD

The entire system includes: liquid sludge storage and pumps; a two-meter belt filter press (or equivalent multiple presses); a proprietary mixing device that mixes dewatered sludge with recycled pellets; the Pelletech dryer unit; a heat generation system (using natural and/or digester gas or waste heat); classification and cooling equipment to separate undersized or oversized pellets and cool them to less than 90°F; and a storage silo that also loads pellets into transport trucks.

The complete facility is typically designed to handle reasonable peaks in sludge feed rate during normal operating hours.

Odor and Air Emission Control

Odors are contained within the process equipment by maintaining slightly negative pressure in materials handling system and the Pelletech dryer itself.

The dryer's exhaust passes through a series of air pollution control steps to condense water vapor, and to remove particulates and soluble gases. The noncondensable gas (saturated air) is then used to make up about 10% of the combustion air for the heat generation equipment. In

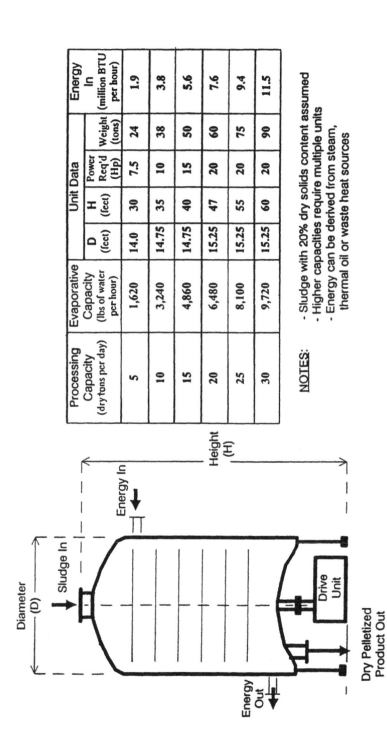

Processing Capacity (dry tons per day)	Evaporative Capacity (lbs of water per hour)	Unit Data				Energy In (million BTU per hour)
		D (feet)	H (feet)	Power Req'd (Hp)	Weight (tons)	
5	1,620	14.0	30	7.5	24	1.9
10	3,240	14.75	35	10	38	3.8
15	4,860	14.75	40	15	50	5.6
20	6,480	15.25	47	20	60	7.6
25	8,100	15.25	55	20	75	9.4
30	9,720	15.25	60	20	90	11.5

NOTES:
- Sludge with 20% dry solids content assumed
- Higher capacities require multiple units
- Energy can be derived from steam, thermal oil or waste heat sources

Figure 11.21 Pelletech unit technical data sheet.

this way, the noncondensable gas with malodorous compounds is thermally destroyed at a temperature of 1800°F or more for over one second, thus effectively afterburning the entire process exhaust. The remaining 90% of the combustion air needed for the heat generation equipment can be drawn from areas needing additional ventilation inside the building (e.g., near dewatering equipment, sludge blenders or mixers, liquid sludge storage tanks) or from inside the building in order to maintain the slight negative pressure.

Process Efficiency

The Pelletech drying facility is extremely efficient with respect to energy usage. Possibly the greatest opportunity for enhancing overall plant energy efficiency is to supplement the primary fuel with digester gas if available. Energy can then be returned from the dryer system in turn to heat digesters.

Over five years of operational experience have demonstrated the practicality, reliability, and economic soundness of the Pelletech dryer as a method for simultaneous thermal drying and pelletization of municipal sludge. Operational costs are competitive compared with other heat-drying or composting systems for PFRP treatment of sludges. Capital costs are also competitive with other heat-drying or composting systems, with comparable practical and environmental features.

IN-VESSEL COMPOSTING OF MUNICIPAL SEWAGE SLUDGE

INTRODUCTION

During the last fifteen years, composting has grown in popularity as a sludge management technique, in part due to the higher costs and greater environmental concerns often associated with alternative methods. Compared to static pile and windrow composting operations, in-vessel systems promise a more stabilized and consistent product, smaller space requirements, and better containment and control of odors. Because in-vessel composting systems are new in this country, however, there is relatively little information available on the operation of actual systems. With funding from the United States Environmental Protection Agency, Camp Dresser & McKee Inc. conducted a technology evaluation of in-vessel composting that was completed in 1989 [12]. This paper is based in part on that study. Included here is a description of the components of any in-vessel composting system, a description of those proprietary systems with full-scale municipal wastewater sludge experience as of 1991, and a case study.

COMPONENTS OF IN-VESSEL COMPOSTING SYSTEMS

The purpose of this section is to provide an overview of the various components of an in-vessel composting facility. A generic process flow diagram showing the components of an in-vessel composting system is shown in Figure 11.22. Three materials (sludge cake, an amendment, and recycled compost) are mixed together and placed into one or several aerated reactors for composting. After composting, the product is removed from the reactor for storage before utilization.

Mechanically and operationally, materials-handling systems are the dominant features of in-vessel facilities. In this regard, in-vessel facilities differ from other more common composting facilities such as static pile systems. In-vessel plants are highly mechanized facilities. Static pile and windrow operations employ manually operated portable equipment such as front-end loaders and windrow-turning machines. In contrast, designers of in-vessel facilities have taken advantage of the fact that the composting is being carried out in a reactor that can be serviced by automated conveyors. There is no inherent technological characteristics that demands the high degree of automation built into current in-vessel systems. It is a tradeoff of capital costs against operating (specifically labor) costs.

Compost Ingredients

There are usually three ingredients to a composting mix: dewatered sludge cake, an organic amendment, and recycled compost. The term "bulking agent" refers to the combination of organic amendment and recycled compost. Dewatered sludge cake and recycled compost are usually

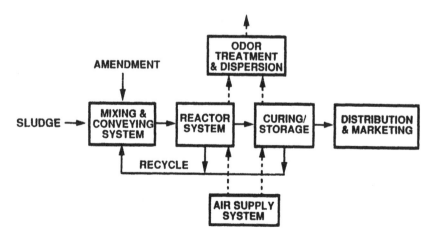

Figure 11.22 Components of an in-vessel composting system.

referred to simply as "sludge" and "recycle," respectively. The mixture of sludge, amendment, and recycle is commonly referred to as the "mix." These conventions will be followed in this paper.

Both mixtures of primary sludge and waste activated sludge, and waste activated sludge alone, are being processed at in-vessel facilities. These sludges are being composted without digestion at most facilities, but with aerobic and anaerobic digestion at others. Although any of several technologies can be used to dewater the sludge, the most common device used in connection with in-vessel systems is the belt filter press.

The bulking agent is mixed with sludge to increase the solids content of the mix, provide structure, and supply some carbon to the composting process. The structure is necessary to maintain the porosity of the mix.

For costs reasons, recycled compost is the preferred bulking agent. The source of recycle can be either discharge material from the bioreactor (first-stage reactor) or finished compost. There is a limit, however, to how much recycle can be used for this purpose. As compost gets older and more stable, average particle size gets smaller. Eventually the recycle particles are so small that they destroy rather than create porosity.

To avoid overuse of recycled compost, an amendment is added to the mix. Unlike static pile and windrow operations, which commonly use wood chips as the amendment, in-vessel systems most commonly use sawdust, wood shavings, or ground-up wood. Shredded bark is also sometimes used. The amendment normally is not recovered, but becomes part of the compost product. This also differs from many static pile facilities that recover and reuse their wood chip amendments.

Usually amendment materials are waste products from local industry, although they can be "manufactured" for this purpose. At Clayton County, GA, for instance, sawdust is made at the plant by grinding wood chips in a hammermill. The wood chips are produced from timber grown on lands irrigated with treatment plant effluent. Another example is Fort Lauderdale, FL, where ground-up brush trimmings and other yard wastes are used for amendment.

Materials Handling Systems

Storage facilities for various materials are needed to provide "slack" in conveying systems so that systems operating at different flow rates or for different times can be matched.

Most compost reactors are loaded for only a portion of each day, while sludge is usually produced on a longer or different schedule. Some plants store liquid sludge in tanks and dewater it at a rate that can be accepted by the compost system. Other plants dewater over a larger portion of the day and store the sludge cake in live-bottom bins or portable containers (if the sludge is dewatered at a site different from the composting plant). Similarly, amendment is brought to the compost plant periodically by the truckload, but is used in a continuous stream only while the reactor is loaded. In this case, storage needed to match these different flow rates. Amendment storage facilities include covered piles, live-bottom bins, or silos similar to those used in the wood products industry.

In some plants, recycle is taken from the reactor at the same time as the reactor is loaded. If taken out at the exact rate that it is needed in the mix, no storage is required. Other plants provide recycle storage to increase the operator's flexibility and loosen the tight constraint that the reactor discharge must exactly match the mixer feedrate. Recycle is commonly stored in live-bottom bins or in piles.

Conveyor systems must be designed to carry a number of different materials—sludge cake, amendment, mix, compost from reactors, and finished compost. Conveyor system designs must account for the different bulk weights, moisture contents, temperatures, and other characteristics of these materials. Conveyor systems must also be designed to work in several different environments, ranging from cold, dry winter conditions outside the plant buildings, to the hot, humid environment above or below the reactors. Site restrictions and the physical design of the reactors force conveyors into a number of different physical configurations— horizontal, vertical, straight, curved, and all combinations. It is not surprising, therefore, to find a number of different kinds of conveyors at a single plant.

Typical kinds of conveyors found at in-vessel facilities include belt, screw, and drag chain. Pneumatic systems and bucket elevators are typically used for conveying sawdust.

Two kinds of mixers are commonly used. Pug mill mixers contain dual rotating shafts fitted with a series of paddles. Plough mixers contain a single rotating shaft with several radial spokes fitted with plough heads on their outer ends.

Reactor Systems

In-vessel composting is a multi-stage process, the first stage of which is accomplished in a reactor, sometimes called a "bioreactor." The detention time in the reactor varies widely from about ten to twenty days depending on system supplier recommendations, regulatory requirements, and costs. Reactor detention time should be based on desired product characteristics, especially stability, and should take into account the detention time in all process phases.

There are three general classes of reactors—vertical plug-flow reactors, horizontal plug-flow reactors, and agitated bin reactors.

Vertical plug-flow reactors are vessels constructed of steel, concrete, and/or fiberglass. The mix of sludge, amendment, and recycle is placed in the top of the reactor. The composting mix is aerated but not agitated or mixed. In theory, it moves as a plug to the bottom of the reactor. There it is removed by a traveling auger discharge device.

Horizontal plug-flow reactors are similar to vertical reactors in that the contents are not mixed within the reactor. The compost is moved through the reactor by the action of a hydraulic door.

Agitated bin reactors differ from plug-flow reactors in the fact that the compost does not move as an unmixed mass through the reactor. Instead, mechanical devices agitate the composting material periodically during its stay in the reactor. Physically, the reactors are open-topped bins with air supplied from the bottom. A variety of methods are used to remove compost from the reactors.

Aeration Equipment

Unlike materials-handling systems, which use a wide variety of conveyors, relatively few different types of air-moving equipment are used. The choice of equipment is generally based on the type of reactor. For air supply, positive displacement blowers are used where the air flow path (and consequent backpressure) is long (greater than ten to fifteen feet). Where the flow path is short, centrifugal blowers can be used. Air supply systems are designed to provide a range of flow rates. Air flow rates are varied by changing the speed of the blower motor, throttling the inlet or outlet, or operating different numbers of blowers. Exhaust blowers are usually constant-speed centrifugal units.

Odor Control Systems

Complete identification of all the odor-causing compounds in composting reactor exhaust has not yet been accomplished. It is known that ammonia and organic sulfides are among the most important compounds. It is also known that constituents of reactor exhaust vary from plant to plant. Exactly what causes this variation is not well-understood. Several techniques are used to control odors produced by dewatering, mixing, and reactor composting operations at in-vessel facilities. Some facilities have tried to control odors by diluting odorous air with outside air. Except for very weak odor sources, this has not been a successful strategy. Some facilities try to reduce odors by bubbling the odorous air through water at an adjacent wastewater treatment plant or passing it through a soil and/or compost biofilter. One or two-stage wet scrubbers are also used. Chemicals typically used in the scrubbers include sulfuric acid, sodium hypochlorite, sodium hydroxide, and various proprietary chemicals.

Not every part of a typical facility is connected to an odor control system. Odor from curing piles and product storage piles, in particular, is not usually controlled. Strong odors from these sources are intermittent, normally being released to the atmosphere only when the interiors of the piles are exposed.

Curing and Storage

Most in-vessel composting systems employ a separate curing and storage step prior to product use. Sometimes curing is carried out in-vessel; however, typical curing is done on impervious surfaces. Curing facilities are sometimes aerated either in a static-pile arrangement with blowers or turned as in a windrow facility. Sometimes covered areas are employed. Detention times for these steps vary considerably at existing in-vessel facilities. However, detention times are more related to regulatory requirements and product utilization rather than to the length of the first stage reactor. Some state regulatory agencies require three to four weeks of curing prior to product use. Storage times usually depend on the demand (especially seasonal) for the compost product.

Product Distribution and Marketing

Every potential compost market requires product quality standards involving such characteristics as the absence of pathogens, particle size distribution, moisture content, stability (oxygen demand), and metals. Examples of "high" quality uses include retail sales of bagged compost to the general public, sales to the nursery industry for potting soil, and sales to landscape contractors and public agencies for use as top dressings. "Lower" quality compost might be used as landfill cover or to reclaim strip mines.

The largest user groups of compost products are topsoil blenders, landscape contractors, and government agencies. Government uses include utilization as a soil conditioner at parks, schools, and highway right-ofways, and as cover material at landfills. Golf courses and homeowners are utilizing significant quantities of compost from some plants. Local nursery industries have generally expressed interest in the product, but typically have purchased only small quantities to date for testing purposes. Compost prices vary from approximately $2 to $20 per cubic yard, depending on local pricing policies and the quantities of compost purchased.

DESCRIPTION OF IN-VESSEL SYSTEMS

This section describes the proprietary in-vessel systems that are currently employed at facilities handling wastewater sludge in the United States.

Vertical Plug-Flow Systems

In the United States there are three major vertical plug-flow system suppliers. The primary differences among them are the configuration of the aeration systems and the discharge devices. Reactor shape, materials of construction, and ancillary equipment also vary, as illustrated in Figure 11.23. The depth of the composting materials is very similar, ranging from approximately twenty-three to twenty-six feet.

Reactors supplied by the Taulman Company are steel cylinders of twenty to forty feet in diameter capped with steel or fiberglass domes. The bottom of the reactor is usually contained in a building. Air is introduced into the bottom of the reactor. It flows up through the compost and is captured under the dome where it is drawn out by an exhaust system. The discharge device is a center-pivot traveling auger screw located in the reactor above the aeration piping. As the auger rotates horizontally around the reactor center, it draws material to a central discharge chute.

Reactors supplied by Purac Engineering, Inc. are enclosed in concrete rectangular structures. Air is introduced into the bottom of the reactor, which measures twenty-six feet wide by seventeen feet long. It flows up through the compost and is captured by a series of exhaust pipes that extend approximately three to five feet into the compost. The rate of gas exhaust is greater than the air supply rate so that some air is pulled into the compost from above as well. Because of this feature, Purac reactors are typically covered with simple metal housings that are not airtight.

The discharge device is an auger powered by end-mounted machinery. It passes horizontally through the compost through a three-foot gap at the bottom of the reactor wall just above the aeration piping. The gap is continuous around the perimeter of the reactor so that the auger can be stored outside the compost bed between uses. To prevent escape of the reactor supply air through the gap, the room housing the discharge device must be airtight. Compost is pulled to one side and dropped onto a conveyor for removal.

Reactors supplied by American Bio Tech are constructed of rectangular steel frames with fiberglass walls. The reactors, which measure twenty-six feet square, are usually housed in buildings for protection from the weather. Air is both fed into and removed from the compost through a series of air "lances"—perforated aeration pipes surrounded by a well screen. The lances hang from headers on top of the reactor and extend to just above the discharge auger. At any one time, half of the lances are supplying air and half are exhausting air. Because of this aeration system, no part of the reactor is airtight. The discharge device is similar in appearance and function to that used in the Purac reactor.

Horizontal Plug-Flow Systems

The second general reactor type is the horizontal plug-flow reactor,

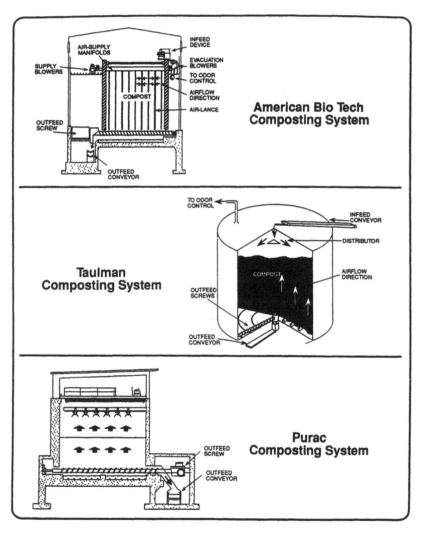

Figure 11.23 Vertical plug-flow systems.

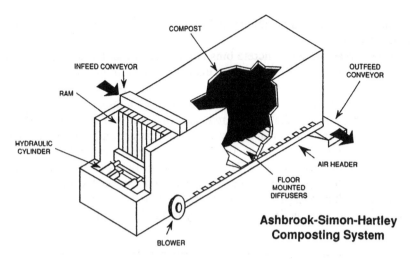

Figure 11.24 Horizontal plug-flow systems.

also known as the "tunnel" reactor because of its shape. Ashbrook-Simon-Hartley is the only supplier of this type of reactor in the United States. The reactor is constructed of concrete with a pneumatically driven steel door at one end and a collecting drag conveyor at the other end. The unit is eighteen feet wide, twelve feet deep, and sixty-five feet long. A typical reactor is illustrated in Figure 11.24. Except when loading, the steel door is positioned flush against the compost. During loading, the door is pulled back away from the compost and fresh mix is placed between it and the face of the compost bed. The door is then closed, which compresses the mix and pushes the whole compost bed longitudinally along the reactor. At the outlet end of the reactor, finished compost sloughs off the bed as it moves forward over the edge of the collecting conveyor. Air is both supplied and exhausted through slots in the concrete floor of the reactor. The aeration system is divided into seven independent zones.

Agitated Bin Reactors

Bin reactors differ from plug-flow reactors in the fact that the compost does not move as an unmixed mass through the reactor. Instead, mechanical devices agitate the composting materials periodically during its stay in the reactor. Physically, the reactors are open-topped bins with air supplied from the bottom, as illustrated in Figure 11.25. The depth of the compost bed is six to ten feet. Bin reactors can be either rectangular or circular. Rectangular reactors are supplied by Compost Systems Company (the Paygro process), International Process Systems, and Royer Industries.

Circular reactors are available from Compost Systems Company (the Fairfield process).

Paygro reactors are concrete bins twenty feet wide, ten feet deep, and up to 730 feet long. The reactors are enclosed in a building. Air is supplied from the bottom and exhausts from the compost bed into the building atmosphere. The aeration system is divided into a number of independent zones along the bottom of the reactor. The compost is both agitated and removed from the reactor by a mobile extraction/conveyor device ("extractoveyor") that spans the compost reactor and rides on rails mounted on the reactor walls. The extractoveyor operates semi-automatically, controlled by an operator in an enclosed cab on the device. It is equipped with rotating toothed drums that agitate the composting material and pull it onto an inclined table conveyor that lifts it out of the reactor.

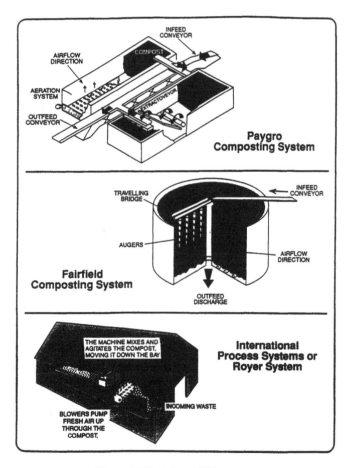

Figure 11.25 Agitated BIN systems.

In the agitation mode, the extractoveyor simply drops the compost off the table conveyor, letting it fall back into the reactor. Agitation is usually done weekly. In the discharge mode, the extractoveyor discharges the compost onto a mobile transfer conveyor that moves behind it. The reactor is loaded by a traveling conveyor similar in construction to the transfer conveyor.

International Process Systems (IPS) supplies a reactor that is similar in style but smaller in size than the Paygro reactor. Like the Paygro system, the IPS reactor is rectangular, aerated from the bottom with independently programmable zones, and enclosed in a building. The reactor is smaller, however, typically measuring six feet wide, six feet high, and 220 feet long. Fresh mix is loaded into the front end of the reactor by a front-end loader. A typical reactor is illustrated in Figure 11.25. The agitation device is also smaller than the Paygro system but is completely automatic. It operates only in the agitation mode. The agitation device makes one pass through the reactor each day. The composting material is dug out and redeposited behind the machine a short distance further along the reactor from where it was picked up. Eventually, the composting material is moved the whole length of the reactor. The finished compost is removed from the open end of the reactor by a front-end loader.

The Royer reactor is similar in size and operation to the IPS reactor. The main difference is that the reactor is supplied with either a cross section of ten feet wide by seven feet deep, or a cross section of seven feet wide by six feet deep.

A Fairfield reactor is a circular bin approximately ten feet deep that is typically covered with an aluminum geodesic dome. Its diameter can vary up to approximately 120 feet. Air is supplied from the bottom of the reactor and exhausts into the space between the top of the compost bed and the underside of the dome, where it is vented away. The aeration system is divided into five independent annular zones. A typical reactor is illustrated in Figure 11.25.

Agitation is provided by vertical augers mounted on half of a rotating bridge that spans the compost bed and rides on a rail mounted on the reactor wall. The augers extend to about six inches above the aeration piping. The augers are mounted at a slight inward angle so that, as they mix the compost, they also move it from the perimeter to the center of the reactor. At the center is a circular discharge chute that accepts finished compost from the innermost auger. Fresh mix is loaded on the perimeter of the reactor by a conveyor on the opposite side of the bridge from the augers. Agitation of the compost bed is not continuous. It occurs only during loading operations.

CASE HISTORY

The Lancaster, Pennsylvania in-vessel composting facility was placed on-line in 1981 and was one of the first in-vessel composting facilities constructed in the United States. The Lancaster facility was selected for this case history because its history is typical of the first generation of facilities. Most of these facilities suffered from early startup problems, experienced odor complaints, and underwent modifications before resuming full-scale operations.

The Lancaster Wastewater Treatment Facility was designed to process 30 MGD. The facility employs the pure oxygen A/O process for nitrification and phosphorus removal. Gravity-thickened primary sludge is blended with dissolved air flotation-thickened waste activated sludge and dewatered by belt filter presses.

The in-vessel composting system was supplied by Taulman Composting Systems Co., Atlanta, Georgia. The facility was designed for thirty-two dry tons of sludge per day at an average dry solids content of 23%. The facility consists of two 42,000 ft³ bioreactors, two 42,000 ft³ cure reactors, and one 19,000 ft³ amendment storage silo. Sludge is transported by belt conveyors from the adjacent dewatering building to sludge storage bins. Amendment (sawdust), sludge, and recycled compost are conveyed by vertical chain conveyors to one of two mixers. The mixture is then deposited into the top of the bioreactors. The compost is transferred from the bioreactors to cure reactors after a fifteen-day detention time. After another fifteen days in the cure reactor the finished product is discharged to trucks. The thirty-day system detention time was based on a volumetric mixture ratio of 1.0:0.5:1.0 (sludge, amendment, recycle). Approximately 15,000 cfm of reactor exhaust gases are ducted to a mist chamber where sodium hypochlorite is added. The composting facility had a capital cost of approximately $10 million when bid in 1984. The local share of the capital costs, after considering EPA grants, was approximately $1.5 million.

During initial operations the facility experienced materials handling equipment problems and odor complaints. In November 1988 the facility was shut down for a nine-month period while repairs were made. Repairs included new screw conveyors to replace drag-chain conveyors and odor control improvements. The major changes to the odor control system included (1) increasing the aeration rate to 24,000 cfm, (2) maintaining continuous aeration rather than intermittent aeration, (3) installing a firststage water scrubber chamber before the mist chamber to reduce ammonia concentrations and cool the reactor gases, and (4) installation of ductwork to direct conveyors exhaust and truck loading area exhaust to a second independent chemical scrubber.

With operations resumed, the facility processes all the sludge produced by the plant—approximately fifteen to eighteen dry tons per day. Dewatered cake solids range between 18 and 21%. Amendment solids average 80 to 95% and are obtained from a company that crushes and grinds wooden pallets. The initial volumetric mix ratio varies but is approximately 1:1:1. Second-stage compost product is approximately 60% solids based on a system detention time of approximately twice design estimates. The composting facility is operated five days per week, two shifts per day. One supervisor and three operators are required per shift. The city estimates the total cost for composting approximately $300 per dry ton, including the local share of capital costs. Compost product is sold to a broker for approximately $2 per cubic yard ($1 per ton). The broker sells to landscape contractors who in turn blend the compost with soil and other materials and sell the final product for $5 to $10 per cubic yard.

REFERENCES

1 United States Environmental Protection Agency (U.S.EPA). 1978. *Current and Potential Utilization of Nutrients in Municipal Wastewater and Sludge, Volume 2.* Office of Waste Program Operations, Contract 68-01-4820.

2 Logan, T. J., B. Harrison and M. D. Che. 1989. "Agronomic Effectiveness of Cement Kiln Dust-Stabilized Sludge," Ohio Edison Program, Ohio Department of Development, Final Report, Columbus, Ohio.

3 Burnham, J. C., N. Hatfield, G. F. Bennett and T. J. Logan. 1990. "Use of Quicklime and Cement or Lime Kiln Dust for Municipal Sludge Pasteurization and Stabilization with the N-Viro Soil Process," *ASTM Symp. on Innovation and Uses for Lime,* pp. 1–28.

4 Goodale, D. M., T. F. Bruetsch and J. Kowal. 1990. "Using N-Viro Soil as a Soil Amendment," Project Report, SUNY, Cobleskill, New York.

5 Gilmour, J. T. and M. D. Clark. 1988. "Nitrogen Release from Wastewater Sludge: A Site-Specific Approach," *J. Water Poll. Cont. Fed.,* 60:494–498.

6 Bennett, G. F. 1989. "Effects of Cement Kiln Dust on the Mobility of Heavy Metals in Treatment of Wastewater Treatment Plant Sludge," Thomas Edison Program, Ohio Department of Development, Final Report, Columbus, Ohio.

7 Logan, T. J. 1990. "Chemistry and Bioavailability of Metals and Nutrients in Cement Kiln Dust-Stabilized Sewage Sludge," *Specialty Conference on Sludge Management.* Alexandria, VA: Water Poll. Cont. Fed., pp. 1–5.

8 Logan, T. J. 1992. "Mine Spoil Reclamation with Sewage Sludge Stabilized with Cement Kiln Dust and Flue Gas Desulfurization Byproduct (N-Viro Soil Process)," *Proc. National Meeting Am. Soc. Surface Mining and Reclamation.* Princeton, WV: ASSAR.

9 Standards for the Disposal of Sewage Sludge; Proposed Rule, 40 CFR Parts 257 and 503, *Federal Register,* February 6, 1989.

10 National Sewage Sludge Survey, 40 CFR Part 503, *Federal Register,* November 9, 1990.

11 Peech, M. 1965. "Hydrogen-Ion Activity," in *Methods of Soil Analysis Part 2.* C. A. Black, ed., Madison, WI: American Society of Agronomy, Inc., pp. 914–926.

12 U.S.EPA. 1989. *Technology Transfer Summary Report: In-Vessel Composting of Municipal Wastewater Sludge,* EPA-125/8-89/016.

CHAPTER 12

Public Policy: The Problem of Public Relations and Acceptance

INTRODUCTION

A LL sludge management projects share two presumptions: one, that they incorporate the best available technology and are at least harmless, if not genuinely beneficial, to the public they intend to serve; two, that they will encounter opposition nonetheless.

Pardon the expression, but one whiff of any proposed sludge management project will precipitate Nimby-ism (from Not In My Back Yard) and bring objections from the very politicians and people to whom it offers a resoundingly wise waste management solution. Engineers and municipal authorities alike can count on antagonism, and, for the public's sake, should plan on doing something sensible about it.

COMMUNICATIONS SENSIBILITY

Where waste is concerned, the public is more immediately emotional than sensible. While this reaction often may be without basis, it certainly isn't without history; we are taught from birth that waste is disgusting— especially somebody else's.

Recent information has taught that some waste is indeed harmful. In fact, the same environmental awareness that gives sewage and sludge management projects a reason for being has also made "waste" synonymous with "toxic" in many people's minds.

A sensible communications plan, therefore, has transition as its goal:

Patricia Hunt, HCI, Inc., Philadelphia, PA.

to move people from a reactive response to sludge to an informed response about sludge management.

If your project is to win acceptance, it has to win it from the *public's* point of view, not yours. This simple truth eludes a great many managers and engineers who are nonplused that their project, so beneficial and attractive from *their* perspective, evokes overwhelming negative reactions from the public.

In that context, sludge—even the cleanest sludge—has some decidedly negative aspects from the public's point of view. Here are some of the forces working against sound utilization of sludge products:

(1) The human waste component—All our lives we're taught that human wastes are dirty and disgusting. Our negative feelings about our own wastes are bad enough, but the notion that we might be exposed to the wastes of others is even more offensive. Therefore, describe the transforming process from waste to useful product.

(2) Waste disposal is a loaded subject—It inevitably smacks of toxic seepage and superfund sites. In fact, even if you drop the word "waste," the "disposal" is likely to sound unattractive if what's being disposed of is headed for your backyard. Avoid both words if possible.

(3) Ignorance isn't bliss—When there's a void of information, it's likely to get filled with bad information and distorted "worst-case" scenarios. Better to fill that void with true, factual information, rather than allow fiction to fill that void.

(4) Your reputation precedes you—The less the public understands about the product itself, the more public opinion will depend on the reputation of the producer. Too often, the public agencies involved in sludge management and utilization have already overloaded public tolerances with other bad news over the years. Water departments that communicate with the public only at times of crisis—a water shortage, for example—or when there's a rate hike, should hardly be surprised when the public looks for the dark side of every announcement made by the utility. Water departments who, like Philadelphia's, have built a reputation for competence and credibility during noncrisis periods, have a distinct advantage in that regard.

REPLACING EMOTIONS WITH FACTS

Recognizing that the current public response toward sludge is characterized by emotionalism identifies an obvious propellant to action: replace the mythology with facts of benefits. Recognize also that the less the public understands about sludge, the less willing it will be to relinquish its emotionalism; and the greater the need for a credible persuader. Ideally, the authority, public agency, or engineering firm leading the project should have a good reputation laid in place long before it asks the public to accept a facility perceived as highly risky or offensive. Ideally, the project advocate will have already proven itself a true public "servant" by previous professional responses to public needs and incidents.

But if, as is often the case, the project advocate has a weak, poor, or nonexistent public image, public resistance to the project will likely remain. A well-laid and assiduously followed public communications plan can, alternatively, facilitate public acceptance by communicating effectively and often that the project advocate is a responsible entity, solidly on the side of public and environmental good.

PLANS FOR SUCCESS

Success begins with knowing the market (yes, market; public acceptance is a selling job). Instead of confronting the public as an unwieldy whole, break it into components. Identify and segregate groups and individuals who will support or oppose the project.

For each of these discrete "markets," prepare a targeted communications strategy. Opinion leaders, the news media, employees, the project's allies, the project's opponents, and the "undecided" people will all have different "hot buttons" that must be discreetly addressed. Understand that public meetings do not provide effective communications forums to most markets; plan direct mail, newsletters, advertising, and smaller meetings for more controlled communications to each group. Enlist the support of employees by keeping them well informed on the project, its rationale, and progress.

It is also highly beneficial to establish an amicable relationship with the media well before the project begins. Do so by giving them background information about sludge in general—major issues, terms and definitions, and basic technology—without the specifics of your pending proposal and its inherent controversy. Later, when they have become more knowledgeable on the subject, prepare a project-specific press kit and invite reporters to tour the facility or proposed site. Reporters often welcome supplied information that helps them fill copy columns and meet deadlines; by supplying it to them, the project advocate can better (though not absolutely) assure the accuracy of the information the public receives on the project. He or she will at least be identified as a "source" on the subject, and, possibly, a dependable resource when the reporter needs clarification of technical matters. All understanding—interpersonal and technical—helps diffuse and adversarialism.

IMPLEMENTING THE PLAN

The ultimate test comes when you actually implement your plan. Every situation is different. However, several *communications tools and techniques* are almost indispensable if you want to win over a skeptical or critical public to a waste management project:

- *A written project description* strategically addresses anticipated public concerns. This becomes the "bible" for communicating about the project. It provides a detailed description of the project that will serve as a consistent story for everyone involved on your team. It might even contain a special glossary of terms for in-house use and external explanations.
- *An introductory booklet* draws on the project description. This is an excellent place to speak about the sponsoring organization's expertise and commitment to environmentally responsible technology and management policies. Don't make the booklet excessively long, but do make sure it tells everything that needs to be told. Interest is likely to be high if your project becomes controversial, so length won't dissuade people from reading most—or all—of the contents.
- *An audiovisual program* can present information explaining the usage process and its environmental and economic benefits in a memorable and convincing fashion. It should address feelings as well as facts, and should *not* contain complex charts and graphs.
- *A sponsor's profile* outlines the knowledge and experience that will ensure that the job is done effectively and safely. Thus, the profile should include a description of the sponsor's record in addressing pertinent environmental problems, its professional credentials, its history of responsiveness to public concerns, and its specific plans for staffing and managing the project.

LET OPPOSITION LIE

One of the most practical requirements of an effective communications campaign on a sludge project concerns budget: don't squander it on the opposition. Consider this: most sludge projects encounter "border people"—that subset of the Nimby population whose homes and businesses are adjacent to the project site or to the roads on which sludge products are transported at either end of the treatment process.

Accept as a given that border people rarely trade their emotions for rationalism; waste neither the project's planning and implementation time, communications budget, nor its manager's energy attempting to placate border people beyond the requirements of the project's siting or operating agreements. Remember that this audience represents a verbal minority.

ADDRESS THE "SWING VOTE"

As in a political election, it is the "swing vote"—the undecided population—that can sink or propel a project. The key to enabling the swing vote to relinquish their emotionalism regarding sludge is to counteract the credibility of the Nimby's arguments. Publicly prove that their claims are founded in emotion, and gain reasonable support from respected public opinion leaders.

While addressing the individual market segment's concerns via targeting communications, use mass media to communicate the fact that the project will benefit 99% of the community; this fact is reassuring to most people and helps mobilize a general consensus. In addition, it provides politicians comfort to know that a large number of constituents support their position.

DEVELOP AN ACTION PLAN

Another key to retaining the viability of a sludge project is to write and follow a flexible action plan. This plan should have a specific timetable and clearly define roles of authority and responsibility for those involved in the project.

Granted, on one level, the specifications contain every justification for the project. This language and approach confuses and alienates the average citizen, however. Project advocates will be more persuasive with communications that make the facts easy for the lay public to understand.

On the other hand, if the project requires acceptance by users, potential buyers will be skeptical of the product if information about it is insufficiently technical. With enough specs to fortify their confidence in the project, horticultural groups, farmers, and engineers engaged in land reclamation can become important proponents of the project.

As with any potentially controversial idea, balance is an important part of how the information is presented. But it goes without saying that communications should be consistent in tone and manner within each audience, and in all cases must be consistently truthful.

IDENTIFY A SPOKESPERSON

The action plan should also identify a single spokesperson for the project to answer questions authoritatively and credibly, and prevent contradictions that might weaken the project's chances of acceptance. Training is available to teach this spokesperson how to effectively represent the project and the organization advocating it. Some key presentational tactics include:

(1) Answer all questions raised, and do so boldly and publicly.

(2) Don't argue with people who ask sensitive questions. Answer, and then elaborate with accepted generalities.

(3) Never reiterate wrong information by phrasing the response to it in a way that repeats the charges, either vocally or in print.

(4) Don't lend credence to the project's opposition by meeting with them in private.

Project spokespersons should also learn this tactic for diffusing hostile questioning at public meetings: after the questioner has spoken, look away from him or her and address the response to the entire audience. This will encourage the hostile one to sit down. The audience will then focus on the spokesperson and be better attuned to his or her words.

COMMUNICATIONS AS AN OBLIGATION

There is no doubt that developing and implementing a public communications plan is a great deal of work—on top of the arduous engineering, administration, and political tasks that typically accompany sewage treatment and sludge management projects.

But communications is important work, because what the public doesn't know about sewage, sludge, incineration, air pollution, and odor, they will fill in with their own fears and myths.

Ignorance can delay public acceptance, which will delay the project in turn, and which can add millions to the bottom-line costs of ventures that already seem costly. Both public officials involved in sludge management and the engineering firms who implement these projects have an obligation to keep them viable for the sake of the environment and the public purse, and to make them understood (any taxpayer's due). It is the project advocate's right to be understood as a benefactor—to tell the public that they are, indeed, being well served by an important and understandable project.

CONCLUSION

Failure to obtain public understanding, and hence public acceptance, of a sludge management program has been known to seriously delay and even scuttle such projects. Public communications, therefore, should be as important an issue as any technical consideration, in planning a sludge management project.

CHAPTER 13

Worldwide Sludge
Management Practices

WESTERN EUROPE—A 1996 PERSPECTIVE

THIS chapter is intended to describe current practices of sludge treatment and disposal in Western Europe. There is a disparity of experience from country to country that reflects cultural and economic history. The European Union (EU) legislation is having a great effect on policies and practices, which both is a reflection of, and an encouragement to, the "greening" of public and political attitudes. This chapter describes the practical consequences of this for sludge disposal managers, particularly in terms of future investment and operations. Practical examples will be given from the United Kingdom and Anglian Water Services in particular, which serves Eastern England and is the largest operating subsidiary of the Anglian Water Group.

INTRODUCTION

The EU has a membership of fifteen countries approaching 370 million residents. Western and central Europe have a number of other countries, Viz., Norway, Switzerland, Hungary, Czech Republic, Slovakia, and Poland, that face similar social and environmental problems. These vary in dimension according to economic and historical circumstances. However,

James W. Bradley, Royds Garden Ltd., Dunedin, New Zealand; Shunsoku Kyosai and Kazuaki Sato, Ministry of Construction of Japan, Tsukuba-shi, Ibaraki-ken, Japan; Peter Matthews, Anglian Water Services, Cambridge, England; Mel Webber, Environment Canada, Burlington, Ontario, Canada.

the principles of environmental management and the techniques that must be employed will be the same for all of these countries. These are clearly demonstrated in the challenge of sewage sludge production and disposal. The EU has made progress in dealing with municipal wastewater in individual countries and as a corporate entity. However, it intends to make still further and substantial progress over the coming years. This will inevitably mean much more sludge to be disposed of. The EU makes legislation and then requires member states to introduce this in national policies, practices, and statistics. To understand this, it is helpful to review the workings of the EU.

THE EUROPEAN UNION

Overall Environmental Policy

The EU takes the view that pollution knows no borders and that polluted air and water can circulate freely throughout the continent. The EU can be a forum for tackling many of these problems, as it represents a "happy medium between a national framework, which is frequently too narrow, and deliberations on a world scale which cannot produce binding measures." Furthermore, when international environmental matters are being planned, the EU takes the view that the fifteen member states will have more influence when speaking with one voice.

The EU is also concerned about the differences in living and working conditions of citizens in different member states. Differing national environmental policies are considered to cause economic disparities that could interfere with the functioning of the "common market," for example by causing distorted competition due to the inequality of business costs. The completion of the internal market in 1992 made the finalization of a proper environmental policy an even more important objective for the EU.

It is considered that one of the most difficult tasks facing the EU in the next two decades will be to manage and improve both the environment and the living standards of the human population without causing either to deteriorate. The EU defines its overall objectives for the environment in five year action programs. The present, and fifth, is in the middle of its term. It has several important ingredients that aim to make environmental policy an essential element of all economic and social policy.

Structure

The EU is comprised principally of a Commission, Council, Parliament, and Court of Justice.

Commission

The Commission is more than the bureaucratic administration of the EU and has many independent powers. It does report to the Council. Only the Commission can propose legislation to the Council, which may adopt the proposal after any suitable modification. The Commission also has the power to take member states before the Court of Justice for failure to carry out obligations, such as the implementation of the Directives.

The Commission itself is made up of Commissioners, each member state providing a defined number; the U.K. has two Commissioners. The Commissioners' allegiance is to the Commission and hence it is quite feasible for a Commissioner to be sponsoring legislation not supported by the interests of his or her home state. Each Commissioner has his or her own private cabinet, but the Commission is supported by Directorates-General. The most important Directorates-General for environmental matters are Environment, Consumer Protection, and Nuclear Safety (DG XI), Science Research and Development (DG XII), and to a lesser extent Internal Market and Industrial Affairs (DG III). The majority of staff within the Commission are permanent, but member states may second staff for periods of time.

Council

Although this is comprised of Foreign Ministers, when particular subjects are being discussed such as the environment, it is normally the relevant government minister who attends. The Presidency of the Council lasts for six months on a rotational basis; when a member state takes over the Presidency it sets the agenda for Council meetings. Hence, if a state is keen to progress a legislative proposal, it will include it on the agenda; if it is not keen it will not. The Council has working groups on various subjects, such as the environment, to discuss proposals. The groups are composed of permanent representatives from each state, who will be supported is discussion by relevant expert officials. Each state maintains a Permanent Representation in Brussels and a committee of these (CORE-PER) prepare matters for Council agendas, and service the working groups, etc.

When a proposal is first made officially by publication as a COM document in the *Official Journal of the Communities,* it is passed by the Commission to the European Parliament and, in certain cases, to the Economic and Social Committee for their Opinions. The Council cannot adopt a proposal until the Opinion of the European Parliament has been given. While a Commission may modify its proposals as a result of these Opinions, it is not bound to do so.

In negotiating within the Council on proposals, the governments of the member states may seek the views of their own democratic systems. Often special government committees may well consider the proposals of the Commission in detail and call national expert witnesses. In this way the national representatives engaged in negotiations within the Commission are properly briefed.

Parliament

This is made up of representatives elected directly every four years. It does not have the right to initiate legislation, but must be consulted before proposals are adopted. In order to achieve this the Parliament has a number of committees, one being on the Environment, Public Health, and Consumer Protection. The reports of those committees are debated by the Parliament before the resolution is adopted for the Opinion. Members of the European Parliament may ask questions and these, with answers, are published in the *Official Journal of the Communities.*

Economic and Social Committee

This is a separate committee within the EU representing employers (one-third), trade unions (one-third), and other interests such as local authorities, consumers, etc. (one-third). It has the right to produce opinions on proposals.

Court of Justice

The judges of the Court are appointed by agreement with the member states. The Court acts as a European forum for legal matters and hence can be important in interpretation of directives, once these have been agreed. Disputes between the constituent parts of the EU can be resolved by the hearing, and one important feature is that the Commission can bring cases against member states for failure to implement Directives.

Legislation

Member states who belong to the EU are signatories of the Treaty of Rome. The Treaty provides for regulations, directives, decisions, recommendations, and opinions. Directives and, to a lesser extent, decisions and regulations, have been relevant to water quality legislation so far. A directive defines a technical requirement, but leaves member states to effect them through national legislation, whereas regulations apply directly. In the case of the U.K., environmental requirements are effected principally through the Water Resources Act and Water Industry Act of 1991. Further legislation on the environment is being implemented through the Environment Protection Act of 1990.

The Directives have a "lead in" period during which time the member states must ensure that there is adequate legislation to ensure implementation. Environmental Directives are binding on member states because of the provision of Article 100 (approximation of laws affecting the Common Market) and Article 235 ("general powers" of the Treaty of Rome). The Treaty of Maastricht has added more requirements. Many organizations may attempt to initiate ideas for the Commission to make proposals. However, the most important sources will be the Commission, Parliament, and Economic and Social Committee. Member states may also raise an issue, and this can be done indirectly by proposals for national legislation which are conveyed to the Commission via the Information Agreement (the Commission's proposal would then take precedence).

Directives relate to the environmental use in which water quality criteria are defined (e.g., use of water for bathing) or a disposal activity (such as waste at sea or titanium dioxide) or for a specific substance in the aquatic environment (e.g., cadmium). The Directives, each related to specific substances taken from a list of priority chemicals, defining what is an acceptable discharge into the environment.

Such definition is achieved by specification of uniform limit values or environmental quality objectives. The "use" Directives and EQOs are consistent with the traditional U.K. approach to water pollution control, the aim of which is to provide water that is wholesome, i.e., fit for the uses required of it. It can be argued that to have both the "use" Directives and EQOs is unnecessary, but the latter is perceived as being important within the context of the general protection of the aquatic environment irrespective of use (which may have special requirements). The nature of the parent "dangerous substances" Directive allows EQOs to be promulgated for any substances and hence to be more flexible.

Several adopted and proposed Directives directly or indirectly relate to sewage sludge disposal and reference will be made to these throughout this text.

The Commission is kept informed of progress on implementation and compliance with the requirements of the Directives by reports produced by the governments of member states.

Research

Environmental protection features prominently in the EC's research program. Topics have included trace organics in water and the treatment disposal use and characterization of sewage sludge have featured in the past. COST 68 lasted from 1972–1990. It has involved countries from outside the EU.

CURRENT SEWAGE SLUDGE DISPOSAL IN WESTERN EUROPE

The collection and treatment of sewage has a long history in western Europe—although the extent of public service varies from country to country. Statistics on such matters are difficult to collect and verify—however the best available data on population and water supply, sewerage, sewage treatment and sludge disposal are given in Table 13.1 [1]. This is taken from the Global Atlas [1], which draws on a number of useful sources such as a report by Lindner [25] and an excellent survey of sludge production, treatment, quality, disposal and reuse by the U.K. Water Research Center for the European Union [26].

Each country has long-term plans to extend or improve the services provided, but the speed of progress has varied according to political priority and availability of money for infrastructure investment. There is no doubt that the oil crisis of the mid 1970s with its inflationary effects starved investment.

However, there has been a "greening" of social and political attitudes and investment has again speeded up. Such a quickening of intention is

TABLE 13.1. Populations Served and Sludge Production (as disposed) in the European Community in 1991–92.

Member State	Total[a] Population (millions)	Population Connected to Sewer (%)	Population Connected to STW (%)	Sludge Disposed tds/y	Sludge Disposed gds/p/d
Belgium	9.9	70	28	59,200	58
Denmark	5.1	93	92	170,300	99
France	56 9	65	50	852,000	82
Germany	79.7	89	83	2,681,200	111
Former West	*62*	*92*	*90*	*2,449,200*	*119*
Former East	*17*	*77*	*58*	*232,000*	*64*
Greece	10 2	45	34	48,200[b]	40[b]
Ireland	3.5	67	45	36,700	64
Italy	57 7	75	60	816,000	65
Luxembourg	0.4	97	87	7,900	62
Netherlands	15.0	97	88	322,900	67
Portugal	9.9	52	20	25,000[b]	35[b]
Spain	39.0	70	59	350,000	42
United Kingdom	57.5	96	85	1,107,000	62
Total (mean)[c]	344 8	(79)	(66)	6,476,400	(78)

[a]Population data for 1991 (Eurostat 1992).
[b]Upper estimate.
[c]Weighted means in parentheses

reflected in the Directive adopted by the EU which will result in virtually every area within the EU being served by sewage treatment facilities of one kind or another by just after the turn of the century. However, changes in economic activity in member states will influence the success in attaining this goal. The Directive is discussed later.

Of course the longer established the infrastructure, and indeed the administration of water management, the more difficult it is to deal with new challenges. An established system needs refurbishment and extension to cope with increased demand as well as tightening environmental targets, and in many ways this can be more difficult than providing a new system.

Given a blank sheet of paper, the manager and designer would often "prefer not to start from here." An established system of management can also represent an entrenched set of attitudes which find adaptation difficult. However, once wastewater management is worked into a culture, there is an appreciation of the challenges involved, and one of the problems encountered is that such cultural attitudes are not developed in some regions and the challenge of dealing with wastewater treatment and sludge disposal may lead to inappropriate decisions particularly at political levels. For instance, the cultural jumps from localized use of night soil in agriculture to a fully developed and supported service for using sewage sludge over a large region may be too much—but it is difficult to make generalizations.

The environment has become a political issue in Europe during recent years. A symbiotic partnership has formed between various interests. New lobby and pressure groups have become political parties, the critical enquiry role of the press has now enhanced the environmental issues, and the overall result has been to encourage the established social system to move towards the green part of the political spectrum. Where sewage is concerned, the responsible authorities have to cope with a deep-seated public antipathy. This can be called a "fecal aversion barrier" and arises from legitimate and essential training given to young children on personal hygiene so as to assist in the prevention of epidemic diseases. It is, therefore, important that managers understand this aspect and respond in an open and sympathetic way to public complaints. It is vital that public confidence is maintained.

To reflect this public and political interest, the EU has taken action with Directives to provide a framework for politicians and managers to act within. Directives have been adopted on the agricultural use of sewage sludge (1986), on groundwater quality (1980) that affects landfills, on urban wastewater treatment (including the prohibition of marine dispersal of sludge) (1991), and on diffuse sources of nitrogen in water (1991). The Directive on incineration of non-hazardous waste (1980) is also relevant. It had been hoped to complete the Directive on Landfills in 1996; however, this was not possible and the commission is considering now how to tackle this issue. It is likely that it will be included in global strategies relating to waste disposal.

The essential point is that due to social and political pressures, sludge production in Europe is going to rise significantly over the next decade. The implications of this are enlarged upon below with particular examples from the U.K. (water companies in England and Wales) and Germany, the two largest sludge producers. In order to provide an understanding of the future sludge production, a summary of the Urban Wastewater Directive is given below.

FRAMEWORK OF AN URBAN WASTEWATER TREATMENT DIRECTIVE

The Directive sets minimum standards which may be improved upon in local circumstances if the environment requires it. The principal requirements of the Directive with respect to sewage treatment are as follows. By dates, varying according to size of works and location, of December 1998–2005 the following degrees of treatment must be provided for municipalities, of 2,000 population equivalent (pe) or more, discharging to fresh water and estuaries and municipalities of 10,000 pe or more discharging to coastal waters.

Standard Treatment

Secondary treatment must meet a standard of 35 mg/L suspended solids and or 25 mg/L BOD (inhibited with allyl-thiourea) and 125 mg/L COD as annual 95 percentile values with specified maxima, but with exceptional samples excluded. An equivalent system of percentile removal may also be used. Monitoring requirements are specified. For high altitudes, the provision of biological treatment may not be appropriate or practical but this is subject to the demonstration of no environmental effects.

Sensitive Areas (Particularly Those Vulnerable to Eutrophication)

The directive sets minimum standards which may be improved upon, in local circumstances, if the environment requires it. Treatment is to be given as the standard treatment as a minimum, but in addition phosphate and total nitrogen will be reduced to specified levels as necessary. These areas were to have been defined by December 1993. The need for treatment at an individual works in a catchment may be assessed against an overall requirement for at least 75% reduction in nutrients.

Less Sensitive Areas

Where comprehensive studies indicate that a discharge would not adversely affect a coastal marine or estuarial environment then primary treatment only is acceptable for those waters for defined sizes of discharges. Primary treatment is defined as settlement or other processes in which BOD is reduced by at least 20% and solids by at least 50%. These areas were to have been defined by treated by secondary processes unless it can be demonstrated that there would be no environmental benefits from the provision of such treatment.

Provision of Facilities

Collection systems should be provided in accordance with best practicable techniques not entailing excessive costs and be designed to limit pollution from storm sewage. Treatment works must be designed for a loading of 60 g BOD per person per day. Discharges of lesser size than those defined above should have "appropriate treatment" by the end of 2005.

Industrial Discharges

Industrial discharges equivalent to sewage of more than 4000 pe are specified and must be treated by the end of 2000.

Sewerage

Adequate collection systems for municipal wastewater must be provided by end of 1998–2005 for all municipalities where the concentration of habitation makes it practical for such systems to be provided. Treatment must be given to all practical and acceptable rates of flow.

Sludge

The Directive also requires that the disposal of sewage sludge to sea should cease by December 1998. In the interim there is to be no increase in volume and a progressive reduction in any toxic persistent and bio-accumulable constituent. Sewage sludge used in agriculture must conform with the provision of the EU Directive passed in 1986.

Others

There are several other requirements. Industrial discharges to sewer should have been authorised by competent authority by December 1993 and reviewed regularly. Finally, the programs should have been defined by December 1993.

INCREASED SLUDGE PRODUCTION

The current investment program and that arising from EU regulations will create substantial increased quantities of sludge for disposal. A survey of EU practices in 1984 by the U.K. Water Research Centre still gives some idea of the changes likely to occur in the next few years [3]. Some populations are expected to rise. By 2020, that of France and the U.K. will have risen to 58.5 M, Ireland to 4.1 M, that of Germany was reported to be likely to about 51 M [4]; however, this does not seem to be happening as Table 13.1 shows. Table 13.2 gives the best information on sludge disposal [1]. However such data are notoriously inaccurate, particularly where the provision of sewerage services is limited; they do indicate the magnitude of activity in each country. Good sludge measurements are essential. However, the comparison of the 1986 survey and the information in Table 13.2 demonstrates the continuing rise of sludge production as expected.

In the U.K. the total impact of the Urban Wastewater Treatment Directive and growth over the period of its implementation will be to cause U.K. production to rise to about 1.5 million dry tonnes by 2006. With the loss of marine dispersal, this means that the demand for land-based disposal methods will double by that date.

Data have been produced that will give some idea of how the increase in environmental standards will affect sludge production. Raw primary sludge is produced in the U.K. at about an average of 52 g/d per person, co-settled activated sludge, 74 g/d per person, and co-settled activated tertiary sludge, 76 g/d per person. If nutrient removal is added, sludge production may be even higher [6].

Many sewage treatment works receive industrial effluent. Where these contain organic waste they will contribute to raw sludge production. Water managers controlling pollution may prefer to have industrial wastes received at central treatment works managed by professions, as opposed to treatment within individual factories and discharge to environmental waters. In these simple terms, it is cheaper for people to use the public sewage systems because the cost of meeting the more stringent treatment standards for discharge to environmental waters will be higher. If there is a charging system for pollutant loads, for discharges to sewers and on to environmental waters this may influence the calculations; the overall net effect is likely to be a declining load to environmental waters and sewage treatment works as a result of better waste management regimes in factories.

An interpretation of the changes occurring is made more difficult by the cycles of decline and change and resurgence in industry, which vary, but observations from circumstances in the U.K. suggest that food processing is

TABLE 13.2. Yield and Disposal of Communal Sewage Sludge in Europe (as at 1992).

	Quantity	Agriculture	Landfill	Incineration	Sea	Other (eg. recultivation, forestry)
			[1000 tonnes dry matter/year (%)]			
Austria	170 (2.3)	30.6 (18)	59.5 (35)	57.8 (34)	—	22.1 (13)
Belgium	59.2 (0.8)	17.2 (29)	32.5 (55)	8.9 (15)	—	0.6 (1)
Denmark	170.3 (2.3)	92 (54)	34 (20)	40.9 (24)	—	3.4 (2)
Finland	150 (2)	37.5 (25)	112.5 (75)	—	—	—
France	865.4 (12)	502 (58)	233.5 (27)	130 (15)	—	—
Germany	2681.2 (36)	724 (27)	1448 (54)	375.2 (14)	—	134 (5)
United Kingdom	1107 (15)	488 (44)	88.6 (8)	77.4 (7)	322 (30)	121 (11)
Greece	48.2 (0.6)	4.8 (10)	43.4 (90)	—	—	—
Ireland	36.7 (0.5)	4.4 (12)	16.6 (45)	—	12.8 (35)	2.9 (8)
Italy	816 (11)	269.2 (33)	449 (55)	16.2 (2)	—	81.6 (10)
Luxembourg	8 (0.1)	1 (12)	7 (88)	—	—	—
Netherlands	335 (4.5)	87 (26)	171 (51)	10 (3)	—	67 (20)
Norway	95 (1.3)	53.2 (58)	41.8 (44)	—	—	—
Portugal	25 (0.3)	2.7 (11)	7.3 (29)	—	0.5 (2)	14.5 (58)
Spain	350 (4.7)	175 (50)	122.5 (35)	17.5 (5)	35 (10)	—
Sweden	200 (2.7)	80 (40)	120 (60)	—	—	—
Switzerland	270 (3.6)	121.5 (45)	81 (30)	67.5 (25)	—	—
Total	7387 (100)	2690.1 (36.4)	3066.2 (41.6)	801.4 (10.9)	380.3 (5.19)	447.1 (6)

The figures are in general agreement with those of Saabye et al. [5].

likely to increase in the future and this will probably result ultimately in more sewage sludge. The factors discussed previously are related to raw sludge production. The treatment and disposal of sludge are related intimately. The treatment of sludge is necessary to achieve environmental goals and to reduce overall costs. Some processes actually increase the quantity of sludge for disposal, such as sludge conditioning and dewatering using inorganic coagulants, whereas other processes decrease it, such as an anaerobic digestion. These are some of the factors influencing the choice of treatment processes.

The introduction of polyelectrolytes and dewatering techniques such as filter belt presses, centrifuges, and membrane pressure filters, have reduced the burden of conditioner chemicals, such as lime and ferrous sulphate. However, it may be necessary to add a small amount of lime to control odors. This is important, as dewatering may become more commonplace as sludges have to be transported greater distances and as incineration may become favored in the future. The development of prefabricated digesters has extended the application of digestion to much smaller treatment works.

From the experience of Anglian Water, much less sludge is dewatered than used to be the case. Dewatering of polyelectrolyte conditioned sludges using belt presses or centrifuges is preferred. This experience is normal elsewhere in the U.K.

Much higher proportions of sludge are dewatered on the continent. It must be emphasized that better management systems require better data, and practical observations suggest that where sludge movement is measured in tanker loads there is a tendency for over-estimation. In other words a tanker can go out under-full but not over-full; so where the total volume is an aggregate of tanker volumes this will be greater than is the volume of the sludge actually present. This problem will be corrected in the future by better automatic measuring devices involving volume and solids removed or discharged, probably connected to computer loggers and even telemetry systems connected to central monitoring computers. It may well be that the amount of sludge predicted as a result of new sewage treatment services may not be as much as predicted because of a combination of all these factors.

SELECTION OF OPTIONS

European waste management policy puts avoidance and minimization first. Avoidance is not possible, but minimization can be achieved with suitable sewage and sludge treatment methods.

The most widely available options in Europe are:

- agricultural utilization
- land reclamation and restoration
- other beneficial uses

- incineration
- marine dispersal
- landfill

The selection of an option on a local basis reflects local or national cultural, historical, geographical, legal, political, and economic circumstances. Pity the waste disposal manager! The degree of flexibility varies from country to country. In the U.K. the choice is made using a practice of common sense now codified in the policy of Best Practicable Environmental Option [9].

Each of the options have criteria for environmental protection to varying extents that must not be exceeded. The observance of these criteria will cost money because of the need to control sludge quality and disposal practice. In planning and implementing a disposal operation, the costs for each option may be calculated and compared. The comparison may be made on the basis of capital and future operational costs discounted to present values. The most favorable option is selected on this basis. This rigorous examination is often reported for the purpose of management in the form of a project appraisal which examines all aspects of the problem, identifies the preferred solution, and provides the outline ready for a design brief. Computer programs have been developed to evaluate these processes—a good example being WISDOM (Water Industry Sludge Disposal Optimization Model) [9].

This approach is encouraged by the Urban Wastewater Treatment Directive, which requires that sludge shall be reused whenever appropriate and that disposal routes shall minimize the adverse effects on the environment. Of course, some options are not examined in detail because prior knowledge or common sense already indicates that the impracticality or high cost of an option preclude the need for further effort. For instance, for a small town in central England, almost a 100 kilometers from the nearest port and surrounded by agricultural land, marine dispersal of the sludge from the sewage treatment works would not be competitive. Equally, some options may not be available for national and regional political reasons. For instance, as explained earlier, marine dispersal will no longer be an option to disposal managers from 1998.

Sludge treatment is considered as "a means to an end." That is, the treatment renders the sludge cheaper to transport and/or more acceptable to the receiving environment. Sludge treatment and disposal should always be considered as an integral part of treatment of wastewater. Comprehensive assessment should embrace the whole process stream. This makes the project appraisal for a new works more difficult, and computer models may be necessary. The danger is that circumstances for sludge disposal are more likely to change over long periods of time. It must also be remembered that the primary purpose of the sewage treatment works is to produce an effluent of a quality acceptable to the receiving waters. For this reason there is a tendency to deal with sewage treatment separately, mindful of the consequences for sludge disposal, and then to deal with sludge disposal as a separate project. Whatever, it must be remembered that, on average, up to almost half of total operational costs for sewage treatment are for sludge treatment and disposal—although this will vary very widely according to local circumstances.

These matters are of interest to managers of individual sewage treatment works, but this management strategy becomes of real benefit when a large region, with many sewage treatment works, is being managed. In this context, water authorities in England and Wales developed the concept of sludge treatment centers that have been inherited by water companies that were established by privatization of the authorities in 1989. Cooperative and regional operations exist elsewhere in Europe, for example the Verband in Germany.

The principal operational options for disposal in practice may be broadly classified into:

- sludge treatment, if any, at an originating works and agricultural use (the use may be local for small works or over fairly substantial areas for large works)
- sludge treatment at a central works and agricultural use
- sludge treatment, if any, at a central works and marine dispersal (some nearby inland works may send sludge for co-dispersal)
- sludge dewatering at a central site and incineration
- sludge treatment, if any, at a central works and disposal to landfill or sacrificial land

The effects of scale are similar in respect to sewage treatment and sludge treatment but, because of the differences in volume, the points at which centralized treatment becomes economical are different. In the Anglian Water Group, the utility that serves eastern England, the 1980/ 81 [6] survey showed there are more than 1100 public sewage treatment works, but sludge is only likely to be taken directly to agricultural land at any time of year from a fraction of these works. Treatment of some sort will only be provided at a very small fraction of works but deal with a high proportion of wastes.

Whereas sewage treatment for populations of less than 500 may be viable, full-scale sludge treatment will frequently not be justified for population equivalents of less than 10,000. In rural areas, it is not uncommon for up to 50% of the sludge treated at works to be tankered in from smaller public works, and private septic tanks, etc. The selection of the site for centralization is particularly important, the cost of tankering varying with the nature of the road network and the density of the works served.

Considerable economies can be achieved, however, by programming to provide for some return loads, primary sludge coming in and treated sludge for disposal going out.

Transport and dewatering costs are linked closely, and geographical location is important in determining the option adopted. The maximum economic radius for liquid sludge disposal varies from location to location but generally does not exceed 15–20 km. Beyond that, it is often most cost-effective to dispose of dewatered sludge. An excellent example of centralization was reported for Cambridge Sewage Treatment Works. By replacing the six filter presses used to dewater digested sludge with one centrifuge, the throughput increased from 180 m^3d^{-1} of sludges from fifteen satellite works from as far away as 30 km [10]. Thickening is one of the most effective ways of extending the economic radius of liquid sludge disposal.

Past studies [11] identified that, while up to half of operational expenditure in the U.K. is consumed by sludge treatment and disposal, only about 15% of capital had been invested in this activity. An analysis of capital and revenue expenditure in sludge treatment and disposal suggested that savings of about 10% on operational costs might be achieved by reallocating the capital investment but not increasing it.

It is essential that the net present value (NPV) of sludge management in the broader context of waste-water management is used to evaluate capital and operational options. Beyond NPV is the even more important understanding of impact on profits and/or customer charges. Simple thickening, modular prefabrication of plants, digestion must all be assessed for the environmental as well as cost benefits. These concepts are still being pursued by the water services companies in England and Wales.

All sludge disposal options are under scrutiny and there are problems—technical, financial, and political—to be faced. The subject of this section is how the water companies, in particular, have found solutions to the challenges of acceptable cost-effective disposal practices.

In order to avoid conflicts between the objectives and constraints of each disposal option, a U.K. national policy-making Standing Committee on the Disposal of Sewage Sludge was formed in 1975. This was serviced by the Department of the Environment (DoE) and National Water Council (NWC) but had representatives from a wide range of interests. It acted as a forum for discussion on environmental protection, operations, and economics for disposal activities including agricultural utilization. It produced many reports and guidelines on all aspects of sludge disposal. With the implementation of the EU Directive on the agricultural use of sewage sludge, this Committee was replaced by DoE by an advisory group. This group had representatives from interested parties including the water industry. Its advice was taken into account in a review of existing guidelines for good practice. A Code of Practice came into force in the autumn of 1989 [12]. Since then, reviews have recommended changes [13,14].

AGRICULTURAL USE

General Considerations

There have been national guidelines in the U.K. since the early 1970s and in a comprehensive form since 1981, hence there was not a significant problem in implementing the 1986 EU Directive on the use of sewage sludge in agriculture [15].

The guidelines were not mandatory, but were relevant to statutory consideration of the impact of operations. However, it has been necessary to introduce regulations implementing the Directive [16]. The guidelines have also been updated to take into account the Directive, operational experience, and research in a "Code of Practice." The regulations and code were published in the autumn of 1989. The code extends the regulations to include additional constraints, and it has the same nonmandatory statistics as the former guidelines.

Similarly, there have been guidelines and statistics in many other European countries, such as France and Germany since 1982, the Netherlands since 1980, and Switzerland since 1980 [17]. These have been summarized in a number of texts. It has been necessary to modify these or introduce new laws to reflect the EU Directive.

In considering the utilization of sludge in agriculture, the following factors are important:

- chemical quality—Limits are set for the receiving soil to prevent phytotoxicity (loss of crop yield) or poor crop quality by virtue of uptake of chemicals, particularly metallic ions, which are thus accumulated in the food chain with serious consequences for human beings. In addition, care has to be exercised, so that surface and groundwaters are not contaminated.
- microbiological quality—While limits cannot be set out on the sludge or receiving soil for species of pathogens, reference can be made to the nature of the sludge (including contributing trade effluents), treatment afforded, and the nature of the receiving situation in the definition of acceptable practice, so that the spread of human, animal, and plant disease is prevented.
- aesthetic quality—This relates to the smells and visual aspects of the operations, and again the definition of acceptable practice takes account of these.
- acceptability and marketability—The usefulness of sludge to a farmer and hence its reputation is important. The requirements of a

farmer will vary from area to area and between types of agriculture. Fertilizer value (NPK), humus value, lime content, and sometimes even water content are factors that interest the farmer, although nitrogen is usually of greatest interest. The sludge must be applied in a manner that is satisfactory to farmers.

Before exploring these matters on a European basis, it was worth noting that the WHO produced reports on the microbiological aspects in 1981 [18] and chemical aspects in 1984 [19]. These were presented to the European Commission in 1985 [20], thus providing an intercontinental perspective.

Recommendations of the World Health Organization

In 1981 and 1984 the Regional Office for the Europe World Health Organization (WHO) convened two Advisory Working Groups to discuss and report on the risk of sewage sludge applied to land. One dealt with the risks of pathogens, and the other dealt with the risk of chemicals present in sludge. In both instances the Working Groups drew heavily on European and North American experiences. This chapter summarizes the main elements of the two reports.

Providing adequate sanitary facilities is a major responsibility facing communities throughout the world. Hence the WHO had a legitimate and extensive role in providing support and guidance on sewage treatment as an important contribution to the good health of Europe and the rest of the world. The Working Groups were charged with consideration of agricultural utilization and similar disposal methods, such as vegetable plots and gardens.

The Working Groups recognized that sludge may be treated prior to disposal in order to reduce disposal costs and to render it more suitable for disposal. Hence, the choice of treatment, if at all, depends on the acceptability of the product and economics.

Principles of Safe Utilization Practice

If sewage sludge is to be disposed of satisfactorily by use as a fertilizer or soil conditioner, it must be wanted, even needed, by customers, particularly farmers. It is a source of nutrients, particularly nitrogen and phosphate, organic matter, lime (especially where this has been used in the treatment of the sludge), and even water. The services provided by the disposer must fit in with good agricultural practice to the extent that there is mutual benefit to disposer and farmer. Sludge only has a modest contribution to make in terms of national fertilizer needs. For instance, in the U.K., with about 80% of the population served by full sewage treatment, if all the sludges were to be used for their nitrogen and phosphate content, only about 5% of N and 10% of P_2O_5 maximum of national needs could be supplied. The inorganic fertilizer industry has, therefore, nothing to fear from this practice. However, the contribution on a local basis around sewage treatment works can be substantial.

Farmers and communities will want to be reassured that the use of sludge is safe. Even with good practice, this presents a difficult public relations problem for disposers because necessary hygiene training at an early age causes a psychological aversion to all fecal matter. This aversion is strengthened by the fact that sewage sludge is the ''sink'' for most of the water-borne waste of the urban community.

Sewage sludge can contain a variety of pathogens, including bacteria, viruses, parasites, and fungi, reflecting the presence of these agents in the human and animal population contributing to the sewage. Consequently, treatment may be required prior to disposal.

The chemical composition of municipal sludge varies widely, particularly in terms of minor constituents. The strictly domestic sources of metals and organic pollutants include a background from fecal matter, as well as contributions from detergent, cosmetics, insecticides, paints, and other materials used in house care. Contributions from corrosion of the buildings and substances applied to gardens, may be included together with aerial deposits from smoke and automobile exhausts if urban runoff is combined with domestic sewage.

Municipal sewage also includes contributions from restaurants, food handlers, and other institutions whose sewage closely resembles that from entirely domestic sources. In addition, there are contributions from small business—painters, printers, dry cleaners, jewellers, dentists, garages, auto workshops, electroplaters, school laboratories, etc.—which can greatly increase the concentration of one or another substance, depending on local conditions.

Industrial effluents may make substantial contributions according to the nature of the industrial processes. In industrialized areas, the concentrations of several metals may be elevated by discharges from metal processing industries, such as electroplating, leather processing, battery and paint manufacturing, etc. Very high concentrations can arise because of discharge from one or more factories.

Industries may discharge large quantities of specific waste, either by historical permission or because of local inattention to pollution problems. In some cases, although pollutants are known to be discharged, there is a reluctance to interfere with companies supplying the source of local payrolls.

It is important, therefore, that industrial effluent control is exercised by the statutory authorities and the manufacturing processes are managed and the industrial effluents are pre-treated to produce acceptable qualities with respect to the sewage effluent and sludge disposed of by the receiving sewage treatment works. This is effective in reducing the excessive loads by individual discharges and in reducing the loads discharged by industry, in general, within a catchment area.

The summary of the two Working Groups' views was that good practice for sludge utilization should have the following features:

- Chemical quality should be controlled in such a way that the receiving soil does not become contaminated so as to cause crops to be hazardous to eat or so as to reduce their yields. Clearly, irresponsible practices can cause insidious long-term problems. Care should be exercised so that pollution of surface and groundwaters does not occur.
- Microbiological quality should be taken into account so that the spread of human, animal, and plant diseases is prevented. It is not possible to control the microbiological quality in quite the same way as the chemical quality (except by the control of animal-derived industrial effluents). Hence the objective of prevention of the spread of disease is achieved by a combination of rigorously controlled utilization practices and treatment of the sludge on the sewage treatment works as required.
- Aesthetic considerations should be taken into account and the operation should not cause nuisance or offense.
- The product and service should be acceptable to farmers.

Chemical Quality

The WHO Working Group considered elemental particularly metallic, and organic chemical residues in sludge and in soils that have received sludge; it also gave some consideration to the way in which water may become polluted by sludge use and, in this context, nitrate received the most attention.

The Group concluded that for heavy metals the following points were of importance. Sludge from areas that have strict discharge control has lower heavy metal concentrations than the sludge from less controlled areas. Heavy metals in sludge, except cadmium, are not expected to affect human health through accumulation in food and fodder plants. There would appear to be little risk to health of applying sludge containing cadmium to forest land or to land producing fodder and seed crops. Tobacco grown on land receiving sludge causes an additional risk to human health from the inhalation of cadmium in tobacco smoke. The uptake by crops of lead from land treated with sludge does not contribute an appreciable amount of lead to the human food chain.

With respect to organic compounds, the Group concluded that information on identified organic pollutants in sludge and their pathways to man is limited. The concentrations reported in the literature appear to be low, and almost always below 10 mg/kg dry solids. The most significant route for organic pollutants from sludge to man is through ingestion of soil and sludge by grazing animals. Transfer through food crops is negligible. The contribution to total human intake of identified organic pollutants resulting from sludge application to land is minor and is unlikely to cause adverse health effects.

From the discussions and conclusions the Group were able to make the following recommendations. Both the short- and long-term health risks, costs, and benefits should be evaluated when selecting a method of sludge disposal. When applying a sludge to land, management practices should consider concentrations of organic and inorganic chemicals in the sludge, maximum allowable accumulation levels in soils, rates of accumulation, local soil conditions (pH, soil organic matter, and cation exchange capacity), climatic conditions, and topography. Sludge should be spread in a manner that meets the needs of the plants and preserves the quality of soil, and ground and surface water. Sludge utilization should be supported by effective industrial effluent control practices and policies. The implications for sewage treatment and sludge utilization should be evaluated before a new chemical is considered for wide use. Sludge should not be applied to growing food crops, such as vegetables and fruits, or in such a way that would contaminate drinking water sources. Sludge should not be applied to land that is used by the public (parks, playgrounds, sports fields) unless adequate precautions are taken to ensure the protection of human health. The lead content should not be above 1000 mg/kg dry solids to protect children against ingestion of lead in soil. Exposure studies should be conducted on worst-case population groups, such as exposed workers, and on adults and children, using home-grown vegetables in acidic soils that have been treated with sludge. Special emphasis should be given in pathway studies to the long-term bioavailability of the different chemical pollutants (both heavy metals and organic compounds).

Concentrations of cadmium in sludge applied to land should be as low as practicable to decrease the risk of its transfer to man through food. Additions of such sludge should be assessed in the light of other cadmium sources, such as fertilizers and atmospheric deposition. Sludge that is excessively contaminated with cadmium should not be used on light and acidic souls because uptake of cadmium is likely to be highest in vegetables grown in these areas. Sludge application rates that are higher than those based on cadmium limitations may be acceptable on land producing plants not used for food, provided there is no risk of contaminating the groundwater or surface water. Tobacco should not be grown on land receiving sludge because of excessive cadmium accumulation by this crop. Further, dietary and fecal cadmium studies should be undertaken to improve knowledge of long-term individual intake and retention of cadmium. Studies should be undertaken to establish (1) how cadmium concentrations in sludge affect potential cadmium uptake by plants and (2) the persistence of cadmium availability to crops. A group of experts should reevaluate the "provisional tolerable weekly intake (of cadmium) per individual" set in 1972 by a joint FAO/WHO Expert Committee on Food Additives. A lot of work has been done on this topic since that time.

More information should be obtained on the organic pollutant contents of sludge, especially where local point sources of industrial pollutants exist. More information should be obtained on the fate and behavior of persistent organic chemicals of significance to health in sludge and soil. Improved methods should be developed for the extraction and preparation for analysis of organic pollutants of health significance in sludges, with a view to undertaking systematic surveys of these compounds in sludge to provide basic information. Substantial progress has now been made.

When applying sludge to land, account should be taken of both the total and soluble nitrogen in the soil, because both can contribute to groundwater pollution.

Microbial Quality

The WHO Working Group discussed the implications of the presence of a wide range of microorganisms in sewage sludge and concluded that the use of sludge results in the distribution of pathogens in the environment. However, a number of other sources and routes of transmission of pathogens contribute to the total risk to human and animal health. The additional risk of a given sludge disposal practice must therefore be considered against this background. It is however, extremely difficult and expensive to measure the risk by microbiological and epidemiological methods. The public health risk from the disposal of sewage sludge on land may nevertheless by reduced by appropriate treatment and use of the land. It would be unrealistic to expect a significant or early reduction in the incidence of disease if other important sources of pathogens remain unaffected.

Acceptable levels of risk in given communities depend on a number of different factors. These include: the health status of the local population; the nature of the soil; the temperature, humidity, precipitation, and groundwater table; the nature of the agriculture and animal husbandry; and the way in which the sludge can be safely transported and spread on the land.

Such acceptable levels cannot be expressed in absolute terms, but the public health risk may be controlled by the use of appropriate guidelines, which would vary from place to place according to local circumstances. *Salmonellae* responsible for food poisoning represent one of the risks to human health that may be increased by the spreading on land of sewage sludge containing these organisms. Although there was no epidemiological evidence available for this (and there is still not), it has been shown that cattle exposed to such sludge may become carriers of *Salmonellae* to a greater extent than controls. *T. saginata* is a specific parasite of man, and where infection occurs the eggs are excreted in human feces. They may be disseminated by sewage sludge and thus infect cattle, the subsequent development of *cysticerci* presenting an infection risk for man. *Salmonellae* and eggs of *T. saginata* in sewage sludge can be eliminated or reduced by various forms of heat treatment, by ionizing radiation, or by long-term storage. In contrast to the *Salmonellae,* the eggs of *T. saginata* are resistant to chemical disinfectants.

The risk to public health from other pathogenic agents appears to be less than those from *Salmonella* and *Taenia* in most areas studied. However, the possible risk from viruses and from parasites such as *Sarcocystis* has not been adequately evaluated. Since then, the view in Europe is that any risks from treated sludge are absent or insignificant.

Measures should be taken to effect a substantial reduction in the concentration of pathogens in sewage sludge before it is allowed to come into contact with crops such as fresh vegetables and fruit that are brought into the kitchen raw. Sludge containing pathogens that may multiply in or contaminate meat, poultry, and dairy products should not be spread on land where food animals are raised unless an adequate interval of time elapses between spreading the sludge on the land and allowing the animals to graze.

The following measures should be taken for the disinfection of sewage sludge so that its use can be relatively unrestricted.

- heat treatment such as 70°C for 30 minutes, thermophilic composting, or heat drying
- ionizing radiation with at least 5 kGy
- extended batch storage for a time that is inversely related to the temperature
- treatment with chemical agents that destroy the organisms in the sludge environment

For restricted use, where direct contact with fresh food that may be brought raw into the kitchen is not involved, anaerobic digestion and other stabilization processes should be utilized if they are carried out in such a way as to minimize recontamination. Uniform regulations should not be imposed over large regions without due regard to local conditions. Carefully planned epidemiological investigations into the relationship between sludge spreading and the public health should be encouraged.

Summary of WHO Recommendations

The Working Groups concluded that land application is a cost-effective method of using sludge and results in minimal risk to health when good management is practiced. Good management practice should be adaptable to new developments and should incorporate monitoring of the sludges, sludged soils, effects on crops, and effects on crop consumers as and when appropriate. The WHO Working Groups, particularly that for microbial risks, did not favor universal recommendations since depending on local circumstances, different measures are needed in different localities and situations to reduce the problem to an acceptable size. The formulation and application of suitable control measures must be a local decision. Several countries have made the decision to formulate national statutes or guidelines. The WHO Working Groups summarized the main features that control measures should take into account.

In the case of chemicals, which may be present in the sludge, it is important that there should be integration or close liaison between management responsible for sewage disposal and utilization and that responsible for industrial effluent control. In determining what can be allowed in discharges, among a wide range of toxicological data available to the control authorities should be data on the hazards arising from dispersion on agricultural soils.

The WHO has not attempted to produce comprehensive guidance on the principles and main features of all aspects of sludge utilization. Good practice in sludge utilization is an important contribution to public health. It assists in the good management of sewage treatment as well as preventing problems in the essential disposal of the sludge.

Improving Quality

All European countries have been exerting greater controls over the quality of sludge. The experience of UK and Anglian Water in particular is given to demonstrate this. An early review of sludge quality was prepared by Berrow and Webber in 1964 [21] and reported in 1972. Six works of various sizes were included for the Anglian Region representing 5–10% of the sludge produced. Whilst this starting point is based on few samples and indeterminate analytical methods—nevertheless it is a very useful comparison for following surveys. Table 13.3 gives a comparison with 1979 data [22] and Figure 13.1 gives the

TABLE 13.3. Improvements in Anglian Sludge Quality (1964–1979).

	50% Concentration Values (range in brackets) mg/kg DS				
	Zn	Cu	Ni	Pb	Cd
1964[a]	2000	800	—	100	
	(1000–4000)	(500–1000)		(300–8000)	
1979[b]	850	500	45	260	5
	(300–2900)	(100–2100)	(5–370)	(50–11000)	(1–97)

[a] Mean.
[b] Used in agriculture.

further improvements up to 1994. These have been achieved by trade effluent control, the use of cleaner industrial practices and the general change in manufacturing. Similar improvements have been observed at a national level with data available from surveys in 1977, 1980, 1982 and 1990 (see Table 13.4) [7,23].

EU Directive

These principles are expressed in the directive and are summarized in Table 13.5. Specified metals are controlled by setting maximum concentrations in the soil (as mg/kg dry solids) and maximum concentrations in the sludge (as mg/kg dry solids) or maximum rates of addition expressed as an annual average over a ten-year period. Limits are set for total zinc, copper, nickel, cadmium, mercury, and lead, and consideration is still being given to chromium. After careful consideration it was decided that limits were not needed for trace organics. Individual member states such as Germany and Sweden are imposing national limits for such substances.

Work in the EC COST research program and in the U.K. indicated that there was no need for specific limits for synthetic organic compounds in sewage sludge or soil. There was an expectation that industrial effluents, discharging in sewerage catchments, would be controlled for specific compounds and relevant information on the potential impact on agriculture through the use of sewage sludge could be used to define limits. However several surveys have been completed for PCBs, furans and dioxins in the U.K. and a report produced [24].

PCB residues were found to be of the order of 200 µg/kg dry sludge, which is generally lower than values from elsewhere in Europe. The mean sum of polychlorinated dibenzo-*p*-dioxins and polychlorinated dibenzofurans has been found to be of the order of 20 µg/kg dry solids. However analytical methods need considerable improvement. The residues tend to

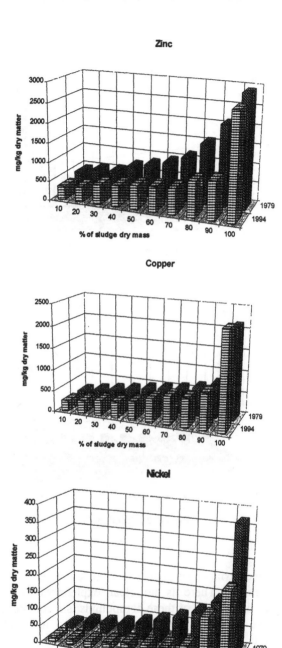

Figure 13.1 Improvements in the quality of sewage sludge used as biosolids in agriculture (1979 to 1994).

Cadmium

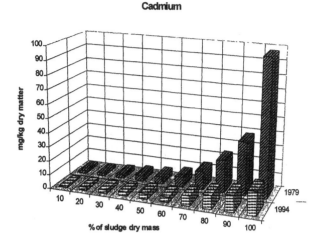

Lead

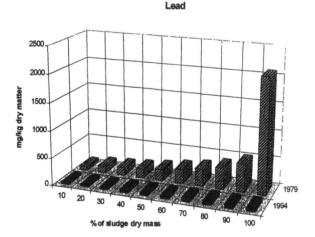

Figure 13.1 (continued)

be higher in industrial areas like PCBs but there do not appear to be temporal variations. There is no clear explanation for the sources. Further studies on the fate of these compounds is recommended although no values have been found which have resulted in a call for urgent action on limits.

The Directive is the minimum standard for all member states, but more stringent requirements can be added for any aspects. Although the U.K. does not have limits for trace organics, it has, for example, added limits for molybdenum and fluoride arising from historical experiences with these substances.

TABLE 13.4. Improvements in National Sludge Quality (1980–1990).

Comparison of Analyses of Sludge's in UK Agriculture, mg/kg, DS				
Zn	Cu	Ni	Pb	Cd
1980				
Min 143	20	5	25	0.1
Med 1002	440	45	260	7.0
Mean 1123	519	101	329	16.3
Max 4920	2900	615	3106	158
1990				
10% 454	215	15	70	1 5
50% 889	473	317	217	3.2
90% 1473	974	225	585	12

The soil limits are set for neutral soil at pH 6–7; a 50% increase is allowable in more alkaline soils (which also contain more that 5% calcium carbonate). Decreased limits must be defined for more acidic soils with no sludge applied to land with a pH value of less that 5.0. The sludge and soil limits are set as ranges, and each member state must choose the appropriate value, taking into account local conditions.

These soil values are to be checked using 25 cm samples composited from subsamples of no more than 25 per 5 ha. If the soil depth is less than 25 cm, samples must be taken to that depth but no less than 10 cm. Analysis of soil must be carried out at a frequency to be determined in each member state for sludges from works serving a domestic population of 5000 people. The analysis for metals must be carried out after strong acid digestion and the reference method must be atomic adsorption spectro-

TABLE 13.5. EU Directive Restrictions on Metals in Sludges/Soils During Agricultural Use.

	Soil (mg/kg DS)[a,b]	Sludge (mg/kg DS)	Rate of Application (kg/ha/yr)[c]
Cadmium	1–3	20–40	0.15
Copper	50–140	1000–1750	12.0
Nickel	30–75	300–400	3.0
Lead	50–300	750–1200	15.0
Zinc	150–300	2500–4000	30.0
Mercury	1–1.5	16–25	0 1

[a]pH 6–7, but Cu, Ni, Zn limits can be increased by 50% in soils of pH > 7.
[b]Where dedicated land used for farming and sludge disposal exceeded these values in 1986 and it can be demonstrated that there is no hazard, and commercial crops are grown and used only for animal consumption, the practice may continue.
[c]10-year average, can be applied in one go
DS = Dry solids.

photometry. Sludges are to be analyzed for pH, the dry and organic matter, nutrients, and metals, but the soil only for pH and metals. The Directive also waives the soil limits for land dedicated to sludge disposal but used for agricultural crops at the time of implementation of the Directive in June 1986. This is with the proviso that there is no hazard to health or the environment and there is restriction of the crops as animal feedstuff where the defined soil limits are exceeded.

Only the nutrient requirement of crops must be provided, and water pollution must be avoided. Sludges that have not been treated by biological or chemical means or by long-term storage in order to reduce fermentability, and health hazards arising from their use, must not be applied to land unless they are ploughed in immediately or injected.

Where treated sludge is applied to grassland or forage crops, grazing or harvesting must not be done is less than three weeks. No sludge must be applied to growing fruit and vegetable crops, except fruit trees, or applied to land used for growing fruit and vegetable crops that are normally in direct contact with soil and are eaten raw, in the period of ten months preceding the harvest and during the harvest.

The disposal authority must keep a register of information on the quantities of sludge produced and supplied, nature and composition (except for works of below 5000 people), treatment, and location of farms receiving the sludge. This information must be available for inspection by competent authorities and used in a consolidated national report to the Commission in 1991 and every four years thereafter. Information on the sludge must also be provided to its users. Member states were given three years from June 1986 to implement the Directive.

Implementation of the Directive and National Guidance

There have been different approaches to implementation of the Directive in individual countries. That used in the U.K. is given as an example. The U.K. government reestablished the national committee to update the national guidelines to take into account the Directive and additional research and operational information. The Code of Practice was produced for the autumn of 1989. The directive itself was implemented by regulations to be effective from September 1. The code supplements the regulations but is not mandatory. It defines these matters within the lien of the member states. In this sense the code and, to a lesser extent, the regulations go beyond the strict requirements of the directive.

Examples of effective sludge treatment processes are described. The additional information enabled the period of storage constituting treatment to be reduced by comparison to the original guidelines (see Table 13.6).

The code also has limits defined for chromium, molybdenum, selenium,

TABLE 13.6. **Examples of Effective Sludge Treatment Processes.**

Process	Description
Sludge pasteurization	Minimum of 20 mins at 70°C or minimum of 4 hours at 55°C (or appropriate intermediate conditions), followed in all cases by primary mesophilic anaerobic digestion.
Mesophilic anaerobic digestion	Mean retention period of at least 12 days primary digestion in temperature range 35°C ± 3°C or at least 20 days primary digestion in temperature range 25°C ± 3°C followed in each case by a secondary stage which provides a mean retention period of at least 14 days.
Thermophilic aerobic digestion	Mean retention period of at least 7 days digestion, with all sludge to be subject to a minimum of 55°C for a period of maturation adequate to ensure that the compost of reaction process is substantially complete.
Composting (windrows or aerated piles)	The compost must be maintained at 40°C at least 5 days and for 4 hours during this period at a minimum of 55°C within the body of the pile followed by a period of maturation adequate to ensure that the compost reaction process is substantially complete
Lime stabilization of liquid sludge	Addition of lime to raise pH to greater than 12 0 and sufficient to ensure that the pH is not less than 12 for a minimum of 2 hours The sludge can then be used directly
Liquid storage	Store retreated liquid sludge for a minimum period of 3 months.
Dewatering and storage	Conditioning of untreated sludge with lime or other coagulants followed by dewatering and storage of the cake for a minimum period of 3 months. If sludge has been subject to primary mesophilic anaerobic digestion, storage to be for a minimum period of 14 days

arsenic, and fluoride. The limits for zinc equivalent and boron in the original guidelines have been dropped, as has the concept of availability for zinc, copper, nickel, and boron. Organic limits have not been imposed although strict control of industrial effluents discharged to sewers is expected. Such control is needed, as well, to protect sewage works processes and sewage effluent quality. Water companies in England and Wales are responsible for industrial effluent control as well as for sludge disposal. The Code states that analysis may be dropped for these additional determinants to once every five years if concentrations do not exceed specified values. All other potentially toxic elements must be analyzed every six months and there is no relaxation for small works, which has been the cause of complaints by water companies. In accordance with U.K. policy, the method of control of sludge use is by soil quality and application rate limits. The exceptions are lead and fluoride limits for sludges applied to grassland. The soil limits are applied to 15-cm samples for arable land. The concentration limits are set for the soil pH bands of 5–<5.5, 5.5–<6, 6–7, >7. For zinc, copper, and nickel, zinc and nickel are the same as the EU upper values, and copper is slightly less. The limits for the other determinants are the same for all soil pH values. Cadmium and lead limits are the same as the EU upper limit, but the mercury limit is the lower EU limit (see Table 13.7).

TABLE 13.7. **Maximum Permissible Concentration of Potentially Toxic Elements in Soil after Application of Sewage Sludge and Maximum Annual Rates of Addition.**

| PTE | Maximum Permissible Concentration of PTE in Soil (mg/kg Dry Solids) | | | | Maximum Permissible Average Annual Rate of PTE Addition over a 10-Year Period (kg/ha)[c] |
	pH[a] 5.0–<5.5	pH[a] 5.5–<6.0	pH 6.0–7.0	pH[b] >7.0	
Zinc	200	250	300	450	15
Copper	80	100	135	200	75
Nickel	50	60	75	110	3
For pH 5.0 and above					
Cadmium	3				0.15
Lead	300				15
Mercury	1				0.1
Chromium	400 (provisional)				15 (provisional)
Molybdenum[d,e]	4				0 2
Selenium[d]	3				0.15
Arsenic[d]	50				0.7
Fluoride[d]	500				20

[a]For solids of pH in the ranges of 5 0 < 5.5 and 5.5 < 6.0 the permitted concentrations of zinc, copper, nickel, and cadmium are provisional and will be reviewed when current research into their effects on certain crops and livestock is completed
[b]The increased permissible PTE concentrations in soils of pH greater than 7.0 apply only to soils containing more than 5% calcium carbonate
[c]The annual rate of application of PTE to any site shall be determined by averaging over the 10-year period ending with the year of calculation
[d]These parameters are not subject to the provisions of Directive 86/278/EEC
[e]The accepted safe level of molybdenum in agricultural soils is 4 mg/kg. However there are some areas in the U.K. where, for geological reasons, the natural concentration of this element in the soil exceeds the level. In such cases there may be no additional problems as a result of applying sludge, but this should not be done except in accordance with expert advice. This advice will take into account existing soil molybdenum levels and current arrangements to provide copper supplements to livestock

For grassland the soil limits apply to 7.5 cm samples but are about 60% higher than the arable land limits for zinc, copper, and nickel. For other determinants there are a variety of restrictions on an increase—for example, in the case of cadmium, none is permitted on permanent grazing land. It is recognized that more information is needed for the control of metals in grassland and further advice should be sought for permanent grassland (see Table 13.8).

As for the annual rates of application of the potentially toxic elements, averaged over a rolling ten-year period, those for copper and zinc are less than the directive, nickel, cadmium, lead, and mercury are the same. However, to minimize ingestion by livestock the application of lead, cadmium, and fluoride should not exceed three times the annual average limit (see Table 13.8).

Where the soil has a pH of less than 5.2, specialist advice should be sought. No allowance is made for background soil concentrations (general assumptions having been made in the derivation of the limits); the exception is molybdenum when specialist advice can be sought to breach the limit in high background soils (and of course there is no statutory inhibition on doing so).

The limits in the code apply to 15 cms (arable) and 7.5 cms (pastoral) samples because government was concerned to ensure that the soil in the "active zones" did not exceed the specified Directive limits. These soil depths had been used in the original guidelines. Because the limits apply in the regulations and Directive to 25 cm samples it would be possible to comply with these and be well above them at a shallower depth—hence the additional restriction in the code. It is therefore possible to deep plough the sludge soil provided that the 25 and 15 cm (or 7.5) limits are not breached.

Before use of sludge and for at least twenty years during sludge use, the soil should be sampled at 25 cm and analyzed. The operational samples should be taken according to local circumstances. Where sites were already in use in September 1989 and there was already some soil information, 25-cm samples should have been obtained and analyzed before December 1991.

The way in which dedicated sites are to be managed are defined in the code and regulations. The sites must have been registered with DoE, and where the limits specified in the code are exceeded, sludge can only be applied in accordance with conditions that are laid down (including monitoring).

As explained earlier, examples of effective treatment are defined in the code. The acceptable uses of treated and untreated sludge are defined in the code in terms of a much wider range of uses than those defined in the regulations and Directive—for instance, turf growing, potatoes, and nurs-

TABLE 13.8. Maximum Permissible Concentrations of Potentially Toxic Elements in Soil under Grass after Application of Sewage Sludge When Samples Are Taken to a Depth of 7.5 cm.

PTE	Maximum Permissible Concentration of PTE in Soil (mg/kg Dry Solids)			
	pH 5.0–<5.5	pH 5.5–<6.0	pH 6.0–7.0	pH[a] >7.0
Zinc[b]	330	420	500	750
Copper[b]	130	170	225	330
Nickel[b]	80	100	125	180
For pH 5.0 and above				
Cadmium[c]	3/5			
Lead	300			
Mercury	1 5			
Chromium	600 (provisional)			
Molybdenum[d,e]	4			
Selenium[d]	5			
Arsenic[d]	50			
Fluoride[d]	500			

[a]See Table 13 7 note b
[b]The permitted concentrations of these elements will be subject to review when current research into their effects on the quality of grassland is complete Until then, in cases where there is doubt about the practicality of ploughing or otherwise cultivating grassland, no sludge applications that would cause these concentrations to exceed the permitted levels specified in Table 13.7 should be made except in accordance with specialist agricultural advice
[c]The permitted concentration of cadmium will be subject to review when current research into its effect on grazing animals is completed. Until then, the concentration of this element may be raised to the permitted upper limit of 5 mg/kg as a result of sludge applications only under grass, which is managed in rotation with arable crops and grown only for conservation In all cases where grazing is permitted, no sludge application that would cause the concentration of cadmium to exceed the lower limit of 3 mg/kg shall be made
[d]These parameters are not subject to the provisions of Directive 86/278/EEC
[e]See Table 13.7 note e.

ery stock growing. Stricter controls for these activities reflect concern for plant and human health protection. Injection of untreated sludge must be carried out according to the WRc *Manual of Good Practice* to be acceptable (see Tables 13.9 and 13.10). No person, user, or supplier should knowingly cause a breach of the regulations—it is an offense to do so.

The other environmental requirements of the directive and additional recommendations are incorporated in the code. These include transmission of weeds, transport, field access, odor control (and in particular the use of spray guns when environmental health officers should be consulted), surface runoff, water pollution, and storage. In the latter cases, advice is given on avoidance of contamination of water by pathogens and nitrate, and sludge should be applied in such a way and a time that ensures maximum crop uptake. The optimum time of application varies with each type of sludge.

TABLE 13.9. **Acceptable Uses of Treated Sludge in Agriculture.**

When Applied to Growing Crops	When Applied before Planting Crops
Cereals, oil seed rape Grass[a] Turf[b] Fruit trees[c]	Cereals, grass, fodder, sugar beet, oil seed rape, etc. Fruit trees Soft fruit[c] Vegetables[d] Potatoes[d,e] Nursery stock[f]

[a]No grazing or harvesting within 3 weeks of application.
[b]Not to be applied within 3 months before harvest.
[c]Not to be applied within 10 months before harvest
[d]Not to be applied within 10 months before harvest if crops are normally in direct contact with soil and may be eaten raw
[e]Not to be applied to land used or to be used for cropping rotation that includes basic seed potatoes or seed potatoes for export.
[f]Not to be applied to land used or to be used for a cropping rotation that includes basic nursery stock or nursery stock (including bulbs) for export.

The code and regulations define the requirements for a Register. However, these go beyond the requirements of the Directive. The date must be recorded for each 5 ha site (or part thereof). In addition, the regulations require that the quality of the elements added must be recorded, the advice given for dedicated sites should be included. The soil analysis is to include the initial estimates prior to December 1991. In addition, the code recommends that information obtained on 15 cm and 7.5 cm operational samples and information on the sludges applied should be included.

However, since 1989 a number of changes have been recommended [13] which have yet to be included as formal revision to the code. Some

TABLE 13.10. **Acceptable Uses of Untreated Sludge in Agriculture.**

When Applied to Growing Crops	When Cultivated or Injected[a] into the Soil before Planting Crops
Grass[b] Turf[c]	Cereals, grass fodder, sugar beet, oil seed rape, etc. Fruit trees Soft fruit Vegetables[d] Potatoes[d,e]

[a]Injection carried out in accordance with WRc publication FR008 1989, *Soil Injection of Sewage Sludge—A Manual of Good Practice* (2nd Edition) [40].
[b]No grazing or harvesting within 3 weeks of application
[c]Not to be applied within 6 months before harvest.
[d]Not to be applied within 10 months before planting if crops are normally in direct contact with soil and may be eaten raw
[e]Not to be applied to land used or to be used for a cropping rotation that includes seed potatoes

TABLE 13.11. 1994 Committee's Recommendations for Changes to U.K. Limit Values for Total Concentrations of Zinc, Copper, and Nickel in Sludge Amended Soil (mg/kg).

	Limit Values			
Soil pH	5.0– < 5.5	5.5– < 6.0	6.0– < 7.0	> 7.0[a]
Zinc[b]	200	200	200	300
Copper[b,c]	80	100	135	200
Nickel[b,c]	50	60	75	110

[a]The increased permissible PTE concentrations in soils of pH greater than 7 apply only to soils containing more than 5% calcium carbonate.
[b]The limit values shown are from the current U.K Regulation [13] However because of the Committee's concern about pH decreasing over time it also recommends that the pH qualification of the limits for nickel and copper should be examined further in the light of decisions to drop the pH qualification for zinc limit values for soil pH 5.0–7 0, in order to protect crops from phytotoxicity from these metals. The Committee recommended that this still needs to be fully addressed
[c]Following on from the Committee's recommendation to examine the pH qualification for limit values for copper and nickel in samples taken to 25 cm the pH qualification should also be examined for the maximum permissible concentrations of copper and nickel in samples taken to a depth of 7 5 cm in grassland

scientists have expressed concern about the potential deleterious effects of zinc in the soil at concentrations approaching current limits, on rhizobia. Although the influence of a variety of other factors needed to be understood better. Concern was also expressed about cadmium. Tables 13.11 and 13.12 summarize recommendations made by a scientific committee established by the government. Although these are challenged by a number of experts, the government has accepted the recommendations to maintain the high level of confidence in the use of sewage sludge as biosolids in agriculture and it published an updated code in May 1996. The government and other interested parties are considering a report by the Royal Commission on Environmental Pollution with respect to the sustainability of soil, which concluded that only fully treated sludges should be utilized in agriculture [14].

Each European state in the Union has taken its own approach in imple-

TABLE 13.12. 1994 Committee's Recommendations for Changes to U.K. Limit Values for Total Concentrations of Zinc and Cadmium in Soil and under Grass after Application of Sewage Sludge When Samples Taken to a Depth of 7.5 cm (mg/kg).

	Limit Values	
	5.0–7.0	> 7.0
Zinc	200	300
	pH 5.0 and above	
Cadmium	3	

menting its local requirements. Tables 13.13–13.20 demonstrate the diversity [25,26]. The limits tend to draw on the precautionary principle, which produces lower values than those in the regulations produced by U.S.EPA, which depend more on risk assessment procedures. The Global Atlas of Wastewater, Sludge and Biosolids Use and Disposal produced by WEF, IAWQ and the EWPCA and published by IAWQ provides valuable comparisons [1].

A good example is Denmark. The objective is to encourage the use of sludge as biosolids in agriculture so quality limits are imposed which will allow the product to be used virtually without practical restriction. Of course, it may mean that some sludges have difficulty in qualifying for use and in order to do so there have to be exceptional restrictions in industrial effluents and on the use of chemicals and materials in homes.

Restrictions have also been implemented on the application of sludge to pasture land in Germany. Individual states within Germany have introduced their own sludge regulations. One aspect of the German regulations that is a step forward from the EU directive is the more detailed consideration of the trace organic pollutants, for example polychlorinated dioxins and furans together with PCBs; there is concern about these substances which has led to restriction on the use of sludge on pasture land(s).

In addition to these, there are other ordinances that may further restrict sludge use—such as the ban on manure spreading in winter in Nordrhine-Westphalia, which has the consequential effect of sludge not being used in winter as well [27].

The Abwassertechnische Vereinigung (ATV) and Deutscher Bauern-Verband (DBV) have been working on joint guidelines for the utilization of manure and sewage sludge. ATV has also prepared guidance on the descriptions of and criteria for disinfection [25,26].

The main way forward is partnership between farmers and disposers. Examples have been ATV and DBV, which have produced guidelines for the use of manure and sewage sludge in Germany, at a national level [27], and Anglian Water and the Agricultural Development and Advisory Service in Eastern England, at a regional level [28]. In Germany and the U.K., many farmers and disposers work together through various kinds of joint operations.

Nutrient Value

Many of the metals present in sludge can be useful as micro-nutrients, but by far the greatest value is in nitrogen, phosphorus, potassium, and organic matter content. Sludge can either be used as fertilizer or as organic soil conditioner.

TABLE 13.13. Sludge Boundary Values (mg/kg) for the EU and Selected European Countries.

		Country						
		Belgium			Germany		Finland	
	EC Directive				Soil		Sewage Sludge	
Parameter	Appendix 1B	Flanders	Walloon	Denmark	pH 5–6	pH > 6	Normal	Improved
Lead	750–1200	600	500	120	900	900	100	150
Cadmium	20–40	12	10	0.8	5	10	1.5	3
Chromium	1000–1500[a]	500	500	100	900	900	300	300
Copper	1000–1750	750	600	1000	800	800	600	600
Nickel	300–400	100	100	30	200	200	100	100
Mercury	16–25	10	10	0.8	8	20	1	2
Zinc	2500–4000	2500	2000	4000	2000	2500	1500	1500
Arsenic	—	—	—	—	—	—	—	—
Fluoride	—	—	—	—	—	—	—	—
Selenium	—	—	—	—	—	—	—	—
Dioxin/furan	—	—	—	—	100[b]	100	—	—
PCBs	—	—	—	—	0.2[c]	0.2	—	—
AOX	—	—	—	—	500	500	—	—

[a] Planned.
[b] ng TE/kg TM
[c] Each for six PCB individual components (Nos. 28, 52, 101, 138, 153, 180)
Heavily contaminated sludge may be diluted with lime, peat, bark, sand, or soil down to normal values (Cd, Hg, Pb)

TABLE 13.14. Sludge Boundary Values (mg/kg) for European Countries.

Parameter	Country							
	France		United Kingdom	Ireland	Italy	Luxembourg		Netherlands
	Recomm'd Value	Boundary Value	Grassland			Recomm'd Value	Boundary Value	
Lead	800	1600	1000	750	750	750	1200	100
Cadmium	20	40	—	20	20	20	40	1.25
Chromium	1000	2000	—	—	—	100	1750	75
Copper	1000	2000	—	1000	1000	1000	1750	75
Nickel	200	400	—	300	300	300	400	30
Mercury	10	20	—	16	10	16	25	0.75
Zinc	2000	6000	—	2500	2500	2500	4000	300
Cr+Cu+Ni+Zn	4000	8000	—	—	—	—	—	—
Arsenic	—	—	—	—	—	—	—	15
Fluoride	—	—	1200	—	—	—	—	—
Selenium	100	200	—	—	—	—	—	—
Dioxin/furan	—	—	—	—	—	—	—	—
PCBs	—	—	—	—	—	—	—	—
AOX	—	—	—	—	—	—	—	—

687

TABLE 13.15. Sludge Boundary Values for More European Countries.

Parameter	Norway	Sweden (due for change)	Switzerland	Spain Soil pH < 7	Spain Soil pH > 7	Austria
Lead	100	200	500	750	1200	500
Cadmium	4	4	5	20	40	10
Chromium	125	100	500	1000	1500	500
Copper	1000	1200	600	1000	1750	500
Nickel	80	50	80	300	400	100
Mercury	5	5	5	16	25	10
Zinc	1500	800	2000	2500	4000	2000
Arsenic	—	—	—	—	—	—
Cobalt	—	—	60	—	—	—
Molybdenum	—	—	20	—	—	—
Selenium	—	—	—	—	—	—
Dioxin/furan	—	—	—	—	—	—
PCBs	—	—	—	—	—	—
AOX	—	—	500 approx. value	—	—	—

Sometimes interest is expressed in constituents such as lime, particularly when it has been used for conditioning prior to dewatering. The most useful and available nutrients are nitrogen and phosphate (although the availabilities do vary from sludge to sludge); application rates are often calculated on the basis of nitrogen. While there can be a need for the recipient farmer to supplement with artificial nitrogen or phosphates according to the quantity of sludge supplied, there is often a need for additional potassium, as sewage sludges are deficient in this nutrient with respect to recommended rates. An important factor is the availability of nutrients to plants during the year of application and this must be taken into account when calculating application rates. However, in a regular program of application there will be residual nutrients available from previous applications, such as from the ''slow release'' organically bound nitrogen.

As described earlier, the EU directive and national controls such as those in Germany and the U.K. restrict the application of sludge so that water pollution does not occur. The preferred application is such that only the right amount of nitrogen is made available to the growing crop; so this has implications for the time and quantity of application. However, concern about nitrate in environmental water, particularly where it is used for water supply, has resulted in increasing restrictions on nitrogen applications particularly in sensitive areas. These restrictions affect sludge disposal and examples exist in Switzerland, the U.K., Germany, and the

TABLE 13.16. Permitted Application Quantities for Sewage Sludge in Some European Countries.

Country	Average Annual Quantity Applied (t dry matter/ha/a)	Period Annual/One-Time Application (years)	Maximum/One-Time Application Quantity (t dry matter/ha)
Austria	2.5	2	5[a]
Belgium	1.4	3	3–12[b]
Denmark	10	10	100
Finland	1	4	4
France	3	10	30[c]
Ireland	2	1	2
Italy	2.5–5	3	7.5–15[d]
Luxembourg	3	1	3
Netherlands	1–10	1	1–10[e]
Norway	2	10	20
Sweden	1	5	5
Switzerland	1.66	3	5

[a]For pasture 50% of the quantity permitted for arable land.
[b]Pasture 3–6, arable land 6–12 t dry matter/ha/3a
[c]In the case of sewage sludge with less pollutants up to 75 t dry matter/ha/10a.
[d]Dependent on cation exchange capacity and pH value.
[e]Dependent on degree of dewatering and on pollutant load

689

TABLE 13.17. Soil Boundary Values (mg/kg) for the EU and Selected European Countries.

		Country						
		Belgium			Germany			
		Flanders						
Parameter	EU Directive 86/278/EC Appendix 1A	Sandy Soil	Clay/Silt	Walloon	pH 5–6	pH > 6	Finland	France pH ≥6
Lead	50–300	50	300	100	100	100	60	100
Cadmium	1–3	1	3	1	1	1.5	0.5	2
Chromium	100–150[a]	100	150	100	100	100	200	150
Copper	50–140	50	140	50	60	60	100	100
Nickel	30–75	30	75	50	50	50	60	50
Mercury	1–1.5	1	1.5	1	1	1	0.2	1
Zinc	150–300	150	300	200	150	200	150	300
Arsenic	—	—	—	—	—	—	—	—
Fluoride	—	—	—	—	—	—	—	—
Molybdenum	—	—	—	—	—	—	—	—
Selenium	—	—	—	—	—	—	—	—

[a]Planned

TABLE 13.18. Soil Boundary Values (mg/kg) in More European Countries.

	Country								
	United Kingdom						Luxembourg		Netherlands (Standard Soil)
	Soil pH				Ireland	Italy	Recommended	Boundary Value	
Parameter	5.0–5.5	5.5–6.0	6–7	>7					
Lead	300	300	300	300	50	100	50	300	85
Cadmium	3	3	3	3	1	15	1	3	0.8
Chromium	400	400	400	400	—	—	100	200	100
Copper	80	100	135	200	50	100	50	140	36
Nickel	50	60	75	100	30	75	30	75	35
Mercury	1	1	15	1	1	1	1	1.5	0.3
Zinc	200	250	300	450	150	300	150	300	140
Arsenic	50	50	50	50	—	—	—	—	29
Fluoride	500	500	500	500	—	—	—	—	—
Molybdenum	4	4	4	4	—	—	—	—	—
Selenium	4	4	4	4	—	—	—	—	—

TABLE 13.19. Soil Boundary Values (mg/kg) in European Countries.

Parameter	Norway	Sweden	Switzerland	Spain Soil pH < 7	Spain Soil pH > 7
Lead	50	40	50	50	300
Cadmium	1	0.4	0.8	1	3
Chromium	100	30	75	100	150
Copper	50	40	50	50	210
Nickel	30	30	50	30	112
Mercury	1	0 3	0.8	1	1
Zinc	50	75	200	150	450
Arsenic	—	—	—	—	—
Fluoride	—	—	400	—	—
Cobalt	—	—	25	—	—
Molybdenum	—	—	5	—	—
Selenium	—	—	—	—	—
Thallium	—	—	1	—	—

Netherlands. An EU directive on nitrogen from diffuse sources introduces restrictions in areas where the use of fertilizers could cause problems. In the U.K., "nitrate vulnerable zones" have been designated under the Water Resources Act of 1991 in which biosolids application is being restricted; general restrictions on nutrient applications are also being recommended [29].

In the debate about the future of agricultural land and agricultural productivity, sludge use may become a casualty. There is a suggestion that there should be a shift from intensive to extensive farming, but in the complex worlds of socioeconomic politics, for the moment the EU has a scheme in which agricultural land can be set aside. Sludge cannot be used in those areas.

There is substantial information on the nutrient content of sludges in the U.K. and the WRc has summarized it to produce operational guidance. Some data are given in Table 13.21 [30].

Services to Farmers

It is very important that agricultural use of biosolids should be viewed as a service to agriculture by the sludge disposal authorities, and throughout Europe there are a variety of arrangements to effect this service. It is very important that the services should have the confidence of the public, farmers, and public health regulatory authorities; they should be supported by extensive literature, advertising, and proper support services including laboratories and control in industrial effluents discharged to sewers. The

TABLE 13.20. Permitted Heavy Metal Loads in Some Countries of the EU.

Parameter	EC Directive 86/275/EC Appendix 1C[a] (kg/ha/a)	Country			
		Great Britain United Kingdom (kg/ha/a)	German (achieved) (kg/ha/a)	France (achieved)[b] (kg/ha/a)	Austria Steiermark (kg/ha/a)
Lead	15	15	1.5	2.4	1 25
Cadmium	0.15	0.15	0.016	0.06	0.025
Chromium	4[d]	15	1.5	3	1 25
Copper	12	7.5	1.3	3	1.25
Nickel	3	3	0.3	0.6	0 25
Mercury	0 1	0.1	0.013	0 03	0.025
Zinc	30	15	2.5	9	5
Arsenic	—	0.7	—	—	0 05
Fluoride	—	20	—	—	Cobalt: 0.25
Molybdenum	—	0.2	—	—	0 05
Selenium	—	0.15	—	0 3	—

[a]Mean value over a period of 10 years.
[b]Based on reference values and an application of 3 t/a
[c]Loads apply for arable land, for pasture half values
[d]Planned.
To be applied only for sewage sludge that exceeds the "High Quality" boundary value
With an application quantity of 10 t dry matter/ha/a and "High Quality" boundary value, an area usage duration/augmentation limit of 100 years results.

TABLE 13.21. Typical Annual Rates of Application for Some Cropping Situations.

Sludge Type	Application Rate	Crop	Available Nutrients (kg/ha) N	P$_2$O$_5$
Liquid undigested	80 m^3/ha	Hay Winter cereals Oil seed rape	48	64
Liquid digested	60 m^3/ha	Grazing	72	48
	80 m^3/ha	Silage Spring cereals	96	64
Cake	50 t/ha	Cereals Grass reseed	55	225
	150 t/ha	Restored areas	165	675

natural aversion that everyone feels towards fecal matter means that communication and language are an essential feature of public acceptance (as well as good operations!). So describing an operation as beneficial recycling of biosolids describes a very different mind-set from that describing the operation as sludge dumping on land. In reference to the right mind-set, from now on in this text, sludge will be described as biosolids when it is used beneficially.

The operation should be well-managed and survive close scrutiny. There is no room for any shortcuts, as these will inevitably jeopardize the long-term future of the operation. An example of such a service is "AWARD/ Service to Agriculture" operated by Anglian Water in the east of England [28].

It is important that methods should be found to optimize the service to farmers and reduce the inequalities between biosolids production and the farmers' needs. Techniques such as the storage of dewatered biosolids on hard standing or liquid biosolids in specially constructed tanks on farms have developed in recent years. In addition, in order to respond to the farmers and public demands for high-quality service, injection of biosolids is now used on a widespread basis. Screening of biosolids (and/or of sewage) has been extended to ensure that litter, particularly of sanitary origin, is not spread on land.

The increasing restrictions on the agricultural use of biosolids are likely to encourage more storage, more dewatering, and more injection of liquid biosolids. However, in the U.K. there is still an overall positive attitude to agricultural use of biosolids, and there have been many political speeches made in which it has been advocated that the additional biosolids produced by the EU Urban Wastewater Treatment Directive, should be used wherever possible.

There is a good relationship between biosolids disposers and farmers in the U.K. The practice is not opposed, and indeed in many instances is promoted by the Ministry of Agriculture, Fisheries, and Food, which is responsible for advising farmers on good agricultural practices. This has been extended in local regions such as Eastern England where Anglian Water has formed a partnership with AOAS as part of MAFF to promote beneficial use of biosolids [28]. The relationship with farmers is also relatively simple. The disposers are responsible for ensuring that good disposal practice is maintained. Contractors may be used for economic reasons, but even then ultimate local responsibility still rests with the disposer under the 1989 Code of Practice. Table 13.22 gives some statistics for Anglian Water showing how a regional operation is organized.

In other countries, such as Denmark, a view is taken that beneficial use as biosolids is best promoted by making the biosolids of such good quality that they can be used almost without restriction. This has implications for limits set (see Tables 13.15–13.20) and consequent restrictions on industrial effluents and use of chemicals and materials in domestic properties.

Methods of Application

In applying biosolids, good agricultural practice must be observed in addition to those constraints to avoiding aesthetic and microbiological problems. Application techniques will vary according to the nature of the biosolids.

TABLE 13.22. Statistics for Anglian Water 1994–1995.

Total disposed	126,400 dry tonnes	
By route	Agriculture	81%
	Sea	8%
	Sacrificial land	9%
	Landfill	2%
By type	Liquid	90%
	Cake	10%
	Fully treated	33%
	Untreated	67%
Agricultural routes	Arable	76%
	Pasture	24%
	Spread by AW	11%
	Spread by contractor	85%
	Spread by farmer	4%
Farm details	Liquid injection	66%
	Liquid surface	34%
	Farm nos	1115
	Lagoons	62

In selecting the machinery for providing the routine services, the water companies take account of a number of criteria. The method of application and incorporation must be free of nuisance, give a good service to the farmer and be efficient, and will be a function of the type of biosolids supplied. This machinery used for dewatered biosolids will have features somewhat similar to that used for solid farmyard manures, whereas that used for liquid biosolids will be somewhat similar to that used for farmyard slurries.

The extent of treatment of the biosolids supplied can influence the technique of application; for instance, anaerobically digested biosolids can be applied to the surface of land without problems, but with raw biosolids, unless the application is in a remote area, there is often a need to plough it in very soon after the application or to inject it. The following features are, therefore, considered when choosing the machinery [31]:

(1) Prevention of nuisance

- Aerosol production must be minimized so that odor nuisance does not occur and there is no risk of disease transmission by droplet drift.
- The equipment must be easy to clean, particularly if it is to go on to a public highway. Dirty machinery will lead very quickly to complaints and criticisms of the services provided.

(2) Service to the farmer

- The spread of sludge should be even, thus providing the farmer with a nutrient or condition of application that will make a reliable contribution to agronomic planning, while also ensuring that the possibility of accumulation of metals in patches is minimized.
- The width and distance of spread should be fully adjustable.
- Where biosolids are being incorporated into the soil as part of the operation, the biosolids consistency should be even, and very little, if any, biosolids should be left visible.
- Good control of driving speed in the field, and hence good control of rate of application, is essential. When coupled with adjustable discharge rate the load applied can be effectively regulated.
- The damage to grassland should be minimized. Soil should not be compacted and grassland should not be rutted or smeared (this is particularly important in wet weather). The requirement should be to keep the tanker well-balanced, when either empty or full, to permit an easy ride over land surfaces.
- If injection is used on grassland, the operation should leave an acceptable sward.

- The choice of application machinery should be appropriate for the type of land or crop being spread (the most fundamental distinction between grassland and arable application).

(3) Efficiency

- The capital and running costs should be reasonable and appropriate.
- The machinery should be robust, reliable, and easy to maintain and service without unduly specialized facilities or expertise being necessary.
- Accessories and replacement parts should be easy to fit.
- The pumps, pipes, and nozzles should be designed to minimize blockage: when a blockage does occur it should be easy to clear.
- The size of machinery used should be appropriate to the circumstances, taking account of width of farm gates and roads.
- Idle time of road tankers and application equipment should be minimized.

Many kinds of equipment are used. Liquid biosolids are often applied by tractor-drawn trailer tankers. With the desire to provide efficient services to agriculture, increasing pressure to ensure that nuisance is minimized, and to keep down costs, there has been substantial interest and developments in the technique of injection, in which biosolids are buried under the surface of land at about 20 cm depth by pumping it through a set of hollow tines. This has developed particularly in the U.K. and the Netherlands for biosolids and farm slurries. Injection of liquid biosolids has increased substantially. In Anglian Water, over half of all biosolids supplied to agriculture as a liquid are injected. A summary of the comparisons of different machinery options is given in Table 13.23 [31,32].

TABLE 13.23. Use of Machinery in the U.K. in 1984.

Equipment	% of Sludge Utilized
Tractor trailer application from works	11
Road tanker surface spread	23
Road and slave tanker surface spread	9
Tractor drawn injector	5
Road tanker injector	5
Rain gun	11
Pipework or reeled drum irrigation	3
Manure spreader	23
Temporary lagoon ploughed in	3
Other	7

The use of road tanker direct to land is the cheapest method. Although at first sight the unit cost of spreading dewatered biosolids using a manure spreader seems high, it is in fact a very cheap application method when compared on the basis of cost/tonne dry solids.

The cost of applying biosolids is, however, generally small compared with the cost of transportation. The biosolids may be delivered to a farm by a medium-sized tanker, which applies it directly to the farmland. It may also be delivered by larger tankers to storage or transfer facilities on the farm and from there it is taken and applied by small self-propelled or tractor-drawn units.

Mechanically dewatered cakes can be applied using normal farming techniques; they are often stored, not only to improve microbiological quality but to achieve a friable structure.

The application rate of liquid biosolids is usually on the order of 1–5 tonnes dry solids/ha (up to 150 m³/ha for wet biosolids) but can be higher for solid biosolids, the actual amount varying with the biosolids type and quality, and the needs of the situation. This should be based ideally on calculations in which nitrogen or phosphorus is the limiting factor not toxic metals. The time of application varies with the type of farming practiced and the availability of land, but to make the fullest use of the nitrogen it should be applied during late winter or preferably early spring.

There have been attempts in the EU to introduce a Directive on landfill but, so far, these have not yet produced an agreed and accepted document. Tables 13.24–13.26 provide information on the current states of agricultural use of biosolids in the U.K. [23]. For comparison the Saabye report [5] provided information on treatment in general in a number of European countries. Table 13.27 summarizes the position.

TABLE 13.24. Quality of Sludge Used in England and Wales for Agriculture 1990–91.

Parameter	10%ile (mg/kg·ds)	50%ile (mg/kg·ds)	90%ile (mg/kg·ds)
Zinc	454	843	1423
Copper	226	479	947
Nickel	14	37	225
Cadmium	1.5	3 2	9.5
Lead	63	204	591
Mercury	1.0	3 0	6.4
Chromium	28	86	489
Molybdenum	2.7	7 5	15.2
Selenium	0.2	1.2	2.6
Arsenic	1 1	3 4	6.8
Fluoride	73	140	308

TABLE 13.25. Quality of Agricultural Land Used in England and Wales for Sludge Disposal, Average Annual Rate of Metal Addition and Area of Land Sludged.

Parameter		10%ile (mg/kg·ds)	50%ile (mg/kg·ds)	90%ile (mg/kg·ds)	Average Annual Metal Application (kg/ha)
Zinc	a	25	58	110	7.10[f]
	b	29	67	124	
	c	32	70	126	
	d	41	74	122	
Copper	a	8	18	36	3.78[f]
	b	8	18	37	
	c	9	18	33	
	d	9	17	35	
Nickel	a	6	14	28	0.03[f]
	b	6	16	30	
	c	6	18	32	
	d	9	21	35	
Cadmium	a	0 16	0.53	1 04	0 03[f]
Lead	e	16	31	68	1 73[f]
Mercury	e	0 04	0.09	0 31	0 03[f]
Chromium	e	11	33	54	0.69[g]
Molybdenum	e	0.49	1 03	6 31	0 06[g]
Selenium	e	0 09	0.32	0.74	0.01[g]
Arsenic	e	3 3	8 2	21 5	0 03[g]
Fluoride	e	24	101	240	0 88[g]

[a]pH range 5 0–<5 5, area of land 2940 hectares
[b]pH range 5 5–<6 0; area of land 5880 hectares
[c]pH range 6 0–7.0, area of land 25480 hectares
[d]pH range >7.0, area of land 14700 hectares
[e]pH range >5.0, 49,000 hectares total area of agricultural land sludged in England and Wales
[f]U K regulation limits prescribe maximum permissible average rate of metal addition over a ten-year period This column covers 1990–91 only and on a site specific basis compliance with the rate of addition limit value will not be known until the end of the ten-year period
[g]Addition of metals covered by the U.K. Code of Practice for Agricultural Use of Sewage Sludge only.

MARINE DISPERSAL

The United Kingdom has been, by far and away, the largest user of the marine dispersal route in Europe. This is controlled in accordance with the requirements of the Oslo Convention for the protection of the North Sea and the northeast Atlantic. Sludges are dispersed in areas in accordance with the terms of licenses issued under national legislation which are renewed annually. Such licenses take account of the quantity and quality of sludge and the nature of the area in which it is proposed that the material should be dispersed [33].

The natural ecology of the area and its commercial uses are protected by the terms of the licence, which regulate the amount and manner of

TABLE 13.26. Treatment of Sludge Used on Agricultural Land in England and Wales.

Sludge Treatment	Agricultural % of Total[a]	Dedicated % of Total[a]
Digestion and dewatering	13.00	0
Mesophilic anaerobic digestion	30 4	93 1
Thermophilic aerobic digestion	0.16	0
Composting	0.56	0
Lime stabilization	0 06	0
Liquid storage	4.30	0
Dewatering and storage	5 23	5.40
Lesser[b]	6 07	1.50
Other	10.3	0
None (raw sludge)	29 4	0

[a]Dry mass of solids
[b]Lesser treatment may include some form of digestion and dewatering, mesophyllic anaerobic digestion, composting or liquid storage not meeting the defined criteria in the U.K. Code of Practice for Agriculture Use of Sewage Sludge

dispersal. Such areas have been monitored very carefully and there is no evidence that these activities have caused any environmental damage. However, such methods have proved to be unpopular with the public and with certain political groups, in spite of the extensive scientific information available. The U.K. government acceded early to the requirements in the EU Urban Wastewater Treatment Directive and anticipated its requirements in 1990 by requiring that all marine dispersal of sludge stop in the U.K. by 1998. Programs to achieve this goal are well advanced [34].

Landfill

Disposal to sanitary landfills is used for more than 40% of sludge in Europe but only about 15% is so disposed in the U.K. where landfill tends

TABLE 13.27. Stabilization Processes in European Countries.

Country	Anaerobic	Aerobic	None	Other
Belgium	40%	21%	38%	
Denmark	40%	23%	30%	7% (lime)
France	49%	17%	32%	2%
Germany	64%	12%	21%	3%
Ireland	19%	8%	73%	
Netherlands	44%	35%	2%	19%
Spain	45%	5%	24%	26% (lime)
Sweden	70%	5%	10%	15% (lime)
United Kingdom	56%	2%	35%	7%

to be used on an ad hoc basis according to economic circumstances, and occasionally because the sludges are heavily contaminated. The sludges are often codisposed with domestic refuse, but in some countries monofill is practiced. Co-disposal is not the subject of specific disposal guidelines. Where problems have existed, this is as a result of the physical nature of sludge in handling and with respect to long-term stability. This and the dwindling availability of landfill sites generally has tended to increase landfill charges. In the U.K. landfill taxes are being introduced.

There are a variety of methods of control of landfill. For instance, in the U.K. sludge is applied to landfill sites in accordance with the terms of licenses issued under pollution control legislation. This will lead to a restriction on the quantity and the quality of sludges being applied, to take account of available space in the landfill and impact on leaching. It is most likely that dewatered sludges rather than liquid sludges will be preferred. There is a trend towards minimum standards for physical properties in some European countries in order to avoid the handling of sludge and to ensure stability. However, in the United Kingdom attention is focused on handling techniques and identification of optimum ratios of addition of sludge to refuse for co-disposal. These pressures and interests have stimulated interest in the direct solidification of sewage sludges after dewatering.

There is increasing evidence that leachate quality may be improved by co-disposal. Methane production is enhanced, probably due to the buffering effect of sludge during the acid phase of anaerobic decomposition of the refuse, allowing methanogenesis to occur many months earlier than without sludge. This effect will undoubtedly make sludge more attractive in co-disposal where commercial recovery of methane as an energy source is envisaged [35].

While landfill disposal has a future, increasing security of economically viable available sites and increasing environmental controls may provide more uncertainty for the future. Any directive on landfill will certainly influence the types of waste being disposed to landfill and the control procedures to be adopted during site operation and for after care. It will be important that the permit system takes account of many factors, including protection of groundwater. In countries such as the U.K. this should not have a significant direct effect on existing practices for sludge disposal, but restricting landfill practices in general could result in many fewer sites being available.

Incineration

There is a considerable amount of sludge incinerated in Europe. However, unless there are particular requirements or restrictions which mean that incineration is the only valid option available, the sludge disposal manager will have in mind that incineration is usually the most expensive option for the disposal of sludge. However, there have been considerable improvements in the technology of incineration in terms of the process engineering, energy efficiency, and compactness of plant. Modern fluidized bed incinerators now appear more attractive both in terms of capital and operating costs compared with the multiple hearth type. In the U.K., incineration is now being considered again after a long period of time in which there were only a few incinerators. With the development of efficient autothermic incineration, requiring no other fuel source except for startup, costs are becoming much more competitive with other disposal options to the extent that incineration is now seen by some as the only solution to the increasingly intractable problems of other sludge disposal options. There is also a renewed interest in the coincineration with other waste materials in power stations and in cement production, but it is unlikely that there will be much takeup of such options at least in the near future [35,36].

However, incineration does not provide complete disposal since about 30% of the solids remain as ash. This is generally landfilled and is regarded in some countries as a highly toxic waste because of its metal content. For this reason the ash materials are considered as suitable candidates for some of the new chemical fixation processes which lock up the mobile metals. However, if the ash is of an appropriate quality not to give rise to problems, it may be used on agricultural land. Incinerator ash is considered difficult to handle, and this will lead to the transfer of technology from the power generation industry.

A major constraint on more widespread use of incineration is the planning difficulty associated with public concern about possible emissions, and this applies in several locations in Europe, including the United Kingdom and Germany.

There is no specific legislation covering the incineration of sludge although at one time it had been proposed that it should be included in the directive concerning the incineration of hazardous waste. This would have disastrous consequences for all sludge disposal and biosolids use. Legislation concerning the incineration of non-hazardous waste is, therefore, relevant.

USES ON NONAGRICULTURAL LAND

Land Reclamation and Restoration

The restrictions applied to biosolids used as an agricultural fertilizer are relevant to some extent to the use of biosolids in land reclamation. The main difference is that biosolids may be applied to land rates very much greater than those permitted for agricultural use because crops are not being grown over the period of time in which biosolids is applied. It therefore seems reasonable that the rate of achievement of the maximum metal concentration may well be much speedier than that for agricultural land, provided that there is an appropriate period of resting once the site has been restored.

This assumes that the site will be used subsequently for agricultural purposes but there can be other uses such as conservation land and recreational land and residential and commercial development. In all cases, care must be exercised so as to ensure that there is no environmental or public health hazard resulting during this subsequent use.

Biosolids make an excellent agent to revitalize the topfill in landfill sites and the soil that remains after substantial building construction or the reclamation of land used for mining and colliery spoil disposal or poor quality scrub land. It can avoid the need to bring in expensive topsoil since biosolids can be mixed with soil-forming materials on site and the resulting mixture used as topsoil. Studies have focused on operational aspect such as methods of application and the avoidance of potential problems. In the U.K. the quality of the soils is driven by the value of agricultural land.

Constraints on expanding this outlet appear to be (1) lack of appreciation of the potential value of derelict land; (2) apprehension of potential environmental problems due to heavy metals pathogens, odor, water pollution, etc.; (3) logistical difficulties of matching continuous biosolids production to a regular and limited availability of restoration sites; and (4) costs of disposal in relation to other local disposal options particularly if additional storage or special application equipment is required [35].

Forestry

Since forests are planted generally on poor soils, tree growth is often limited by nutrients, and consequently biosolids can provide a good source of nutrients and organic matter. Trials in Europe and the United States have shown that when sewage sludge is used as biosolids it can be an effective alternative to conventional forest fertilizers. Ongoing research is studying growth response, environmental impact, and operational aspects, but experiences show that sludge can be utilized satisfactorily in commercial coniferous forests either to prepare the ground before planting using dewatered sludge or to fertilize growing trees using liquid sludge.

In these circumstances there should be no problems due to pathogen transmission or food chain contamination. Environmental concerns relate to the possibility of runoff since forest land is often on a steep slope in water catchment areas, and to the acid soils which may cause metal mobility. European experience matches that of the United States in that it has been identified that forest soils are well-suited to biosolids applications. Forest soils have high rates of infiltration, which reduce runoff and ponding; large amounts of organic material, which immobilize metals in the biosolids; and perennial root systems, which allow year-round application in mild climates.

There is concern, however, that if such forests are accessible to the public for recreational purposes, there may be a potential hazard, and so restrictions have been placed in a number of countries in Europe such as Germany, on the use of biosolids in commercial forests with public access. If European rules are evolved they should, however, be sensible. Guidelines were issued in the U.K. in 1994 [30]. These recommended that maximum soil concentrations are based on those used for agricultural land.

Sacrificial and Dedicated Land

Sacrificial land is land that has been selected conveniently close to a sewage treatment works and can receive large quantities of sludge without causing environmental problems, including water pollution. Sludge is applied in large quantities and ploughed in. The biological activity of the soil breaks down the sludge organic matter. This technique is not common, but as the costs of sludge disposal rise it may well become more popular. There are a number of sites in Europe that used to be sewage farms, but now this treatment technique has been replaced by a modern treatment works and the land is used to dispose of the sludge.

Land is used for the cultivation of forage crops or small grain crops for animal feed and is a low-risk option for historically contaminated soils in which metal levels may have already exceeded maximum permissible concentration. The EC directive permits the continued use of such dedicated land already in use in 1986, provided it is used to grow crops exclusively for animal consumption with no resulting hazard to either human health or the environment. Stricter management along these lines can permit productive farming to continue on such land which might otherwise be condemned to permanent dereliction. The big advantage of this option is the low cost [32]. As the restrictions on disposal become greater, the vulnerability to external factors increases and one of the consequences is a reawakening of interest by disposal managers in the use of directly owned sites, be it for agricultural purposes in accordance with the 1986 Directive or as sacrificial land. It may well be that new sacrificial sites could be unpopular in some areas of Europe.

Horticultural Land and Gardens

Use of biosolids in gardens and for horticultural crops could be viewed as a subsidiary class in agricultural use, and indeed the statistics in Table 13.2 include them together. However, it represents a comparatively high-risk outlet in terms of effects of metals and potential pathogen transmission because there is likely to be little control over how the biosolids are used. If the biosolids are to be marketed in this way, substantial controls have to be applied to the quality of the biosolids both in terms of metals and pathogens. This causes the price to rise, and it is an expensive method of disposing of sludge. However, such methods of disposal tend to be high-profile in terms of public relations and are becoming increasingly attractive in view of the environmental concerns about the damage caused by the use of peat. If biosolids are to act as a peat replacement, then the most obvious way forward is in composting.

In spite of the fact that composting has received a considerable amount of attention, only a relatively small amount has been composted in the past in Europe either alone or in combination with domestic waste or municipal refuse. Composting is often portrayed as being an alternative method of disposal, whereas in fact it is one of the methods of treatment to render biosolids suitable for use in agriculture or horticulture. For this reason the composting has not been a major activity in European countries because it is more expensive than the established uses of liquid biosolids in agriculture, particularly of digested biosolids. However, there are signs that this is changing.

A clear distinction must be made between composted biosolids used in agriculture (for which rules exist already) and composts, which contain a derivative of biosolids and which are used in horticulture and gardens. The environmental concern is that it is possible for crops such as salad vegetables to be grown in biosolids.

If interest increases in the future it is quite clear that there will be a need for guidelines on the use of composted materials including biosolids in horticulture and gardens. There have already been extensive discussions within the context of the Economic Community COST 68 program. One way forward would be for compost to have the same metal quality as that of the soil limits in the EU Agricultural Use Directive and to be disinfected. There are regulatory and investment implications in these requirements.

A number of biosolids treatment processes have been developed to produce earth supplements. An example is the N-Viro process and this had been marketed as a commercial bagged product. There are no indication that this is moving out into the full agricultural market but it could provide a service and product with a more professional appearance. Similar changes are developing with granulated or pelletised dried biosolids. These are products more acceptable in smaller scale situations and may fill some needs in agriculture although the more traditional forms of biosolids will still fulfil a major need for farmers.

A number of states within the EU already have laws and regulations relating specifically to composts or in general to marketing and customer protection for facilities, etc. [39]. It is understood that the EU is formulating a directive on fertilizers and compost, but it is not possible to assess effects yet. It has been in progress for several years. The exclusion from the definition of organically grown crops of those fertilized with biosolids was not helpful.

Minor and Novel Uses

There are a wide range of other uses for sludge which exploit its energy or chemical content. A major problem is the complex nature of sludge making the extraction or manufacture of useful substances difficult or prone to contamination. Various types of sludge have been used as an animal feed stuff but with the possible exception of fish this has not been successful. A range of potentially valuable constituents can be extracted from sludge such as protein, grease, vitamin B_{12}, metals, and phosphorus, but none has been found to be commercially viable in Europe, partly because of the disposal problems of residues. The economics are unattractive, being very susceptible to changes in world commodity prices.

Production of fuel oil by low-temperature pyrolysis have been demonstrated elsewhere in the world and there is some interest in Europe in developing the process, particularly under the current circumstances where the price of energy appears to be increasing.

Other successful minor outlets appear to be developing such as the incorporation of sludge into building materials—an example is building brick. Sludge has also been turned into fiber board and sludge incinerator ash is used as a lightweight aggregate and filler for clay pipes and tiles. In some instances the ash could be melted to produce slag, which in turn is used as a construction material from which there is less opportunity for metals to leach. None of these outlets have been found to be of more than local importance, and reflect personal interest. At present an entrepreneurial spirit is required on the part of the sludge producer to promote and unusual outlet for his products. However, as the pressure on conventional sludge disposal routes becomes greater and costs rise, and sludge disposal authorities are inculcated with a commercial attitude, it may well be that these processes will develop in Europe [35].

There are also techniques available that are designed to produce sludges that are suitable for specialized uses. Examples are the N-Viro process and vermiculture (another product of which is worms, which may be a useful source of protein and fishing bait).

The methods of disposal in future are likely to be the classic options, but minor and novel uses, including nonagricultural use of compost, are likely to fill niche markets, which play important but minor roles in overall schemes within Europe.

The management of the Common Agricultural Policy within the European Union requires the agricultural land be set aside on a rotational basis to control food production. There is a grant support system. This land cannot be fertilized and it is regretted that biosolids cannot be added.

TRENDS AND THE OUTLOOK FOR THE FUTURE

A starting point is to look at the baseline situation that existed in 1991 when the urban wastewater treatment directive was passed (Table 13.1 [1]).

The Global Atlas [1] predicts that sludge production will rise to 10 million tonnes by 2005 with utilization and incineration being the prime routes for disposal (see Table 13.28) [25]. This fits the 10 year prediction made back in 1984. At that time, it did not foresee the loss of marine dispersal and it predicted 7 million tonnes in 1994 with 37% recycled to the land, 40% disposed to landfill, 16% incinerated and 6% disposed at sea.

The WRc report [24] predicted that by the end of 2005 landfill will become the smallest outlet, at about 17%, with recycling the greatest at 45% and incineration at 35%. The report recognises that these reports are influenced by the fact that Germany will be producing 38% of the sludge in the EU and the overall strong decline in dispersal to landfill will be in the main due to the anticipated implementation of the German policy to restrict the landfilling of organic matter-rich wastes. At this point it is worth mentioning that one of the great mysteries of the data concerning sludge disposal in Europe is the fact that Germany has been consistently much higher in terms of per capita production in many of the surveys over recent years. There seems no obvious reason for this. If the per capita production in Germany for treated sludge disposed is dropped to figures comparable to those found elsewhere in the European Union, not only

TABLE 13.28. Future Development of Quantity and Disposal of Communal Sewage Sludge in the Old 12-Member Countries of the EC [1000 tonnes dry matter/year (%)].

Year	1984	1992	2000	2005
Utilization	2057 (37)	2504 (39)	3617 (40)	4576 (45)
Incineration	518 (9)	715 (11)	2088 (24)	3872 (38)
Landfill	2988 (54)	3257 (50)	3200 (36)	1615 (17)
Total quantity	5563 (100)	6476 (100)	8906 (100)	10063 (100)

will the total sludge production in Europe fall, but this will affect the statistics for trends in future disposal. Clearly stabilization of sludge may well increase, although this may not be true for incineration. Denmark and Italy are expected to be incinerating greater proportions of their sludge and disposing of smaller proportions to landfill although the actual quantity going to landfill in Italy is expected to increase due to the great increase in total sludge production. A greater proportion of the sludge increase would be recycled at the expense of landfilling, though again the actual quantities to both outlets will increase. Germany predicted a substantial increase in the use of incineration, with less sludge being recycled. Increases in the proportions of sludge undergoing incineration and recycling were expected in the U.K. with decreases in both the proportions and quantities of sludge disposed of at sea and to landfill. It was expected that less sludge would be used in agriculture in Luxembourg due to problems caused by heavy metal contamination. Spain had predicted a decrease in the quantity of sludge disposed of at sea and a corresponding increase in the use of recycling. Little change was foreseen in the proportions of sludge going to each outlet in the Republic of Ireland.

Not all the EU member states were able to predict the quantities of sludge that would go to the various disposal routes over the coming years, so it was not possible to give an accurate prediction of future disposal practices for the whole of the EU.

With the changes being implemented by the EC it is likely that substantial quantities of the sludges being disposed of at sea will be incinerated and some will go to agricultural land. On balance it would appear that the long-term growth area for sludge disposal in Europe is going to be incineration, with a steady supply to agriculture, where there is an already proven success for this outlet. The uncertainty is how much biosolids will be acceptable in agriculture, where it has not been used before. As stated earlier, this would indicate that the most likely form of sludge to be disposed of in future will be dewatered sludge because of the need to transport sludge to agriculture land over greater distances, and the need to have properly dewatered sludges for incineration and for landfill disposal. The provision of dewatering increases operational flexibility; dewatering equipment can be switched off so that liquid sludge can be supplied to customers if necessary and appropriate, but if there is no equipment it cannot be switched on if liquid sludge disposal becomes impossible. The question to be answered by planners is how much backup dewatering capacity is required in a dual option operation.

It may well be that sewage treatment processes with lower sludge production may become more favored, although these are often more expensive to run and hence may not be employed extensively. However, this may be counteracted by the additional sludge production due to chemical treatment of sewage to remove phosphate. Certainly a whole new pholosophy on treatment processes is bound to emerge to achieve the best quality, minimum quantity sludges while still treating sewage satisfactorily.

In terms of agricultural utilization, there will be continuing pressure for sludge to be properly treated, and it is likely that this form of treatment will be anaerobic digestion. In the 1984 survey, 79% of the EC sludge was treated prior to disposal, with 56% undergoing anaerobic digestion (mainly mesophilic digestion) and 16% aerobic digestion (mainly cold digestion). Table 13.28 shows that this position was initially unchanged in 1992 [5]. However, if the sludges are incinerated subsequently, it is likely that there will be less digestion, because the need to provide the treatment is less and indeed, may even be not beneficial because the calorific value of digested sludge is less than undigested sludge. Processes such as composting and N-Viro may well find a bigger role than they have at the moment. In this way sludge disposal managers can extend the suite of products available to their customers ranging from the domestic to the large-scale farm market. Drying and pelletization might fall into this category; these may help with restrictions on the times of application (requiring more storage), improved customer care image, and flexibility of services.

It is quite clear that data are difficult to collect and collate, but modern responsible management dictates that records should be kept and therefore on this basis information should be more readily available in the future. For instance the EU directive on the agricultural use of sludge requires the maintenance of registers.

With the greening of social and political attitudes, there will be ever-increasing stringency on the controls applied to sewage sludge, which will make life difficult for sludge disposal managers throughout Europe. For instance, in spite of the fact that incineration is probably going to be the most appropriate method of disposing of sludge in the long-term future, this relies on the fact that the necessary planning and development approvals can be obtained for siting incinerators. Disposal managers are likely to find that no matter how well-designed and operated the plants will be, there will be opposition in principle from local people on the basis of "not in my back yard" [27]. Arising from the fecal aversion barrier, guidance from central government on planning issues can be helpful. The U.K. government has issued guidance on this matter.

It is not possible to review all of the developments in each European country. Indeed even within countries there are varying developments for region to region. The Global Atlas gives a good insight into some of these variations. It is then, for example, that there are varying degrees of difficulty. The tight limits controlling biosolids use in the Netherlands have had the effects of forcing disposers to choose incineration [42].

Concerns about heavy metals and organics is making biosolids use more difficult in Denmark and Sweden, but there are high hopes for the future. The Swedish government is discouraging landfill and incineration. In Denmark the rules for biosolids use are very stringent. The reasoning is that more controls will sustain confidence, particularly with farmers and this will make use easier. In the Netherlands and Belgium one of the problems has been that of the vast quantities of animal manure which must be disposed. Some ideas of current views for the future can be obtained from the Global Atlas. In France, landfill will be phased out within a few years. In Germany beneficial reuse is being limited in essence to arable land but incineration is banned in some Lander and landfill is likely to be discouraged as explained earlier and this had been a primary option for the future. Even these statements are likely to be too sweeping and there will be many local variations. There is a need for continuing close collaboration and co-operation between EU countries.

The "precautionary principle" takes precedences over "risk assessment." As a basis for policy and practise in several European countries. The argument is that because society takes an antipathetic view of sewage sludge (or even biosolids if that word has been introduced) probably due to the fecal aversion barrier—everything should be done to inspire confidence but avoiding any risk. Into this melting pot of psychology and environmental science must now be added the concept of sustainability. Success depends on being able to devise a policy which recognizes public concern, is sustainable and affordable but does not give in to unreasoning paranoia.

So where is sludge in England and Wales going over the next 20 years? The consequences of the EU Urban Wastewater Treatment Directive are being addressed in the Water Company investment programs for 1995–2000. As explained, total sludge production could rise by at least 50%. More pessimistic views, drawn from a 1992 survey, suggest a doubling [24]. The disposal of this sludge will not only require solution in its own right but will impact on existing operations. All of this might be achieved against a background of tightening constraints and legislation; problems will be exacerbated by the loss of marine dispersal; plans have been announced for coping with this loss [28]. In terms of change to the proportions of disposal, agricultural use will remain the principal option—the increased requirement for land counter balanced to some extent by increased constraints. Incineration, will increase in terms of quantity and proportion—due principally to the loss of marine dispersal. Landfill will increase after the loss of marine dispersal but this will decline due to increasing constraints and loss of available sites. Finally the development of other methods of use will rise as the traditional outlets become restricted. Examples are use in land restoration, forestry, deciduous and coniferous forestry and in horticulture. However, this is likely to be minor in terms of sludge disposal; Biosolids use in this context is a service to the user and not a facility to the disposer; it could have big impact on the user of activities but not so much on biosolids use. This is summarized in Figure 13.2.

The choice of solution on a local or regional basis will be made initially and continuously reevaluated using economic models. The most obvious model in the UK is WISDOM but others are available. Sludge disposers are more likely to have to demonstrate that the most appropriate options have been adopted to regulatory agencies and even the general public. It is possible that such efforts will be associated with environmental impact assessment and audit.

It is envisaged that the current practice of storage in farms to balance supply and demand may have to be phased out due to the pressure of public opinion. There will be greater quantities of biosolids dewatered and more biosolids will be treated. Anaerobic digestion, pasteurisation processes, drying and pelletisation and composting will all increase within the context of agricultural use. Whilst there is a current demand for biosolids compost as a peat replacement it is difficult to predict how successful this will be. The replacement is not straight forward.

Injection of biosolids into farmland is likely to develop even more. Although public relations is already a very important feature of successful operations, public opinion is likely to require an even greater effort by disposers. There will be increasing controls over industrial effluents discharged to sewers and it is inevitable that there will be pressure to use "clean" technologies to result in even cleaner biosolids. Whatever happens, the future of sludge disposal in the U.K. will be primarily a function on what happens in a European context.

Cleaner biosolids will aid another development. Disposal managers have always been able to market sludge as an organic alternative to chemical fertilisers. However, this is rather different to the concept of organically grown crops for which chemicals must be absent. Only "clean sludge's" would qualify for use in growing crops marketed as organically grown. Whilst this market is likely to be restricted the growing "green" interest in organic crops is bound to have some influence. This could also lead to a restriction of uses of household chemicals and this is what is happening in Scandinavia. However the environmental benefits and consequences need to be assessed, for example what would happen if household disinfectants were to be phased out? One other trend is worth examining. Over recent decades, there has been a concentration of effort on developing what could be called the basic options of sludge disposal. There has only been peripheral interest in exotic treatment processes such as oil, protein and vitamin production. These were never economically

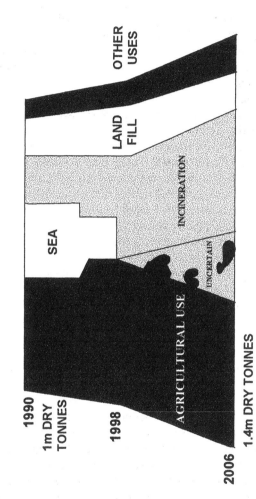

Figure 13.2 Expected changes in sludge disposal (ENGLAND and WALES, 1990–2006).

attractive. However the new economic and regulatory order is bound to focus attention on these again.

CONCLUDING OBSERVATION

It is very clear that the treatment of urban wastewater is going to be of key significance in the development of the environment throughout the European Union. This will create more sludge. It is recognised that very large sums of money will be involved and this has attracted the attention of the Council of Ministers because these sums will have impact on national budgets. Whether or not the timescales of the Urban Wastewater Treatment Directive are practicable in engineering and scientific terms is a separate issue from whether or not individual States can afford the necessary expenditures. Legislation containing sludge disposal continues to extend in Europe as a whole and in individual countries. When sludge is used in agriculture, controls over fertilizers and manure's will also have an effect. The costs of sludge treatment and disposal are a significant proportion of the costs of treating urban wastewater and hence this topic will be significantly influenced by any decisions taken on investment. It is absolutely vital that the consideration of sludge treatment and disposal is integrated in technical as well as economic terms, with the total requirement for urban wastewater treatment. An arrow pointing off the side of sewage treatment works' design schedule, as often happened in the past, is not acceptable!

Once the works are constructed, careful and responsible management of the total works including sludge treatment disposal facilities is vital. As the technology of the works processes and environmental requirements increase, so the skills of the operators will have to increase and vocational training will become even more important. Several initiatives are in hand on this matter.

Public acceptance is now crucial as the genuine risk aspects are contained by proper controls and management what remains is perception. Of course perception is reality and there is and will continue to be a need for information and education to sustain on-going policies and practices.

There is a need for fully integrated policies which balance structure and local flexibility. European waste water groupings such as the European Water Pollution Control Association and the European Waste Water Group have recommended strategies to the European Commission. The European standards organisation CEN has been established to develop standards for sludge quality but not standard sludge's or biosolids which would not be helpful. New initiatives are being taken to produce integrated policies for waste and water management which ensure good connections between initiatives such as in the Groundwater Action Programme.

These organizations also all promote exchange of information and experience. Such exchange is developing to embrace transatlantic comparisons. We should all benefit. It is clear that in Europe, just as in the U.S. there are now "votes in sewage" as wastewater management is considered to be a high social and political priority.

Such political attention which includes the costs of services as well as the environment issues has led to a reevaluation of the way the services are provided. Contracting out on outsourcing and even privatisation is on the agenda in Europe for all water and wastewater services. This is particularly evident in England and Wales where the privatised water utilities manage all municipal wastewater. The expertise developed in doing this in some companies such as Anglian Water is a major driver in establishing global businesses.

Whatever system water pollution managers work in they have "a duty of care" and have taken their rightful place in protecting the environment of Europe.

JAPAN—AN OVERVIEW OF TREATMENT, UTILIZATION, AND DISPOSAL

This section will first review the changes in sludge generation and disposal that have occurred in Japan over the past ten years. A discussion of the present status and technology development of sludge treatment processes will be presented. The methods of turning sludge into resources and the beneficial uses of sludge in Japan will be explained. Finally, the recent trend in the centralized treatment of sludge will be shown, introducing the regional sewage treatment project (ACE plan) by the Japan Sewage Works Agency.

INTRODUCTION

Japan is an island country with a population of around 122 million and a land area of about 378,000 km². More than 70% of the area is mountainous and the remainder is habitable area with a ground gradient of less than 10%. All production and consumption activities are conducted within this habitable area; therefore, the activities per km² of the habitable land area have been greater than those in any other country. The population has steadily concentrated into the urbanized areas. In 1960, the ratio of the urban population to the total population was 43.7%, but by 1985 it had increased to 60.6%. Twenty-one percent of the total population lives in Tokyo and the eleven designated cities, which comprise only 1.5% of the total land area.

Sewered population in Japan has increased mainly due to the promotion of the Five-Year Program for Sewerage Construction, and included 44% of the total population at the end of fiscal 1990. The Seventh Five-Year Program for Sewerage Construction, with fiscal 1991 as the first year, is a plan for achieving 55% sewered population by the end of fiscal 1995. In parallel with the spread of sewerage facilities, treated effluent has increased from year to year, and the generated sludge has increased at almost the same rate. The treatment and disposal of the sludge generated becomes very problematic not only in large cities where disposal sites are scarce, but also in less populated cities where sludge treatment is newly started. Sludge treatment and disposal is not only an urgent problem in the field where sludge is generated daily, but is also an important problem that requires a permanent future solution.

SLUDGE GENERATION AND DISPOSAL

The number of wastewater treatment plants operated in Japan at the end of fiscal 1988 was 736. A breakdown of the treatment processes that are used is shown in Table 13.29. Secondary treatment, which achieves a BOD concentration of less than 20 mg/L in the treated effluent, is generally applied in Japan with conventional activated sludge process, including step aeration, being used in more than 80% of all the plants. In terms of volume of treated effluent, more than 90% of the wastewater is treated by the activated sludge process. Thus, the sludges generated from the wastewater treatment plants in Japan are primary sludge and excess activated sludge.

Figure 13.3 shows the annual trend in the volume of wastewater treated and the amount of sludge generated based on the data from *Sewerage Statistics Report* issued annually by the Japan Sewage Works Association [46]. The generated sludge is the measured value of the sludge removed from the liquid treatment system and transferred to the sludge treatment system. The solids concentration of these sludges is in the range of 0.5 to 2.0%, depending upon the treatment plant source. The volume of generated sludge has increased each year at a growth rate of about 5%. This is higher than the observed growth rate of treated effluent. Causes of this are probably: (1) increase in the strength of raw wastewater, (2) upgrading the degree of wastewater treatment, and (3) decreased concentration of the sludge generated. Recently, the sludge being generated has shown poorer thickenability than previous years.

On the other hand, the quantity of sludge being disposed has decreased in recent years as indicated in Figure 13.4. This is mainly due to the fact that incineration is more widely used, and sludge conditioning for

TABLE 13.29. Number of Wastewater Treatment Plants in Japan Classified as to Treatment Process (FY 1988).

Treatment Process	Design Daily Maximum Flow Rate (Thousand m³/day)						
	<5	5–10	10–50	50–100	100–500	<500	Total
Plain sedimentation	—	—	3	—	1	—	4
High-rate trickling filter	2	2	5	—	—	—	9
High-rate aeration	2	—	9	3	2	—	16
Activated sludge							
Conventional	52	50	226	90	116	11	545
Step aeration	3	4	16	16	16	7	62
Extended aeration	13	1	2	—	—	—	16
Contact stabilization	1	—	—	—	—	—	1
Pure oxygen	—	1	3	1	2	—	7
Nitrification-denitrification	—	1	—	2	1	—	4
Sequentional batch	3	—	—	—	—	—	3
Oxidation ditch	31	11	—	—	—	—	42
Rotating biological contactor	13	4	6	1	—	—	24
Contact aeration	2	—	—	—	—	—	2
Others	1	—	—	—	—	—	1
Tertiary	(11)	(6)	(3)	(8)	(7)	(1)	(36)
Total	123	74	270	113	138	18	736

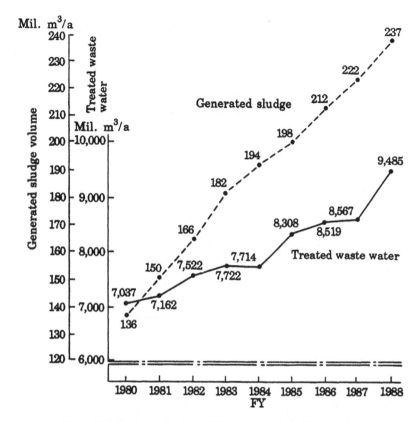

Figure 13.3 Yearly trend of treated wastewater and sludge generation.

dewatering is being converted from inorganic coagulants, ferric chloride and lime, to organic coagulants. The general situation in Japan may be summarized as one where the generated sludge is steadily increasing, but where effective measures for reducing the final sludge volume in the sludge treatment stage are aggressively being taken.

The methods of sludge disposal, including beneficial uses and the types of sludges being disposed, are summarized in Table 13.30 in terms of the dry solids of generated sludge. Based upon this data, 1.36 million tons of solids-based generated sludge was treated and disposed of in fiscal 1988, of which 58% was incinerated and 15% was beneficially utilized for agriculture, green spaces, and construction. The data presented in Table 13.30 was developed by determining the solids-based generated sludge for the individual plants using the data of thickened sludge and its concentration from the 1988 edition of *Sewerage Statistics Report*.

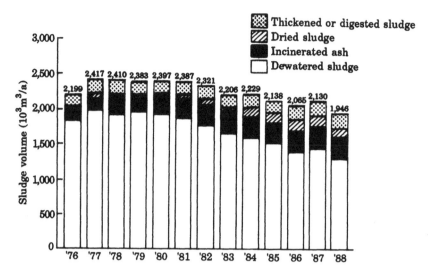

Figure 13.4 Yearly trend of disposal and utilized sludge volume.

SLUDGE TREATMENT

Sludge Treatment Systems

A generalized flow diagram of the sludge processing and disposal schemes used in Japan is shown in Figure 13.5. These processes are employed for stabilization and volume reduction. Volume reduction is achieved by decreasing the moisture content during the dewatering processes, by moisture removal and gasification of organic matter during incineration and by reducing void ratios during the melting process. Since the solids concentration of generated sludge from wastewater treatment systems is 1 to 2%, the reduction ratio of dewatered sludge with a solids concentration of 20 to 30% is about 15 times. This will be further reduced about 10 times by incineration, due to the gasification of organic matter. The stabilization processes include anaerobic digestion, aerobic digestion, and composting, and about one-half of the organic in the sludge is decomposed. A further reduction in volume occurs in composting due to the evaporation of moisture.

Table 13.31 shows the number of treatment plants in Japan, classified by sludge treatment and solids-based treated sludge. From this it can be seen that the prevailing sludge treatment schemes are raw sludge dewatering (236 plants), digested sludge dewatering (210 plants), sludge dewatering combined with incineration (129 plants), and digested sludge dewatering followed by incineration (67 plants). These four flow schemes account

TABLE 13.30. Sludge Disposal and Utilization (Generated Sludge Solid Basis FY 1988).

Sludge Type	Method of Disposal or Beneficial Use					Total	
	Landfill	Agricultural and Landscaping	Construction Works	Ocean Dumping	Other	DS t/a	%
Liquid	—	—	—	7127	195	7322	0.5
Dewatered	334,027	45,957	—	2931	3835	436,750	32.0
Composted	10,587	95,989	—	—	3166	109,742	8.0
Dried	7997	6153	—	—	387	14,537	1 1
Incinerated ash	711,727	8581	50,213	—	18,956	789,477	57.9
Metal slag	2785	—	3454	—	—	6239	0.5
Total DS—t/a	1,117,123	156,680	53,667	10,058	26,539	1,364,067	—
Total %	82.0	11.5	3.9	0.7	1.9	—	100.0

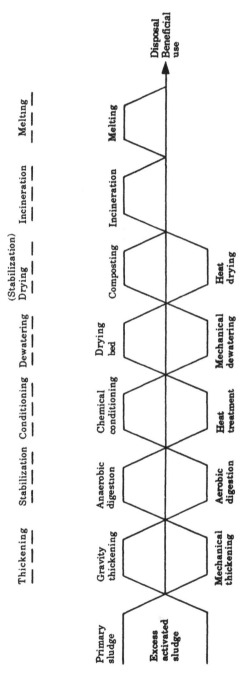

Figure 13.5 Generalized sludge processing and disposal flow diagram.

TABLE 13.31. Present Status of Sludge Treatment Flow (FY 1988).

Final Sludge Type	Sludge Treatment Flow	Number of WTPs	Treated Sludge Solids	
			Dry Weight (1000 t/a)	Percent
Liquid sludge	Without treatment	2	0.1	0.01
	Thickening	6	0.1	0.01
	Thickening—anaerobic digestion	4	7.1	0.52
	Subtotal	12	7.3	0.54
Dewatered cake	Thickening—dewatering	236	163.3	11.97
	Thickening—anaerobic digestion—dewatering	210	261.4	19.16
	Thickening—aerobic digestion—dewatering	12	4.0	0.29
	Thickening—heat treatment—dewatering	10	8.1	0.59
	Subtotal	468	436.8	32.02
Dried sludge	Thickening—drying	12	0.3	0.02
	Thickening—anaerobic digestion—drying	11	1.4	0.10
	Thickening—dewatering—drying	9	7.3	0.53
	Thickening—anaerobic digestion—dewatering—drying	20	5.6	0.41
	Subtotal	52	14.5	1.07

(continued)

TABLE 13.31. (continued).

Final Sludge Type	Sludge Treatment Flow	Number of WTPs	Treated Sludge Solids Dry Weight (1000 t/a)	Percent
Composted sludge	Thickening—dewatering—composting	40	44.7	3.28
	Thickening—anaerobic digestion—dewatering—composting	43	65.0	4.76
	Thickening—aerobic digestion—dewatering—composting	1	0.1	0.01
	Subtotal	84	109.7	8.05
Incinerated ash	Thickening—dewatering—incineration	129	436.2	31.98
	Thickening—anaerobic digestion—dewatering—incineration	67	296.1	21.70
	Thickening—aerobic digestion—dewatering—incineration	3	0.1	0.01
	Thickening—heat treatment—dewatering—incineration	8	48.6	3.56
	Thickening—c.g. process—incineration	1	0.7	0.05
	Thickening—wet oxidation—dewatering	1	7.8	0.58
	Subtotal	209	789.5	57.88
Melted slag	Thickening—dewatering—melting	5	3.2	0.23
	Thickening—anaerobic digestion—dewatering—melting	1	2.6	0.19
	Thickening—dewatering—incineration—melting	2	0.5	0.04
	Subtotal	8	6.3	0.46
	Total	833	1364.1	100.0

for treatment of 85% of the total treated solids. In addition, a new sludge treatment system using a melting process in conjunction with other treatment processes shows promise as a process for producing construction materials.

Thickening

As a result of the spread of separate sewer systems and improved lifestyle in Japan in recent years, the wastewater properties have changed. The organic content of the sludges generated in the treatment plants has increased, and problems of sludges with poor thickenability have developed in many of the treatment plants. Excess activated sludge tends to have a poor thickenability, and recently many of the treatment plants have adopted the use of mechanical thickening for these sludges, as shown in Table 13.32. Separate thickening, i.e., gravity thickening for primary sludge and mechanical thickening for excess activated sludge, has been increasing.

The flotation method currently being adopted is mostly dissolved air flotation, which is able to obtain a sludge concentration of 3 to 4% for excess activated sludge. Centrifugal thickening is carried out in two types of centrifuges, either solid bowl or basket type, and is able to obtain a higher concentration than flotation thickening but at a higher power cost.

The separate use of gravity thickening for primary sludge and excess activated sludge has the effect of improving the conventional gravity thickening method as compared to its use on combined sludge. The sludge concentration obtained by separate gravity thickening is in the range of 3 to 5% for primary sludge and 1.5 to 2.5% for excess activated sludge. The thickening and putrefying properties are considerably different between primary sludge and excess waste activated sludge, and it may be desirable to further develop the use of separate gravity thickening considering these differences in characteristics.

A remarkable development in thickening techniques is a pelletizing thickening process recently developed by the Japan Sewage Works Agency, jointly with a private enterprise [47]. This method coagulates and pelletizes sludge by the combined use of a metal coagulant, iron or aluminum-based, and an amphoteric polymer, and it is used as a sludge thickening and conditioning process for belt press filter. Figure 13.6 shows the process flow diagram of a pelletizing thickening system. The Japan Sewage Works Agency started technical evaluation of efficient sludge thickening methods, including the pelletizing method, in fiscal 1990.

Anaerobic Digestion

As stated previously, anaerobic digestion is being used at more than

TABLE 13.32. Increase of Treatment Plants Using Mechanical Thickeners.

Fiscal year	Flotation Thickening		Centrifugal Thickening		Other Mechanical Separate Thickening	Without Thickening	Gravity Thickening	Total
	Combined[a]	Separate[b]	Combined[a]	Separate[b]				
1981	2	17	1	6	1	65	402	494
1982	4	20	1	8	1	68	420	522
1983	5	24	1	10	1	73	434	548
1984	8	29	2	16	2	69	461	587
1985	9	31	2	21	2	75	491	631
1986	15	37	5	23	2	62	518	662
1987	10	40	3	30	2	98	520	703
1988	8	42	5	41	2	103	535	736

[a]Primary sludge and secondary sludge are thickened together.
[b]Gravity thickening for primary and mechanical thickening for secondary sludge.

724

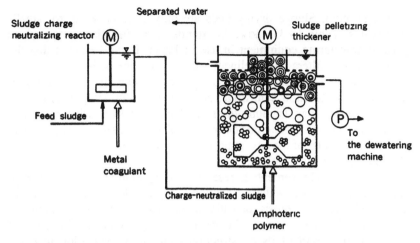

Figure 13.6 Schematic flow diagram of the pelletizing thickening system.

40% of the wastewater treatment plants in Japan. The characteristics of the anaerobic digestion process as sludge treatment are:

- decrease in the solids in the sludge
- stabilization of organic matter in sludge

Other auxiliary advantages include:

- use of the digestion gas generated
- storage function of the digestion tank

As a stabilization process, it is more efficient than similar aerobic digestion or composting processes. The process can be made energy-saving by the use of the digestion gas generated.

Examples of recent technical improvements in the anaerobic digestion process include highly concentrated sludge digestion, with use of mechanical thickening, and construction of egg-shaped digesters for improving the agitation and efficiency. In the Comprehensive Research and Development Project executed by the Ministry of Construction (Biofocus WT) conducted in fiscal 1986 to 1990, the application of the immobilization method to methane fermentation and liquification pretreatment using heat treatment and enzymes, were reported to be effective for improving the digestion efficiency [48].

Dewatering

The main method of sludge dewatering is some type of mechanical dewatering device (vacuum filter, filter press, belt press, screw press, centrifuge, etc.). Sludge drying beds are only of limited use for small-scale plants. Table 13.33 shows the trends in the installation of various types of dewatering equipment in use in Japan. It can be noted that the use of vacuum filters has declined, whereas the use of belt presses has increased, so that presently belt press installations outnumber vacuum filter installations. In addition, centrifuges and belt presses utilizing polymers are more numerous than vacuum filters and filter presses using inorganic coagulants, such as ferric chloride or lime. The increased use of polymers is probably because of the advantage of less sludge volume due to chemical addition when using a polymer.

Some recent achievements in research and development for dewatering include a two-agent dewatering method using two kinds of polymer and also an electro-endosmosis dewatering process for removing moisture by moving liquid by means of electrical energy. Also, the Ministry of Construction as part of its construction technology evaluation in fiscal 1989, carried out studies with respect to the development of automatic moisture measuring systems for dewatered sludge. Several instruments were found to be in the stage of practical use for the control of the dewatering process.

Incineration

There are several types of sludge incinerators, such as: multiple-hearth, fluidized-bed, rotary-kiln, and stroker. The number of incinerator installations has greatly increased in recent years, as shown in Table 13.34. While the multiple-hearth incinerators still comprise the majority of installations (seventy-two in 1988), the installation of fluidized-bed incinerators has greatly increased and is now the type of choice. With respect to incineration technology, aggressive improvements have been made over the last ten years in both energy-saving processes and environmental measures for air pollutants and offensive odors. The top level technology in the world now seems to have been completed in the area of incineration in Japan.

With respect to energy-saving processes, the conditions of autogenous combustion of sludge were clarified in the report. "Technical Evaluation of the Autogenous Combustion Systems," performed by the Japan Sewage Works Agency from 1985 to 1987 [49]. By the combination of optimum control of the amount of combustion air and prestage drying process, the heat balance of incinerators has been recently improved. In order to achieve autogenous combustion, these incinerators require a water content of 75% for dewatered sludge with organic coagulant and 60% for that with inorganic coagulant. That is, dewatered sludge with a low calorific value of 400 to 600 kcal/kg is said to be required for the autogenous combustion operation.

TABLE 13.33. Trends in the Use of Existing Dewatering Machines.

Fiscal Year	Vacuum Filter	Filter Press	Centrifuge	Belt Press	Screw Press	Others	Total
1980	604	220	241	124	6	10	1205
1981	577	240	306	189	7	11	1330
1982	560	262	320	245	10	18	1415
1983	552	284	315	289	13	10	1463
1984	541	294	306	334	13	1	1489
1985	517	325	311	378	17	0	1548
1986	486	333	324	428	20	20	1611
1987	472	324	332	484	23	3	1638
1988	452	326	342	561	24	8	1713

TABLE 13.34. Trend in Incinerator Usage in Japan.

Type	Number of Installations						
	1982	1983	1984	1985	1986	1987	1988
Multiple hearth	54	55	55	57	61	71	72
Fluidized bed	27	27	36	36	44	52	53
Rotary kiln	14	15	15	16	16	20	20
Pyrolysis	2	3	3	3	2	2	1
Rotary bed	3	3	3	3	0	0	1
Melting furnace	1	1	1	1	2	2	3
Wet oxidation	1	1	1	1	2	2	1
Total	102	105	114	117	127	149	151

In Japan, exhaust gas from the combustion of sludge is restricted by diversified provisions in different regulations and standards. The Air Pollution Control Law prescribes the concentration of emission of SO_x, dust, hydrogen chloride, and nitrogen oxides. With respect to SO_x, total pollution load regulations are imposed for certain areas. The controlled value of emission concentration for sludge incinerators is shown in Table 13.35.

Offensive odor is regulated by the Offensive Odor Control Law, and the concentration of eight substances with offensive odor is restricted at the property lot lines as indicated in Table 13.36.

Control of the exhaust gases from sludge incinerators is normally performed by applying a series of treatment processes, including cyclones, gas washers, and electrostatic precipitators to the exhaust gas. Figure 13.7 shows a typical system flow diagram for a fluidized-bed incinerator.

Melting

If sludge is heated to 1200 to 1500°C, the organic material in the sludge is decomposed and burned, and the remaining inorganic material becomes melted liquid. The liquid is cooled, solidified, and melted slag is formed. The process of melting sludge was developed recently in Japan and has the advantages of a greater volume reduction than incineration and also of making metals insoluble. At present, from the viewpoint of utilizing the melted slag in various construction materials, this process is now drawing much attention.

Shown in Table 13.37 are the models and characteristics of the melting furnaces that have been developed and put to practical use for wastewater sludge. Feed sludge to the melting furnace can be incinerated ash and/or dewatered sludge. Table 13.38 summarizes the characteristics of melting furnaces presently in operation.

Slag obtained by melting generally has different physical properties depending on the cooling method, and the melting method is selected depending on the intended use of the melted slag. The relation between cooling method and the physical properties of slag is shown in Table 13.39.

BENEFICIAL USE OF SLUDGE

Present Situation

The beneficial use of sludge can be realized by utilizing the sludge itself and/or by using the methane gas generated in the sludge treatment process as energy. These beneficial uses are classified by form and shown in Figure 13.8. The uses were mostly established in the 1980s, except the uses in agriculture and green spaces, which had been practiced before then.

TABLE 13.35. Discharge Standards of Air Pollutants from Wastewater Treatment Facilities.

Regulated Substance	Sludge Incinerator		Gas Turbine	Diesel Facilities
	Off Gas ≥40,000 Nm³/h	Off Gas <40,000 Nm³/h		
Sulfur oxides (Nm³/h)	K value regulations[a] K = 3.0–17.5 (K = 1.17–2.34)[b]		K value regulations[a] K = 3.0–17.5 (K = 1.17–2.34)[b]	K value regulations[a] K = 3.0–17.5 (K = 1.17–2.34)[b]
Dust (g/Nm³)	0.15(0.08)[b]	0.50(0.15)[b]	0.05(0.04)[b]	0.10(0.08)[b]
Hydrogen chloride (mg/Nm³)		700	—	—
Nitrogen oxides (ppm)	250	700	70	950

[a]SO_x is regulated by the following formula: $q = K \times 10^{-3} He^2$, q is SO_x discharge, Nm³/hr; K is coefficient defined at each location, He is compensated stack height, m.
[b]Values in parentheses are applied for special districts.

TABLE 13.36. Range of Permissible Emission Concentrations of Odor Causing Compounds.

Compound	Air Content (ppm)
1. Ammonia	1–5
2. Methyl mercaptan	0.002–0.01
3. Hydrogen sulfide	0.02–0.2
4. Methyl sulfide	0.01–0.2
5. Methyl disulfide	0.009–0.1
6. Trimethyl amine	0.005–0.07
7. Acetaldehyde	0.050–0.5
8. Styrene	0.4–2

With respect to the amount of beneficial use of sludge itself, it is about 15% of the sludge generated (fiscal 1988) as indicated in Table 13.40. The amount of sludge for beneficial uses is about 450,000 m^3/year, which is about 23% of the total amount of disposed sludge of 1.95 million m^3/year. Its breakdown shown in Table 13.40 is: 417,000 m^3/year used for agriculture and landscape planting, and 37,000 m^3/year for construction materials.

Utilization for Agriculture and Landscape Planting

Since wastewater sludge contains large amounts of organic matter, and components of nitrogen and phosphorus, its utilization for agriculture and landscape planting is generally by spraying the sludge on farmland and green spaces as fertilizer or soil conditioner. Sludge has been used for many years in Japan for agriculture and landscape planting. In fiscal 1988, the total amount or part of the wastewater sludge was used for agriculture and landscape planting by 157 local public bodies. The forms of sludge used for agriculture and landscape planting are varied and include: dewatered sludge, incinerated ash, dried sludge, digested sludge, and most recently composted sludge. Most of the dewatered sludge delivered to fertilizer companies is composed within the companies.

With respect to the utilization of sludge for agriculture and landscape planting, the following problems are pointed out:

- mixing-in of pathogens, parasites, and seeds of weeds
- difficulty in handling due to offensive odor and high water content
- contamination of soils by heavy metals
- determination of adequate sludge amount of application

Among these problems, the first two can mostly be solved by composting. From the viewpoint of the use as an organic fertilizer for checking the

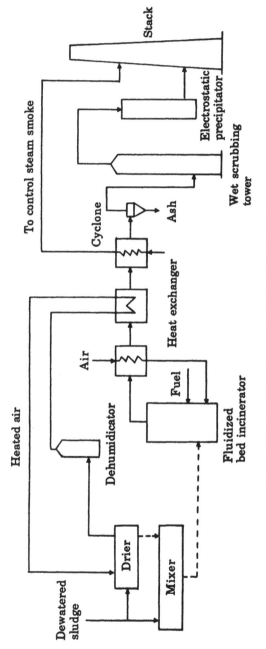

Figure 13.7 Typical flow diagram of a fluidized bed incineration system.

732

TABLE 13.37. **Types of Sludge Melting Furnaces.**

Type of Melting Furnace	(Type of Energy) Heating Method	Melting Subject	Outline of Furnace
1. Reverberatory melting furnace	(Fuel) Gas Heavy oil Powdered coal	Dried cake	Dried cake is placed around the core in a doughnut shape. When the fire is lit, the inside upper wall of the furnace is heated, forming a kind of reflector to maintain melting. The burner is used until the furnace reaches a steady-state. Once the steady-state has been reached, the furnace temperature is controlled by adjusting the cake input through changes in the ceiling height.
		Ash	The ashes are placed inside the furnace and the furnace is heated by the burner so that its upper part forms a kind of reflector in order to maintain melting. The heat to maintain melting is entirely supplied by the fuel combustion heat from outside.
2. Cyclone melting furnace	Same as above	Dried cake	With the cake air, a revolving flow along the inside wall of the cylindrical furnace is generated to lengthen the retention time of the cake. The cake combustion in turn heats the furnace wall to maintain melting.
		Ash	While the burner heats the furnace, a revolving flow of ashes and air is generated along the inside wall of the cylindrical furnace to maintain melting. The heat to maintain melting is entirely supplied by the fuel combustion heat from outside.

(continued)

733

TABLE 13.37. (continued).

Type of Melting Furnace	(Type of Energy) Heating Method	Melting Subject	Outline of Furnace
3 Coke bed melting furnace	(Fuel) Coke	Dried cake	Dried cake and coke are thrown into the furnace in turn to assure the permeability and energy supply required for continuous melting.
		Ash	Ashes and coke are thrown into the furnace in turn to assure the permeability and energy supply required for continuous melting.
4. Electric arc melting furnace	(Electricity) Arc heat to melting subject	Dried cake	When electricity is turned on to the carbon electrode in the furnace and the base metal at the bottom of the furnace, arc discharge occurs, to generate the heat to maintain melting
	Arc heat and reflected heat	Ash	Same as above
5. Microwave melting furnace	(Electricity) Friction heat caused by oscillated rotation of molecules	Dried cake	Theoretically possible
		Ash	When the microwave (electromagnetic) radiation is applied to the ash (which is a dielectric substance) the ashes commence electrical oscillation and melt by their self-generated heat. The thermal conditions can be regulated by the microwave output.

TABLE 13.38. Melting Furnaces in Operation or under Construction as of March 1991.

Name of WTP	Type of Furnace	Operation Starting Time	Treating Capacity		
			DS/day (t/day)	Wet Cake/day (t/day)	Water Content of Cake (%)
Ai River WTP, Osaka Prefecture	Coke bed	July 1985	8	40	80
Kase WTP, Kawasaki City	Electric arc	September 1980	12.8	32	60
Oyabe River WTP, Toyama Prefecture	Reverberatory	August 1988	5.3	22	76
Onoe WTP, Kakogawa City	Cyclone	August 1987	15	50	65
Chuou WTP, Chiba City	Cyclone	June 1987	3	15	80
Konanchubu WTP, Shiga Prefecture	Cyclone	April 1990	12	40	70
Nanbu Sludge Plant, Tokyo Metropolis	Cyclone	April 1991 (proposed)	32	160	70
JSWA ACE Plan:					
Hyogo Western Sewage Sludge Regional Treatment Plant	Coke bed	November 1989	40 × 2	150 × 2	75
Osaka Northeastern Sewage Sludge Regional Treatment Plant	Coke bed	November 1989	10	45	78
Osaka Southern Sewage Sludge Regional Treatment Plant	Reverberatory	December 1990	25 × 2	115 × 2	78

TABLE 13.39.   Effect of Cooling Method on Slag Characteristics.

Cooling Method		Slag Characteristics
Rapid cooling	Cooling by water	Vitreous, low strength, fine grained
Gradual Cooling	Cooling by air without temperature control	Mostly vitreous, low strength, massive or crushed
	Cooling by air with temperature control	Crystallized, high strength, massive or crushed
Reheating	Reheating of quenched fine grained slag at 900–1000°C	Crystallized, high strength, massive or crushed

fatigue of soils and recycle of resources, the promotion of the utilization of sludge for agriculture and landscape planting is much desired.

With respect to the heavy metals, the maximum concentrations of mercury, arsenic, and cadmium in sludge to be used for agriculture and landscape planting were established by the Fertilizer Control Law. Also, a control standard for zinc concentration in soils in farmland was established by the Environment Agency. These values are summarized in Table 13.41.

Shown in Figure 13.9 is the list of composting facilities for wastewater sludge possessed by municipalities. With respect to the composting equipment, a technical evaluation was made by the Japan Sewage Works Agency, and "Design Manual for Sewage Sludge Composting Facilities" (draft) was formulated in fiscal 1987. Figure 13.10 shows a basic flow diagram for the composting process. The basic design requirements developed for the composting process are shown in Table 13.42 [50].

The Ministry of Construction formulated "Guide for Applying Wastewater Sludge in Urban Landscape Planting" (draft) in March 1987 and indicated the standards for the amounts of composted sludge, dried sludge, and digested sludge cake to be applied to parks and landscape planting. Table 13.43 shows the amount of sludge to be applied in terms of nitrogen content of the sludge, based upon the quality of the soils for landscape planting and depth of application.

Utilization for Construction

Trials of reuse of incinerated ash and melted slag as construction materials is aggressively being promoted by several cities and the ACE plan of the Japan Sewage Works Agency. The ACE plan will be explained in detail later; its name came from "the sludge utilization for Agriculture, Construction works, and Energy." Some examples of use of sludge as

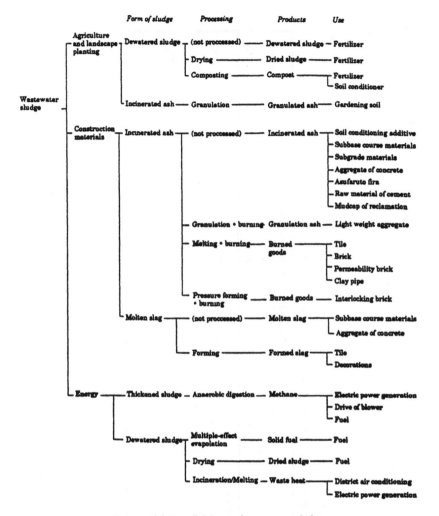

Figure 13.8 Beneficial use of wastewater sludge.

construction materials, includes: incinerated ash of lime-conditioned sludge utilized as a soil conditioning additive, and melted slag utilized as roadbed materials.

The Ministry of Construction carried out surveys and research for five years, from fiscal 1981 to 1985, on the theme of "Development of Waste Materials Utilization Techniques in Construction Projects" as part of the Comprehensive Research and Development Project. As a result, several new techniques related to the wastewater sludge were proposed and included:

TABLE 13.40. Status of Beneficial Use of Wastewater Sludges in Japan (as of Fiscal Year 1988).

Classification		Volume of Sludge (Thousand m³/year)					
		Dewatered cake	Incinerated Ash	Dried Sludge	Composed Sludge	Digested Sludge	Total
Agricultural and landscape planting	Conducted by municipalities	85	4	4	17	0	110
	Delivery to fertilizer companies	230	3	1	72	1	307
	Subtotal	315	7	5	89	1	417
Construction material		0	37	0	0	0	37
Total		315	44	5	89	1	454

TABLE 13.41. Regulation Standards of Heavy Metals and Toxic Organics for Agricultural Use.

Compound	Standard	Comment
As	50 mg/kg DS	Sludge content, based on
Cd	5 mg/kg DS	Fertilizer Control Law
Hg	2 mg/kg DS	
Alkyl Hg	Not detected	Leachate test (10% v/v),
Hg	0.005 mg/L	based on Waste Disposal
Cd	0.3 mg/L	and Public Cleansing Law
Pb	3 mg/L	
Organic phosphorus	1 mg/L	
Hexavalent Cr	1.5 mg/L	
As	1 5 mg/L	
CN	1 mg/L	
PCB	0.003 mg/L	
Zn	120 mg/kg DS	Soil content, for agricultural soil, notification of Environment Agency

(1) "Pavement Design and Construction Manual Utilizing the Incinerated Ash from Wastewater Sludge" (draft)

(2) "Concrete Design and Construction Guide Utilizing Slag Aggregate from Wastewater Sludge" (draft)

(3) "Quality Standards for Incinerated Sludge Aggregate (artificial lightweight aggregate) for Light Weight Concrete" (draft)

In response to these results, several municipalities started construction works using the above as construction materials. Among them, an artificial lightweight aggregate was initially developed aggressively by the Tokyo Metropolitan Government, and the outline of its production flow is shown in Figure 13.11. The pelletized lightweight aggregate can be adjusted to any grain size within the grain size range of 0.3 to 3.5 mm, and its bulk density is 0.8 to 0.95 kg/L. Its use as aggregate for lightweight concrete for high-rise buildings and other structures is expected.

Another form of utilization of sludge as construction materials is to produce building products utilizing incinerated ash or melted slag as part or all of the raw material. The production of clay pipes and water-permeable concrete blocks from incinerated ash and production of concrete aggregates from slag are examples of this utilization. Recently, a process for manufacturing calcined bricks with 100% of incinerated ash was developed. The flow chart is shown in Figure 13.12. Table 13.44 summarizes the development and application of various construction materials from incinerated ash and melted slag.

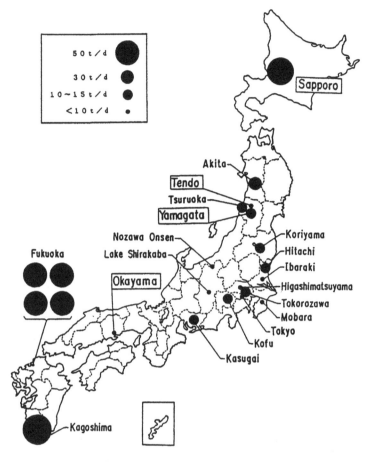

Figure 13.9 Sewage sludge composting plants in Japan.

In order to promote the utilization of wastewater sludge as resources, the Association for the Utilization of Sewage Sludge was founded in 1977 by relevant municipalities, and the Association has performed various activities related to the beneficial use of sludge. The Association is now preparing a "Utilization Manual of Wastewater Sludge for Construction Works" (draft) in order to fully collect knowledge of the utilization of sludge as construction materials by many cities, as explained herein before.

Utilization as Energy

Digester gas, which is mainly comprised of methane, is obtained from the anaerobic digestion of wastewater sludge. Conventionally, this gas

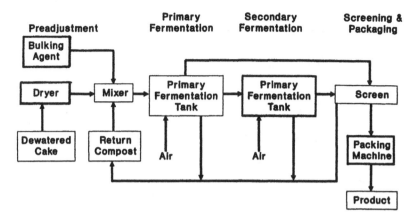

Figure 13.10 Basic flow diagram for composting.

was used as a heat source for heating a digestion tank or as a supplemental fuel for an incinerator.

In addition, digester gas can be utilized as an energy source for diversified purposes through electric power generation. Digester gas can be used as fuel for generating electric power that can be utilized as part of the electric power used in wastewater treatment plants. In European countries, electric power generation by digester gas has been widespread since the 1950s. In Japan, people paid attention to energy conservation after the occurrence of the oil crises, and a number of electric power generating facilities utilizing digester gas were constructed in the 1980s. The Public Works Research Institute of the Ministry of Construction formulated "Design Manual of Power Generation by Digester Gas" (draft) in 1985 by collecting and editing all the achievements in previous surveys and researches. At present, electric power generation by digester gas is being carried out at seventeen plants in Japan, as shown in Table 13.45.

Another method of utilizing the digester gas is the use of the kinetic energy from a gas engine directly as power. Such power is being utilized for driving gas compressors for digester mixing in both the Chubu and Tarumi Plants in Kobe City.

Another form of energy utilization, other than digester gas, is the use of waste heat from sludge incineration. Conventionally, this waste heat was used for sludge drying prior to incineration or as a heat source for warm-water swimming pools in the neighborhood. Recently in Kobe City, a more diversified utilization of the waste heat has been in progress. The waste heat is utilized as a heat source for a hot water supply system for a newly developed housing zone on Rokko Island, reclaimed land in Kobe.

Metropolitan Tokyo is generating electric power by developing a new

TABLE 13.42. Basic Flowchart for Composting.

	Preadjustment	Primary Fermentation	Secondary Fermentation	Screening and Packaging
Process Objectives	Preparation of environment suitable for microbiological reaction • Nutrients (C/N) • Water content • Gas permeability (grain size) • pH • Inoculation with seed bacteria	Aerobic fermentation • Decomposition of easily degradable organic matter • Deodorization • Disinfection • Removal of water	Aerobic fermentation • Decomposition of degradable organic matter • Removal of water • Enhancement of preservability	Enhancement of merchandisability • Improvement of handling • Preservability
Operations	• Mixing of feedstock (dewatered cake, additives, return compost) • Setting of target moisture content • Adjustment of pH (neutralization) • Return of compost • Improvement of aeration	• Aeration • Turnover and transshipment of mixture • Separation of return compost	• Turnover • Aeration • Separation of return compost • Storage	• Size control and classification • Separation of return compost • Packing • Storage

TABLE 13.42. (continued).

	Preadjustment	Primary Fermentation	Secondary Fermentation	Screening and Packaging
Major	• Hoppers • Mixers • Weigher • Dryer	• Fermentation tank • Blower • Turnover equipment • Return conveyer • Deodorizer	• Fermentation tank (yard) • Turnover equipment • Return conveyer • Blower	• Classifier • Packing machine • Storage tank • Return conveyer • Weigher
Remarks	• In some plants, the drying of dewatered cake and the mixing of return compost and additive are omitted.	• In some plants, compost is not returned.	• Primary fermentation is used as secondary fermentation tank. • Usually, aeration is omitted. • Piling type is employed mostly. • Secondary fermentation tank often is used as a storage tank.	• In some cases, the compost product is trucked in bulk.

TABLE 13.43. Guideline for Sludge Application to Landscape Planting.

| Application Depth | Allowable Nitrogen g/m²ª | | |
	For Excellent Grade Soil	For Good or Fair Grade Soil	For Poor Grade Soil
< 10 cm	10–15	30–45	45–60
< 20 cm	20–30	60–90	90–120
< 30 cm	30–45	90–135	135–180

ªThe numerical values shown in this table are given by total nitrogen weight in the applied sludge

technique for making a solid fuel from wastewater sludge utilizing a multiple-effect evaporation system to remove the water.

CENTRALIZED TREATMENT OF SLUDGE

The centralized treatment of sludge was normally performed in large cities such as Tokyo and Nagoya. Recently, Yokohama City has developed a plan for centralized sludge treatment, as shown in Figure 13.13, which is considered a typical one [51]. Sludge from the existing eleven treatment plants will be centralized at two sludge treatment centers using twelve pipelines for transporting the sludge to the centralized facilities. The sludge treatment flow diagram at the Hokubu Sludge Treatment Center in Yokohama City is shown in Figure 13.14. The organic matter in sludge will

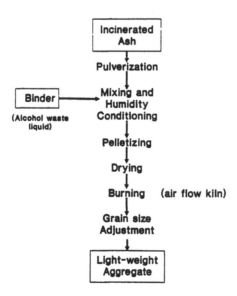

Figure 13.11 Manufacturing flowchart of artificial lightweight aggregate.

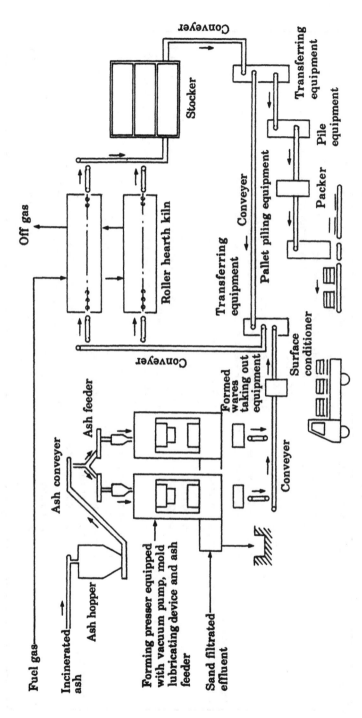

Figure 13.12 Manufacturing flowchart of interlocking brick made from incinerated ash.

745

TABLE 13.44. Technology Developing and Practicing Circumstances of Sludge Ash and Slag Use for Construction Works.

Form of Sludge	Use	Organization
Incinerated ash of lime-conditioned sludge	Soil conditioning additive	Yokohama City, Nagoya City, Shiga Prefecture, Kyoto City, Public Works Research Institute (PWRI)
	Subgrade materials	Sapporo City, Saitama Prefecture
	Subbase course materials	Sapporo City, Saitama Prefecture, Hamamatsu City, Japan Sewage Works Agency (JSWA)
	Interlocking brick	Kyoto City
	Tile	Shiga Prefecture, Kyoto City
	Aggregate for concrete	Sapporo City, Shiga Prefecture, Kyoto City
	Raw material or earthenware	Shiga Prefecture, Osaka Prefecture, Hyogo Prefecture, Kyoto City
Incinerated ash of polymer-conditioned sludge	Subgrade materials	Ibaraki Prefecture, Kyoto City, Kobe City, Nara Prefecture
	Subbase course materials	Ibaraki Prefecture, Kawasaki City, Kyoto City, Nara Prefecture
	Interlocking brick	Niigata Prefecture, Kyoto Prefecture, Kobe City, PWRI
	Interlocking brick	Sapporo City, Tokyo Metropolitan Government, Saitama Prefecture, Yokohama City, Kyoto Prefecture, PWRI
	Tile	Sapporo City, Niigata Prefecture, Utsunomiya City, Nagoya City, Kyoto City, PWRI, JSWA
	Lightweight aggregate	Tokyo Metropolitan Government, Nagoya City, Kyoto City, Nara Prefecture
	Raw material of earthenware	Sapporo City, Yokohama City, Nagoya City
	Permeability brick	Nagoya City
Molten slag	Subgrade materials	Gifu City, JSWA
	Subbase course materials	Chiba City, Tokyo Metropolitan Government, Kawasaki City, Toyama Prefecture, Gifu City, Osaka Prefecture, Osaka City, PWRI
	Interlocking brick	Toyama Prefecture, JSWA
	Aggregate for concrete	Kawasaki City, Toyama Prefecture, Shiga Prefecture, Osaka City, PWRI, JSWA
	Formed slag	Tokyo Metropolitan Government, Osaka City, PWRI

TABLE 13.45. List of WTPs Performing Digester Gas Power Generation.

| Treatment Plant | Operation Starting Time | Gas Engine | | Electric Power (kW) | Digester Gas Generation (FY 1989) $\times 10^3$ m^3 | Used Gas Volume (FY 1989) $\times 10^3$ m^3 | Generated Electric Power (FY 1989) $\times 10^3$ kWh |
		Engine Type	Engine Power (PS)				
Seibu WTP, Asahikawa City	April 1984	Dual fuel	750	500	2105	1430	3291
Nishimachi WTP, Tomakomai City	April 1982	Dual fuel	650	400	1939	805	1645
Nambu WTP, Hakodate City	April 1989	Spark igntion	750	500	2288	531	954
Tonan WTP, Kitakamigawa-joryu RSS, Iwate Prefecture	January 1990	Spark ignition	202	135	1334	73	133
Yamagata WTP, Yamagata City	September 1988	Spark ignition	265	178	604	593	975
Odai WTP, Tokyo Metropolis	September 1988	Spark igntion	1100 × 3	680 × 3	5028	4999	11,548
Chubu TWP, Yokohama City	November 1983	Spark ignition	195	130	968	418	740
Hokubu Sludge Treatment Center, Yokohama City	April 1989	Spark ignition	1350 × 4	920 × 4	11,625	8975	16,132

(continued)

747

TABLE 13.45. (continued).

| Treatment Plant | Operation Starting Time | Gas Engine | | | Digester Gas Generation (FY 1989) $\times 10^3$ m³ | Used Gas Volume (FY 1989) $\times 10^3$ m³ | Generated Electric Power (FY 1989) $\times 10^3$ kWh |
		Engine Type	Engine Power (PS)	Electric Power (kW)			
Nambu Sludge Treatment Center, Yokohama City	November 1989	Spark ignition	1743 × 2	1200 × 2	2609	1870	3839
Chubu WTP, Yamato City	March 1983	Spark ignition	240	160	694	639	1151
Asano WTP, Kanazawa City	April 1984	Spark ignition	92	60	1111	228	369
Seibu WTP, Otagawa RSS, Hiroshima Prefecture	February 1988	Spark ignition	300	200	2550	757	1510
Hiakari WTP, Kitakyushu City	April 1984	Spark ignition	300 × 2	200 × 2	4969	1563	2589
Shinmachi WTP, Kitakyushu City	February 1984	Spark ignition	240 × 2	160 × 2	1333	572	1028
Chubu WTP, Fukuoka City	April 1984	Spark ignition	360	240	2881	541	1651
Naha WTP, Chubu RSS, Okinawa Prefecture	June 1981	Spark ignition	410	270	1716	1104	1651
Nago WTP, Nago City	January 1983	Spark ignition	41	25	33	0	0

WTP = wastewater treatment plant; RRS = regional sewerage system

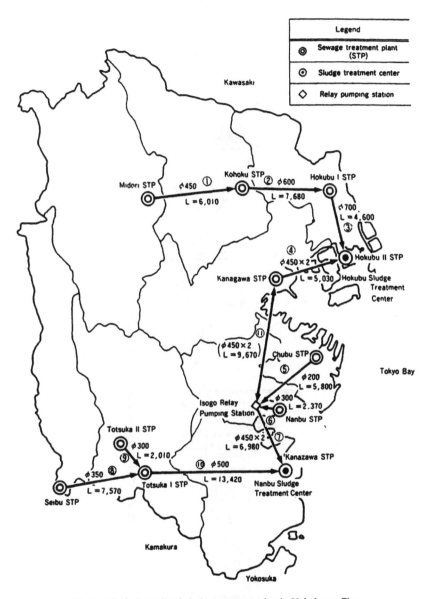

Figure 13.13 Centralized sludge treatment plan in Yokohama City.

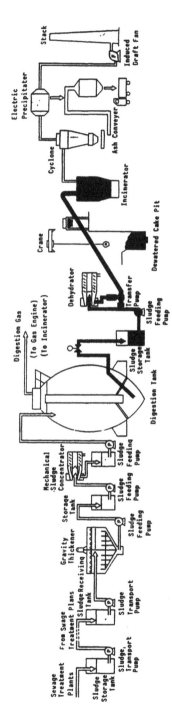

Figure 13.14 Sludge treatment flowchart at Hokubu sludge treatment center in Yokohama City.

be recovered as methane gas through anaerobic digestion, and used as an energy source for electric power generation. The digested sludge will be incinerated by fluidized-bed incinerators.

The regional sewage sludge treatment project (ACE plan) sponsored by the Japan Sewage Works Agency is an area-wide sludge treatment plan passing over multiple administrative jurisdictions [52]. Under this project, the construction and maintenance of the treatment facilities are carried out by the Japan Sewage Works Agency after receiving requests from the participating municipalities. Four regional wastewater sludge treatment plants have been constructed and are being operated in the Kinki District as shown in Figure 13.15. Sludge treatment under the ACE plan is characteristic in that incineration and/or melting processes are provided in every plant, thereby prescribing the utilization of incinerated ash or melted slag as construction materials within the master plan for the project. Table 13.46 shows the treatment flow and the scale of the project of two regional treatment plants for the Hyogo area.

As indicated in Table 13.47, fifty-five cities are performing some kind of centralized sludge treatment as of fiscal 1988, which is about 60% of the ninety-five cities that own multiple wastewater treatment plants. In comparison, centralized sludge treatment was performed in thirty-five cities in fiscal 1984, which was about 60% of the cities possessing multiple wastewater treatment plants. The centralized treatment of sludge has been steadily expanding in recent years.

The reasons for requiring centralized sludge treatment are:

(1) Economic sludge treatment

(2) Effective environmental measures

(3) Promoting the use of sludge as a resource

One of the important themes in centralized treatment in the future is probably the establishment of regional treatment incorporating the administrative jurisdictions for medium and small cities as in the case of the ACE plan. With respect to this, the Ministry of Construction is requesting, from fiscal 1990, the formulation of a comprehensive plan for wastewater sludge treatment to the prefectural governments.

NEW ZEALAND SLUDGE MANAGEMENT PRACTICES

INTRODUCTION

Geography

New Zealand consists of two main islands and a number of smaller islands located in the southwest Pacific Ocean. The combined area of the country,

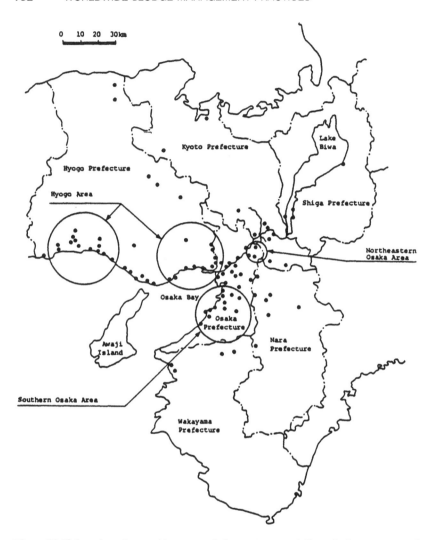

Figure 13.15 Location of area-wide sewage sludge treatment and disposal plants constructed according to the ACE plan.

270,000 square kilometers, is similar in size to Japan or the British Isles. New Zealand is over 1600 kilometers long, 450 kilometers wide at its widest point and has a coastline length of 5650 kilometers. The country is mountainous, more so in the South Island than the North Island. A mountain chain, the Southern Alps, runs along the length of the South Island, with 223 named peaks higher than 2300 meters and 360 glaciers. The highest peak is Mt. Cook—3764 m, and the longest glacier is the Tasman (29 kilometers). New

TABLE 13.46. Process Flow and Design Capacity of the ACE Plan for the Hyogo Area.

Item	Eastern Hyogo	Western Hyogo
Applicant municipalities (wastewater treatment plants)	Hyogo Prefecture (Muko River, Upper and Lower Regional Wastewater Treatment Plants); Amagaski City (Hokubu and Tobu Wastewater Treatment Plants)	Hyogo Prefecture (IBO River Basin Regional Wastewater Treatment Plant); Himeji City (Chubu, Seibu, Tobu, Shikama, Takagimae, Fukuimae, Takagigawanishime, and Shigomae Wastewater Treatment Plants)
Process flow	Sludge receiver → Centrifugal thickener → Centrifugal dewatering machine → ↓ Incinerator (fluidized bed and pyrolysis furnaces) ↑ Disposal ↑ Recycling and harnessing facilities → Products made from sludge	Sludge receiver → Centrifugal thickener → Centrifugal dewatering machine → ↓ Incinerator (melting furnace) ↑ Disposal ↑ Recycling and harnessing facilities → Products made from sludge
Design raw sludge processing capacity	9700 m³/day (water content: 99%)	13,500 m³/day (water content: 99%)
Disposal method	Effective use and landfill	Effective use and landfill
Project cost (FY 1986-FY 2005)	Approximately ¥37,000 million	Approximately ¥ 58,000 million
Scheduled date for treatment startup	FY 1989	FY 1989

TABLE 13.47. The Number of Municipalities Which Had Centralized Sludge Treatment in 1988.

Number of Municipalities Having More Than Two WTPs	Number of Municipalities Practicing Centralized Sludge Treatment	Transporting Method			
		Pipe Line	Truck Transport of Dewatered Sludge	Truck Transport of Liquid Sludge	Other
95	55	19	11	24	1

Zealand's rivers are mainly swift and difficult to navigate. They are important as sources of hydroelectric power, and substantial artificial lakes have been created as part of major hydroelectric schemes.

Population

New Zealand's population is 3.3 M (1986 census), 74% of whom reside in the North Island. For the last hundred years the trend has been for a northward drift of people [53]. The islands of New Zealand have been ethnically and culturally connected to Polynesia for at least 1000 years. Less than 200 years ago, its population and cultural heritage was wholly that of Polynesia, but now New Zealand is dominated by cultural traditions that are mainly European, emanating especially from Britain.

About four-fifths of New Zealanders are of European origin, predominantly from the British Isles, but also including people from the Netherlands, Yugoslavia, Germany, and other nations. The indigenous Maori population make up the next largest group of the population, about 12.4% in 1986. The third main ethnic group is the Pacific Island Polynesians, who made up around 3.5% of the population at the time of the 1986 census.

Customs

The two predominant ethnic groups in New Zealand, namely Caucasian (European) and Maori, by and large are the primary determinants of New Zealand culture and customs, although significant minority populations of various other Pacific and Asian peoples contribute on a local scale.

With regard to wastewater treatment and disposal (including sludge management), a significant constraint given prominence and legal recognition nowadays is the nature of certain beliefs held by the Maori, given stature by the signing of a formal treaty between England and Maori tribes in 1840, which guaranteed the latter certain rights (the Treaty of Waitangi). Amongst these beliefs is one that holds that human waste must be purified by land, not water. As a consequence, new sewerage schemes in New Zealand now take this into consideration during options evaluation, more so than in the past, where water-based disposal systems (especially coastal outfalls) were employed. Being a country with a large coastline and extensive natural features of significance, New Zealanders spend a lot of time "outdoors," either working or for recreation.

Government

New Zealand is a monarchy with a parliamentary government. The Crown is vested in the same person as the British Crown, and Queen Elizabeth II has the title Queen of New Zealand. Although an independent state today, New Zealand's constitutional history can be traced back to 1840 when, by the Treaty of Waitangi, the Maori people exchanged their sovereignty for the guarantees of the treaty and New Zealand became a British colony. A constitution is concerned with the establishment and composition of the legislative, executive, and judicial organs of government, their powers and duties, and the relationship between these organs.

General elections are held every three years. Presently there are two main political parties, National and Labour (with the government elected in 1990 being National) and a number of smaller parties. Proportional representation is presently an issue, for which a nationwide referendum is soon to be held.

Other Notable Features

A significant part of New Zealand's export earnings derive from primary production. As a consequence, increasingly stringent international standards with regard to levels of contaminants in meat, fish, dairy products, and fruit reflect significantly in local growing and associated activities. Similarly, a significant new industry in New Zealand is tourism, particularly taking advantage of the large areas of undeveloped and sparsely populated country in the South Island, including Fiordland, the Southern Alps, and other coastal and forested national parks. Consequently, New Zealand needs to maintain and indeed enhance its international reputation for clean waters, pollution-free areas, and low levels of any contamination in export products. With such an image to foster and protect, the inappropriate management of sludge from either small or larger communities or indeed the primary industries themselves is an obvious inconsistency, one that must be avoided. New Zealand consequently is "nuclear free," and a signatory to most international pollution-prevention conventions.

QUANTITIES OF SEWAGE AND SLUDGE

Sewered Population

The sewered population in New Zealand is now estimated to be approximately 85% although once new sewage treatment facilities now planned come online, including in particular Wellington (150,000 people) and other cities like Wanganui (40,000 people), the sewered and treated population will rise to approximately 96%.

Most communities with populations from approximately 250 upwards now have reticulated community sewerage schemes, unless the ground conditions are particularly conducive to on-site septic tanks and ground soakage systems. The central Government's Department of Health former subsidy scheme provided up to 40% of the capital costs for small communities, and using a sliding scale, down to 10% for large urban areas. This was terminated in 1986, and its pending termination prompted many small communities then still unsewered to install schemes.

Domestic Contribution

Typical per capita figures used in New Zealand are 250 liters per head per day of domestic sewage with suspended solids and biochemical oxygen demand (BOD_5) loadings of 65 to 70 grams per capita per day.

Industrial Contribution

New Zealand has a very significant industrial wastewater load from primary based industries that are intensively carried out in New Zealand, particularly farming and forestry. The major industrial effluents are accordingly from meat processing, woolscouring, tanneries, the dairy industry, and timber processing. Lesser amounts of industrial effluents result from a range of normal manufacturing industries.

By far the largest proportion of the industrial wastewaters, predominantly those stemming from the agricultural sector, are not located in urban areas, and consequently have their own wastewater treatment and disposal systems, including sludge management and disposal systems. Wastewaters from the meat processing plants are the largest group, having a per annum biochemical oxygen demand (BOD_5) equivalent of some 160 million persons.

WASTEWATER TREATMENT PROCESSES

Since the early 1950s there has been a wide range of treatment methods developed as septic tank systems were phased out. Initially much of the practice followed British technology, but in more recent years a full spectrum of development including a number of New Zealand innovations have occurred.

The predominant processes used involve oxidation ponds, activated sludge, primary plants, and trickling (biological) filtration. The most common method of treatment in terms of population, principally because the two largest plants, Auckland (Manukau) and Christchurch (Bromley) operate it, is biological filtration with plastic media biofilters, followed by oxidation (polishing) ponds.

Of the communities with a census population greater than 1000 persons, the 165 plants serving these communities are distributed as shown in Table

13.48 [54]. Additionally, there are over 200 smaller community sewerage schemes that predominantly use oxidation ponds where siting considerations have favored their use.

Treatment processes more recently installed include the carrousel type extended aeration plant at New Plymouth and Porirua, a number of wetland schemes (particularly for small communities), UV disinfection, some rotating biological contractors (Keri Keri, Duvauchelles), sequencing batch reactor in Lake Taupo area, and (recently commissioned) a modified Bardenpho plant for phosphorus and nitrogen removal prior to land disposal of the effluent at Rotorua [55]. Wellington City has recently called tenders for a secondary treatment plant with UV disinfection.

Of considerable interest in New Zealand has been the development and application of rotating drum milliscreens, with the Hutt Valley plant being the first of some seven major plants now in operation, in all cases as a first stage of treatment prior to ocean outfall discharge. Rotating drum milliscreens use a cylindrical construction of a fine wedgewine screen mesh of 0.5 to 3 mm aperture size. Raw wastewater is distributed tangential to the inside of the rotating drum, which screens out the solids and conveys them to the end of the drum by a series of inclined guide vanes fixed to the inside of the drum. The screened wastewater passes through the screen mesh. The Hutt Valley milliscreening plant was the first major municipal plant of its type in the world.

Land disposal of treated effluent is practical and its use is increasing, particularly for smaller communities, and wetlands for effluent polishing are also being used to an increasing extent.

SLUDGE PRODUCTION

It is estimated that currently 40,000 dry tonnes/year of domestic source sludge is produced and this is likely to rise to around 55,000 dry tonnes/

TABLE 13.48. **Methods of Treatment Employed by Communities with Population Greater than 1000.**

Treatment Method	Number of Plants
None (all in planning stages)	7
Ocean outfalls with fine screening (milliscreening)	8
Imhoff	5
Primary	5
Trickling filter	16
Activated sludge	19
Oxidation ponds	102
Other	3

year within the next decade with the new, proposed, and upgraded domestic treatment facilities producing increased sludge amounts. Typically a figure of approximately 55 grams per capita per day has been taken for a combined primary and secondary (trickling filter plant sludge) but with more extended aeration and other plants these figures are now more variable relative to process. These figures do not include the industrial waste component from municipal treatment. Although this has dropped relative to the domestic population in recent years, is still likely to increase the current municipal-sourced amount by 25% or more, that is, another 10,000 dry metric tons.

It must also be noted that these figures exclude the sludge slowly accumulating on the bottom of the many oxidation pond systems in New Zealand. No comprehensive surveys have been undertaken of industrial sludge produced by industries' own wastewater treatment and disposal systems. In many cases, however, industries' objectives include maximizing reuse of waste materials (e.g., in the meat industry, paunch, tallow, protein, and rendering materials to fertilizers) and disposing of wastewater onto land with a minimum of treatment, hence minimizing sludge production.

SLUDGE TREATMENT TECHNOLOGIES

Thickening (Prior to Sludge Stabilization/Heat Processes, etc.)

Of the plants with sludge treatment systems, it is estimated that by dry sludge weight approximately 15% use continuous gravity thickeners, 2% use mechanical dewatering, and the remainder achieve some limited thickening as part of clarifier operation.

Dewatering (After Sludge Stabilization Processes)

Filter belt presses and centrifuges are being used to an increasing extent with some evidence of a preference in larger plants of filter belt presses over centrifuges. In extended aeration plants the first stage of dewatering of the stabilized sludge from the aeration basins/clarifier is frequently by a continuous gravity thickener prior to other mechanical dewatering or drying processes.

Biological Stabilization and Thermal Processes

Of the 40,000 dry tonnes/year of domestic-sourced sewage sludge it is estimated that currently:

- 5% is subject to cold (unheated) anaerobic digestion.
- 84% is subject to heated (mesophillic) anaerobic digestion.

- 1% is composted.
- 6% is incinerated (the Dunedin City plant).
- 4% is disposed of in a non-stabilized form.

These figures are only estimates of current practice and new plants and changing practices are expected to alter the balance of these figures. Chemical stabilization of sludge is not practiced to any significant degree. Of the stabilized sludges some 30% (by weight) are then stored with natural dewatering/consolidation in open sludge storage lagoons.

Milliscreenings

The development and application in New Zealand of fine (down to 0.5 mm aperture size) rotating drum milliscreening plant for municipal sewage treatment has also resulted in the need for milliscreening disposal. The most commonly used method is landfilling using the Department of Health guideline of burial with at least 300 mm of fine grained soil. Lime addition prior to burial makes the screenings less offensive and provides some degree of pathogenic destruction; however, this practice has not generally been followed. Incineration is now being practiced at Hutt Valley and composting trials have been carried out [56].

REGULATION OF SLUDGE MANAGEMENT PRACTICES

The following existing legislation covers various aspects of sludge management, including disposal.

Present Legal Constraints

(1) Under the present Water and Soil Conservation Act (1967) a waterrights (i.e., permit) is required to place material that is likely to result in contamination of natural water onto land. Strictly therefore, sludge disposal onto land requires a water right, although in some instances, particularly for small quantities more randomly disposed of, water rights are not always applied for.

(2) Under the present Town and Country Planning Act, "land use" approval in district planning schemes is needed where the land use is changed from the present zoning. Where, however, it can be shown that the land use is compatible with the sludge disposal (e.g., sludge as a fertilizer for crops) then planning approvals are not required.

(3) Where a "nuisance" or "public health" problem may arise, approval is needed under the Health Act for the specific practice. This approval must be given by the Medical Officer of Health covering that region.

(4) The Clean Air Act covering air emissions from boilers, incinerators, etc., where a "best practicable means" approach has been developed.

The above and other legislation is soon likely to be replaced by one comprehensive new piece of legislation (Resource Management Act) as introduced below.

Existing Department of Health Guidelines for Disposal of Sewage Sludge onto Land

The first guidelines were prepared in 1973 with the current guidelines prepared in 1984 following a 1983 nationwide survey of sludge disposal practices [57]. That survey showed that while the larger local authorities were generally conservative in their approach and in most cases sludge was stored for at least two years before use, many of the smaller local authorities were more liberal in releasing sludge for use. The 1984 guidelines:

- provide local authorities with more flexibility in the ultimate disposal of sludge
- acknowledge the low-grade fertilizer value of sludges and the fact that the spreading of sewage sludge on agricultural land and private gardens can provide an economically attractive disposal method for local authorities
- took into account that levels of heavy metals were then high in several New Zealand sludges, this according to the level and type of industrial activity in treatment plant catchments (these have since dropped in most cases with less industry and tighter trade waste controls)
- recognized the importance of agricultural exports to the New Zealand economy by adopting a more conservative approach than that suggested by the EEC and U.K. guidelines

The guidelines set recommendations on disposal of sewage sludge onto land by:

(1) Categorizing the sludge type according to treatment method and the land after-use and restrictions—giving storage, grazing time and other limits for each matrix category—these criteria principally relating to the pathogenic potential of the sludges; Table 13.49 reproduces the 1984 guidelines.

(2) Giving recommended permissible additions of elements in sludges to uncontaminated soils over a period of thirty years or more. Tables 13.50 and 13.51 reproduce the 1984 guidelines. These figures are based on sludge being thoroughly incorporated into the top 200 mm

TABLE 13.49. Recommendations for the Disposal of Sewage Sludge on Land.

Sludge Type	Land After-Use and Restrictions					
	Grazed Crops, Hay, Silage	Orchards, Turf	Human Food Crops (Cooked)	Human Food Crops (Uncooked)	Home Gardens, Parks, Reserves, Playing Fields	Forestry, Crops for Conservation, Land, Recreation
Mesophillic anaerobic digestion with <40% reduction of volatile solids	a	b	c	d	cde	f
Aerobic digestion from extended aeration plant or mesophillic anaerobic digestion with <40% volatile solids reduction or cold anaerobic digestion	ae	be	ce	de	cdh	f
Lagooned septic tank sludge	h	bh	ch	dh	cdh	f

a No grazing by cattle or pigs within six months of sludge application, other animals four weeks
b Fruit or turf is not to be harvested within three months of sludge application
c Crops for human consumption (cooked) should not be harvested within six months of sludge application.
d Crops for human consumption (uncooked) should not be sown for twelve months after sludge application: other crops may be sown in the interim.
e Store sludge for twelve months prior to use.
f Conditions of public access to be determined by the Medical Officer of Health
h Lagoon or store for two years prior to disposal
NB: If restrictions a, b, c, and d cannot be reliably ensured, restriction should be applied.
Source: New Zealand Department of Health, 1984 Guidelines

TABLE 13.50. Recommended Maximum Permissible Additions of Elements in Sewage Sludge to Uncontaminated Non-Calcareous Soils Over a Period of Thirty Years or More.

Element	Uncontaminated Soil Background Concentration (mg/kg) "Extractable"	Maximum Permissible Addition of Element (kg/ha) "Total"
Zinc	2.5	560
Copper	5	280
Nickel	1	70
Zinc equivalent	20.5	560
Boron	1	4.5 (1st year)
		3.5 (subsequent years)
	"Total"	
Chromium	100	1000
Cadmium	1	5
Lead	50	1000
Mercury	0.1	2
Molybdenum	2	4
Arsenic	5	10
Fluorine	200	600
Selenium	0.5	5

Note. Additions of zinc, copper and nickel are subject to the overriding limitation of the zinc equivalent, because these elements appear to be additive when present in high concentrations. The zinc equivalent assumes comparative toxicities of zinc, copper and nickel of 1 2 8, but these may not be valid for all crops grown under all conditions.
"Extractable"—Metals extracted by EDTA, boron extracted by hot water.
"Total"—Metals extracted by strong acid, fluorine determined by fusion and ion selective electrode, boron extracted by hot water.

of the soil; also, up to 20% of the total addition (kg/ha) may be made in a single year, provided that time is allowed for the running average rate of addition to fall to the thirty-year average before the next application is made.

Proposed New Legislation

A new Resource Management Act was passed in July 1991 and became operative in October 1991. This replaces many existing outdated statutes in New Zealand including those referred to previously. The new legislation requires an integrated approach to pollution control, being based on a "Best Practicable Option" approach and having sustainable management of resources as its fundamental premise. It integrates the "mixed media effects" of land/water and air and require "discharge permits" for such activities as sludge disposal onto land, into landfills and other procedures.

In addition to the new legislation, Regional Waste Management Plans

TABLE 13.51. Provisional Maximum Permissible Concentrations of Elements in Arable Soils to Be Reached in Thirty Years or More.

Element	Non-Calcareous	Calcareous
	(mg/kg dry solids)	
"Extractable"		
Zinc	280	560
Copper	140	280
Nickel	35	70
Zinc equivalent	280	560
Boron	3.25	3.25
"Total"		
Chromium	600	600
Cadmium	3.5	3.5
Lead	550	550
Mercury	1	1
Molybdenum	4	4
Arsenic	10	10
Fluorine	500	500
Selenium	3	3

are in the process of being adopted throughout New Zealand's thirteen regions. These plans identify sludge as a special waste, and accordingly set procedures for its safe handling, reuse, and/or disposal in that region.

SLUDGE MANAGEMENT AND DISPOSAL PRACTICES

Table 13.52 gives estimated totals of the municipal domestic-sourced sludge. In reviewing this table the 1991 "stockpile" increase results in a

TABLE 13.52. Municipal Sewage Sludge Disposal Practices.

Disposal Method	Dry Tonnes Disposed of Per Year		
	1985	1991	Expected by Year 2000
Farmland/gardens/landscaping	15,000	7200	14,300
Forestry	500	800	9900
Land reclamation	500	800	3300
Landfill	14,000	11,800	22,000
Sea	0	0	0
Thermal destruction (incineration)	2000	2400	2750
Other· stockpile	3000	17,000	2750
Total	35,000	40,000	55,000

Note All figures are estimates only.

large part from the present storage regime at the largest New Zealand plant, Auckland, and the increase landfill for year 2000 is based on the present proposal to landfill the sludges from the proposed Wellington plants which will be operational by then.

With much more attention being paid to sludge management in the future, it is expected that beneficial use technologies will be further developed and implemented. This is referred to in the "by year 2000" figures given in Table 13.52.

The event of the anticipated Resource Management Act coupled with the emphasis being placed on waste minimization, resource reuse and integrated environmental approaches to waste management projects is expected to result in greater reuse of sludge.

Composting is expected to increase, as are other soil enhancement procedures. Appropriate standards and controls will be needed for such procedures to ensure agricultural practices, so important to New Zealand's economy, are not jeopardized.

Composting of waste solids is increasingly being seen by the New Zealand meat industry as a means of turning a waste material into a product that has value, and of reducing the overall cost of disposal. A number of plants are now actually promoting and marketing the material under locally identifiable produce names and offering customers the option of purchasing 25 kg bags or trailer loads in bulk. Pressure on landfill operations will see an increasing recycle/reuse approach being adopted, particularly with respect to organic solid wastes.

SUMMARY REVIEW OF MAJOR URBAN AREA SEWAGE TREATMENT SLUDGE MANAGEMENT PRACTICES AND PROPOSALS

The following summary covers the major municipal plants in operation or for which planning is in progress. The populations given are the populations connected to the municipal treatment facility in discussion. This review includes the new and more innovative applications (on New Zealand standards) being considered for Rotorua, Christchurch, and also Auckland.

Auckland (Manukau Sewage Purification Works) (Population 650,000)

This plant, being New Zealand's largest, is a secondary treatment plant involving plastic media trickling filters, followed by oxidation (polishing) ponds. The dewatered anaerobically digested sludge has previously been stored on-site for more than two years before being given away (free) to the public and landscape contractors. The mixed sludge has been treated in anaerobic digesters at 37°C for twenty to twenty-five days. The digested sludge is pumped to six storage lagoons that cover 33.5 ha total area at 2 m depth. This gives a sludge detention of two to four years. Gravity thickening takes place in the lagoons with removal of supernatant from the lagoons. Sludge is pumped to dewatering beds, which are earthen basins 600 mm deep where the retention is fifteen to twenty-four months and a final product of 50% solids is achieved.

Sludge Use

Until recently, the material excavated from the dewatering beds was made available to members of the public and soil contractors without charge. The dried sludge was apparently used for the following purposes:

• backfill material
• landscaping of new sections and other developments
• topdressing of lawns and gardens
• bulk amendments for gardens, tree planting, etc.

This program generally saw all the sludge being removed from the site, although seasonal demands varied. There are no records available to show the fluctuations in demand over the year. A local soil/organic mix commercial operator noted that the freely available sludge material cut into his retail business significantly.

In September 1990, this free, uncontrolled distribution of the dried sludge stopped as a result of concerns about heavy metals in the sludge and the unknown and unmanaged use of the material under the free distribution program. The material excavated from the dewatering beds is currently being stockpiled at the treatment plant.

New Mechanical Dewatering Facility

As a result of the sludge lagoons being at or over capacity, a decision was made in 1988 to divert a portion of the sludge outflow from the digesters for dewatering by mechanical belt filter presses. The facility, now under construction, will include:

• belt filter presses
• sludge conditioning facilities to improve the dewaterability
• an ability to increase the solids content from 2% to 25%
• a capacity to dewater up to 30 tonnes (dry solids) per day (one shift)
• capacity to install additional presses in the future to dewater up to 64 tonnes per day

The sludge will be digested for twenty-one days but will have had no further storage at the time of dewatering. Thus utilization would be far more restricted under the Department of Health Guidelines than with the current, aged sludge material. Distribution of the mechanically dewatered sludge would not be permitted without severe restrictions.

Future Sludge Use

Long-term decisions have yet to be made, but at present the Auckland Regional Council is likely to initiate a conceptual design of a 10 tonne per day aerated static pile composting facility and a 10 tonne per day lime stabilization facility [58]. If successful, this plant will be expanded. It is hoped that this product will be marketed in the longer term.

Hamilton (Population 96,000)

This primary treatment plant incorporates heated anaerobic sludge digestion (35°C) at twenty-four days average detention. Recent operational changes have incorporated pipework to keep solids content in the primary digester as high as possible by returning digested sludge to the lead digester from a thickening tank. This maintains solids in the lead digester at between 2.5 and 3%. This has resulted in a much better sludge being processed in the subsequent sludge press. The pressed sludge is stockpiled on-site. A local fertilizer manufacturing firm uses some quantities and rehabilitation of a refuse landfill site has also been recently carried out using the aged sludge.

Rotorua (Population 50,000)

The Meat Industry Research Institute of New Zealand (MIRINZ) is working with the Rotorua District Council to develop a process for composting the sludge from Rotorua's Secondary Bardenpho Wastewater Treatment Plant. Four trials using a temperature-controlled aerated pile technique were completed in 1989. Rapid stabilization of primary sewage sludge was achieved using sawdust and recycled compost as a bulking agent. The compost product proved to be an excellent plant growth medium and soil conditioner.

The next stage of the project, which is currently underway, is to evaluate a prototype composting bin. This bin was designed by MIRINZ and has the capacity to stabilize up to 70 m^3 of primary and waste activated sludge and bulking agent. Results so far show that the bin and its aeration system enable much better and easier control of compost temperature and air distribution than the open piles used in the earlier trials. The current trials will be completed by mid 1991 with the view to establishing a full-scale plant in the near future.

New Plymouth (Population 65,000)

The New Plymouth treatment plant commissioned in 1984 is an extended aeration plant incorporating a carrousel aeration basin configuration. The sludge (at an age of twenty days) is still rather unstable and has in the past given rise to unpleasant odors when stockpiled after dewatering. The latter process uses filter belt presses and is a relatively difficult exercise.

A range of filter and treatment disposal methods has been used, including trials of facultative lagooning. Present disposal techniques involve a Soil Enrichment Program where dewatered sludge (at 18% dry solids) is mixed with sandy soil and then the low-grade fertilizer soil material is used for nurseries and other such uses. The sludge has low levels of heavy metals and its use meets the New Zealand Department of Health Guidelines as well as the U.S.EPA and EEC regulations.

Palmerston North (Population 65,000)

This is a primary treatment plant with heated anaerobic digestion (35°C) followed by aerated lagoons that are operated seasonally in accordance with effluent discharge conditions into the receiving river.

The digested sludge is stored in large open storage lagoons. As these become full they are excavated and the sludge stockpiled and naturally dried on-site. The Council's Parks and Gardens Department uses the sludge in a controlled way for gardens and parks around the city area. More recently, use has also been made of the sludge as a topsoil type material in developing the finished areas of the Council's adjacent refuse landfill.

Wellington City (Population 150,000)

Wellington City has recently called tenders for a compact type secondary treatment plant on the confined site near the airport. Sludge management options have been studied and the proposal under further consideration is to pump the active sludge direct from the treatment plant some 8 km to the Happy Valley landfill site where heated anaerobic digestion will be carried out before landfill disposal of dewatered sludge. Supernatant liquor would be returned to the sewerage system and excess electricity from sludge gas generation fed into the local electrical supply grid.

Christchurch (Bromley Treatment Plant) (Population 300,000)

Since the late 1960s, the Christchurch City Council has disposed of stabilized sludge onto 190 ha of farmland surrounding the treatment plant.

The land is owned by the Council, and is operated as a farm with beef stock being grazed on pasture grown on the sludge-conditioned soils. The sludges are derived from primary sedimentation, biological filtration (plastic media) and subsequent secondary sedimentation, and are stabilized by conventional mesophillic anaerobic digestion (35–39°C) and lagooning for eighteen weeks (with decant liquor returned for treatment).

After more than twenty years of sludge application, the capacity of soils of the farm to accommodate the loadings is becoming exceeded, particularly for certain trace metals. As a consequence, the Council is proposing to carry out major trials to investigate the possibility of sludge application to forest lands as a long-term management solution. These trials will probably commence in late 1991–1992. At present the Town Planning and Water Rights applications are being heard to enable the proposed trial to take place.

Dunedin (Tahuna Water Pollution Control Plant) (Population 85,000)

This primary treatment plant incorporates New Zealand's only sludge incinerator. The incinerator is a Dorr-Oliver fluidized bed. The capacity is normally 9 tonnes (ds) of sludge per day operating autogenously (except for startup) at a temperature of 900°C.

Gravity picket fence thickening forms the first phase of sludge thickening, followed by filter belt presses. The filter belt presses replaced the originally installed centrifuges, the replacement decision being made on the basis of significantly lower operating costs.

Invercargill (Population 40,000)

The existing primary treatment plant with heated anaerobic sludge digestion is presently being upgraded to a secondary plant by installation of plastic media biological filtration. Highly concentrated woolscour flow-down effluent is also treated in the heated anaerobic digesters. This has been a very successful operation over the last eleven years. Digested sludge is fed directly to long-term storage sludge holding lagoons that have supernatant drained off and returned to the plant.

CONCLUSION

Sludge treatment and disposal in New Zealand has in the past been characterized by heated anaerobic digestion and landfill or land disposal for the larger plants with a variety of methods, some rather informal, being used at the smaller plants. Increased public awareness and interest in waste minimization and resource reuse coupled with the requirements of the new

Resource Management Act are now resulting in increased focus being placed on all aspects of residuals management.

In new wastewater management projects for sludge, as that for the Hutt Valley area, sludge management objectives are being drawn up from the outset so that processes and procedures can be formulated around these objectives. The objectives of sludge management adopted for the Hutt Valley project, that typify a number of others also under planning in New Zealand, are:

- to adopt relatively simple, flexible, and where possible multiple sludge treatment and disposal/reuse options
- to reuse or otherwise dispose of the sludge or end products arising from its treatment in an environmentally safe manner that protects public health and the ecosystem
- to acknowledge the reuse potential of municipal sludge and adopt practices that allow reuse as possibly economical and otherwise desirable

Increasing application is expected to be made of sludge reuse technologies, particularly those incorporating disposal onto land. Guidelines and regulations are needed to ensure such practices are appropriately applied and controlled.

CANADIAN APPROACH FOR LIMITING METALS ON LAND FROM MUNICIPAL SLUDGES

A summary of Canadian information concerning the production and agricultural utilization of municipal sludge is presented. Sludge quantities (1981) are estimated and waste guidelines for agricultural utilization are compared. Research findings on waste metal uptake by agricultural crops are reviewed. The future of agricultural utilization is considered.

INTRODUCTION

Sewage sludge is a nutrient-rich, largely organic by-product of municipal wastewater treatment. It must be removed from the treatment facility and the disposal options are ocean dumping, incineration, landfilling, and utilization on agricultural land. Recent high energy costs and environmental awareness have increased interest in sludge disposal by utilization on agricultural land. Managed properly, this practice minimizes environmental risks, takes advantage of the fertilizer and soil conditioning value of sludge, and is frequently the least expensive disposal option.

There are two types of environmental risk (temporary and persistent) associated with sludge utilization on agricultural land. Temporary risks disappear within one year or at most a few years following sludge application and include malodor, pathogens, groundwater contamination with nitrate-nitrogen, and phytotoxicity due to soluble salts or toxic biodegradation products from inadequately stabilized sludge. Persistent risks remain long after the temporary ones have disappeared and include increased concentrations of industrially produced organic compounds such as polychlorinated biphenyls (PCBs) and waste metals in soil. It is estimated that the half-life of persistent organics such as PCBs in soil is about ten years [59] while that of most waste metals is about one thousand years [60]. "Waste metals" as used in this paper refers to metals and non-metals (e.g., As, B, F) of environmental concern.

It is widely accepted that sewage sludge utilization on agricultural land must be regulated to minimize the risks associated with heavy metal buildup in soil. Consequently, guidelines for this purpose have been developed in many countries, including Canada. The objectives of this paper are to present Canadian information on:

- municipal sludge production and agricultural utilization
- waste metal guidelines for agricultural utilization
- research findings on waste metal uptake by agricultural crops
- future agricultural utilization

MUNICIPAL SLUDGE PRODUCTION AND AGRICULTURAL UTILIZATION

Based on 1981 data (Table 13.53), a majority of the Canadian population is served by sewers. The remainder, mainly rural, is served by septic tanks. Approximately one-half of the Canadian population is served by sewage treatment. In British Columbia, Alberta, Saskatchewan, Manitoba, Ontario, Prince Edward Island, and the Yukon Territory, the populations served by sewage treatment approximate those served by sewers. However, in Québec, New Brunswick, Nova Scotia, Newfoundland, and the Northwest Territories, many fewer people are served by sewage treatment than by sewers. With the exception of Québec and the Northwest Territories, these jurisdictions are maritime, and raw sewage is released to rivers and the Atlantic Ocean. In Québec, raw sewage released to rivers and lakes has caused serious water pollution problems and a very large program of sewage treatment plant construction has been undertaken. In the Northwest Territories, one-half of the sewage does not receive treatment.

Twenty-nine percent of the estimated 1981 total sludge production in Canada was utilized on agricultural land, and 37% was disposed of by landfilling, incineration, and ocean dumping (Table 13.53). The remaining 34% represented sludge that would have resulted from sewage treatment

TABLE 13.53. Canadian Population [61] and Estimated Municipal Sludge Quantities for 1981.

Jurisdiction	Population (10³ people)		Sewage Treatment	Stabilized Sludge (10³ tonnes dw)		
	Total	Sewered		Total[a]	Disposal[c]	Agricultural Utilization[c]
British Columbia	2.72	1.64	1.57	28	10	
Alberta	2.21	1.54	1.49	63[b]	3	40
Saskatchewan	0.96	0.57	0.57	15.4[b]	6.2	3.8
Manitoba	1.02	0.79	0.79	20[b]		17
Ontario	8.55	6.58	6.37	192[b]	125	61
Québec	6.38	5.48	0.41	82	6.7	
New Brunswick	0.69	0.37	0.19	7 1	3.3	
Nova Scotia	0.84	0 45	0.11	7.5	2.5	
Prince Edward Island	0.12	0.06	0.05	0 9		
Newfoundland	0.56	0.32	0.06	5.1	1.1	
Yukon Territory	0.02	0.02	0.02	0.3		
Northwest Territories	0.04	0.04	0.02	0.7		
Canada	24.1	17.8	11.6	422	158	122

[a] Estimated for the sewered population assuming 40 and 64 g capita^{-1} day^{-1} of anaerobically digested sludge, respectively, for population served by primary and activated sludge plants [62], 50 g capita^{-1} day^{-1} for population served by lagoons and 40 g capita^{-1} day^{-1} assuming anaerobic digestion of primary sludge for sewered population with no sewage treatment.

[b] Adjusted to reflect increased sludge quantities resulting from tertiary treatment (e.g., chemical addition for phosphorus removal)

[c] Based on sludge removal from primary and waste activated plants only. Data for Alberta, Manitoba and Ontario are measurements whereas other data are estimates.

(e.g., in Québec) and sludge contained in sewage treatment lagoons, which is removed irregularly, if ever. However, this breakdown differed widely among jurisdictions within Canada. For example, most of the sewered population in Ontario was served by mechanical wastewater treatment, and agricultural utilization and disposal accounted for 97% of the sludge production. Agricultural utilization was practiced widely in Ontario and accounted for 32% of the sludge production. By contrast, there was very little wastewater treatment in Québec in 1981 and sludge disposal accounted for only 8% of production. Agricultural utilization of sludge in Québec was prohibited by legislation. Large proportions of the sludge produced in Alberta, Saskatchewan, and Manitoba were utilized on agricultural land. In Alberta, land-applied sludge originated mainly from the cities of Calgary and Edmonton, and in Manitoba from the city of Winnipeg. The Atlantic provinces and British Columbia had expressed interest in agricultural utilization. However, this practice did not account for significant quantities of sludge. In British Columbia there was considerable interest in composting sludge to produce horticultural potting media and soil conditioners.

WASTE METAL GUIDELINES FOR AGRICULTURAL UTILIZATION

Agricultural utilization of sludge in Canada comes under both federal and provincial jurisdiction. The sale of sludge and sludge-based products as fertilizers and soil amendments is within the purview of the Fertilizers Act and Regulations administered by Agriculture Canada, and a federal trade memorandum [63] has been developed to regulate this activity. However, utilization of sludge supplied free of charge is within provincial jurisdiction and there is variable regulation as follows. Ontario [64], Alberta [65], Nova Scotia [66], and Québec [67] have adopted guidelines; British Columbia [68] and Saskatchewan [69] have developed draft guidelines; Manitoba [70] has adopted a guideline for Winnipeg sludge utilization; and the remaining provinces have no guidelines. The waste metal guidelines for agricultural utilization of sludge in Canada vary with jurisdiction as indicated in Table 13.54.

Ontario

Waste metal guideline development for agricultural utilization of sludge in Canada originated in Ontario during the early 1970s following initiation of the program of phosphorus removal from wastewater undertaken in the lower Great Lakes region. The phosphorus removal program resulted in both increased sludge production and increased pressure for sludge disposal by utilization on agricultural land.

TABLE 13.54. Waste Metal Guidelines for Agricultural Utilization of Municipal Sludge in Canada.

Waste Metal	Maximum Acceptable Concentrations in Sludge (mg kg⁻¹ dw)				Maximum Acceptable Loadings to Soil (kg ha⁻¹)							Maximum Acceptable Concentrations in Soil (mg kg⁻¹ dw)	
	Canada[a]	ON[b]	BC[c]	PQ[d]	Canada	ON	BC[d]	AB[e]	SK	MB[f]	NS	PQ	ON
As	75	170	75	(20)	15	14	15		12	0.85	9	7.5	14
B				(200)						580			
Cd	20	34	20 (25)	(15)	4	1.6	4	5–10	1.2	1.1	1	2	1.6
Co	150	340		(100)	30	30	30	0 8–1.5	30	1.4	19	15	20
Cr		2800	150	(1000)		210		50–100	100	205	130		120
Cu		1700		(1000)		150		100–200	120	45	95		100
Hg	5	11	5 (10)	(10)	1	0.8	1	0.2–0.5	0.4	0.15	0.5	0.5	0.5
Mn				(1500)									
Mo	20	94	20	(25)	4	4	4		4	<2.8	2.5	2	4
Ni	180	420	180 (200)	(180)	36	32	36	12–25	30	4.8	20	18	32
Pb	500	1100	500 (1000)	(500)	100	90	100	50–100	80	60	55	50	60
Se	14	34	14	(25)	2.8	2.4	2.8		2	0.15	1.5	1.4	1.6
Zn	1850	4200	1850 (2500)	(2500)	370	330	330	150–300	300	225	200	185	220

[a] Values are for sludge and sludge-based products containing ≤5% N dw basis and represented for sale as fertilizers or soil supplements.

[b] Values are for aerobically digested sludges and dewatered anaerobically digested sludges.

[c] Values are for agricultural high grade and, in parentheses, low grade sludges.

[d] Values are recommended maxima and, in parentheses, obligatory (i e., not to be exceeded) concentrations

[e] Choice of value depends upon site suitability for sludge application

[f] Values derived assuming 56 t ha⁻¹ dw of Winnipeg sludge applied to land and the 1980 mean heavy metal concentrations in sludge as follows: As—15; B—10,400, Cd—19.5; Co—25; Cr—3700; Cu—770; Hg—2 6; Mo—<50, Ni—85, Pb—1050; Se—2.6; Zn—4050 mg kg⁻¹ dw [71].

ON—Ontario, BC—British Columbia, PQ—Québec, AB—Alberta, SK—Saskatchewan, MB—Manitoba, NS—Nova Scotia.

Following a thorough review of the information available on sludge utilization in agriculture, background waste metal concentrations were determined for more than 200 Ontario agricultural soils (0–15 cm depth). Maximum acceptable concentrations in soil were defined assuming a doubling of the background levels and maximum acceptable loadings as follows: As, 14; Cd, 1.6; Co, 10; Cr, 30; Hg, 0.2; Mo, 4; Ni, 32; Pb, 30; Se, 0.8; and Zn, 110 kg ha^{-1} were calculated. As further experimental information and experience were obtained, the limits for Co, Cr, Hg, Pb, Se, and Zn were increased to their present levels (Table 13.54).

Liquid anaerobically digested sludge is the principal form of sludge available for land application in Ontario. However, maximum acceptable metal concentrations were not defined for these sludges. They contained appreciable amounts of nitrogen and it was recommended that their application rates be based upon plant-available nitrogen concentration. Thus, ammonium plus nitrate-nitrogen to heavy metal ratios were defined to determine acceptability for use in agriculture [64].

By contrast, maximum acceptable waste metal concentrations were defined for aerobically digested sludges and dewatered anaerobically digested sludges (Table 13.54). These sludges contain much less plant-available nitrogen than liquid anaerobically digested sludges, but they contain substantial amounts of phosphorus, organic nitrogen, micronutrients, and organic matter, which are valuable to agriculture.

Canada and British Columbia

With the exception of Cd, Cr, and Cu, the maximum acceptable waste metal loadings to soil in sludge represented for sale as fertilizers or soil supplements in Canada were chosen to approximate the Ontario values (Table 13.54). No limits were defined for Cr and Cu due to lack of evidence that buildup of these metals in soil represents a significant environmental risk. The limit for Cd was set at 4 rather than 1.6 kg ha^{-1}; based on considerable evidence [72,73] that the 1.6 kg ha^{-1} value is unnecessarily conservative. Maximum acceptable waste metal concentrations in sludge (Table 13.54) were calculated assuming the maximum acceptable waste metal loadings to soil and a 200 t ha^{-1} dw cumulative sludge loading. The Canadian limits were adopted for the draft British Columbia Guidelines.

Alberta

The Alberta limits for waste metal loadings to soil (Table 13.54) vary depending upon site suitability for sludge application. Sites are classified according to soil pH, texture, slope, and depth to groundwater (Table 13.55). The metal loading limits increase with increasing site suitability

TABLE 13.55. Classification of Sludge Application Sites in Alberta [65].

Site Characteristic	Suitable for Sludge Application			Not Suitable for Sludge Application
	Class 1	Class 2	Class 3	Class 4
pH	≥6.5	≥6.5	≥6.5	≥6.5
Texture[a]	CL, SiCL, SiL, Si, SiC, L, SCL, SC	C, HC	LS, SL	Sand and gravel
Slope (%)	0–2	2–5	5–9	>9
Depth to potable aquifer (m)	>5	3–5	2–3	<2

[a]C-clay, L-loam, S-sand and H-heavy.

and parallel but are more conservative than the Ontario limits. Agricultural land in Alberta is easily available for sludge spreading and it was decided that waste metal buildup in soil should be minimized.

Saskatchewan

Maximum acceptable waste metal loadings to soil (Table 13.54) defined in draft guidelines for sludge use in agriculture are generally smaller than the Ontario values.

Manitoba

In Manitoba, agricultural utilization of Winnipeg sludge is regulated by a Clean Environment Commission Order. Currently, land may receive one 56 t ha^{-1} dw application of sludge and the maximum acceptable waste metal loadings to soil (Table 13.54) were derived assuming 1980 mean metal concentrations for the sludge [71].

Quebec

Recommended and obligatory (not to be exceeded) waste metal concentrations in sludge for agricultural utilization, have been defined (Table 13.54). Official approval and soil and plant monitoring are required for utilization of sludges exceeding the recommended values. The maximum acceptable waste metal concentrations for Québec soils were derived from the Canadian "maximum acceptable loadings to soil" (Table 13.54). It was assumed that the weight of one hectare of soil, 15 cm deep, is two million kg and that, for example, adding 15 kg ha^{-1} As would result in a 7.5 mg kg^{-1} As concentration in soil. In general, the maximum acceptable waste metal concentrations for Québec soils are smaller than for Ontario soils.

Nova Scotia

Guidelines were developed to regulate agricultural utilization of sludge in Pictou County. In general, the limits for waste metal loadings to soil are similar to the Alberta maxima.

CANADIAN RESEARCH ON HEAVY METAL UPTAKE

Canadian research on heavy metal uptake from sludge treated soil has been conducted in the greenhouse, in lysimeters, and in the field. However, experience has shown that greenhouse data may be misleading because uptakes are larger than in lysimeters or the field. The following is an abstract of lysimeter and field experimental information.

Yields of orchard grass, bromegrass, wheat, and corn grown in Ontario following land application of sludge were as large as, or larger than, yields following application of commercial fertilizer [74]. There was no evidence of yield reduction from sludge even where applications greatly exceeded the recommended rate [64]. Cadmium loadings in sludge ranging from 4 to 7.4 kg ha^{-1} caused minor increases in the Cd concentrations in orchard grass, bromegrass, wheat straw and grain, and corn grain. However, the Cd concentration in corn stover increased from approximately 0.4 to 1 mg kg^{-1} dw. Zinc loadings ranging from 341 to 1032 ha^{-1} caused small increases in the Zn concentrations in orchard grass, bromegrass, wheat straw, wheat grain, and corn grain. The Zn concentrations in corn stover increased from approximately 40 to 160 mg kg^{-1}. The Cd and Zn concentrations, respectively, in corn stover were related to the amounts of nitrilotriacetic acid-extractable Cd and Zn in soil and to the cumulative amounts of Cd and Zn added to soil in sludge [75]. In general, including soil pH as an independent variable in regression analysis improved the above-mentioned relationships. Copper loadings ranging from 240 to 320 kg ha^{-1} caused approximately 2 mg kg^{-1} dw increases in the Cu concentrations of all plant materials except corn grain which exhibited no increase. Nickel loadings of 910 and 624 kg ha^{-1}, respectively, increased the Ni concentrations in orchard grass from approximately 2 to 21 mg kg^{-1} and of bromegrass from approximately 1 to 6 mg kg^{-1}. A loading of 536 kg ha^{-1} increased the Ni concentration in corn stover and grain approximately 1 mg kg^{-1}. Lead and Cr loadings up to 405 and 736 kg ha^{-1}, respectively, did not significantly increase the concentrations of these metals in crops.

The Cd and Zn concentrations in plant materials (mainly corn leaves) taken from old sludge application sites in Ontario, generally exhibited small increases over those in control materials [76]. Increases for Cd ranged up to 0.7 mg kg^{-1} with a median value of 0.02 mg kg^{-1}, and for Zn up to 100 mg kg^{-1} with a median value of 7 mg kg^{-1}. There was no consistent effect of sludge application on the Cu, Ni, Pb, Mo (except at Stratford), and Cr concentrations in plant materials. The Stratford sludge added much more Mo to soil than the other sludges and increased the Mo concentration in corn leaves from 0.7 to 32 mg kg^{-1}. The largest estimated metal loadings to these sites were Cd, 3.1; Zn, 595; Cu, 111; Ni, 118; Pb, 90; Mo, 19.4; and Cr, 628 kg ha^{-1}.

Wheat grain and straw grown on a calcareous soil treated with 400 to 800 t ha^{-1} dw of Winnipeg sludge exhibited no increase in Pb concentration, minor increases in Cd and Cu concentrations, and moderate increases in Zn concentrations [77]. With one exception each, the Cd and Zn concentrations were <0.5 mg kg^{-1} dw and <60 mg kg^{-1} dw, respectively. Estimated metal loadings to the soil assuming the following sludge composition: Cd, 18; Zn, 2060; Cu, 495; and Pb, 370 mg kg^{-1} dw and a 600 t ha^{-1} dw sludge application rate were: Cd, 11; Zn, 1240; Cu, 300; and Pb, 220 kg ha^{-1}.

In Alberta, sludge application generally increased yields of alfalfa grown on a solonetzic soil at Lethbridge [78] and on a calcareous chernozem soil at Edmonton [79], and of barley grown on a calcerous chernozem soil at Calgary [80]. However, the effects of sludge application on heavy metal concentrations in the plant materials were minor. Alfalfa exhibited occasional increases in Zn and small decreases in Mn. Barley straw exhibited increases in Zn, occasional increases in Pb, and small decreases in Mn. Barley grain exhibited occasional increases in Cd and Zn. Frequently, the differences in concentrations between years greatly exceeded the differences due to sludge treatment. Copper, Ni, and Cr concentrations in the plant materials were not affected by sludge treatment. The Calgary, Edmonton, and Lethbridge sludges exhibited low to moderate waste metal concentrations, and although the maximum sludge loadings to experimental sites ranged from 30 to 78 t ha^{-1} dw, the maximum metal loadings were only: Cd, 0.7; Zn, 53; Cu, 28; Ni, 4; Pb, 48; Cr, 75; and Mn 19 kg ha^{-1}.

FUTURE AGRICULTURAL UTILIZATION

Agricultural utilization of sludge is a well-established practice in Canada and is usually much cheaper than disposal where there is relatively easy access to land. It accounts for 29% of the estimated total sludge production and 44% of the estimated sludge removal from wastewater treatment facilities and is expected to remain an important future sludge management option.

TABLE 13.56. Waste Metal Guidelines for Agricultural Utilization of Sludge [81].

Waste Metal	Maximum Acceptable Loadings to Soil (kg ha^{-1})										All Countries	
	Canada	Denmark	Finland	France	Germany	Netherlands	Norway	Sweden[a]	United Kingdom	United States	Range	Median
As	15			5.4	8.4	2	0.2	0.075	10		2–15	10
Cd	4	0.2	0.1			2	0.4	0.25	5	5, 10, 20	0.1–20	5
Co	30					—					0.4–30	
Cr				360	210	100	4	5	1000		4–1000	210
Cu				210	210	120	30	15	280	125, 250, 500	30–500	210
F									600			
Hg	1			2.7	5.7	2	0.14	0.04	2		0.14–5.7	2
Mn							10					
Mo	4								4			
Ni	36			60	60	20	2	2.5	70	50, 100, 200	2–200	60
Pb	100			210	210	100	6	1.5	1000	500, 1000, 2000	6–2000	210
Se	2.8						58		5		2.8–5	
Zn	370			750	750	400	60	50	560	250, 500, 1000	60–1000	500

[a]Five-year loadings can be repeated. Values are not included in "All Countries—Range and Median."

779

The risks associated with waste metal application to agricultural land are well recognized, and guidelines to limit loadings to soil, have been developed for several Canadian jurisdictions. There is variability between the limits (Table 13.54). However, they generally correspond with the mid-range of values adopted by a large number of other countries (Table 13.56). Canadian research on waste metal uptake by plants from sludge treated soils indicates that even the maximum suggested values are not likely to cause significant crop production or animal and human health problems. Further research and practical experience may indicate that some relaxation of limits is warranted.

Agricultural utilization of sludge according to the Ontario guidelines has been practiced for several years and has proven satisfactory for both the agricultural and wastewater treatment communities. The guidelines have prompted considerable reductions of waste metal concentrations in several sludges. These reductions were necessary to maintain sludge acceptability for agricultural utilization and in most cases were accomplished by improved management of industrial processing or pretreatment of industrial effluents. Only a few sludges have been declared unacceptable for agricultural utilization due to high metal concentrations.

Although agricultural utilization of sludge in Canada is based on sound scientific information there are pockets of public resistance usually due to incorrect information or unsatisfactory past experience. Efforts to increase public acceptance have included the development of two information packages [82,83] and audio-visual presentations.

REFERENCES

1 IAWQ, EWPCA, WEF *Global Atlas of Wastewater and Biosolids Use and Disposal.* IAWQ Scientific and Technical Report Number 6, 1996, P. Matthew, ed., London, UK.

2 European Community Council, "Council Directive of 21 May 1991 Concerning Urban Wastewater Treatment," *J. Euro. Committees* 30 May 1991.

3 Newman, P. J., A. V. Bowden and A. M. Bruce. 1989. "Production Treatment and Sewage Sludge," in *Proc. EC/EWPCA Conf. Sewage Sludge Treatment and Use. New Developments, Technological Aspects and Environmental Effects.* Amsterdam, Sept., 1988, Dirkzwager and L'Hermite, eds., Elsevier, London.

4 EC. 1987. EEC Statistical Office, Population Statistics, Brussels.

5 Saabye, A., and H. D. Schwinning, 1994. "Treatment and beneficial use of sewage biosolids in the European Community." *WEF Speciality Conference Series, International Management of Water and Wastewater Solids for the 21st Century. A Global Perspective.* Washington, June.

6 Bruce, A. M. 1981. "A Note on Measurement of Sludge Production. Treatment and Use of Sewage Sludge," COST 69 BIS Final Report III. Technical Annexes p. 59, Commission of the European Communities, May.

7 Department of the Environmental/National Water Council (DoE/NWC). 1983. Stand-

ing Committee on the Disposal of Sewage Sludge (SCDSS). Sewage Sludge Survey, 1980 Data, August.

8   Water Research Centre WRc. September 1990. "A Methodology for Undertaking BPEO Studies of Sewage Sludge Treatment and Disposal," Medmenham.

9   Bruce, A. M. 1990. "Sewage Sludge: Making the Best of It," *Inst. Wat. Environ. Manag. Handbook,* 1989–90.

10   Beeching, A. 1981. "Centrifuge Operation at Cambridge STW," presented to East Anglian Branch IWPC, February, Cambridge.

11   Calcutt, A. T. and J. Moss. 1984. "Sewage Sludge Treatment and Disposal. The Way Ahead," *Water Pollution Control,* 83:163.

12   Department of the Environment. 1989. "Code of Practice for Agricultural Use of Sewage Sludge," HMSO. London, December.

13   Maff/Department of the Environment. 1994. Review of the Rules for Sewage Sludge Application on Agricultural Land. Fertility Aspects of Potentially Toxic Elements. Report of the Independent Scientific Committee. HMSO, London.

14   Royal Commission on Environmental Pollution. 1996. 19th Report on Sustainable Use of Soil. HMSO, London.

15   European Community. 1986. "Council Directive on the Protection of the Environment and in Particular of the Soil When Sewage Sludge is Used in Agriculture," *Official J. Euro. Comm.,* L 187/6, 4 July.

16   HMSO. 1989. Sludge (Use in Agriculture) Regulations 1989.

17   House of Lords. 1983–84. Select Committee on the European Communities. "Sewage Sludge in Agriculture," Evidence presented at EWPCA. HMSO 1983.

18   WHO. 1981. "The Risk to Health of Microbes in Sewage Sludge Applied to Land," Report of a WHO Working Group. Stevenage, January. EURO Reports and Studies No. 54. WHO Regional Office for Europe, Copenhagen.

19   WHO. 1985. "The Risk to Health of Chemicals in Sewage Sludge Applied to Land, Report of a Working Group," R. B. Dean and M. J. Suess, eds., *Waste Management and Research,* 3:251–278.

20   Matthews, P. J., E. Lund and R. Leschber. 1986. "Health Risks of Microbes and Chemicals in Sewage Sludge Disposed to Land—Recommendations to the World Health Organisation," in *Proc. EC Int. Symp. Processing and Use of Organic Sludge and Liquid Agricultural Waste,* Rome, October 1985, P. L'Hermite, ed., Pub Reidel, Dordrecht.

21   Berrow, M. L., and Webber, J. 1972. "Trace Elements in Sewage Sludge," *J. Sci. Fd.* April 93.

22   Matthews, P. J., and Barnden, A. D. 1980. Report on Metals in Sludges used in Agriculture in the Anglian Water Authority April 1979–March 1980.

23   Department of the Environment. 1993. UK Sewage Sludge Survey. CES, London.

24   Foundation for Water Research. 1993. The Examination of Sewage Sludge for Poly-chlorinated Dibenzo-*p*-Dioxins Polychlorinated Dibenzo-furans. Laboratory of the Government Chemist for the Environment. D/009.

25   Linder, K. H. 1996. Current Developments in the Field of Sewage Sludge at the European Level. Korrespondenz Abwasser, February.

26   Water Research Centre. 1995. Report NOs. EC3646 Part I Survey of Sludge Production Treatment Quality and Disposal in the European Union, and 3757 Part 2 Quality Criteria Classification and Strategy Development.

27   Matthews, P. J. and W. Schenkel. 1989. "Modern Sludge Management. The Managers Choice," Ibid., Ref. [3].

28 Garnett, P., Dudman, S., Andrews, D. 1996. "Marketing of Biosolids," *WEF Speciality Conference Series. 10 Years of Progress and a Look Toward the Future.* Denver, August.

29 Ministry of Agriculture Fisheries and Food. 1991. Code of Good Agricultural Practice for the Protection of Water.

30 WRc. 1985. *The Agricultural Value of Sewage Sludge. A Farmers Guide.*

31 Matthews, P. J., D. A. Andrews and R. F. Critchley. 1986. "Methods for the Application and Incorporation of Sludge Into Land," in *Proc. EC Int. Symp. Processing of the Sewage Sludge,* Brighton, September 1983 P. L'Hermite, H. Ott, eds., Pub. Reidel, Dordrecht.

32 Hall, J. E. 1989. "Methods for Applying Sludge to Land. A Review of Recent Developments," Ibid., Ref [3].

33 Institute of Water Pollution Control. 1978. *Sewage Sludge III. Utilisation and Disposal.* Maidstone.

34 Maff. March 1991. Press release, John Gummer announces plans to phase out sewage sludge disposal at sea.

35 Hall, J. E. 1990. "Soil and Sludge Management," *Treatment and Use of Sludge. Concerns and Options.* WRc Medmenham, June.

36 Colin, F., P. J. Newman and L. Spinosa. "Thermal Treatment of Sludge," *Proc. EC (1989) Workshop Bari,* June 1988.

37 HMSO. 1994. Waste Management Licencing Regulations.

38 HMSO. 1993. *A Manual of Good Practice for the Use of Sewage Sludge in Forestry.* Forestry Commission Bulletin 107. London.

39 Budlingmaiur, W. and P. L'Hermite. 1989. "Compost Processes in Waste Management," *Proc. EC Workshop,* Neresheim, Sept., 1988.

40 HMSO. 1991. DoE/WO Circular. Water Industry: UK Planning No. 17/91. London.

41 Anon. 1990. New Disposal Options for Sewage Sludge. ENDS Report 191. London.

42 Anon. 1993. Dutch Farm Sludge to End by 1995. *World Water and Environmental Engineer.* January.

43 Oake, R. 1991. *Sludge Disposal—The Options.* Water Services Association.

44 *European Workshop on Disposal Routes for Wastewater Sludge.* 1991. WSA/SPDE/ FEDERGASAQUA. Brussels, July.

45 Department of Trade and Industry. 1991. *Proc. Seminar Urban Water Pollution Management in Europe,* University of Sheffield, June.

46 Japan Sewage Works Association. 1990. *Sewage Works Statistics FY 1988.* Vol. 45.

47 Watanabe, Y. et al. 1990. "Ultra-Compact Sludge Thickening System—Innovative Sludge Handling through Pelletization/Thickening," presented at the *63rd WPCF Annual Conference.*

48 Matsui, T., S. Kyosai and M. Takahashi. 1990. "Progress in the Research on Application of Biotechnology to Wastewater Treatment," *Proc. 4th WPCF/JSWA Joint Technical Seminar on Sewage Treatment Technology,* pp. 55–75.

49 Murakami, T. and K. Murakami. 1986. "Autogenous Combustion of Sewage Sludge," *Proc. Tenth United States/Japan Conference on Sewage Treatment,* EPA/600/9-86/ 015a, pp. 19–66.

50 Ishida, T. 1988. "Existing Conditions for Agricultural Utilization of Sewage Sludge Compost in Japan," presented at the *Sewage Treatment and Use* held by EC and EWPCA, Amsterdam.

51 Yoshida, S. 1986. "Centralized Sludge Treatment in Yokohama," *Proc. Tenth United States/Japan Conference on Sewage Treatment,* EPA/600-9-86/015a, pp. 91–126.

52 Tamaki, T. 1988. "Areawide Sewage Sludge Treatment and Disposal Project— ACE Plan," *Proc. Eleventh United States/Japan Conference on Sewage Treatment Technology,* EPA/600/9-88/010, pp. 51–74.

53 Department of Statistics. *New Zealand Official 1990 Year Book.*

54 Fitzmaurice, John R. 1989. "Wastewater Disposal in New Zealand," presented at the *WPCF Asia/Pacific Rim Conference on Water Pollution,* Hawaii, October 1989.

55 McNeill, K. P. and J. W. Bradley. 1988. "Alternative Waste Treatment Systems in Southern New Zealand," in *Alternative Waste Treatment Systems,* R. Bhamidimarri, ed., Elsevier Applied Science.

56 Hedgeland, R. November 1986. "Disposal of Milliscreening," *Transactions of the Institute of Professional Engineers of New Zealand,* Volume 13, 3/CE.

57 New Zealand Department of Health. 1984. "Guidelines for the Disposal of Sewage Sludge on Land," Circular Memorandum (General) No. 1984/93.

58 Auckland Regional Council. 1989. "Operation and Research Group for Drainage— 1989 Report."

59 Fries, G. F. 1982. "Potential Polychlorinated Biphenyl Residues in Animal Products from Application of Contaminated Sewage Sludge to Land," *J. Environ. Qual.,* 11:14–20.

60 Bowen, H. J. M. 1975. "Residence Times of Heavy Metals in the Environment," in *Proceedings of International Conference on Heavy Metals in the Environment,* Toronto, Ontario, 1:1–19.

61 National Inventory. 1981. "National Inventory of Municipal Waterworks and Wastewater Systems in Canada, 1981," Minister of Supply and Services Canada, Cat. No. En 44-10/81, Ottawa, Ontario.

62 Schmidtke, N. W. 1981. "Sludge Generation, Handling and Disposal at Phosphorus Control Facilities in Ontario," in *Characterization, Treatment and Use of Sewage Sludge,* P. L'Hermite and H. Ott, eds., Eur 7076, D. Reidel Publishing Co., pp. 190–225.

63 Standish, J. F. 1981. "Metal Concentrations in Processed Sewage and By-Products," Agriculture Canada, Trade Memorandum T-4-93, Ottawa, Ontario.

64 Ontario Ministries of Agriculture and Food, Environment and Health. 1978; revised 1986. *Guidelines for Sewage Sludge Utilization on Agricultural Lands.*

65 McCoy, D., D. Spink, J. Fujikawa, H. Regier and D. Graveland. 1982. *Guidelines for the Application of Municipal Wastewater Sludges to Agricultural Lands.* Alberta Environment, Edmonton, Alberta.

66 Departments of Agriculture, Environment and Health. 1983. *Guidelines for Utilization of Digested Sewage Sludge on Agricultural Lands in Nova Scotia.*

67 Ministére de l'Environnement Québec. 1987. "Valorisation Agricole des Boues de Stations d'Epuration des Eaux Usées Municipales—Guide de bonnes pratiques."

68 British Columbia Ministries of Agriculture and Food, Health and Environment. 1982. *Guidelines for Use with the Regulations under the Waste Management Act for Control of the Discharge of Sludge to Land—Draft.*

69 Saskatchewan Environment. 1987. *Saskatchewan's Guidelines for the Use of Sewage Sludge on Agricultural Lands—Draft.*

70 1986. "Order No. 1089 VO of the Clean Environment Commission under the Clean Environment Act."

71 1981. "Sludge Drying Beds and Land Application Operation," a brief presented to the Manitoba Clean Environment Commission, Winnipeg, Manitoba.

72 Ryan, J. A., H. R. Pahren and J. B. Lucas. 1982. "Controlling Cadmium in the Human Food Chain: A Review and Rationale Based on Health Effects," *Environ. Res.*, 28:251–302.

73 Davis, R. D. and E. G. Coker, 1980. "Cadmium in Agriculture with Special Reference to the Utilization of Sewage Sludge on Land," Water Research Centre, Technical Report TR 139, Stevenage, Herts., England.

74 Webber, M. D., Y. K. Soon and T. E. Bates. 1981. "Lysimeter and Field Studies on Land Application of Wastewater Sludges," *Water Sci. Technol.*, 13:905–917.

75 Soon, Y. K., T. E. Bates and J. R. Moyer. 1980. "Land Application of Chemically Treated Sewage Sludge: III. Effects on Soil and Plant Heavy Metal Content," *J. Environ. Qual.*, 9:497–504.

76 Webber, M. D., H. D. Monteith and D. G. M. Corneau. 1983. "Assessment of Heavy Metals and PCB's at Selected Sludge Application Sites," *J. Water Poll. Control Fed.*, 55:187–195.

77 Zwarich, M. A. and J. G. Mills. 1979. "Effects of Sewage Sludge Application on the Heavy Metal Content of Wheat and Forage Crops," *Can. J. Soil Sci.*, 59:231–239.

78 Graveland, D. N. 1983. "The Effect of Digested Bed Dried Sewage Sludge on Brown Solonetzic Soil and Associated Plant Growth," Alberta Environment, Technical Report, Edmonton, Alberta.

79 Lutwick, G. and D. N. Graveland. 1983. "Effects of Anaerobically Digested Sewage Sludge on Crop Yield and Elemental Accumulation in Alfalfa and Soils near Edmonton, Alberta," Alberta Environment, Technical Report, Edmonton, Alberta.

80 Graveland, D. N. 1983. "A Seven Year Study of Soils and Crops Following Application of Calgary Sewage Sludge," Alberta Environment, Technical Report, Edmonton, Alberta.

81 Webber, M. D., A. Kloke and J. Chr. Tjell. 1984. "A Review of Current Sludge Use Guidelines for the Control of Heavy Metal Contamination in Soils," in *Processing and Use of Sewage Sludge*, P. L'Hermite and H. Ott, eds., EUR 9129, D. Reidel Publishing Co., pp. 371–386.

82 Black, S. A., D. N. Graveland, W. Nicholaichuk, D. W. Smith, R. S. Tobin and M. D. Webber. 1984. "Manual for Land Application of Treated Municipal Wastewater and Sludge," Environment Canada Report EPS 6-EP-84-1, Ottawa, Ontario.

83 Webber, M. D. 1984. "Land Utilization of Sewage Sludge—A Discussion Paper," Expert Committee on Soil and Water Management Report, Agriculture Canada, Ottawa, Ontario.

Index

785

T - #0230 - 101024 - C0 - 229/152/43 [45] - CB - 9781566766210 - Gloss Lamination